Cell Structure and Function

TRANSFUSION &
TRANSPLANTATION SCIENCE

edited by Robin Knight

BIOMEDICAL SCIENCE
PRACTICE

EXPERIMENTAL & PROFESSIONAL SKILLS

edited by Nessar Ahmed, Chris Smith & Qiuyu Wang

CYTOPATHOLOGY

edited by Behdad Shambayati

CLINICAL BIOCHEMISTRY

edited by Nessar Ahmed

DATA HANDLING
AND ANALYSIS

Andrew Blann

HAEMATOLOGY

Andrew Blann, Gavin Knight,
& Gary Moore

MEDICAL
MICROBIOLOGY

edited by Michael Ford

CELL STRUCTURE
& FUNCTION

edited by Guy Orchard & Brian Nation

HISTOPATHOLOGY

edited by Guy Orchard & Brian Nation

fundamentals OF
biomedical science

Cell Structure and Function

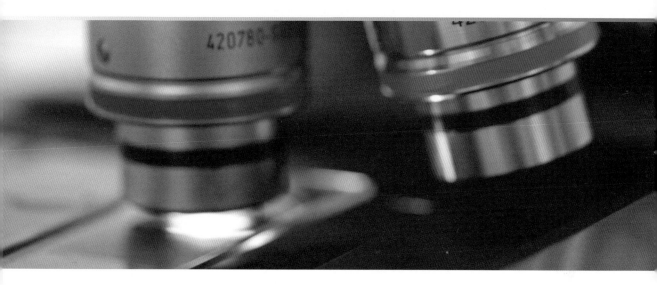

Edited by

Guy Orchard
Consultant Grade Biomedical Scientist and Laboratory Manager
Viapath (formerly GSTS Pathology),
St John's Institute of Dermatology, London

Brian Nation
Editor,
British Journal of Biomedical Science

OXFORD
UNIVERSITY PRESS

OXFORD
UNIVERSITY PRESS

Great Clarendon Street, Oxford, OX2 6DP,
United Kingdom

Oxford University Press is a department of the University of Oxford.
It furthers the University's objective of excellence in research, scholarship,
and education by publishing worldwide. Oxford is a registered trade mark of
Oxford University Press in the UK and in certain other countries

Published in the United States of America by Oxford University Press
198 Madison Avenue, New York, NY 10016, United States of America

British Library Cataloguing in Publication Data
Data available

Library of Congress Control Number: 2014933294

ISBN 978-0-19-965247-1

Printed in Great Britain by
Bell & Bain Ltd, Glasgow

Acknowledgments

We would like to thank the contributing authors for all their hard work in the preparation of the individual chapters that comprise this textbook, and also acknowledge the help, support and guidance received from staff at Oxford University Press, in particular Jonathan Crowe, Dewi Jackson, Alice Mumford, Sarah Broadley and Angela Butterworth.

Special thanks are also due to our respective families, especially Sarah, Ross and Kim, for their understanding and forbearance during the many hours devoted to the completion of this, our second volume, in the Fundamentals of Biomedical Science series.

Guy Orchard

Brian Nation

Foreword

Of the trillions of cells that make up the human body, it's quite amazing to think that there exists absolute order within this vast sea of cellular diversity. A fundamental point to remember is that every cell belongs to a cell line, which in turn belongs to a system. The organs of the body play a pivotal role in maintaining the body systems and offer their own fascinating structural complexities.

As a student of biomedical science, you are taught early on that understanding the normal structural and functional features of the human body should always form the basis for understanding disease states. Put more simply, it's about how disease affects the normal cellular and body systems. In order to comprehend this, it is essential to appreciate fully what 'normal' looks like, and that's what this book is about.

Unlike textbooks and colour atlases that currently exist, this book attempts to link the systems and cross reference the interlinking themes. The early chapters set the scene and give the reader the fundamentals of cell structure and how they form systems and organs. There is also a chapter on the tools that enable us to study cells. The book then explores the body systems and organs chapter by chapter, from the inside out.

When looking at children's plastic interlocking bricks, one can see how clever the designer had been and how clearly thought through were the minor details of how to construct any number of different models. As a concept, this is not a million miles from the workings of the human body, in that every cell is part of a larger system or organ and has its own role to play within that matrix, just like the parts of a children's model.

Finally, this book is designed to complement the other volumes in the Fundamentals of Biomedical Science series; indeed, it sets the foundations for the series and cross references to other volumes throughout. As with the other books in the series, it follows a similar style that emphasizes key points, key terms and self-check questions to support and encourage the reader to check their understanding as they progress through the text.

'An investment in knowledge pays the best interest'
Benjamin Franklin

Guy Orchard
Brian Nation

An introduction to the Fundamentals of Biomedical Science series

Biomedical scientists form the foundation of modern healthcare, from cancer screening to diagnosing HIV, from blood transfusion for surgery to infection control. Without biomedical scientists, the diagnosis of disease, the evaluation of the effectiveness of treatment, and research into the causes and cures of disease would not be possible. However, the path to becoming a biomedical scientist is a challenging one: trainees must not only assimilate knowledge from a range of disciplines, but must understand—and demonstrate—how to apply this knowledge in a practical, hands-on environment.

The Fundamentals of Biomedical Science series is written to reflect the challenges of biomedical science education and training today. It blends essential basic science with insights into laboratory practice to show how an understanding of the biology of disease is coupled to the analytical approaches that lead to diagnosis. Produced in collaboration with the Institute of Biomedical Science, the series provides coverage of the full range of disciplines to which a biomedical scientist may be exposed.

Learning from this series

The Fundamentals of Biomedical Science series draws on a range of learning features to help readers master both biomedical science theory, and biomedical science practice.

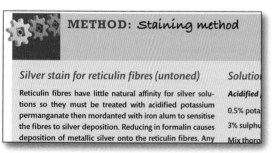

Method boxes walk through the key protocols that the reader is likely to come across in the laboratory.

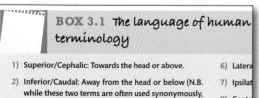

Additional information to augment the main text appears in **boxes**.

Clinical Correlation boxes emphasize at a glance how the material sits in a clinical context.

> **CLINICAL CORRELATION**
>
> *Clostridium difficile*
>
> This is an anaerobic, Gram-positive bacillus which forms part of the norma
> around two-thirds of children under two years old and in at least three per cent
> may be much higher in the elderly or hospitalized. Some strains can produce
> rise to diarrhoea, and this often seems to be triggered by the disruption of n
> by the use of broad-spectrum antibiotics.

Further features are used to help consolidate and extend students' understanding of the subject

Key points reinforce the key concepts that the reader should master as they work through each chapter, while **Summary** points act as an end-of-chapter checklist for readers to verify that they have remembered the principal themes and ideas presented within the chapter.

> **Key points**
>
> Most cells share a common organization, containing similar organelles an
> features, but the relative amounts of different organelles vary in each
> type according to their function.

Key terms in the margins provide instant definitions of terms with which the reader may not be familiar; in addition, each title in the series features a **glossary**, in which the key terms featured in that title are collated.

> **Apoptosis**
> This is an inflammation-independent mechanism through which controlled cell death is initiated.

protein produced primarily by th
and acts as a hormone. Erythrop
production of EPO is stimulated b
improve oxygen delivery to meta
a range of other growth factors t
and increasing proliferation and

Erythropoietin must bind to speci
cell survival, proliferation and ma
binding partner (**ligand**), an intra

Self-check questions throughout each chapter provide the reader with a ready means of checking that they have understood the material they have just encountered; answers to self-check questions are presented in the book's Online Resource Centre.

aries between these signals and the actin networks. Espins also help in the fo
bundles in hair-cell stereocilia and in testes. Plastins/fimbrins also act as actin-b
in intestinal brush border epithelia and hair cells. EPS8 and related proteins can
of actin bundles and regulate their activity. The control of the actin cytoskeleto
dynamic, and motile and sedentary cells both depend greatly on this functiona

> **SELF-CHECK 1.12**
>
> In which type of cell is actin found in the greatest quantity in the human bod

Discussion questions are provided at the end of each chapter, to encourage the reader to analyse and reflect on the material they have just read. Hints and tips for answering the discussion questions are provided in the book's Online Resource Centre.

 Discussion questions

5.1 What different types of glial cells are there in the CNS and what are th
5.2 Why is it essential for neurons to have an efficient system of intracellula mechanisms exist for this?
5.3 Describe the structural arrangement of large nerve fibres in the PNS.

Cross-references help the reader to see biomedical science as a unified discipline, making connections between topics presented within each volume, and across all volumes in the series.

f cells and have numerous cilia on their upper
or beating motion and move mucus and par-

ting goblet cells, so named because of their
Goblet cells exude large quantities of mucus,

al cells believed to have the ability to undergo
und in the epithelium.

> **Cross reference**
> To find out more about the mechanism of cilia motility, se Chapter 11

Online learning materials

online resource centre

Each title in the *Fundamentals of Biomedical Science* series is supported by an Online Resource Centre, which features additional materials for students, trainees and lecturers.

www.oxfordtextbooks.co.uk/orc/fbs

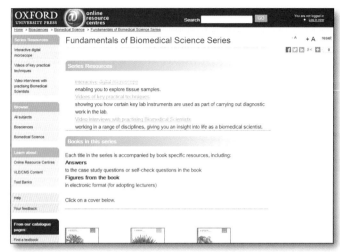

Guides to key experimental skills and methods

Video walk-throughs of key experimental skills are provided to help you master the essential skills that are the foundation of biomedical science practice.

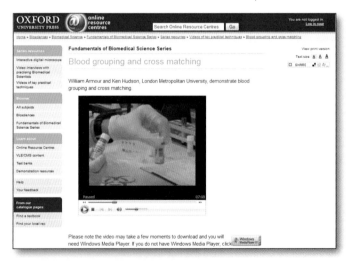

Biomedical science in practice

Interviews with practicing biomedical scientists working in a range of disciplines give a valuable insight into the reality of work in a Biomedical Science laboratory.

Virtual microscope

Visit the library of microscopic images and investigate them with the powerful online microscope, to help gain a deeper appreciation of cell and tissue morphology.

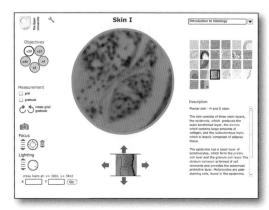

Lecturer support materials

The Online Resource Centre for each title in the series also features figures from the book in electronic format, for registered adopters to download for use in lecture presentations, and other educational resources.

To register as an adopter visit **www.oxfordtextbooks.co.uk/orc/** and follow the on-screen instructions.

Any comments?

We welcome comments and feedback about any aspect of this series.
Just visit **www.oxfortextbooks.co.uk/orc/feedback/** and share your views.

Contributors

Andrew Blann,
Consultant Clinical Scientist and Senior Lecturer,
University Department of Medicine, City Hospital,
Birmingham, UK

Judy Brincat
Histopathology, Melbourne Pathology, Australia

Suha Deen
Queen's Medical Centre Campus, Nottingham University
Hospitals, UK

Andrew Evered
Principal Lecturer in Biomedical Science,
Cardiff Metropolitan University, Wales, UK

David Furness
Professor of Cellular Neuroscience, Institute for Science
and Technology in Medicine and School of Life Sciences,
Keele University, UK

Carole Hackney
Professor, Advanced Imaging and Microscopy, UK

Rosalind King
Department of Clinical Neurosciences, Institute of
Neurology, UCL, UK

Gavin Knight
Senior Lecturer, School of Pharmacy and Biomedical
Sciences, University of Portsmouth, UK

Ian Locke
Principal Lecturer, Department of Biomedical Sciences,
University of Westminster, UK

Tony Madgwick
Department of Biomedical Sciences, University of
Westminster, UK

Richard Mathias
Cellular Pathology, Salford Royal Hospital, UK

Joanne Murray
Senior Lecturer, Department of Human and Health
Sciences, University of Westminster, UK

David Muskett
Histopathology, East Lancashire Hospitals
NHS Trust, UK

Brian Nation
Editor, *British Journal of Biomedical Science*

Kathy Nye
Consultant Medical Microbiologist, Public Health
Laboratory Birmingham, Public Health England,
Birmingham Heartlands Hospital, UK

Guy Orchard
Consultant Grade Biomedical Scientist and Laboratory
Manager, Viapath (formerly GSTS Pathology), St. John's
Institute of Dermatology, London

Alberto Quaglia
Lead Consultant Histopathologist
Institute of Liver Studies, King's College Hospital,
London, UK

Anne Rayner
Institute of Liver Studies, King's College Hospital,
London, UK

Behdad Shambayati
Consultant Clinical Cytologist-Specialty Lead,
Cytopathology Department, Surrey Pathology Service,
Ashford and St. Peter's Hospitals Foundation NHS
Trust, UK

Tony Warford
Department of Biomedical Sciences, University of
Westminster, UK

Contents

1

Introducing the cell: the unit of life

Carole Hackney and David Furness

Learning objectives

After reading this chapter you will have gained knowledge and understanding of:

■ The range and diversity of cells in the three main domains of life.

■ One possible route by which eukaryotic cells may have evolved from prokaryotic cells.

■ The overall organization of the eukaryotic cell and the structure and function of subcellular organelles.

■ Protein synthesis and the role of ribosomes in translating the genetic code.

■ Membrane flow between organelles and its role in protein sorting.

■ Lipid synthesis and the role of the endoplasmic reticulum.

■ Metabolism and the role of mitochondria.

■ The composition and organization of the cytoskeleton.

■ Cell division and specializations.

■ Cell death pathways.

The diversity of life is evident everywhere we look in our environment. Even in a tiny region of the world's biosphere, such as a pond or a few grams of soil, there is a huge richness of organisms; both visible and microscopic life abound and find ways to fill every ecological niche. Taxonomically, these organisms have until recently been classified into five major kingdoms: Animalia, Plantae, Fungi, Protista and Monera. The latest classifications divide life into three domains: the archaea (primitive bacteria-like forms), the bacteria and the eukaryota.

Despite the diversity, and the many differences between organisms, there are fundamental characteristics that all share. They are composed of specific types of organic molecules which can be subdivided primarily into proteins, lipids, carbohydrates and nucleic acids, arranged in various molecular structures (Box 1.1). Firstly, each organism carries the instructions for building and maintaining its structure in a

BOX 1.1 The molecules of life

Proteins are chains of amino acids that have a primary structure (sequence of amino acids) which then fold into secondary and tertiary structures, forming a variety of shapes. Many form subunits that then assemble into other macromolecular arrangements such as channels in the plasma membrane and cytoskeletal filaments. Specific proteins can occur as different isoforms, in which there are usually small changes in the amino acid sequence that modify to some extent the protein's properties. The basic amino acid structure is shown in Figure 1.1.

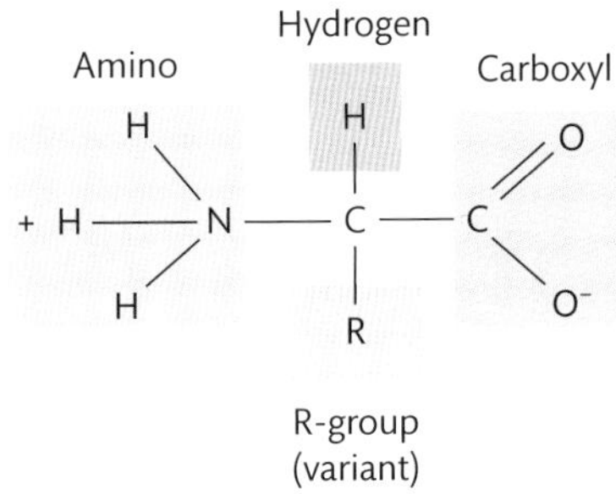

Carbohydrates are composed of carbon, hydrogen and oxygen (the hydrogen and oxygen are roughly in the same proportion as in water [i.e. H_2O], hence explaining the 'hydrate' part of the name). Carbohydrates are present commonly as polymers of saccharides (polysaccharides) forming structural elements such as the cellulose that makes up a plant cell wall. The major fuel in the cell used for respiration (the process that releases energy) is a carbohydrate, glucose ($C_6H_{12}O_6$) (see Figure 1.2).

Lipids (Figure 1.3) are chains composed primarily of carbon and hydrogen coupled to other components that together determine their properties. They are generally hydrophobic or amphiphilic (meaning they have both hydrophobic and hydrophilic parts) and can be found in membranes or stored for utilization as a nutritional source. The hydrophilic part of an amphiphilic lipid is a polar molecule such as a phosphate group (as found in the plasma membrane) or carboxylic acid (as found in fatty acids).

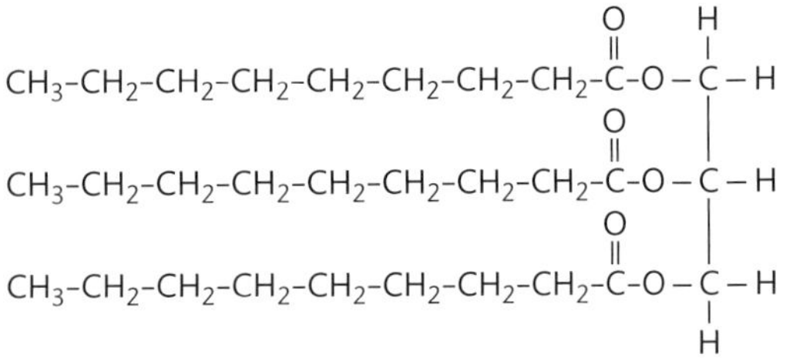

Nucleic acids (Figure 1.4) are large polymers composed of nucleotides, a complex of a base (a purine or pyrimidine molecule), a sugar and a phosphate group. The sugar-phosphate group forms the backbone of the nucleotide polymer. Single-stranded RNA differs from double-stranded DNA in that one of the four bases, uracil (U), replaces thymine (T) in DNA, while the other three (adenine [A], guanine [G] and cytosine [C]) are common to both. It also has a difference in the sugar: in DNA it is deoxyribose, whereas in RNA it is ribose. The double strand of the DNA is formed by the coupling of two single strands by bonds between complementary pairs of bases (A+C, G+T). The combinations of A+G or C+T are energetically unfavourable and therefore do not occur.

FIGURE 1.1
Diagram showing the basic structure of an amino acid.

Carbohydrate = Carbon + Water

$C_n(H_2O)_n$
or
$C_6H_{12}O_6$

FIGURE 1.2
A 3D model of glucose.

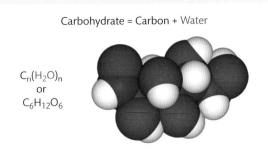

FIGURE 1.3
An example of a triglyceride lipid.

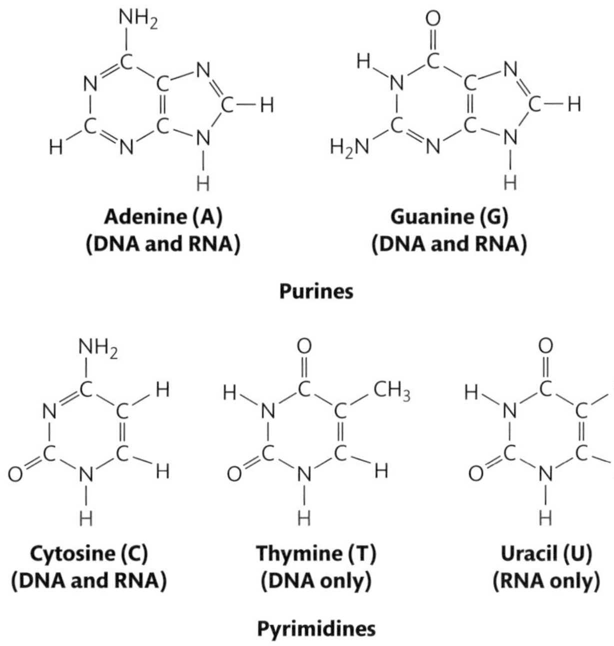

FIGURE 1.4
Nucleotides: the basic building blocks of nucleic acids.

molecule–deoxyribonucleic acid (DNA)–that codes for all the complex materials of life, and forms genes that drive the synthesis of each of the organism's constituents. Second, with a few specialized exceptions, all organisms are made of enclosed units called cells. The exceptions are viruses, which are not cells but strands of ribonucleic acid (RNA) or DNA in a protein jacket, capable of hijacking cellular machinery to reproduce themselves.

Cells can be found living singly or organized in various levels of colonial or multicellular forms. The diversity of organisms is matched by an even greater diversity of cellular composition. In the human body, as in the vast majority of multicellular organisms, highly specialized cells carry out distinct functions. They are so adapted to their place within the body that they cannot normally survive alone in the natural environment; they can live outside the body only if grown in culture in very precise conditions, and often only in modified form.

Some cells can also inhabit other cells, as will be commented on further, and many kinds of single prokaryotic and eukaryotic cells are found as parasites and disease-producing pathogens in multicellular organisms (Clinical correlation 'Disease-producing prokaryotic and eukaryotic single cells').

SELF-CHECK 1.1

How might multicellularity have arisen from single-celled organisms?

Key points

Virtually all forms of life are composed of cells, either singly or arranged in groups. The main exceptions are viruses, which, according to some definitions of life, are not truly living.

CLINICAL CORRELATION

Disease-producing prokaryotic and eukaryotic single cells

■ A wide variety of single-celled pathogens cause disease.

■ Cholera, typhoid, tuberculosis, methicillin-resistant *Staphylococcus aureus* (MRSA) and tetanus are just a few examples of diseases caused by bacterial infection, but there are many others.

■ Bacterial diseases can generally be treated by antibiotics, although rapid growth and reproduction rates enable organisms to mutate and evolve resistance to the drugs if they are overused.

■ Mycoplasms are disease-causing bacteria that lack a cell wall and are generally resistant to antibiotics that often work by targeting the cell wall.

■ Eukaryotic single-celled organisms cause diseases such as sleeping sickness, amoebic dysentery, toxoplasmosis and malaria.

1.1 Development of cell theory

Historically, the concept of cells only became possible with the invention of the optical microscope. Robert Hooke in the seventeenth century is credited with the first observation of cellular architecture in cork. The microscopic appearance of cork is a collection of tiny thick-walled compartments. In cork, the contents of these compartments are missing and the tissue is dead, but nevertheless this gave the early microscopists a concept of the cellular nature of life. Some years later, in 1838–9, the cell theory was suggested independently by Schleiden and Schwann, who stated that cells make up most living organisms. This is the origin of the modern-day idea that cells are the fundamental unit of life.

Since the development of cell theory, our understanding of cells has grown along with new developments in methodology and instrumentation. The first key step was clearly the development of the microscope (Box 1.2), in particular the electron microscope, which enabled the substructure of cells to be seen clearly for the first time. Subsequently, there was development of biochemical methods for analysis of cell constituents, genetic techniques and, over the past few years, genomics, where an individual's genes can be 'fingerprinted', and the entire human genome (the collection of genes that are contained within the genetic material) is now known. Most recently, the development of the protein equivalent of genomics (proteomics) and bioinformatics for searching and comparing genes and proteins means the cell biologist has an enormously powerful range of tools with which to study cells (see Chapter 2).

Cross reference

For more information about microscopy, see the *Histopathology* volume of this Fundamentals of Biomedical Science series.

Key points

Robert Hooke first identified cell structure in cork, and the cell theory was later independently formulated by Schleiden and Schwann in the 1830s.

BOX 1.2 Types of microscopy available to the cell biologist

- **Light microscopes** were first invented in the seventeenth century when they utilized ambient light or light from a candle to observe samples. Now, they normally have an inbuilt light source. They have a maximum resolution (ability to discern detail in a sample) of 0.25 μm, which is limited by the wavelength of the light used. They can be used to observe small living samples or histological sections that have been stained with dyes (see the *Histopathology* volume in the Fundamentals of Biomedical Science series). There are a number of variants in which optical effects are used to enhance the image, such as phase contrast or Nomarski optics.

- **Fluorescence microscopes** take advantage of the fact that some biological molecules fluoresce naturally under ultraviolet (UV) light. They contain a set of filters that enable the fluorescence to be observed safely (UV light can damage the retina) and also allow different dyes to be used simultaneously. Fluorescent staining methods and immunofluorescence (a technique that allows fluorescent dyes to be attached to specific molecules in cells and tissue) have enhanced the usefulness of fluorescence microscopes.

- **Confocal microscopes** are a derivative of the fluorescence microscope where a laser is used as the source of illumination and scanned across the sample, resulting in the emission of fluorescent light. The emitted light passes through a pinhole that excludes out-of-focus light so the image is made clearer. By stepping the microscope focus down through the sample, stacks of images can be obtained that are serial optical sections, enabling the sample to be reconstructed in three dimensions (3D). Examples of confocal images can be seen in Figure 1.25

Recent developments in confocal microscopy have resulted in better resolution than can be obtained by light microscopy.

- **Electron microscopes** (EM) utilize a source of electrons (commonly a heated tungsten filament) as an illuminant. Owing to the dual particle-wave properties of electrons, they can be treated like a light beam, and the much shorter wavelength of an electron permits a resolution of 0.2 nm, with magnifications correspondingly higher than obtained by light microscopy (see the *Histopathology* volume in the Fundamentals of Biomedical Science series). There are two types of EM, scanning EM (SEM) where the electron beam scans over the surface of the sample causing electrons to be emitted to form an image (see Figures 1.5 and 1.7), and transmission EM (TEM) where the electron beam passes through a thin section of a sample, stained with heavy metals, generating a contrast image (see Figures 1.8 and 1.9a).

- **Transmission EM** can also be used to examine negatively stained small particulate samples (e.g. viruses, bacteria—see Figure 1.6) and freeze-fractures, where a sample is frozen and fractured so that membranes are cleaved along their inside surface. A metal film is used to make a replica of the fractured surface that can be viewed in the TEM. This enables protein particle distribution in membranes to be visualized (see Figure 1.11), while, with etching, structural arrangements of the cytoskeleton can also be observed.

- **Scanning EM** can also be used to examine freeze-fractured samples, in this case looking at the original sample so that the 3D architecture inside cells and organelles can be observed (see Figures 1.7 and 1.9b).

So, what is a cell? The simplest forms, which include archaea, bacteria and simple blue-green algae (cyanobacteria) are prokaryotic organisms (Figure 1.5). These are microscopic organisms that live mostly as independent single cells, although they are often gathered into masses and colonies, and they contain all that is needed to function as a discrete organism. In a typical bacterium, the DNA strand floats free in a watery solution called the cytoplasm that contains ribosomes, small granular organelles needed to synthesize proteins, enclosed in a thin layer of lipids and proteins, the plasma membrane (Figure 1.6). Typically, the membrane is enclosed in a thick wall that gives the cell protection against adverse environmental conditions. The wall may be decorated by hairs (pili) and motility may be provided by a whip-like appendage, the motile flagellum.

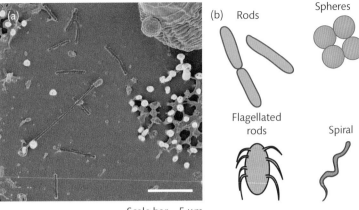

Scale bar = 5 μm

FIGURE 1.5
(a) SEM of pond bacteria showing two of the major forms, rods and spheres. (b) Diagrams of different forms of bacteria.

It is assumed that the earliest forms of cells were prokaryotes, but that cells of this type then evolved into a more complex form, the eukaryotic cell (Figure 1.7). Eukaryotic cells can also exist as free-living, independent organisms, in the kingdom Protista, but also form the basis of all complex multicellular organisms in the other kingdoms. Eukaryotic cells are more elaborate and typically larger than prokaryotes, some free-living Protista being easily visible to the naked eye, although mostly they are microscopic like prokaryotes.

The main differences between prokaryotic cells and eukaryotic cells is that the genetic material (the DNA) is gathered into a membrane-enclosed body, the nucleus (Greek: karyon) and that the cell cytoplasm contains a range of subcellular macromolecular complexes, termed organelles. Organelles are found in three main forms: membrane structures that help compartmentalize

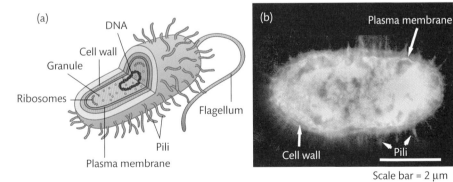

Scale bar = 2 μm

FIGURE 1.6
(a) The structure of a bacterium: DNA does not reside inside a nucleus but is located as a free-floating strand. The plasma membrane lies inside a tough cell wall and the cytoplasm contains ribosomes and granules. (b) Negatively stained TEM of a bacterium. The pili, cell wall and membrane are clearly visible.

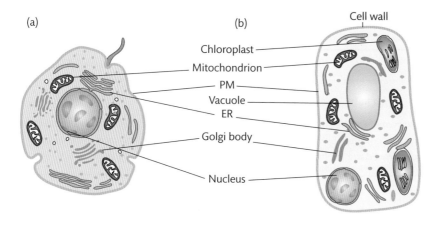

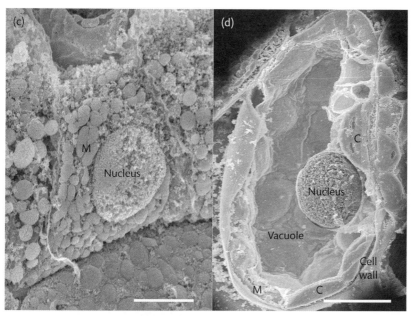

Scale bar = 5 μm   Scale bar = 10 μm

FIGURE 1.7

Eukaryotic cells. (a) Animal cell showing the principal organelles. (b) Plant cell for comparison. PM: plasma membrane; ER: endoplasmic reticulum.

Comparison between plant and animal cells by SEM. (c) Animal cell from the kidney, which is about 10 μm across; a central nucleus is surrounded by a packed cytoplasm containing organelles. (d) The plant cell from a leaf (shown at lower magnification than in c) is larger than the animal cell. A vacuole occupies the middle of the cell and the whole is surrounded by a thick wall. Mitochondria are present in both (M) but chloroplasts (C) are only found in the plant cell.

the cell into different regions, cytoskeletal structures that form scaffolding to support its shape and confer movement when required, and granular organelles that perform a variety of functions and act as storage or package molecules for release from the cell. The principal organelle types will be considered in greater detail in this chapter.

Which key invention enabled the development of cell theory?

Key points

Prokaryotes are the simplest forms of cells and exist as unicellular forms, although they may form colonies. Eukaryotic cells can exist as single-celled organisms but their appearance allowed the evolution of multicellular organisms of considerable complexity.

1.2 Origins of cells

Prokaryotic cells are already quite complex assemblages of mutually interacting, interdependent biological molecules. To consider in detail how the first life forms might have arisen to create these complex assemblages is beyond the scope of the present chapter. However, it appears likely to have been via a process of abiogenesis (i.e. organic life arising from non-living inorganic and organic reactions in the early environment). Some of the organic molecules needed to make a cell, including quite complex amino acids that make up proteins, can assemble spontaneously in the right conditions, and have been detected even in the harsh environment of space.

It is also assumed that cell membranes must have arisen through a spontaneous process. Phospholipids, molecules that combine a phosphate group (hydrophilic) and lipid chains (hydrophobic) are vital parts of the modern-day cell membrane, forming the lipid bi-layer where the polar phosphate groups are exposed on the outside on either side of the bi-layer, and the hydrophobic lipids lie inside away from the external aqueous environment. When phospholipids are mixed with water, they spontaneously form micelles, small droplets with the hydrophobic lipid chains inside the droplet and the polar heads forming the surface, and bi-layers that presumably could roll up into little packets. While spontaneously occurring phospholipids are not known, similar amphiphilic molecules can form in the environment and so produce bi-layers spontaneously. It seems likely that micelles or droplets formed from these bi-layers could trap and also protect other useful macromolecules, giving rise to early cellular forms. This may be how ancestral prokaryotes appeared, but many questions remain over this process: no clear idea has emerged about how self-replicating molecules like DNA that code for other biomolecules might have appeared spontaneously.

Better understood is how eukaryotic cells might have arisen from earlier, more primitive cells. While not universally held to be true, one assumption has been that eukaryotic cells evolved by combining elements of independent prokaryotic cells into a form of 'colony' or mutualistic aggregation that over time became an integrated single cellular system. In its most widely accepted form, this process is represented by the serial endosymbiosis theory (SET). The best examples of how this process might work are the probable origins of two of the main membranous organelles found in cells, mitochondria and chloroplasts (Figure 1.7). Mitochondria are bacterium-sized, usually rod-shaped organelles that are involved in producing the chemical fuel that provides the cell with its energy, and they are virtually ubiquitous. Chloroplasts are found in cells that photosynthesize (extract energy from sunlight to create biomolecules) such as plant cells.

Mitochondria and chloroplasts resemble prokaryotic forms. They contain their own DNA which floats in their internal matrix and self-replicate, although in present-day organisms they cannot be synthesized independently of the nuclear genome. Thus, it is postulated that mitochondria are derived from a free-living bacterium that became integrated with other prokaryotes into a common cell membrane where it provided some advantage to the other 'organisms' within the same membrane, a form of endosymbiosis. Similarly, chloroplasts could be a blue-green alga that has formed a symbiotic relationship, providing a fuel source utilizing photosynthesis. In fact, many present-day animals (e.g. *Hydra*, a common coelenterate organism living in ponds and streams) incorporate algae into their cells to derive an advantage from the by-products of photosynthesis. An endosymbiotic relationship with bacteria has also been observed in some eukaryotic cells. For example, bacteria can be found living and replicating in the cytoplasm of ciliated eukaryotic protozoa in the stomach of ruminants (Figure 1.8).

Details of the SET remain to be determined; there are some anomalies, such as the fact that the DNA strands in mitochondria and chloroplasts are rather shorter than those of typical bacteria or blue-green algae, and the nucleus contains genes that are involved in their replication, even though they cannot be made entirely from instructions in nuclear DNA. Thus, gene transfer seems to have taken place over time. Nevertheless, it seems highly likely that through aggregation and mutual advantage via a symbiotic process, eukaryotes evolved from prokaryotes and then gave rise to the more complex forms possible from a more sophisticated cellular design.

The SET gives a reasonable explanation of how eukaryotic cells may have evolved, but there are other theories; for example, those suggesting that prokaryotes evolved by simplification of eukaryotes, or that archaea, bacteria and eukaryotes all arose at about the same time. Evidence from the fossil record suggests that prokaryotic forms did pre-date eukaryotic forms, but the precise pathway for the evolution of eukaryotes continues to be a fascinating and still developing topic.

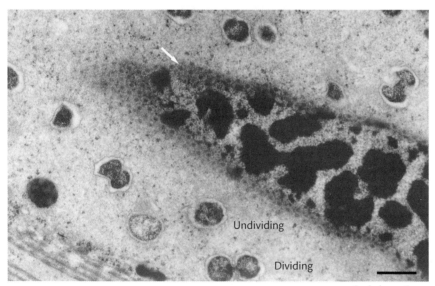

Undividing

Dividing

Scale bar = 1 µm

FIGURE 1.8

Transmission electron micrograph of endosymbiotic bacteria inside an endosymbiotic ciliate living in a sheep's rumen. The bacteria can be seen in various stages of cell division. Note also the protozoan nucleus filled with dense chromatin bodies and the envelope, here seen in tangential view, studded with nuclear pores (white arrow).

SELF-CHECK 1.3

What advantages might be gained by prokaryotic organisms becoming integrated into a single cellular unit?

Key points

Eukaryotic cells are thought to have evolved from simpler prokaryotic cells by serial endosymbiosis.

1.3 Diversity of prokaryotic cells

There are three main forms of prokaryote: bacteria, archaea and cyanobacteria. The basic structure of a bacterium has been described earlier (Figure 1.6). Bacteria vary in shape; they can be rods, spheres or spirals (Figure 1.6), range in size from 1 μm to 10 μm and are present in vast numbers in all ecosystems. They have ribosomes that are biochemically distinct from those of eukaryotes.

Archaea share similarities with, but are also distinct from, the bacteria and cyanobacteria in a number of ways. Many are extremophiles living in harsh environments where other prokaryotes and eukaryotes cannot survive. They tend to be extremely small (<1 μm) and possess cell membranes based on phospholipids that are different from those in bacteria and that, in some cases, appear to form single layers rather than bi-layers. Some use sunlight as an energy source (phototrophic) although they do not generate oxygen like photosynthetic organisms. Some are chemotrophic, using inorganic compounds such as sulphur or methane as an energy source. They contain eukaryote-like ribosomes, and have motile flagella that are similar to those of bacteria.

Cyanobacteria are distinct from archaea and other bacteria because they possess an elaborate system of internal membranes containing pigment that facilitates photosynthesis. This process, as in higher plants, generates oxygen; the pigment gives them their blue–green colour. They also contain ribosomes like those found in bacteria. The cyanobacteria typically form filamentous colonies composed of individual cells ranging from <1 μm to 20 μm in length. They have a bacterium-like cell wall and lipid/protein-based cell membrane. They lack flagella but can glide through the water in which they live.

1.4 The general structure of eukaryotic cells

Eukaryotic cells (Figure 1.7), like prokaryotic cells, are bounded by a plasma membrane. Unlike prokaryotes, however, they have multiple internal membrane systems that segregate the interior of the cell into compartments (Figure 1.9). The characteristic feature of the eukaryotic cell is the large, often spherical, membrane-enclosed nucleus containing the DNA of the cell. This is surrounded by flattened, sometimes very extensive stacks of membrane forming a network throughout the cell, the endoplasmic reticulum (ER), which is involved in the synthesis of

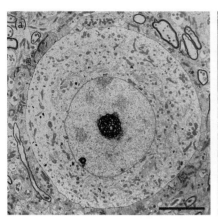
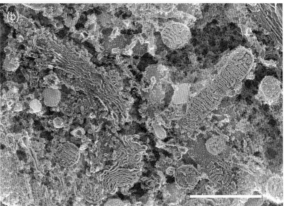

FIGURE 1.9

(a) A spherical neuron from the brain stem with its large central nucleus and dense nucleolus, and cytoplasm filled with organelles. Scale bar: 10 µm. (b) Various membranous organelles can be found in eukaryotic cells. Colour enhanced SEM showing ER (yellow), Golgi (blue) and mitochondria (red) in the cytoplasm (green). Scale bar: 1.5 µm.

cellular materials. Interspersed with these there are smaller, more discrete membrane assemblies called the Golgi apparatus, which helps to sort materials synthesized in the ER, the rod-like membrane-enclosed mitochondria, which are the energy supply for the cell, and a variety of membranous bodies such as endosomes, lysosomes and vesicles. Unique to plant cells is the chloroplast, a membrane-enclosed body with its own internal membranes that is required for photosynthesis. Comparison of animal and plant cells shows some other differences; for example, plant cells are bounded by a tough cellulose-based cell wall and commonly have a central fluid-filled vacuole (Figure 1.7).

In addition to membranes, all eukaryotic cells possess a range of filamentous organelles that cooperate with each other to form an internal cellular skeleton (cytoskeleton). The main forms of cytoskeletal filament are microtubules, intermediate filaments and microfilaments, each of which is a polymer form of a specific protein, but these filaments also have a wide variety of associated proteins that enable them to perform a range of different functions. These include providing structural support to maintain the shape of the cell and its different structures; positioning the membranous organelles in the cell and transporting them around; facilitating various motile functions that enable cells to move; and forming the basis of the cell division process that creates new cells.

Interspersed among these filaments and membranes, also contained in the cytoplasm is a range of granular organelles. These often contain secretory materials prepared by the cell for export, or used, for instance, to store fuel such as glycogen, a molecule that can be converted to glucose for respiration, a series of reactions that release energy for cellular processes.

The components of cells described here are, of course, not fixed, unvarying objects; each has its dynamic properties that change as the cell performs its own functions. Specialized cells show varying numbers of these organelles depending on their function; for example, cells that provide mechanical support often have an abundance of cytoskeletal proteins. Cells engaged in considerable protein synthesis show substantial amounts of ER. Cells that use a lot of energy contain many mitochondria. Thus, individual cells show considerable diversity in their structure and contents.

What are the principal types of organelle found in eukaryotic cells?

> ### Key points
>
> Prokaryotic cells lack nuclear envelopes, their DNA floating as strands in the cytoplasm. Eukaryotic cells have a membrane-enclosed nucleus that segregates the DNA from the remainder of the cell.

1.5 Membrane systems

Membrane systems permeate the cytoplasm of the cell and surround the nucleus, with the whole cell bounded by the plasma membrane. Most of the membrane systems flow into each other, creating aqueous compartments within the cell that are distinct from the cytoplasm, and there are import and export processes that involve the membrane systems. The mitochondria and, in plant cells, chloroplasts are excluded from the flow of membranes.

1.5.1 Plasma membrane

The plasma membrane provides a boundary around the cell that protects it from, and provides an interface with, the environment. According to the current model of the plasma membrane, the fluid mosaic model (Figure 1.10a), it is composed of proteins floating in a bi-layer of

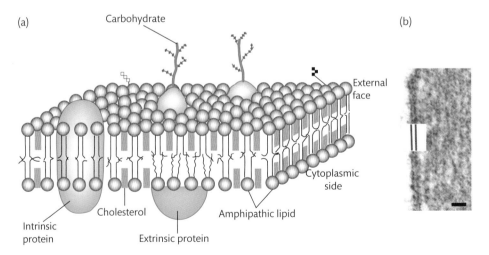

(a)   Carbohydrate   (b)

External face

Cytoplasmic side

Intrinsic protein   Cholesterol   Extrinsic protein   Amphipathic lipid

FIGURE 1.10

(a) Fluid mosaic model of the plasma membrane showing the lipid bilayer made of two layers of phospholipids tail-to-tail and various associated proteins. An external coat of carbohydrate, the glycocalyx, is also present. (b) The plasma membrane in a transmission electron micrograph with the two dense lines of the phospholipid heads separated by a transparent area representing the tails (as indicated in the inset). Scale bar: 25 nm.

phospholipids visible as a pair of thin parallel lines under the transmission electron microscope (TEM; Figure 1.10b), decorated by an external coating of carbohydrates forming the glycocalyx.

> ## Key points
> The plasma membrane is an interface between the cell and its environment with a semi-permeable structure that allows important substances to enter or exit the cell.

Lipid component of the plasma membrane

The phospholipids that form the bi-layer are amphiphilic molecules. This means they have a hydrophilic (water-loving) and hydrophobic (water-hating) part; they also have a polar phosphate group that is hydrophilic, attached to two hydrocarbon chains which are lipids and are hydrophobic. These are packed side by side and the two layers form because the hydrophobic lipid tails lie inside while the heads lie outside where they can interact with the aqueous environment, with the molecules arranged tail to tail (Figure 1.10). On average, a membrane comprises 50 per cent proteins and 50 per cent lipids, but this varies significantly: lipid-rich membranes are found, for example, in the myelin sheath that surrounds nerve axons. The opposite, a membrane with high protein content, is found, for example, in hair cells of the inner ear where the membrane has large numbers of prestin (a motor protein) molecules.

Cross reference
Nerve cells are considered in greater detail in Chapter 5.

The lipid bi-layer primarily provides a barrier as only small non-polar molecules can cross the hydrophobic central part. Thus, it excludes ions, which are charged particles, and proteins or carbohydrate molecules. However, many lipids, other small hydrophobic molecules, and gases can also cross the membrane. In addition, to some extent, water molecules can pass through the membrane, although primarily they pass through aqueous protein pores (see 'Proteins of the plasma membrane' which follows). The bi-layer is not necessarily symmetrical; different forms of phospholipid can be present on the internal surface than on the external surface. For example, the inner leaflet often contains phosphatidyl serine head groups that have a negative charge. This causes an intrinsic charge difference (membrane potential) to occur across the membrane, making the inside of the cell more negatively charged. This provides an electromotive force that tends to attract positive ions inwards and negative ions outwards. As the ions cannot cross the hydrophobic part of the membrane, the cell controls influx and outflow by other means (see 'Proteins of the plasma membrane' which follows). Other lipids found in the membrane include cholesterol, the presence of which can stiffen the membrane by reducing its fluidity. The lipid components confer various permeability and mechanical properties on the membrane.

> ## Key points
> On average, the plasma membrane comprises 50 per cent lipid and 50 per cent protein, the former being arranged in two layers (the lipid bi-layer) with hydrophilic heads.

Proteins of the plasma membrane

The proteins in the plasma membrane can be embedded in the bi-layer, in which case they are intrinsic (Figures 1.10a and 1.11), or associated with the membrane on either its internal or external surface, in which case they are extrinsic. Extrinsic proteins can be anchored in the membrane by small attached lipid tails that project into it, or by association with intrinsic proteins.

FIGURE 1.11

Freeze-fracture electron microscopy reveals the protein constituents of the plasma membrane as little particles. Here, the fracture plane has gone through the membrane where gap junctions communicate between two cells. The patches of particles represent the gap junction proteins, but in between there are also other particles representing different proteins. Scale bar: 100 nm.

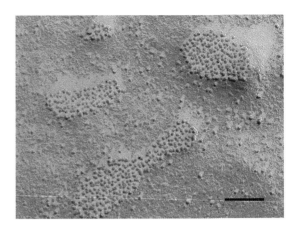

The main functions of a membrane are determined primarily by its protein content. Intrinsic proteins typically provide an opportunity for internal or external materials to cross the membrane. In this case, the two main forms of protein that allow this are channels and carriers. Channels are molecules composed of several subunits that collectively form a protein complex with a hydrophilic pore running through; this allows water, ions and small molecules to pass through. The pores can be opened or closed (gated) by a specific mechanism; these include chemical ligands that bind to the channel and cause it to open or close, such as neurotransmitters in the nervous system, voltage changes (see Chapter 5) or mechanical deformation by a physical structure. There is a huge variety of such channels: one, aquaporin, is involved in water transport or osmoregulation and is found in many places across all divisions of life; for example, in bacteria, in plant roots, and in kidney, intestine and lung in vertebrates (see Chapters 6 and 12). Mutations and other factors that affect channel function are the basis of a number of major diseases in humans (Clinical correlation, see below).

CLINICAL CORRELATION

Diseases caused by ion channel or transporter failure and other membrane protein abnormalities

- Cystic fibrosis is a genetically inherited disease caused by abnormality of chloride channels. Failure to distribute chloride and associated sodium ions results in thicker than normal mucus in the respiratory system causing poor lung function which leads to inflammation and infection.

- One inherited form of epilepsy (generalized epilepsy with febrile seizures) is caused by abnormality in a voltage-gated sodium channel.

- Type 1 episodic ataxia (periodic bouts of muscle non-coordination) can be caused by genetic abnormality in a voltage-gated potassium channel.

- Failure of glutamate transporters is considered to be a cause or contributory factor in various neurological disorders, such as stroke, amyotrophic lateral sclerosis and idiopathic epilepsy. Not only can inward transport of glutamate stop, but it may also go into reverse, causing neurotoxicity as a result of the build up of extracellular glutamate.

- Multiple sclerosis, where scar tissue in the form of plaques occurs in various regions of the central nervous system, is caused by an autoimmune response to myelin (the substance that wraps around nerve cell axons) and associated proteins.

Aquaporin specifically excludes ions from its pore, but other channels allow both positive and negative ions into or out of the cell with varying degrees of selectivity. There are sodium-, potassium- and chloride-specific ion channels, for example. Some of these can be gated by ligands, others by voltage, and they are often associated with specific activities such as the generation of action potentials in the nervous system. Cation or anion channels are less selective, allowing the passage of several positively or negatively charged ions, respectively.

Other proteins that facilitate the passage of substances across the membrane are called carriers or transporters. Some act passively, helping substances to travel down their concentration gradients (facilitated diffusion). The act of binding produces a change in the structure of the transporter protein, resulting in the transfer of the substrate across the membrane. This does not require the release of energy, hence it is passive, but it is facilitated transport as the substrate cannot cross the membrane without help. An example of facilitated diffusion is the release of glucose stored as glycogen inside the cells of the liver; facilitated diffusion allows the glucose to be released into the bloodstream.

Cross reference

The structure and function of liver cells is described in Chapter 11.

Active transport utilizes chemical energy to move molecules across the membrane against the concentration gradient (i.e. in the opposite direction to diffusion). In this case, the substance binds to the carrier and then, using an energy source, it pumps the molecule across the membrane. Two forms of pump, the symporter and antiporter, use energy derived from the electrochemical gradient of one ion to drive the transport of a second ion. For example, the glucose symporter SGLT1 co-transports one glucose molecule into the cell for every two sodium ions it imports into the cell—in other words, the molecules move in the same direction (hence symport). It is found, for example, in the small intestines and trachea. An example of an antiporter is the sodium–calcium exchanger that pumps calcium ions out of the cell using sodium import to drive the process—the ions move in opposite directions (hence antiport).

Other energy-utilizing transporters are enzymes that hydrolyze a molecule of adenosine triphosphate (ATP) to drive the target molecule from one side of the membrane to the other. Calcium can also be pumped by this means, using the plasma-membrane calcium ATPase. Another important molecule is sodium–potassium ATPase, which transports sodium out and potassium in against their concentration gradients.

Other types of plasma membrane protein act as receptors at the cell surface. Receptor proteins are a class that responds to the binding of a signalling molecule. This functional type overlaps with the ligand-gated ion channels noted above that can also be classed as receptors for the signalling molecule that causes them to open. Other receptor proteins may trigger a cellular response by initiating a 'second messenger' system within the cell; this is where activation of the receptor in turn causes activation of an intracellular molecule, often a G-protein, which can bind the guanine nucleotide in either the activating (triphosphate or GTP) or inactivating (diphosphate or GDP) form. When the receptor activates the G-protein, it binds GTP and the complex then affects some processes inside the cell through a cascade of chemical reactions.

SELF-CHECK 1.5

What component of membranes forms the bi-layer?

Key points

The protein composition of membranes is variable and primarily determines the membrane function in most cases. Protein receptors and ion channels are required for cells to interact with their environment and to respond to signals from other cells.

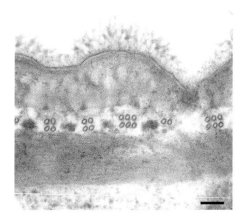

FIGURE 1.12
The glycocalyx, which in some cells is particularly prominent. This ciliate from a sheep's rumen has thick protective glycocalyx-like bushes sprouting from its pellicular surface. Other interesting structures include the bundles of subpellicular microtubules, seen as little circles in cross section. Scale bar: 100 nm.

The glycocalyx

The carbohydrate component of the membrane is found on its outer surface (Figure 1.12). It is composed of polysaccharides (polymers of sugar-like molecules) that are attached to, and extend from, the membrane proteins. The combination means that these molecules are called glycoproteins or proteoglycans and differ in the relative content of carbohydrate to protein and also in the type of carbohydrate present. The glycoproteins/proteoglycans associated with the membrane can be intrinsic molecules, passing through the membrane, or extrinsic molecules, associating with the extracellular part of intrinsic proteins. Another way in which carbohydrates are associated with the membrane is as glycolipids, a combination of lipid and carbohydrate.

The functions of the glycocalyx are varied; for example, it provides protection from chemical or physical injury, mediates immune reactions, provides cellular recognition characteristics and allows cell-to-cell adhesion. During fertilization, it allows the sperm to recognize the egg cell, and during development it enables cells to recognize where they are and where they are going during migration.

1.6 **Nucleus**

The nucleus (Figures 1.7, 1.9 and 1.13) in an animal or plant cell is typically a spherical structure (Figure 1.9), although it can have other forms (Figure 1.13) and is generally located centrally, provided the cell is not in the process of dividing. It consists of nucleoplasm bounded by a double layer of membrane (the nuclear envelope). This envelope is punctuated by pore complexes; these are not simple openings but have a defined structure that allows the selective passage of materials into or out of the nucleus. Inside the nucleus, the electron microscope reveals dense bodies (chromatin) and a specific subregion called the nucleolus (Figure 1.9a) in a lighter ground substance, and small particles in the lighter substance.

The chromatin contains DNA wrapped up into complexes with nuclear proteins (histones). There are two forms of chromatin: euchromatin, which is loosely packed to allow access to the chemical machinery that reads (transcribes) the genetic instructions in the DNA; and heterochromatin, which is more densely packed and in a form in which the DNA is not available for transcription. The transcription of DNA results in the production of a

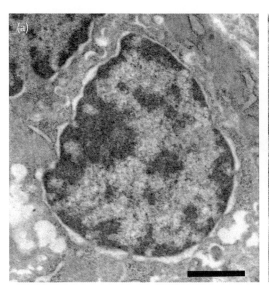 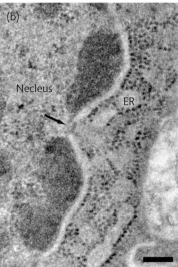

FIGURE 1.13

(a) The nucleus contains dense chromatin, lighter nucleoplasm and is surrounded by a nuclear envelope composed of two membranes with a gap in between (light halo). Scale bar: 1 μm. (b) The envelope contains pores (arrow) and is continuous with the ER. Scale bar: 200 nm.

messenger molecule (messenger RNA [mRNA]). The RNA molecule is similar to DNA in some respects in that it consists of nucleotides coupled by a sugar-phosphate backbone. Unlike DNA, which has a helically wound, double-stranded structure, RNA is typically single stranded. The mRNA is an instruction released from the nucleus to drive protein synthesis in the cytoplasm.

The nucleolus is the site of ribosome production, and the ribosomes ultimately end up in the cytoplasm. They are made of two different-sized components, the 60S and the 40S subunits. Each comprises ribosomal RNA (rRNA) and proteins. The ribosomal subunits do not form a complete ribosome within the nucleus but pass through the nuclear pore complex to form ribosomes in the cytoplasm, and these function in protein synthesis. Many different molecules are needed to make this process work correctly. The ground substance contains the ribosomal subunits, water and other molecules such as mRNA.

The nuclear pore complex regulates the passage of molecules into and out of the nucleus and therefore separates the mechanisms of transcription of the DNA code in the nucleus from translation of the code into protein synthesis in the cytoplasm. It has a structure consisting of nucleoporins, protein subunits that form an octagonal arrangement around the rim of the pore, with some central gating structures also evident, forming the basket. The gating of the nuclear pore is selective, requiring interaction between phenylalanine-glycine repeats that are part of the nucleoporin subunits and a transporting molecule that can carry another molecule (the cargo) in to or out of the nucleus through the pore. Small molecules do not need this gating process to pass through the pore, but larger cargo requires a nuclear localization sequence for inward passage, or a nuclear export sequence to leave the nucleus. These sequences, a group of amino acids with a specific signature, allow cargoes to be targeted and transferred by specific transport molecules (Figure 1.14).

(a)   (b)

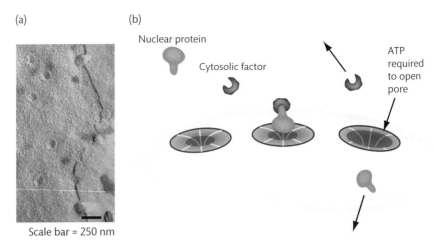

Scale bar = 250 nm

FIGURE 1.14

(a) Nuclear pores seen by freeze fracture electron microscopy. (b) Operation of the nuclear pore complex. The pore consists of six subunits and is 80–100 nm across. Gating of the pore requires ATP and proteins targeted for the nucleus, a signal peptide to identify them, and be guided in by cytosolic factors.

SELF-CHECK 1.6

How do instructions encoded in the DNA in the chromosomes get out of the nucleus?

Key points

The nucleus not only segregates the DNA from the rest of the cell, but also controls the process of protein synthesis by release of mRNA.

1.7 Cytoplasm

1.7.1 Endoplasmic reticulum

The endoplasmic reticulum (ER) is a membrane system that is concentrated around the nucleus but extends in a network throughout the cell. Like the plasma membrane, the ER membrane has a lipid bi-layer with proteins associated with it. The membranes are arranged in sacs called cisternae, bounded on either side by the membrane, separating an internal compartment (the lumen) from the external cytoplasm. There are two forms of ER found in cells: rough ER (RER) and smooth ER (SER).

The RER is so called because in electron micrographs its surface is seen to be studded with ribosomes, giving the appearance of a granular coating (Figures 1.15 and 1.16). Typically, the cisternae of RER are flattened and lie in stacks. In places, the lumen of the RER and its enclosing membranes are continuous with those of the nuclear envelope, which also can have ribosomes on its outer surface. The SER forms sacs and tubules that are less flattened than those of RER and lack the ribosomes (Figure 1.16).

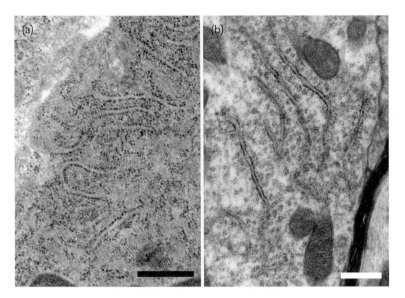

FIGURE 1.15

Rough endoplasmic reticulum (RER) in two different kinds of cell (a) Intestinal cell. (b) Neuron. The RER in both cases consists of stacks of double membranous cisternae outlined on the either side by ribosomes (black spots). Groups of ribosomes (polyribosomes) are also floating in cytoplasm. Scale bars both 0.5 μm.

Functionally, the RER is where protein synthesis occurs, with ribosomes becoming attached to its surface as they translate the mRNA instructions produced in the nucleus (Box 1.3). As a ribosome in the cytosol starts to translate the mRNA strand into a protein, the initial stage of the protein contains a specialized region called the signalling peptide that binds to a receptor on the ER membrane. The protein inserts into the ER membrane so that the ribosome becomes associated with its surface, and as the protein grows it passes through the membrane

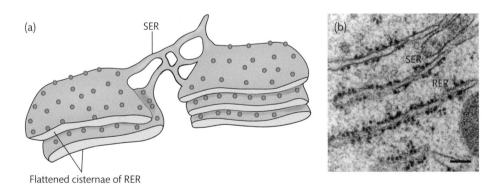

FIGURE 1.16

(a) Schematic diagram of rough (RER) and smooth (SER) endoplasmic reticulum. The flattened stacks of cisternae of RER are studded with ribosomes and continuous with SER membranes. (b) Detail of the continuity between RER and SER in a TEM image. Scale bar: 100 nm.

BOX 1.3 *The synthesis of proteins*

- Protein synthesis, summarized in Figure 1.17, requires mRNA from the nucleus, which binds to a large and a small ribosomal subunit.

- In order for proteins to be synthesized in the ER, a signal peptide is first transcribed that attaches the ribosome to the ER membrane.

- When proteins are destined for the cytoplasm, the ribosomes do not attach to the ER but float free.

- The RNA moves along the ribosome, attaching amino acids in sequence to produce the protein; more ribosomes join the first part of the RNA to make more copies of the protein, forming either attached or free-floating polyribosomes.

- In the ER, as the protein grows it is translocated through the membrane to be internalized.

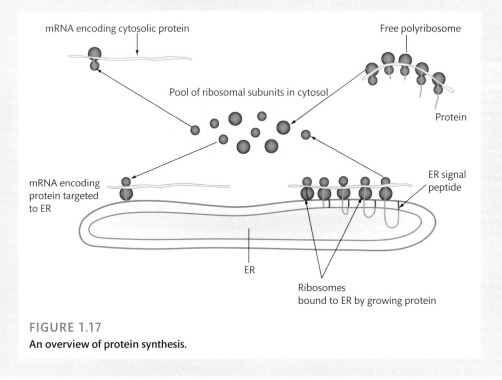

FIGURE 1.17
An overview of protein synthesis.

into the lumen. The mRNA strand moves along the ribosome as it is translated and the initial translated part attracts another ribosome, and this second ribosome forms an association with the surface, next to the first. As more ribosomes are added, up to about six single ribosomes, the string of ribosomes along the strand curls into a spiral, producing the characteristic appearance of a polyribosome. Within the lumen of the ER there are chaperone and stress proteins that help to regulate the process of protein synthesis. Calcium is also stored in the ER and takes part in this process.

The SER lacks the ribosomes that are a feature of RER but is involved in various synthetic processes, in particular the production of lipids and in the metabolism of carbohydrates. It can also act, with RER, as an intracellular calcium store, involved in calcium homeostasis. The SER is also associated with the detoxification of harmful substances.

> **Key points**
>
> Endoplasmic reticulum functions both in protein and lipid synthesis, the former in conjunction with ribosomes.

1.7.2 Golgi apparatus

Another membrane structure that takes part in the production of molecules in the cell is the Golgi apparatus, also known as Golgi body (Figure 1.18a and 1.18b). Golgi bodies are discrete structures that lie in the cytoplasm, interspersed with the ER and in some respects are continuous with ER membrane systems. In fact, the nuclear envelope, ER and Golgi apparatus take part in exchange of membranes and the continuity of their internal lumens through membrane flow, which in some cases results in release of materials out of the cell (Figure 1.19). Vesicles, which are small packets bounded by membrane, bud off from the ER and join the Golgi apparatus in a well-defined region of the latter. The content of these vesicles are the proteins that have been translated in the ER and which then undergo modification and sorting in the Golgi apparatus.

The Golgi body consists of stacks (cisternae) of membranes enclosing a lumen. There is a well-defined entry point for ER membranes, the *cis*-face of the Golgi membrane stack, and an exit point, the *trans*-face of the stack. Vesicles join on or bud off at the edges of the cis and trans cisternae, respectively, and transfer the contents between adjacent stacks in the middle of the Golgi body (Figure 1.20). They can also return to the ER from the Golgi cisternae. The vesicles and their contents are guided through the membranes by resident proteins that recognize parts of the system, some of which guide them back to the ER, while others guide them to

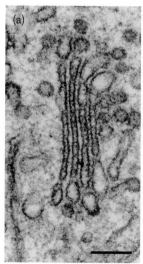
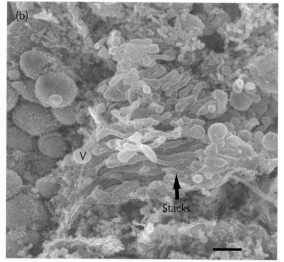

Scale bar = 1 μm   Scale bar = 1 μm

FIGURE 1.18

The Golgi apparatus seen by (a) TEM and (b) SEM. Flattened stacks of membranous cisternae form the centre of the apparatus, with vesicles (V) budding from them.

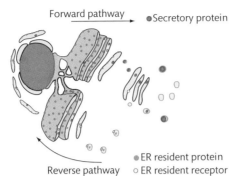

Forward pathway
●Secretory protein

ER resident protein
○ ER resident receptor
Reverse pathway

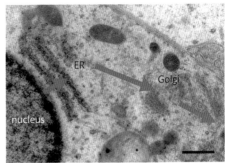

Scale bar = 2 μm

FIGURE 1.19

The diagram shows the pattern of membrane flow guided by resident proteins and receptors that target the different cisternal and vesicle membranes and their contents along the pathways. Secretory proteins synthesized in the ER pass through the forward pathway. Membranes return via the reverse pathway. The transmission electron microscope (TEM) image shows the close association of the nuclear, ER and Golgi membranes in a cell.

the output of the Golgi apparatus where the modified proteins are enveloped in a vesicle or membranous body.

The contents of the vesicles produced by the Golgi apparatus vary: some contain proteins destined to be incorporated into the plasma membrane, while others contain enzymes that are packaged into a lysosome. Lysosomes fuse with debris or pathogens and digest them, and are used to destroy damaged organelles in the cell. Some vesicles are secretory.

SELF-CHECK 1.7

Give three key ways in which the ER differs from the Golgi apparatus.

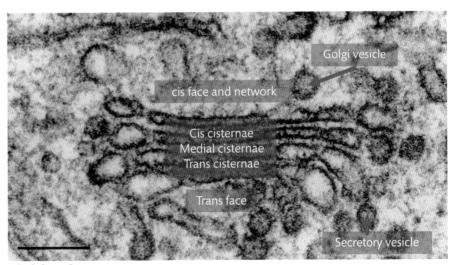

Scale bar = 1 μm

FIGURE 1.20

The Golgi apparatus and the organization of its cisternae.

Key points

The Golgi apparatus sorts proteins synthesized in the rough endoplasmic reticulum and determines their ultimate fate.

1.7.3 Exocytosis and endocytosis

Exocytosis

The flow of membranes through the membranous compartments of the nuclear envelope, ER and Golgi body also involves the plasma membrane. The contents of secretory vesicles generated by the Golgi apparatus are released from the cell in a process called exocytosis. There are many examples of exocytosis, for example: (i) hormones such as insulin are released by secretion from cells in the pancreatic islets of Langerhans into the bloodstream to travel to their target cells and organs; (ii) neurotransmitter is released from secretory vesicles in nerve terminals (Figure 1.21) into the synaptic cleft between two contacting neurons—neurotransmitter vesicles differ, however, in that they can be recycled by local membrane flow; and (iii) mucus is secreted by goblet cells into the intestinal lumen to protect the intestinal wall from degradation (see section 1.10).

The basic process of exocytosis requires the secretory vesicle to fuse with the plasma membrane so that the contents are released outside the cell. Neurotransmitter vesicles, for example, are targeted to the membrane through a variety of proteins in the nerve terminal that bind to them, carry them to the membrane and then trigger their release.

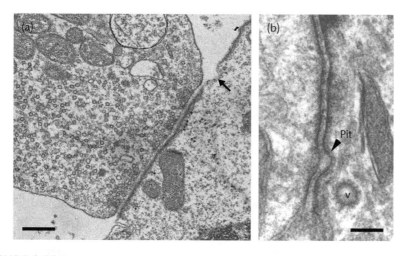

FIGURE 1.21

(a) Secretory vesicles at a synapse. The vesicles are located in the nerve ending and fuse with the membrane to release their contents. This image also shows an example of probable endocytosis (arrow) in the post-synaptic cell. Scale bar: 500 nm. (b) Endocytosis via clathrin-coated pits and vesicles (v) at a synapse. Scale bar: 100 nm.

In the case of many secretory events, if this process were to continue without compensatory action by the cell then the plasma membrane would grow and grow by membrane addition, so the cell would swell. To avoid this, the cells also have an endocytotic mechanism.

Endocytosis

Endocytosis (Figure 1.21) fulfils two functions: (i) it retrieves the internal membranes from the plasma membrane, thus preventing the continued growth of the latter; and (ii) it permits the uptake of external molecules and materials by a means independent of the channels and carrier proteins of the plasma membrane. Endocytosis is used at synapses, for example, to retrieve neurotransmitter, although this occurs in concert with other local mechanisms (see Chapter 5).

Endocytosis can occur in several different ways; for example, pinocytosis, macropinocytosis, receptor-mediated endocytosis (also known as clathrin-mediated endocytosis), via caveolae or phagocytosis. Pinocytosis is the inward budding of a small area of membrane to form a pit, which is then internalized as a vesicle. It is associated with the uptake of solutes and liquids. In macropinocytosis, a larger area of membrane reaches out and surrounds small particles. Both forms of endocytosis are non-specific in that they do not target a particular molecule or substance. Caveolae are flask-shaped areas of membrane, larger than pinocytotic pits, with stable lipid-protein domains. Although their functions are not well understood, they do appear to be involved in uptake of materials. In receptor-mediated endocytosis, specific external target molecules bind to receptor proteins on the cell membrane, activating the formation of a small clathrin-coated pit, followed by internalization to form a clathrin-coated vesicle. On a larger scale than any of the former is phagocytosis. Here, the plasma membrane bulges out, stretching around and enclosing a target object usually the size of a bacterium, or in the case of protistan, organisms living in ponds or the sea, or other small eukaryotic organisms.

SELF-CHECK 1.8

What is the main difference between exocytosis and endocytosis?

Key points

In addition to the semipermeable membrane, substances can enter or exit the cell via endocytosis and exocytosis.

1.7.4 Mitochondria

Mitochondria are a major membrane-bound cellular organelle, but they do not take part in the exchange of membrane occurring in the secretory pathway. Mitochondria are typically considered to be about the size of a bacterium and are rod shaped. They do, however, display various forms; in particular, they can be quite long and branched. The significance of the branching is unclear—it may reflect part of the replication process where new mitochondria form by division of an existing one.

The mitochondrion is bound by an external (outer) membrane surrounding an internal (inner) membrane. The inner membrane is thrown into folds, called cristae, which project into the

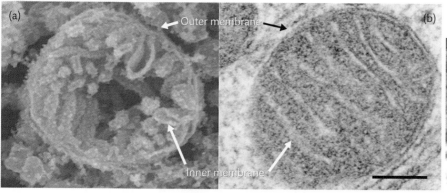
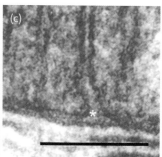

FIGURE 1.22

Mitochondrial structure. (a) SEM and (b) TEM at the same magnification. The mitochondrion has an outer membrane surrounding the whole structure and an inner membrane that lies inside the outer membrane and folds into the interior of the organelle to form cristae. A point of folding can be seen in (c) (*). Scale bars: 500 nm.

matrix inside the organelle (Figure 1.22). They contain their own DNA strand and replicate semi-autonomously. To achieve this they also have internal ribosomes, which are smaller (55S) than those of the cytoplasm (80S) or bacteria (70S) and contribute most of their proteins. However, they also rely on nuclear genes to contribute to replication.

The surface of the cristae facing the matrix, and the matrix itself, contain proteins that are primarily enzymes associated with the citric acid cycle (Krebs cycle; Figure 1.23), which, among other reactions, leads to the generation of ATP for use as fuel for cellular reactions. Hence, the mitochondrion is often referred to as the 'power house' of the cell, providing the energy it needs for life.

In summary the:

- composition of the matrix, the intramembrane fluids and the external cytoplasm are all different; the first two are controlled by transmembrane proteins in the inner and outer mitochondrial membranes that regulate the transfer of molecules for import or export and involvement in the Krebs cycle.
- outer membrane contains transport molecules called porins.
- inner membrane contains ATPases and specialized transport proteins.
- matrix contains enzymes of the citric acid cycle.
- intermembrane space contains several enzymes that use ATP formed in the matrix to phosphorylate (add a phosphate group) other proteins.
- Krebs cycle 'starts' with the conversion of pyruvate and fatty acids into acetyl coenzyme A (Figure 1.23); it produces nicotine adenine dinucleotide hydride (NADH) and flavine ADH (FADH2), which contain energetic electrons that are used to pump H^+ out of the matrix and this drives adenosine triphosphate (ATP) synthesis.

As well as ATP generation, mitochondria possess a number of enzymes involved in lipid metabolism and can synthesize membrane components such as phosphatidic acid, CDP-diacylglycerol, phosphatidyl glycerol, and cardiolipin. However, they cannot synthesize phosphatidylcholine, phosphatidylserine, phosphatidylinositol or sterols so these need to be

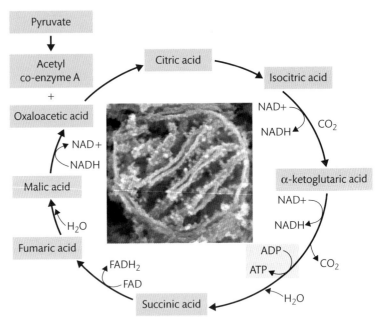

FIGURE 1.23

Simplified view of the Krebs cycle (also known as the citric acid cycle or the Szent-Györgyi-Krebs cycle), which takes place in the mitochondrial matrix and generates ATP as fuel for the energetic processes of a cell. The enzymes of the cycle (coloured green in the central image of a mitochondrion) are attached to the cristae (membrane folds), coloured blue. NADH and $FADH_2$ assist in the production of more ATP.

imported from other organelles. Nevertheless, mitochondrial lipid synthesis may contribute substantially to overall cellular phospholipid components.

SELF-CHECK 1.9

What is meant by semi-autonomous replication of mitochondria?

Key points

Mitochondria are organelles that provide the main site for the generation of ATP, the energy source for the cell.

1.8 Membrane systems and disease

The previous sections have discussed how membranes provide the main ways in which cells are subdivided into compartments and that mitochondria are required to generate fuel in the form of ATP for cellular activities. Consequently, membrane systems and mitochondria are present in the vast majority of eukaryotic cells. Diseases which target these membrane systems (Clinical correlation) therefore tend not to be specific to individual cell types or organs, but are heterogeneous.

CLINICAL CORRELATION

Diseases of membrane systems

- The main functions of the endoplasmic reticulum (ER) are synthesis, folding and sorting of proteins. Functional defects could occur at any of the transport and sorting steps involved in this function. For example, mutations in the signal peptide affecting membrane translocation or signal processing are thought to cause familial hypoparathyroidism, coagulation factor X deficiency, inherited central diabetes and pancreatitis. Misfolding of a chloride channel (see Clinical Correlation 1.2) is thought to underlie cystic fibrosis.

- Diseases affecting the Golgi apparatus include polycystic kidney disease, an autosomal dominant condition affecting adults. Transport of proteins from the trans-Golgi network to the membrane of proximal tubular cells is impaired by alterations in cell structure. Other diseases involve inappropriate glycosylation (addition of carbohydrate groups) to proteins in the Golgi apparatus, causing, for example, mental abnormality and liver disease.

- Nuclear envelope deficits, such as those caused by mutations in lamins, produce fragile nuclear membranes that are less resistant to mechanical stress. An example of a disease caused by this is progeria.

- Mitochondrial diseases are commonly caused by mutations in mitochondrial DNA. They often affect multiple organs and tissues because mitochondria are present in virtually all cells (except red blood cells); two mitochondrial diseases include mitochondrial myopathy (muscle weakness) and 'diabetes mellitus and deafness'. Inherited mitochondrial disorders are maternal as the organelles are semi-autonomous and so require the original maternal mitochondria to replicate (the sperm cell does not contribute mitochondria during fertilization).

1.8.1 Cytoskeleton

The cytoskeleton is a complex array of macromolecular filaments that spreads throughout the cytoplasm, helping to position organelles and conferring motile properties either on subcellular structures or on the cell as a whole; it also plays a vital role in the cell division process. The three major classes of filamentous structure are, in decreasing order of diameter, microtubules, intermediate filaments and actin filaments (microfilaments) (Figures 1.24 and 1.25). Each of these three classes of filament is accompanied by a range of associated proteins that enable each to self-interact, to interact with the other filaments, or to associate with membranes of various organelles.

> ### Key points
> The cytoskeleton consists of three main classes of polymerized protein filaments and a range of associated proteins that provide structural support, motility and transport.

Microtubules

As the name implies, microtubules are tubular filaments, typically about 24 nm in diameter. They are ubiquitous in eukaryotes, but tubular organelles with structural and molecular similarity to eukaryotic microtubules have also been found in some prokaryotes.

The major constituent of eukaryotic microtubules is a globular protein called tubulin, which is transcribed as three different isoforms, α, β, and γ. A typical microtubule is composed of dimers of α- and β-tubulin. Under the right conditions, the dimers polymerize into chains called protofilaments (Figure 1.26). Polymerization requires the presence of GTP, which stabilizes the

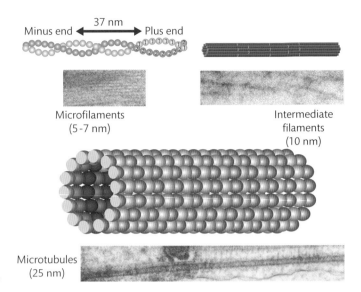

FIGURE 1.24
The three main types of cytoskeletal filament.

joining of two dimers, then, with another molecule of GTP, another dimer is added so that the protofilament lengthens. As the protofilament forms it then joins with adjacent protofilaments to form a sheet that curls up into the complete tubule. The most common types of microtubule contain 13 protofilaments, but there are variations in this number between different cell types and within a single cell (Figure 1.27). One of the known prokaryotic microtubules has 12 protofilaments.

The presence of GTP helps to form a stable cap at the growing end of the microtubule (termed the plus end), on to which new dimers can add. At the other end (termed the minus end), the microtubule can depolymerize, by hydrolysis of the GTP to GDP, making it unstable so that the dimers disengage. Thus, a microtubule shows dynamic instability—it can grow or shrink according the rate of addition or loss of dimers. This enables it to move or to change position, or to move other organelles that are attached. Thus, as well as having a structural role, microtubules can also be highly dynamic.

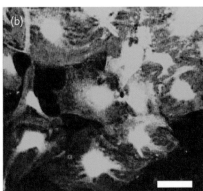
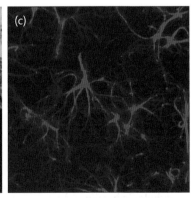

FIGURE 1.25
(a) Actin and (b) tubulin staining in cultured fibroblasts. (c) Glial fibrillary acid protein (intermediate filament) in astrocytes from the brain, viewed with a confocal microscope. The actin forms bundles/cables clearly visible as lines in the cells. The tubulin surrounds the nucleus and forms finer strands spreading through the cell body. Scale bar: 100 nm.

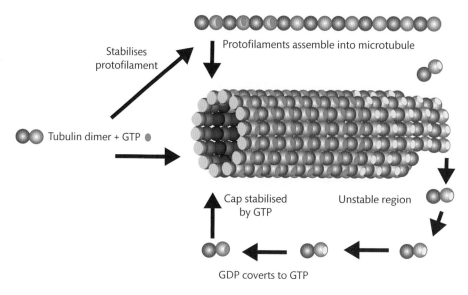

Stabilises protofilament

Protofilaments assemble into microtubule

Tubulin dimer + GTP

Cap stabilised by GTP

Unstable region

GDP coverts to GTP

FIGURE 1.26

Polymerization of microtubules from tubulin dimers. Addition of GTP to the dimer allows them to assemble into protofilaments, which in turn join up to form a microtubule. The GTP keeps the microtubule stable, but conversion to GDP causes instability and depolymerization.

There is a range of other proteins associated with microtubules that contribute to their functions. One family is called microtubule-associated protein (MAP) and includes MAP1, MAP2, MAP4, and also MAPT (tau protein). While MAP1 and MAP2 are found in nerve cells, MAP4 is common to a wider range of cells. These proteins are involved in stabilizing or destabilizing microtubules and also, potentially, in linking them to other structures. There are also cytoplasmic linker-associated proteins (CLASPs), which facilitate interactions with membranes, and other proteins, such as kinesin and dynein, that utilize ATPase to move organelles along microtubules, or slide adjacent microtubules with respect to one another.

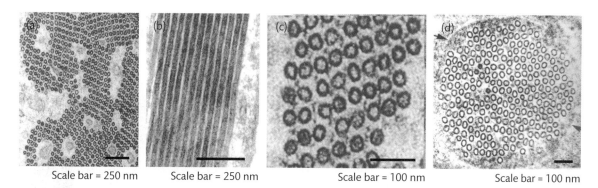

Scale bar = 250 nm   Scale bar = 250 nm   Scale bar = 100 nm   Scale bar = 100 nm

FIGURE 1.27

Microtubules in bundles. (a) Transverse and (b) longitudinal section of a thick bundle of microtubules from an inner ear supporting cell. These microtubules are the same diameter and are linked together by intermicrotubular links. (c) Detail from (a), illustrating the typical ring-like profiles in cross section. (d) A bundle of microtubules from the spindle of a rumen ciliate. In this example, microtubules of several different diameters, and presumably therefore different protofilament number, can be seen.

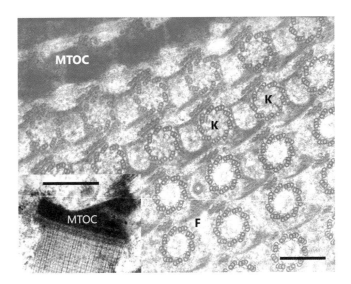

FIGURE 1.28

A complex microtubular network in a ciliated protozoan. As well as microtubules forming rows of kinetosomes (K, also known as basal bodies), there are various other fibrils (F) and interacting materials. The kinetosomes sit on dense bars that act as MTOCs. In the inset, an MTOC can be seen in side view with a thick microtubule bundle coming from it. Scale bars: 200 nm.

Cross reference

Cell division is discussed in Chapter 13

Microtubules are organized into various bundles (Figure 1.27) and networks (Figure 1.28). Partly this is a function of the associated proteins and crosslinkers, but also these arrays are organized by microtubule-organizing centres (MTOCs). In addition, there are microtubule-nucleating sites (MTNCs), which can initiate polymerization of microtubules. A ubiquitous MTOC is the centrosome, consisting of a pair of centrioles. The centriole is a structure composed of nine triplets of microtubules and containing dense material. The centrosome is typically located close to the nucleus and is a self-replicating body containing RNA, which can organize the microtubular cytoskeleton and the spindle that is used for separating chromosomes during cell division. Some MTOCs consist of a relatively large electron-dense area when seen under the electron microscope (Figure 1.28), with microtubules extending away from it, and they can contain the γ-tubulin that is involved in nucleating microtubules.

The functions of microtubules are varied; for example, they are found in the cilia (see 'Structure and function of cilia' that follows) that are also common to virtually all cell types, at least at some point in their development. Cilia are motile protrusions that cover the surfaces of cells, often where movement, either of the cell itself or of materials in the external medium, is required. Cilia line the airways where they contribute to the mucociliary escalator that removes dust and other particulate material from inside the lungs (see Chapter 6), and also the internal surface of the Fallopian tubes where they move along ova (egg cells) that have been released into the tube for fertilization. Cilia also form the tails of sperm cells, enabling them to swim towards the ovum. They are the main motile organelles of ciliated protozoa.

Structure and function of cilia

The structure of a cilium is highly conserved, meaning it is similar in all circumstances in which it is found, provided there is no disease or damage. It is also closely related to the flagellum, which is a longer version found in various protozoa, plants and simple animals. The ciliary

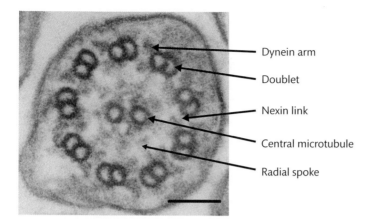

Dynein arm

Doublet

Nexin link

Central microtubule

Radial spoke

FIGURE 1.29

Structure of a cilium with two central microtubules and nine peripheral doublets. Various features can be recognized within the axoneme (central core) of the cilium. Scale bar: 50 nm.

(and flagellar) axoneme, the central structure that supports the body of the protrusion, is composed of a group of nine microtubule pairs (doublets) and a central pair of single microtubules (Figure 1.29). The doublet microtubules consist of one standard 13-protofilament microtubule (subfibre A) with a second partial microtubule (subfibre B) of ten independent protofilaments, and an additional four to five being shared with subfibre A.

Various linking proteins occur regularly along the shaft of the axoneme: dynein arms shaped like wellington boots regularly extend from each doublet towards the adjacent doublet; nexin links connect one doublet to the next; the central microtubules are surrounded by sheathing material, and radial spokes project from them towards the outer doublets (Figure 1.29).

The cilium moves with a whipping motion, the stroke produced by sliding of the doublets along one another on one side of the axoneme. The sliding occurs because the dynein arms, which are composed of an ATPase, bind to the adjacent doublet microtubules and release energy by converting ATP to ADP, in the process producing a power stroke that causes the two connected doublets to slide against each other. The whipping action is not symmetrical; the main stroke produces force against the surrounding fluid, while the recovery stroke does not, so that propulsion occurs in one direction. In some circumstances, in particular in ciliated protozoa such as *Paramecium*, in which the cilia are the main motile agent, cilia beat together synchronously in metachronal waves to produce a smooth, gliding action.

Other microtubular structures and functions

The cilium is an example of a major organelle that is composed of microtubules. However, the microtubules often appear as networks where a level of organization is less apparent. The dynamic nature of polymerization and depolymerization means that such networks are constantly shifting and changing in response to the needs of the cell.

There are also circumstances in which microtubules form thick, stable bundles. An example of this is in the inner ear where mechanical struts are needed to support a vibrating tissue (Figure 1.27). Other examples include the thick microtubular bundles of certain protists (Figure 1.28).

Another circumstance in which the microtubular network is a compromise between order and dynamic activity is in the spindle that is involved in cell division. As will be seen later, the process of cell division separates genetic material, in the form of chromosomes, into two daughter cells. The separation is achieved by microtubules attaching to the chromosomes, drawing them apart from their sister chromosomes.

SELF-CHECK 1.10

Microtubules are found in virtually every eukaryotic cell, but are they also found in prokaryotes?

Key points

Microtubules are scaffold-like structures that can be highly dynamic, changing length and transporting other organelles around inside the cell, as well as providing motile structures such as cilia and the cell-division spindle.

Intermediate filaments

As their name implies, intermediate filaments are intermediate in size between microtubules, which are larger, and microfilaments, which are smaller (Figure 1.24). They are about 10 nm in diameter and are more varied in composition than are either microtubules or microfilaments as various different proteins can make intermediate filaments. They form tough, durable polymers supporting the nuclear envelope and cells in general. In epithelia, they can form transepithelial networks that link cells, via desmosomes, into a strong interlinked sheet.

There are five classes of intermediate filament:

- type I: acidic keratin produced by epithelial cells
- type II: basic keratin produced by epithelial cells
- type III: includes vimentin (produced by endothelial cells and leucocytes), glial fibrillary acid (nervous system glial cells—Figure 1.25) and desmin (muscle cells)
- type IV: neurofilaments—heavy, medium and low molecular mass forms (nerve cells)
- type V: lamins (intranuclear), which support the nuclear envelope.

Intermediate filaments are composed of rod-like monomers that polymerize initially into a coiled dimer, which then form pairs (tetramer) consisting of four monomers. Multiple tetramers coil around one another to form a long chain, or filament. These filaments are very stable but can show a degree of dynamic activity, with polymerization and depolymerization.

SELF-CHECK 1.11

How do intermediate filaments differ from microtubules?

Key points

Intermediate filaments are less dynamic than other cytoskeletal filaments and serve primarily to provide structural support in cells and in the nucleus.

Microfilaments

Microfilaments represent a third class of fibrillar organelle, which is composed of a 49 kDa molecular mass globular protein called actin. Actin exists as monomers in the cytoplasm,

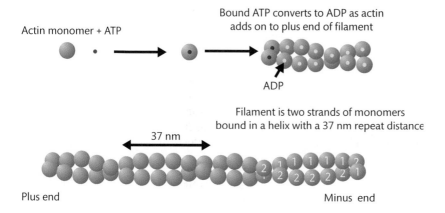

FIGURE 1.30
Polymerization of monomeric actin to form microfilaments.

which can polymerize, as tubulin dimers do, into filaments. They are different from micro-tubular filaments, however, in a number of ways. First, a complete microfilament is composed of two chains of actin polymers wound around one another in a helix, with an overall diameter of 5–7 nm, and a helical periodicity of 37 nm. The filaments polymerize by addition of mono-mers at the plus end, and depolymerize by loss of monomers at the minus end. To stabilize the filament, the actin monomer binds to a molecule of ATP, which then converts to ADP as the monomer joins the filament (Figure 1.30). The relative rate of polymerization and depolymer-ization determines whether the filament lengthens or shortens, and equal addition and loss of monomers results in treadmilling, as is the case with microtubules. Actin functions in multiple cell activities primarily involving structural support or movement. It is a major component of muscle cells where, together with another protein that forms filaments, myosin, it forms the contractile element that shortens muscles during their contraction. Actin has three main iso-forms (α, β and γ). α-actin is the muscle-type actin, while β- and γ-actin are present in most, if not all, non-muscle cells, although γ-actin is implicated in the development of muscles. The isoforms all have relatively similar amino acid sequences, with about 97 per cent homology, and β- and γ-actin may be able to substitute for each other. Mice and humans lacking γ-actin are relatively normal overall but do show progressive hearing loss.

Actin interacts with various other proteins, some of which regulate polymerization and depoly-merization, such as profilin and thymosin, respectively, and others that allow actin filaments to connect together in a range of networks, called actin-associated proteins. Within cells, three kinds of actin-rich structures can be defined by the way in which the actin and their associated proteins interact (Figure 1.31). There are gel-like networks in which actin filaments are not aligned with each other but instead cross over in a network. There are actin bundles where a number of actin filaments lie in parallel all with the same direction (i.e. all aligned so that their plus ends are at the same end), crosslinked by short protein bridges such as plastins. There are also actin bundles in which the filaments lie in anti-parallel arrays (i.e. the plus and minus ends of adjacent filaments are at opposite poles of the bundle).

An example of an actin network includes the elastic submembranous lattice that allows red blood cells to deform without damage as they pass through the narrowest of capillaries. This network is composed of various proteins including spectrin (an elastic molecule), ankyrin (a molecule that anchors actin to transmembrane proteins), and the membrane proteins them-selves that enable the actin-spectrin lattice to act on the membrane. Actin gels can have dif-ferent properties depending on cooperation between different linking proteins. For example,

Cross reference

See Chapter 10 for more detail on muscle tissue

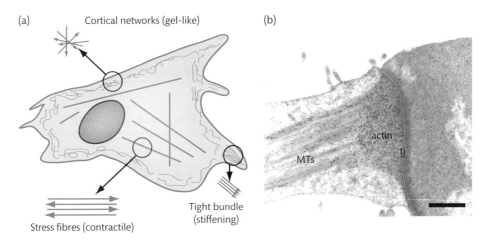

(a) Cortical networks (gel-like)   (b)

actin

MTs   tj

Stress fibres (contractile)

Tight bundle
(stiffening)

FIGURE 1.31

(a) Different types of actin network in cells. Cortical networks are like gels, stress fibres form contractile cables, and tight bundles form stiffening structures. (b) Actin in a cortical network, interacting with microtubules (MTs). This TEM image also illustrates immunogold labelling. The actin has been labelled with gold particles (visible as black dots) that allow its precise location to be seen. The actin gel is located next to a tight junction (tj) between two cells. Scale bar: 200 nm.

filamin and α-actinin, both actin-associated linking proteins, work together to make an actin gel much stiffer.

Actin bundles are common in a number of cell types. They form the stiff core of microvilli, such as those found in intestinal epithelial cells forming the brush border, and the enlarged micro-villi (stereocilia) of hair cells that detect mechanical vibrations. Actin bundles, called stress fibres, are found in fibroblasts of ligaments and are also commonly seen in cultured mammalian cells *in vitro*. As noted earlier, they also occur in muscle cells were they are interspersed with myosin filaments. Actin bundles can be strengthened by increasing the number of bridging proteins that hold the filaments of the bundle together.

The formation and regulation of the actin cytoskeleton is under the control of a number of proteins in response to extracellular signals. In many cases, Rho GTPases can act as intermediaries between these signals and the actin networks. Espins also help in the formation of actin bundles in hair-cell stereocilia and in testes. Plastins/fimbrins also act as actin-bundling proteins in intestinal brush border epithelia and hair cells. EPS8 and related proteins can affect the length of actin bundles and regulate their activity. The control of the actin cytoskeleton is complex and dynamic, and motile and sedentary cells both depend greatly on this functionality

SELF-CHECK 1.12

In which type of cell is actin found in the greatest quantity in the human body?

Key points

Actin forms cable-like structures which can confer various properties, from stiffening projections of the cell and supporting the plasma membrane to providing movement of intracellular organelles and also of whole cells.

The above description of the cytoskeleton illustrates the importance of cytoskeletal components and their associated proteins in maintaining cell structure and powering dynamic activities, both intracellular and extracellular. Of necessity, such an important component of cells can go wrong in various ways resulting in a number of associated diseases and disorders (Clinical correlation).

CLINICAL CORRELATION

Diseases caused by cytoskeletal abnormalities

■ Microtubules are almost ubiquitous organelles in the human body. Necessarily, if they go wrong, cells which are heavily dependent on them for their functions are affected. A major example of this in the central nervous system is Alzheimer's disease, where degeneration of cells is accompanied by the collapse of the microtubules due to defective tau protein.

■ Disorders caused by defective intermediate filaments include progeria, where lamin is mutated, and various skin and hair disorders where different forms of keratin are genetically abnormal. One example, monilethrix, causes beading and fragility of hair; another, epidermolysis bullosa simplex, causes blistering of the skin.

■ Disorders of actin filaments also present in many different forms, and vary according to which actin isoform is affected. Mutations in α-actin, the main isoform in muscle cells, results in over 140 different myopathies; disorders of γ-actin primarily affect hearing and the intestine, in which tissues this isoform is abundant.

Granular organelles

The main types of granular organelle are ribosomes (which have already been described) and storage organelles. These can be used to stock up fuel reserves of, for example, glycogen, which is found as granules in the cytoplasm of a range of cell types. Glycogen granules form a rapidly mobilizable pool of material that can be used to obtain glucose quickly when needed. Plant cells also contain starch grains, another type of storage material, often within chloroplasts.

1.9 **Cell division**

One of the most important activities a cell can undergo is cell division, where the cell replicates itself to produce two offspring. In one form of cell division, mitosis, the offspring are genetically identical, containing a full complement of the original cell's genetic material. However, in the other form of cell division, meiosis, genetic material is reduced by half during the process to produce a gamete. In males, the gametes are sperm cells, while in females they are eggs; when the former fertilizes the latter, the resultant offspring (zygote) contains a novel combination of genes from the two gametes, producing a completely new organism (see Chapter 13).

Mitosis is used as a means to replace cells that die and also during development. Many adult mammalian cells are terminally differentiated, meaning they have reached a final specialized state and cannot perform this activity. However, stem cells or progenitor cells in various tissues and also during embryogenesis have to divide to produce new cells, in the former because of ongoing cell death (which may be a perfectly normal condition of the tissue) and in the latter because new cells are needed to make up the huge number that constitute the body of an adult organism.

An example of a constantly renewing tissue is the blood. It is composed of red cells (erythrocytes) and white cells (leucocytes) at a proportion 99 per cent red cells to 1 per cent white cells. Red cells in mammals are unusual in that once they are mature they lack a nucleus (this is not the case in birds or reptiles) and cytoplasmic organelles, including mitochondria. The absence of mitochondria means that the erythrocytes, which still require active membrane pumps and other ATP-dependent activities, eventually run out of fuel and begin to die, typically living about 120 days. Thus, they need to be replaced constantly, and the source of new erythrocytes in adults is primarily bone marrow, although other tissues such as the spleen and thymus take part in maturation.

Cross reference

Red and white cell families and blood cell genesis are covered in more detail in Chapter 4.

Key points

Cells cannot arise *de novo* (assemble spontaneously), they have to be derived from a pre-existing cell through cell division.

1.9.1 Mitosis

All cells arise from a preceding cell division cycle. In the somatic cells of most animals (those making up an organism's body), each cell contains a double set of chromosomes (diploid state) in which virtually all the cell's genetic code is located (with the exception of mitochondrial DNA), one set from the father and the other from the mother. In humans, there are 46 chromosomes (i.e. 23 from each parent). Two of the 46 are the pair of sex chromosomes, designated X and Y. Females contain two X chromosomes (XX) while males contain an X and a Y chromosome (XY). The remaining chromosomes are called autosomes, and they are also paired, so that there are 22 homologous (containing equivalent genes) pairs of autosomes (chromosomes 1 to 22).

The majority of cells spend most of their time in **interphase**, a state in which they are not ready for cell division, and carry out their day-to-day activities. If a cell is destined to undergo mitosis then cell division proper takes place (Figure 1.32) in the following sequence:

- *Interphase gap 1 (G1)*: The cell grows, forms organelles and synthesizes proteins. As the cell readies itself for cell division, the centrosome replicates and the two centrosomes remain together at one side of the nucleus.
- *Synthesis phase (S)*: DNA is replicated to form a complete copy.
- *Interphase G2 (occurs in many, but not all, cells)*: The cell prepares for cell division.
- *Prophase*: The DNA begins to coil up into chromosomes. The DNA synthesis results in all chromosomes being duplicated, each to contain identical copies called **chromatids**. The two centrosomes also separate to either side of the nucleus and begin to form a spindle. The chromatids acquire kinetochores, a region to which microtubules can attach. As prophase nears completion, the nuclear envelope begins to break down and the nucleolus disappears.
- *Metaphase*: The nuclear envelope disappears, the chromosomes line up together, along a plane called the metaphase plate, and the spindle microtubules emanating from the centrioles attach to the chromatids.
- *Anaphase*: The chromatids are pulled apart and move along the spindle to opposite poles of the cell, thus producing a complete copy of all the chromosome pairs on either side.
- *Telophase*: The spindle disintegrates and the cell undergoes cytokinesis, whereby the cell body divides into two by means of a contractile ring of actin and myosin (in animal cells) around the cell's equator. New nuclear envelopes form to produce a pair of daughter cells.

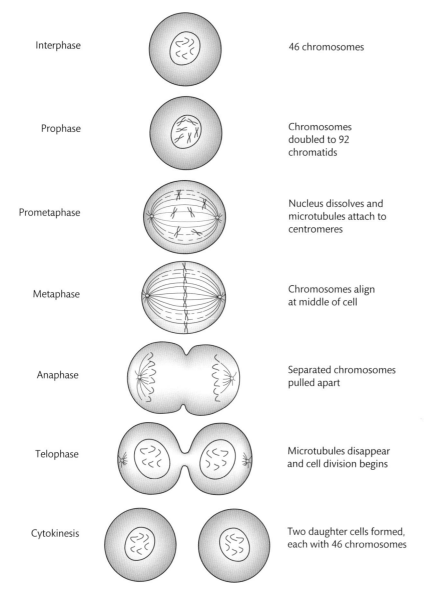

Interphase		46 chromosomes
Prophase		Chromosomes doubled to 92 chromatids
Prometaphase		Nucleus dissolves and microtubules attach to centromeres
Metaphase		Chromosomes align at middle of cell
Anaphase		Separated chromosomes pulled apart
Telophase		Microtubules disappear and cell division begins
Cytokinesis		Two daughter cells formed, each with 46 chromosomes

FIGURE 1.32
Mitosis.

Key points

Mitotic division produces two exact copies of the parent cell and is used to generate new somatic cells (body cells that make up tissues and organs) during growth and repair.

1.9.2 Meiosis

Meiosis initially is somewhat similar to mitosis. However, it is also called a reduction division because, from a single cell, the end result of the process produces four daughter cells,

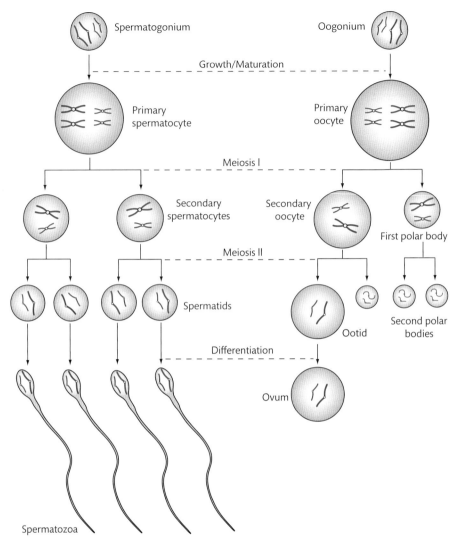

FIGURE 1.33
Meiosis.

each containing only half of the normal chromosome complement (Figure 1.33; see also Chapter 13). Hence, the number of chromosomes in a human cell is reduced from 46 (diploid) to 23 (haploid). In this process, the chromosome pairs are assigned randomly to daughter cells so that the new genome is not identical to either parent. In addition, when the initial pairing of homologous chromosomes occurs, the chromatids can intertwine and swap parts with one another, hence recombining genes from different parents into new 'mixed' chromosomes. This recombination leads to new combinations of genes and characteristics, which ultimately helps to cause natural variation in organisms, a process that contributes to natural selection and evolution. As ecological systems change, a highly variable population may yield individuals who are more likely to survive the changes.

The stages of meiosis have similar features to those of mitosis but with some important differences:

- *Prophase I*: As in mitosis, DNA synthesis results in each chromosome duplicating and forming pairs of chromatids. The nuclear envelope starts to break down and the spindle starts to form. The pairs of homologous chromosomes come together and their chromatids inter-twine and undergo recombination.

- *Metaphase I*: The chromosome pairs now align along the metaphase plate and spindle microtubules begin to attach to the kinetochores on the chromatids. Here, there is a differ-ence with mitosis—the kinetchores of each chomatid pair align together and attach to spin-dle microtubules emanating from the same pole. Those of the homologous chromosome align towards the opposite pole.
- *Anaphase I*: The spindle microtubules now pull the homologous pairs apart (unlike in mitosis, where the sister chromatids for each chromosome are separated at this stage).
- *Telophase I*: The separated chromosomes gather at either end of the cell and a cytokinetic furrow forms. Both offspring are now in the haploid state but each chromosome contains two chromatids.

In some organisms a nuclear envelope briefly forms at this stage (called interkinesis), but it is normal for meiosis to continue without. The daughter cells then each undergo a second division:

- *Prophase II*: The chromosomes remain condensed and a new spindle forms.
- *Metaphase II*: The chromosomes align along the metaphase plate and the spindle microtubules attach to the chromatids, this time in opposite directions for each chromatid pair.
- *Anaphase II*: The chromatids are pulled apart to either side of the cell.
- *Telophase II*: Cytokinesis separates the two daughter cells and each now contains a haploid set of chromosomes with a unique combination of genes.

Meiosis is used in males to create male gametes (sperm) and in females to produce female gametes (eggs). However, in human females, gamete production is complicated. Meiosis starts in the fetus to produce primary oocytes, but it is halted until puberty when eggs start to mature into secondary oocytes and are released for fertilization each month. The secondary oocyte is also arrested in meiosis until fertilization has occurred. In contrast, after puberty a male con-tinues to produce sperm cells by meiosis throughout his life (see Chapter 13).

During sexual reproduction the haploid sperm and egg fuse together (fertilization) and create a diploid zygote. Owing to the fact that it is made from two completely new combinations of genes, derived from the parents but recombined, the offspring will have a new combination of characteristics. Subsequently, the zygote divides mitotically so that all the cells of the offspring are genetically identical, except for their own gametes which are produced when maturation is complete. In some circumstances, the fertilized egg divides into two separate identical zygotes producing identical (monozygotic) twins—dizygotic twins are produced when two separate eggs are each fertilized by two different sperm.

1.9.3 Abnormalities in cell division

Mistakes in cell division give rise to a number of different conditions in humans. One example is Down's syndrome, which is caused by a person having three copies of chromosome 21 (trisomy 21). In this case, chromosome 21 has failed to separate from its counterpart during the second stage of meiosis to produce gametes. The result is two copies in one gamete and none in the other gamete. The former then takes part in fertilization, and the resultant zygote has three copies of the chromosome, leading to the developmental abnormalities associated with this condition.

1.9.4 Cell death: programmed versus unprogrammed

Cells die in a number of ways throughout the lifetime of an organism. The cause can be traumatic (e.g. from a wound or infection) in which case it is called necrosis because the cells die in an uncontrolled manner and disintegrate to release their contents, including enzymes that damage other nearby cells and tissues. The debris from this process is phagocytosed by macrophages, white blood cells whose role is to remove the dead cells and materials before they cause too much damage or promote infection (e.g. septicaemia).

Cross reference

You can read more about controlled cell death, or apoptosis, in Chapter 15.

Other cells die as part of a natural replacement or developmental process, or to maintain a stable population. A good example is skin cells which under normal conditions divide in the lower layers of the skin but then are moved to the outer layers where they lose their organelles and are keratinized (i.e. filled with the protein keratin, which is also the basis of hair and nails). The cells of the skin are shed constantly so they have to be replaced to maintain a stable population and a complete protective layer. This is an example of controlled cell death in the adult animal.

Key points

Cells die not only because of damage and disease, but also because they need to be refined in number and type.

During embryonic development, programmed cell death (apoptosis) is used to shape the final organism. For instance, the human hand is initially developed with webbed fingers and toes. Under normal circumstances, the webbing dies away in an organized manner to allow the digits to separate and operate independently. The nervous system also contains too many cells during development and those that are not needed in the adult are removed using controlled cell death.

In apoptosis, the cells tend to go through some specific morphological changes (Figure 1.34). The cell surface starts to become convoluted and the chromatin in the nucleus condenses.

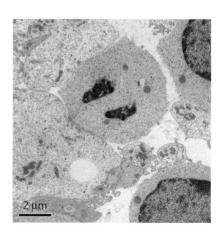

FIGURE 1.34

Apoptosis in a mutant mouse where inner ear development is halted. The nuclear chromatin has condensed and the nucleus fragmented into two. There are few organelles remaining and the cytoplasm is condensed. Scale bar: 2 μm.

The cytoplasm also starts to condense. The nucleus begins to produce 'blebs' (small, round chromatin-filled bodies) and the cell splits into small lumps, called apoptotic bodies, which can then be phagocytosed. As the membrane remains intact around these bodies, the cell contents do not escape and cause damage, as occurs in necrosis. The process is triggered by soluble signalling factors (growth factors, hormones and even intercellular contacts) and these signals activate a cascade of biochemical events that leads to the morphological changes, at the core of which are enzymes called caspases. As well as caspases, a number of other genetically programmed factors function in apoptosis. Thus, this is a genetically controlled process leading to programmed elimination of cells.

When programmed cell death fails to occur as required, in adults as well as during developmental stages, then cells may continue to replicate and give rise to tumours. Thus, we need programmed cell death to ensure the useful replacement of organs, but unprogrammed cell death can lead to serious health consequences.

SELF-CHECK 1.13

Give two main ways in which apoptosis differs from necrosis.

> **Key point**
>
> Apoptosis (or programmed cell death) is the controlled destruction of cells preventing escape of potentially damaging intracellular enzymes.

1.10 **Different cells and tissue formation**

Tissues are collections of cells of various different kinds that work together to fulfil the tissue's function and, together with other tissues, make up organs of the body. To make a tissue, cells have to be connected so that they do not fall apart. To facilitate this they are usually embedded in an extracellular matrix that provides support and helps to maintain the integrity of the tissue. In addition, connections between cells can allow materials to pass directly from one cell's cytoplasm to its neighbour (gap junctions), tie cells together (adherens junctions), or prevent the passage of materials between cells (tight junctions). Intermediate filaments also form intracellular networks that associate with the adjacent cell's intermediate filaments via adherens junctions (desmosomes) and so form transcellular networks.

A feature of cells in tissues is that there may be several types the structure and composition of which are adapted to different functions. An example of this is in intestinal epithelium (Figure 1.35) where columnar cells are absorptive (taking in the digested nutrient) and goblet cells secrete mucus to line the intestinal wall and protect it from physical damage as food is moved along. Goblet cells are so named due to their goblet shape. The cup-like cell body is filled with a high concentration of RER, with mucus secreted into the central cavity. In contrast, columnar epithelial cells are roughly cylindrical, with no central cavity, and show a brush border of tiny hair-like projections, called microvilli, that increase their surface area.

Another adaptation is that cells which use higher amounts of energy, such as liver cells (hepatocytes) that are involved in detoxifying the blood (see Chapter 10), will contain relatively more mitochondria. In hepatocytes, mitochondria account for about 40 per cent of the cell

Cross reference

The alimentary canal is considered in more detail in Chapters 7 and 8.

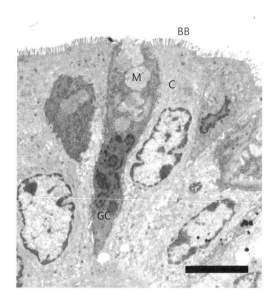

FIGURE 1.35

Goblet cells (GC) and columnar (C) cells in the intestinal epithelium. The goblet cell secretes mucus (M) while the columnar cells absorb digested materials and have a brush border (BB) of microvilli to increase their surface area. Scale bar: 5 μm.

membranes. In contrast, in a secretory pancreatic cell, mitochondria respresent about 20 per cent of the cell membranes. The opposite is true for the RER, which is required by the pancreatic islet cell for synthesis of insulin, where the proportion of RER is 65 per cent of the membranes compared with 35 per cent in the hepatocyte.

SELF-CHECK 1.14

If a cell contains a large amount of RER, what do you think its main function is likely to be?

Key points

Most cells share a common organization, containing similar organelles and cytoskeletal features, but the relative amounts of different organelles vary in each individual cell type according to their function.

1.10.1 Specialization of specific cell types

Not only do cells vary considerably between different organs and within organs and tissues, according to their specific functions, but also even the same cell type can be varied systematically according to specific need. Examples of this are found in several organs, but it is demonstrated elegantly in the cochlea (the hearing organ of the inner ear), the retina (where light is detected in the eye) and on the surface of the body among skin cells.

In humans and other mammals, the cochlea is a spiral structure with a long internal sensory epithelium, the organ of Corti, sitting on a membrane, the basilar membrane. Located in the epithelium are two types of sensory hair cell (inner and outer hair cells) that detect sound-induced vibrations of the basilar membrane, and these cells are covered by specialized

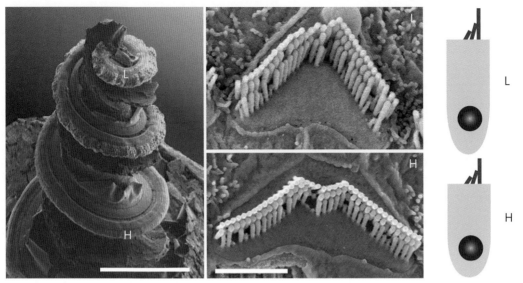

Scale bar = 1 mm

FIGURE 1.36

Cochlear hair cell specializations. The spiral organ of the cochlea (left) shows rows of hair cells with tiny hairs (stereocilia) that emerge from their tops (middle). The height of the stereocilia and the width of the rows (middle and right panels) and the length of the cell body (right) all become shorter towards the bottom of the spiral, corresponding to changes in the frequency to which the cells respond best, from low (L) to high (H). Scale bar in middle image: 2 μm.

actin-containing protrusions called stereocilia (actually these are enlarged microvilli). Both the outer hair cell body and its stereocilia become systematically shorter from the top to the bottom of the spiral, while maintaining essentially the same cellular organization (Figure 1.36; see also Chapter 5), an adaptation that is thought to reflect the ability of the cells to respond more effectively to increasingly higher frequencies of sound.

Another example of this is the rod and cone cells of the vertebrate retina. These cells have essentially the same organization: an outer segment contains membrane disks studded with molecules of photopigment, an inner segment includes the nucleus and cell organelles, and it ends in an axon. The rod outer segment is much longer, providing a larger membrane surface, as it is designed to detect lower light levels, and the cone cell is shorter with a smaller membrane surface area carrying variations in the photopigments, allowing it to detect different wavelengths of light for colour vision.

A third example is keratinization of skin cells. These cells are organized in layers, the lower living layers gradually giving way to increasingly keratinized upper layers of skin cells, with those on the surface being dead. The degree of keratinization is greater on friction-bearing surfaces, such as the skin on the soles of the feet and the palms of the hand, than elsewhere on the body surface.

Key points

Cells specialized for a particular activity also show specialization between them which adapts them to specific aspects of their overall function.

Chapter summary

Although cells of the body share common organelles and generally have a similar composition, they are also highly modified compared with each other, in order to perform specific functions. By and large, all cells in a multicellular organism share the same genetic instructions. Clearly, therefore, it is how that code is read, and how gene expression is controlled in different cell types during development and beyond, that is the key to producing the various types in the community of cells that is an organism.

- Cells are the basic unit of life and, with the exception of certain particulate forms that can reproduce under some conditions (viruses), make up all living things.

- The two main types of cell are prokaryotes, which lack a nucleus, and eukaryotes, which possess a nucleus.

- Eukaryotes may have evolved from prokaryotes acting together in a communal way initially, until eventually they became associated together within a single membrane.

- The modern eukaryotic cell is encapsulated in a plasma membrane, which contains a watery cytoplasm within which are a number of organelle types and a nucleus containing the genetic material of the cell.

- The organelles occur in three forms: membranous, filamentous or granular.

- Membranous organelles, with the exception of mitochondria and, in plant cells, chloroplasts, take part in membrane flow where they interchange membranous components during protein synthesis and sorting in the cell, and give rise to secretory materials that are destined for release from the cell.

- Cytoskeletal organelles provide the internal skeleton for the cell that gives it its form and architecture, and also provide structures for cell motility, intracellular transport and cell division.

- Granular organelles are involved in protein synthesis (ribosomes) and the storage of materials.

- Cells arise from previously existing ones through a process of cell division called mitosis, which produces exact copies. Gametes that are needed to produce new organisms arise from previously existing cells through a reduction division, meiosis, which halves the number of chromosomes present and permits a new combination of genes to be produced.

- Excess or damaged cells are destroyed through two main pathways: necrosis, which has side effects that can be damaging to other cells; and apoptosis (programmed or controlled cell death) that removes unwanted cells efficiently to maintain stable populations or reduce to the final number of cells during development.

- Cells of different types are specialized in various ways, and cells of the same type can show variations that fit them precisely for their specific function.

Further reading

- **Alberts B, Johnson A, Lewis J, Raff M.** *Molecular biology of the cell* 5th edn. New York: Garland Science, 2008.

- James Morré D, Mollenhauer HH. Microscopic morphology and the origins of the membrane maturation model of Golgi apparatus function. *Int Rev Cytol* 2007; **262**: 191–218.

- Khurana S. *Aspects of the cytoskeleton. Advances in molecular and cell biology.* London: Elsevier Science, 2006.

- Margulis L. Serial endosymbiotic theory (SET) and composite individuality. Transition from bacterial to eukaryotic genomes. *Microbiology Today* 2004 Nov; **31**: 172–4.

 # Discussion questions

1.1 How might mitochondrial genes become transferred to the nucleus, if the SET theory is correct?

1.2 Can viruses be considered a life form?

1.3 What determines how a cell becomes one specific type?

1.4 From where does endoplasmic reticulum in a cell arise if it is needed for its own synthesis?

1.5 Why should proteins have different isoforms?

Answers to self-check questions and discussion questions, hints and tips are provided on the book's Online Resource Centre.

 Visit www.oxfordtextbooks.co.uk/orc/orchard_csf/.

2

Studying cells: essential techniques

Gavin Knight

Learning objectives

After studying this chapter you should confidently be able to:

- Describe the historical aspects of cells and cell theory.
- Describe the application of various techniques to the study of cells.
- Compare and contrast the different types of microscopy and their applications.
- Discuss the principles underpinning the different methods of cell enumeration.
- Compare and contrast the different technologies available for the investigation of a patient's cytogenetic status.
- Describe the application of molecular biology to cytogenetic analysis.
- Discuss the role of molecular techniques in the study of cells.

As biomedical scientists, we are interested in how the body works in health and disease. Whether our specialty is histopathology, cytology, haematology, blood transfusion, microbiology, biochemistry, or even molecular biology, it is essential to have an understanding of the structure and function of cells and the technologies available to help answer fundamental questions about the human body. This chapter examines the role technology plays in furthering our understanding of cells, with particular emphasis on those found in the human body. As demonstrated throughout this chapter, our understanding of the human body depends on the technology available at the time. Over the course of 400 years, our understanding of the human body has increased exponentially, largely facilitated by the development of microscopy. At this point we begin our journey into studying cells.

2.1 Where did the study of cells begin?

The first light microscope was developed by the spectacle makers Hans and Zacharias Jansen in Middleburg, Holland, in 1595. Zacharias took credit for this development, although many believe he received some help from his father. By placing two convex lenses in a set position within a tube, the Jansens produced the first compound microscope. Although this microscope is no longer in existence, it was described in a letter by William Borel, the Dutch ambassador to the court of King Louis XIV of France, as a vertical brass tube almost two and a half feet in length with a lens fixed at each end, stabilized by a dolphin-shaped tripod.

It was not until the use of microscopy by Robert Hooke, curator of experiments at the Royal Society in London, that the composition of organisms and their fine structure was reported. In 1665, Hooke published *Micrographia*, containing his original microscopy investigations. Importantly, in his examination of cork, Hooke was the first scientist to identify a pocketed honeycomb structure. Each pocket he described as a cell (from the Latin *cella* meaning 'a small room'), a term we use routinely today.

The first in-depth scientific study using microscopy was performed in 1661 by Marcello Malpighi, who visualized the structure of the frog lung and demonstrated that blood flows in a system composed of capillaries. These findings were published in *De Pulmonibus Observationes Anatomicae* and confirmed the structure of the circulatory system originally proposed by William Harvey, a prominent physician of his time, in his 1628 publication *Exercitatio Anatomica de Motu Cordis et Sanguinis in Animalibus*.

Significant progress in the study of cell structure was made by the Dutch draper Antonie van Leeuwenhoek, who, in 1678, improved upon microscope lens design, and then identified and examined bacteria, protozoa and spermatozoa in fluids. In a series of publications, van Leeuwenhoek described these microorganisms as 'animalcules' based on his observations. Following confirmation of his findings by Hooke, van Leeuwenhoek was invited to join the **Royal Society**.

It was not until 1838 that Matthias Jakob Schleiden, a German botanist, stated that cells or their products make up every part of a plant. Subsequently, in 1839, the zoologist Theodor Schwann formulated the same conclusions for organisms in general, stating that cells are elementary parts of organisms. These published comments led to the beginnings of 'cell theory'.

In 1861, Max Schultze provided a definition of a cell as a small mass of nucleated protoplasm, and in 1865 he described, for the first time, four of the five major types of white blood cell (neutrophils, eosinophils, monocytes and lymphocytes), with only basophils missing from his studies. Incredibly, in his study, Schultz visualized the granules in neutrophils and eosinophils, and demonstrated the phagocytosis of small particles by neutrophils (Figure 2.1).

Some 340 years after van Leeuwenhoek's advances, microscopy remains a fundamental tool in the investigation of cells and tissues. Whereas initial applications of microscopy were purely in the research arena, it is now used as an essential diagnostic and research tool. In the next section of this chapter, microscopy and its applications will be considered in more detail.

Royal society

Founded in 1660, the Royal Society has the mission statement to 'recognise, promote and support excellence in science and to encourage the development of science for the benefit of humanity'. The world's oldest scientific journal *Philosophical Transactions of the Royal Society* is published by this society.

SELF-CHECK 2.1

What was the difference in the microscopy studies of Hooke, published in 1665, and Leeuwenhoek, in 1678?

2.2 **Introduction to microscopy**

Microscopy involves the resolution of structures beyond the ability of that possible with the naked eye. Since the introduction of microscopy by Jensen in 1595, huge technological advances have seen progress made, from the typical compound microscope to the assembly of small, powerful machines that help us visualize the ultrastructure of cells. In some situations, these powerful microscopes can even allow us to visualize molecular complexes.

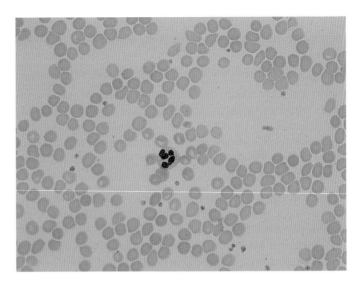

FIGURE 2.1
Peripheral blood photomicrograph using brightfield microscopy. A neutrophil is in the centre, surrounded by numerous red blood cells. The small purple fragments are platelets.

Technological advances in microscopy are at the forefront of our comprehension of human biology, and understanding the essence of normal and diseased cells and tissue can only really be appreciated by visualizing these structures under the microscope. Some of the different types of microscopy are outlined below.

2.2.1 Light microscopy

Light microscopy covers any type of microscopy that uses a source of illumination to aid the visualization of a specimen. There are various different types of light microscopy, including:

- brightfield
- phase contrast
- fluorescence.

Brightfield microscopy

Brightfield microscopy is the method used most commonly to investigate cell structure in health and disease, and generally involves the examination of chemically stained specimens, as shown in Figure 2.1. Notice the colour imparted on the specimen by the chemical stain. Depending on the magnification employed, light microscopy can have a multitude of applications, ranging from the assessment of the structure of small organisms, as was reported in Hooke's *Micrographia*, to the identification of intracellular organelles. Light microscopes contain the following important components, the relative positions of which are shown in Figure 2.2:

- light source (lamp)
- condenser
- stage
- objective lens
- ocular lens.

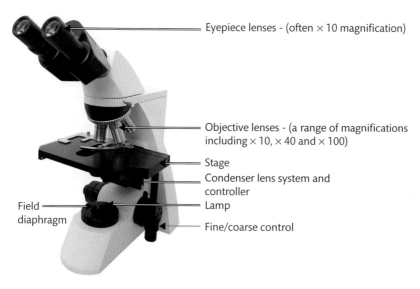

Eyepiece lenses - (often × 10 magnification)

Objective lenses - (a range of magnifications including × 10, × 40 and × 100)

Stage

Condenser lens system and controller

Field diaphragm

Lamp

Fine/coarse control

FIGURE 2.2
The basic structure of a light microscope.
© Dragan Trifunovic/istockphoto.com

The light source allows illumination of the specimen to enable visualization. It is necessary to use a thin section of tissue as the light source is positioned below the specimen, and light must pass through before reaching the objective lens and ultimately the observer. If the specimen is too thick, insufficient light will pass through to allow visualization to occur. The specimen is placed on the stage above the light source and light is focused onto the specimen using the condenser. The condenser contains an iris diaphragm to control the amount of light directed through the specimen. The objective lens has a short focal length and the eyepiece ocular lens provides additional magnification to allow the specimen to be focused for visualization.

> **Key points**
>
> When examining a specimen using light microscopy, two degrees of magnification must be considered. Magnification obtained using the objective lens is often ×10, ×40, or ×100, and must be multiplied by the magnification of the eyepiece ocular lens (often ×10). Therefore, true magnification will be in the range ×100 to ×1000.

SELF-CHECK 2.2

List the five important components of a light microscope and briefly outline the role of each.

One of the most important areas of consideration in light microscopy is the **resolving power** of the microscope. The resolving power, or resolution, is the smallest distance required between two points on a specimen to allow those points to be seen as separate; this can be seen in Figure 2.3. The resolving power is inversely proportional to the wavelength of the light passing through the specimen (i.e. the shorter the wavelength, the greater the resolving power)

FIGURE 2.3

Resolution is the ability to distinguish two objects as
separate entities. Poor resolving power showing two points
in close proximity overlapping is depicted on the left; good
resolution, or resolving power, where the two points are
seen as discrete entities is depicted on the right.

2 objects
visualised
with poor
resolving
power

Same 2
objects
visualised
with good
resolving
power

and the **numerical aperture** of the objective lens and the condenser. The numerical aperture
describes the ability of the lens to gather light and, as a consequence, the greater the numerical
aperture, the better the resolution obtainable at a fixed distance.

Phase-contrast microscopy

First outlined in 1934 by Frits Zernike, phase-contrast microscopy is particularly useful for the visualization of unstained specimens. This is important when live, unfixed specimens are required
for scientific investigation. While for the majority of diagnostic applications, fixed and stained
histological and cytological specimens are used, it is sometimes necessary to examine live cells.

As its name implies, phase-contrast microscopy relies on the production of a high-contrast
image without the need for special stains. The principle of this technique is based on differences in the **refractive index** of different regions within the same specimen. Light passes
through a condenser annulus and, in regions with a low refractive index, straight through
the specimen into the objective. The main elements of a phase-contrast microscope and
their spatial relationships are shown in Figure 2.4. Light passing through areas with a high

Refractive index

Refractive index is the ratio
of the speed of light in one
material compared to that in
a second material of greater
density. An illustration
demonstrating refraction is
provided in Figure 2.5. For each
pair of materials, the figure
calculated is a constant. In this
ratio, the numerator represents
the incident light and the
denominator represents the
refracted light.

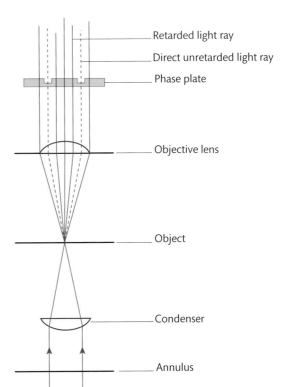

Retarded light ray

Direct unretarded light ray

Phase plate

Objective lens

Object

Condenser

Annulus

FIGURE 2.4

**An illustration of the light
path used in phase contrast
microscopy. Note the use of
the annulus and phase plate in
producing phase contrast.**
© Oxford University Press 2012

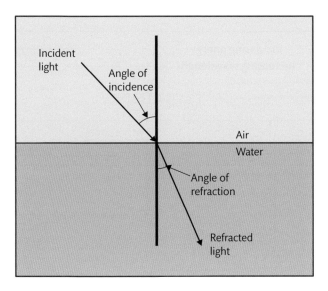

FIGURE 2.5

Refraction occurs when the speed of light changes due to differences in the materials though which light passes. The angle of incidence and the angle of refraction differ at this interface.

refractive index will have an altered wavelength and will therefore be out of phase. This out-of-phase light will interact with a phase plate incorporated in the objective lens, leading to a reduction in the wavelength of light. Refocusing the diffracted light with the light that passed unimpeded through the specimen produces a phase-contrasted image against a dark background.

SELF-CHECK 2.3

Explain the term resolving power.

Key points

Visible light forms a small part of the electromagnetic spectrum. According to electro-magnetic theory, light is transmitted in a wave motion. In quantum theory, light is con-sidered as energy quanta or photons (i.e. discrete packages of energy). Taken together, these two opposing perspectives form what is known as **wave-particle duality**. The light wavelengths visible to the naked eye range from 400 nm (violet) to 740 nm (red), as shown in Figure 2.6. It is the wave-like properties of light that lead to the limits of resolution. Wave-like properties inhibit our ability to use light microscopy to visualize the fundamental building blocks of cells (i.e. atoms and molecules), and use of standard methods of light microscopy means that resolution is limited to half the wavelength for visible light (i.e. approximately 200 nm).

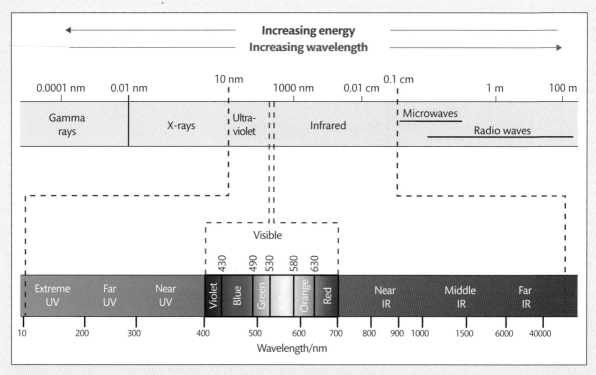

FIGURE 2.6

The electromagnetic spectrum. Of particular interest in light microscopy is the visible spectrum (centre). UV and IR are used in laser capture microdissection, as outlined later in this section.

© Oxford University Press 2011

Fluorescence microscopy

Fluorescence microscopy is an important technique for the investigation of specimens stained using fluorochromes, which are substances that absorb light at a specific wavelength and then emit light at a lower wavelength. Fluorescence was first identified by Sir George Stokes in 1852 while Lucasian Professor of Mathematics at the University of Cambridge. The utilisation of fluorescence in microscopy ensured fluorescence microscopy has become one of the most important technologies in cell biology.

Fluorescence microscopy utilizes light of known wavelengths to excite fluorochromes added to specimens, leading to the emission of light of a shorter wavelength. This emitted light can then be detected and visualized. Separation of light from the excitation and emission spectra can be achieved using filters, ensuring that the user can only see the fluorescent light emitted from the fluorochromes added to the specimen. Separation or differentiation between excitation and emission spectra is facilitated through the selection of specific fluorochromes that have narrow excitation wavelengths and emit light at a specific and different wavelength to that of the excitation light, as demonstrated in Figure 2.7. The longer wavelength, lower-energy light that follows excitation is called the **Stokes shift**, which is a consequence of a loss of excitation energy.

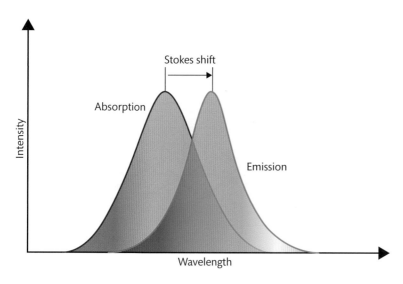

FIGURE 2.7
Stokes shift occurs as a consequence of energy loss during the transition between the higher-energy excitation wavelength and the lower-energy emission wavelength, resulting in fluorescence.
Cepheiden/CC-BY-SA

Key points

Fluorescence occurs following the interaction of light of a certain wavelength with a molecule. This molecule may absorb the light (in the form of photons) and then emit light of a longer (lower energy) wavelength which will be of a different colour. Initially, the absorbed light will cause ground-state electrons (i.e. those in their lowest energy state) within the molecule to gain energy and move to an excited state. Some of this energy will dissipate through collisions with surrounding molecules, although in order to reduce its energy to the ground state, the residual energy will be emitted as a short burst of light. The emission of this light energy facilitates the return of the electron to its ground state. The emitted light is fluorescence.

SELF-CHECK 2.4

Define the term Stokes shift.

Total internal reflection fluorescence (TIRF) microscopy is used to examine regions of cells close to and including the membrane (Figure 2.8). Of particular interest are the processes of cellular adhesion, movement and protein trafficking, and TIRF imaging relies on the following important properties:

- excitation light must pass through a medium with a high refractive index, perhaps a specifically designed coverslip or oil, to a medium with a low refractive index, such as an aqueous mounted specimen.

- the angle at which the excitation light strikes the interface between the two media must exceed the angle (called the critical angle) at which the excitation light can pass into the lower refractive index medium, which leads to total internal reflection.

- an electromagnetic field vector is generated and exponentially decays into the area of lower refractive index; this evanescent wave, identical in frequency to the excitation light, will

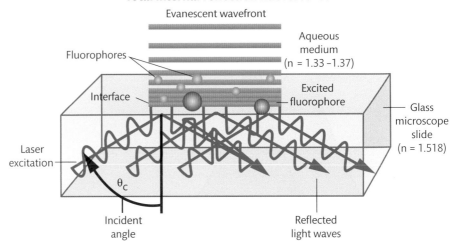

Total internal reflection fluorescence

FIGURE 2.8
For TIRF to succeed, light must pass from an area of high refractive index to an area of low refractive index. Rather than passing through the specimen, the critical angle is exceeded, leading to total internal fluorescence. The evanescent wave generated excites fluorochrome in the periphery of the specimen, generating fluorescence.

then transiently excite fluorochromes added to elements of interest within the specimen adjacent to the membrane, leading to fluorescence.

An illustration of the principles of TIRF microscopy is provided in Figure 2.8. Notice how the evanescent wave enters the cell and generates fluorescence.

Confocal microscopy was first designed in 1955 by Marvin Minsky, the son of an ophthalmologist and a junior fellow at Harvard University. The confocal microscope can use either reflected light or fluorescence to visualize specimens. By possessing a high signal to noise ratio, the confocal microscope permits the production of very clear images which can, if taken sequentially on a series of thin slides, be assembled to provide a three-dimensional visual representation of the scanned specimen. The appearance of a modern confocal microscope is shown in Figure 2.9.

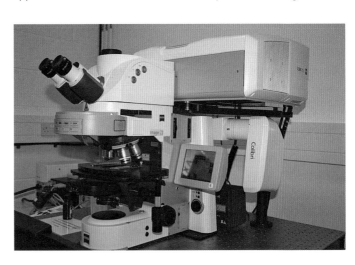

FIGURE 2.9
A state-of-the-art Zeiss Imager Z2 confocal microscope, in use at the University of Portsmouth.

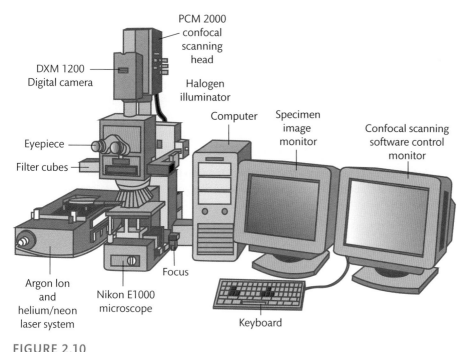

FIGURE 2.10

Typical requirements of confocal analysis. In this case, Nikon equipment is used.
© Nikon Corporation. All rights reserved.

Confocal microscopy was developed in an attempt to remove excess scattered light which otherwise reduces the quality of the image. The removal of scattered light was largely achieved by utilizing a 'pinhole' to allow the illumination of a single point on the specimen at any one time. The light reflected from this is selected by its ability to traverse an additional 'pinhole', further reducing scattered light. Consequently, the background lighting common to conventional fluorescence microscopy is largely avoided.

In order to visualize a whole section, the confocal microscope must be attached to an appropriate computer system, as illustrated in Figure 2.10. As only one point on a specimen is illuminated at any given time, an entire image needs to be assembled for analytical purposes.

Cross reference

More detail on light microscopy can be found in the *Biomedical Science Practice, Histopathology, Haematology* and *Medical Microbiology* volumes in this Fundamentals of Biomedical Science series.

2.2.2 Electron microscopy

Substituting visible light with electrons provides a method of improving significantly the resolving power of the microscope. This substitution forms the basis of the electron microscope. By increasing the velocity of electrons through a specially designed microscope system, it is possible to reduce the wavelength of the electron beam, thus significantly increasing resolving power. In a similar way to the way that light in a light microscope is focused using lenses, the electron beam is focused by electromagnetic lenses. In order to ensure that the electrons can follow the correct path to the tissue and be focused accurately to achieve the appropriate resolution, specimens in electron microscopes must be maintained within a vacuum. Failure to achieve a vacuum will result in the presence of additional particles in the electron beam and collisions between the electrons and the foreign particles. These collisions will reduce the velocity of the electrons and lead to a failure of optimal beam focusing.

The two important types of electron microscope are:

- transmission electron microscope (TEM)
- scanning electron microscope (SEM).

Cross reference

More detail on TEM can be found in the *Histopathology* volume in this Fundamentals of Biomedical Science series.

By examining the way in which electrons interact with the components of the specimen, TEM is used to provide information about tissue structure. These electrons are essentially able to 'carry the image' of the specimen and develop the image on a fluorescent screen (or similar medium), providing an accurate, high-resolution image of the specimen's structure. An overview of the structure of a TEM is shown in Figure 2.11.

The SEM has a lower resolving power and is used to scan surfaces of a specimen and build an accurate representation of the surface appearance of a gold- or palladium-coated specimen.

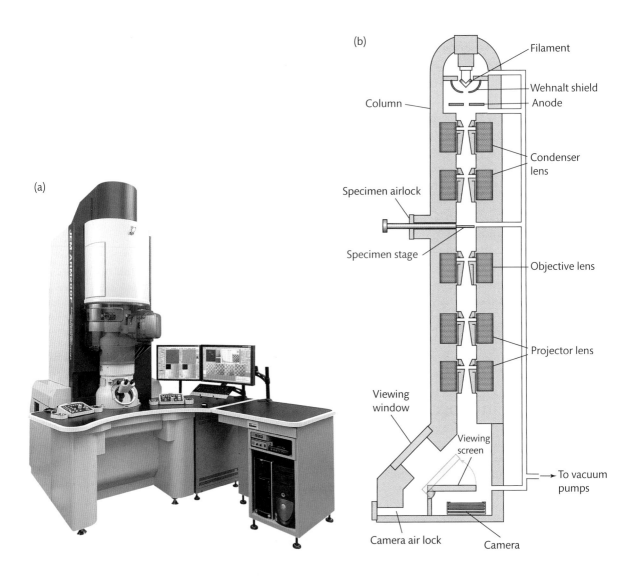

FIGURE 2.11

The main components required for transmission electron microscopy. The internal components of the microscope are also shown and are required to focus electrons onto the specimen for imaging.

This metal coating is necessary to permit the production of secondary electrons from the surface of the specimen to produce an appropriate topographical representation of its architecture which is suitable for analysis.

SELF-CHECK 2.5

Briefly compare and contrast transmission electron microscopy (TEM) and scanning electron microscopy (SEM).

Electron probe microanalysis

Electron probe microanalysis (EPMA), developed by Raimond Castaing in 1951, is a technique very similar in principle to SEM, used for the assessment of small, homogenous regions within a sample. The region of the specimen is bombarded with an electron beam, and the intensity of the X-rays emitted by these regions of interest is measured and compared with a standard of known composition. The ratio of X-rays emitted from the specimen compared to the standard forms the k-ratio, which can be converted to provide a concentration of the chemicals or compounds at the point of interest. Biological applications of EPMA include examination of cellular drug uptake, intracellular calcium distribution analysis and ion transportation.

Key points

Whether considering the resolving power of microscopes or the thickness of sections cut prior to analysis, an awareness of the definitions and significance of the small sizes we are utilizing is essential. Along with all other scientific disciplines, it is important to compare sizes accurately. In histopathology we deal with sizes in the micrometre (μm) and nanometre (nm) ranges. The quick reference guide below demonstrates a comparison of these different sizes:

1 metre is the base unit for the international system (SI)

1 centimetre (cm) = one hundredth of a metre (1×10^{-2})

1 millimetre (mm) = one thousandth of a metre (1×10^{-3})

1 micrometre (μm) = one millionth of a metre (1×10^{-6})

1 nanometre (nm) = one billionth of a metre (1×10^{-9})

Remember that the wavelength of visible light is 400-740 nm and the resolution of a light microscope is limited to half the wavelength of light (i.e. 200 nm). For comparison, the wavelength of an electron is approximately 0.005 nm; hence the enhanced resolution obtained using electron microscopy.

2.2.3 Laser capture microdissection microscopy

Laser capture microdissection (LCM) microscopy is a technique developed by the US National Institutes of Health in 1996 to replace methods of manual cell extraction. This technique enables the dissection of specimens, while visualizing areas of interest under the microscope, in an accurate, precise and timely manner, provided the specimens have been dehydrated adequately. A wide range of specimens can be interrogated using LCM, including:

- paraffin-embedded tissue
- frozen sections

METHOD Specimen processing and visualization for light and electron microscopy

In the same way that light microscopy uses stains (or dyes) to facilitate the visualization of different structures, electron microscopy requires the use of additives to permit visualization of the structure and architecture of a tissue. Whereas formaldehyde is used predominantly to fix tissues for light microscopy, glutaraldehyde is used in electron microscopy to fix proteins, and osmium tetroxide is used to fix lipids. It is essential that the fixatives do not form precipitates or alter the tissue as this could impede the electron beam.

Should the tissue need to be embedded (solid tissues require embedding, whereas liquid preparations such as blood do not), paraffin wax is often used for light microscopy. A rigid support medium should be used for electron microscopy in order not to impede the path of the electrons through the specimen.

Once embedded, sections of the specimen need to be taken. The thickness of the specimen depends on the method of visualization used. For light microscopy, section thickness between 1 µm and 20 µm is appropriate; for electron microscopy, however, a section thickness of just 50-100 nm is required.

Once tissue is embedded and sectioned, it is necessary to add contrast to the specimen. Very thin sections are usually almost transparent and therefore require contrast to enable differentiation between components within the tissue. Although a large number of stains are available, two commonly used for light microscopy are haematoxylin and eosin (H&E). Haematoxylin is a basic stain that reacts with negatively charged groups (e.g. DNA) to produce a blue colour. Conversely, eosin is an acidic stain and reacts with positively charged cellular components (e.g. cytoplasmic components), imparting a red colouration on the tissue.

While H&E staining is appropriate to use for adding contrast to tissues for light microscopy, this dye combination will not add value in electron microscopy. It is important to add 'stains' that will impede the flow of electrons through tissue in a specific and predictable way to allow architectural interpretation. By adding elements with a high atomic number (e.g. lead or uranium), differential staining of the tissue section is possible. Lead interacts with lipids such as those found in the cell membranes, while uranium binds to DNA and proteins. These elements increase the density of the target cellular components, impeding electron flow through the tissue. This alters the way in which the tissue image develops on the fluorescent screen.

Cross reference

More details on cytochemical stains can be found in the *Histopathology* and *Haematology* volumes in this Fundamentals of Biomedical Science series.

- touch preparations
- smears (including from peripheral blood)
- cytospin preparations
- cell blocks.

Water retained in the specimen prevents the sample fully adhering to the instrumentation, resulting in inadequate tissue capture. Depending upon the downstream investigations to be employed, samples can be fixed with formaldehyde (good for investigating DNA), acetone or ethanol (a good fixative for investigating RNA).

Several different LCM technologies are available, including:

- infrared (IR)
- ultraviolet (UV)
- IR/UV.

These LCM technologies can be combined with various microscope technologies, including brightfield, fluorescence and phase contrast, to allow visualization of the tissue of interest. Importantly, LCM enables scientists to separate complex, heterogeneous tissues into homogeneous, specific cells for subsequent downstream analysis with the minimum of contamination.

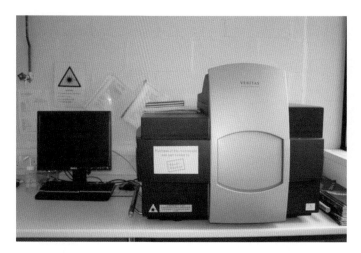

FIGURE 2.12

The Arcturus Veritas laser capture microdissection microscope at the University of Portsmouth. All microscopy and dissection features are enclosed within the housing. Visualization and processing is controlled through the attached computer.

Downstream analysis may include molecular techniques such as the polymerase chain reaction (PCR) following DNA or RNA extraction, and western blotting or northern blotting, respectively.

The basic instrumentation on which LCM is based includes a specially designed computer program, a typical microscope set-up, the presence of one or two lasers, depending upon the nature of the technology (either one IR and/or one UV laser), and a specialized collection chamber within the housing. As shown in Figure 2.12, the appearance of modern LCM is rather unassuming with the internal workings retained within the housing.

Although the internal workings are contained within the housing, in the Arcturus Veritas the hatch situated at the front allows samples and caps to be added and collected as necessary. This is shown in Figure 2.13.

Ultraviolet LCM uses tissue mounted on a membrane and placed on a glass slide. Once the cells of interest have been identified using microscopy, the UV laser ablates the unwanted tissue and a margin surrounding the target cells. The separated target cells are then catapulted into a cap ready for collection and downstream processing.

Conversely, the IR laser in IR LCM heats a thermoplastic cap, which is placed over the target cells by a robotic arm, until it melts (this takes a fraction of a second). Melting of the cap allows

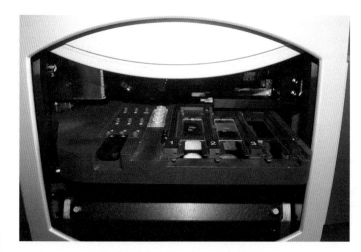

FIGURE 2.13

Internal workings of LCM. Three slide ports are available, and a supply of caps is available to enable specimen capture. Once specimens have been microdissected, used caps can be retained for collection in the area on the left.

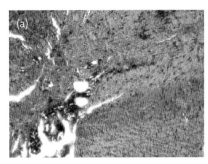
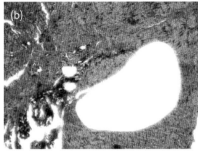
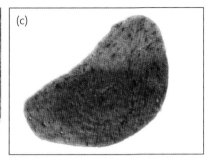

FIGURE 2.14

A sample of palmar fibromatosis before (a) and after (b) microdissection. Note the captured tissue (c), which has been collected for further analysis.

From Wang & Zhu (2006). Clonal analysis of palmar fibromatosis: a study whether palmar fibromatosis is a real tumor. *Journal of Translational Medicine* 2006; **4**: 21. © 2006 Wang and Zhu; licensee BioMed Central Ltd.

the target cells to adhere to it prior to cooling. On cooling, the cap contracts and shears the target cells from the tissue. The cells can then be processed using the UV collection protocol.

In the case of IR/UV LCM, once the tissue has been visualized and the appropriate cells or regions have been selected, the cap—made of a thermoplastic membrane—is placed, by an analytical arm, over the cells of interest. The low energy IR (capture) laser melts the thermoplastic membrane over the target cells and adheres to them. The UV (cutting) laser then ablates a narrow margin around the target cells or tissue, preventing contamination and allowing separation of the target cells from the tissue section. These target cells can then be retained and used for specific analyses. Microdissection is demonstrated in Figure 2.14. Notice the neat margins on the dissected tissue.

Cross reference

More detail on laser capture microdissection can be found in the *Cytopathology* volume in this Fundamentals of Biomedical Science series.

2.3 **Visualization of microscopic structures**

In microscopy, to enable visualization of the structure of cells and tissues, specimens must be treated with fixatives. Navigation and interpretation of unstained sections of tissue is very difficult as all elements appear uniform, with no distinguishing features (i.e. there is no contrast). Pretreatment in the form of fixation and processing is beyond the remit of this chapter, although we will discuss the application of stains, dyes and chemistry in the visualization of tissues using light microscopy.

Dyes and stains have been applied in microscopy for over a century following their original applications in the textile industry, make-up and artwork. Both Hooke and Leeuwenhoek used cochineal (carmine) in their studies during the seventeenth century to provide contrast to their tissue sections. It is now known that dyes and stains work through chemical attraction, and, depending on the chemical structures of the tissues, the stain will combine in a reproducible way, leading to **tinctorial staining**. Scientists have control over the appearance of tissues, and can interpret the components of cells and tissues according to the way in which stains and dyes interact with the chemical groups within these structures. Acidic dyes are negatively charged anions in solution and bind to, and are retained by, tissues containing

Tinctorial staining

These methods use dyes to demonstrate components of cells and tissue structures.

cationic, or positively charged, structures. These structures may be referred to as **acidophilic**, or acid-loving, due to their affinity for acidic dyes. Conversely, structures that bind to basic (cationic) dyes have acidic components and are called **basophilic**. In addition to basic and acidic dyes, neutral, natural, synthetic and amphoteric dyes are in use in various staining applications.

We have now established that the specificity of dyes is obtained through chemical interaction. These dyes may occur naturally or be synthesized. Synthetic dyes are often derived from colourless benzene. In itself, benzene is not effective in enhancing contrast. To enhance contrast, two groups must be added—a **chromophore** and an **auxochrome**. Chromophores absorb light at a visible wavelength and can introduce colour to a tissue component, thereby enhancing contrast. However, in order for dyes to bind selectively to particular chemical groups, dyes must possess the appropriate complementary chemical structure, often in the form of an ionizable group. This group, called the auxochrome, enables the dye to bind in a complementary fashion to certain tissue structures in a permanent or semi-permanent fashion. In a number of circumstances, metal salts, called **mordants**, are added to facilitate binding between the tissue and the dye, in essence forming an ionic bridge. Various mordants are available, each influencing the colour produced from the tissue-mordant-dye interaction.

The most frequently employed stains are **haematoxylin** and **eosin**, and may be used in combination (H&E stain) or separately. In a well-preserved section using appropriate H&E staining, the architecture and the tissue and cell **morphology** can be easily established. In approximately 80-90 per cent of histology cases, H&E-stained slides provide sufficient information to enable formulation of a diagnosis.

Morphology
Shape or structure of a cell or tissue.

Haematoxylin, often used as a nuclear stain, is derived from the logwood *Haematoxylum campechianum*. Prior to use, haematoxylin must be oxidized to haematein and combined with mordants to allow binding to tissue. Depending on the mordant added, a colour will be generated on staining ranging from a blue nuclear stain (with potassium alum or aluminium alum) to black when iron is added. A counterstain used alongside haematoxylin, eosin, is used to stain cytoplasmic components of cells, providing a number of different shades of red to aid the interpretation of tissue architecture.

Should H&E staining prove insufficient for diagnostic purposes, special stains may be used. Special stains are not used routinely, rather they are applied purely to explore specific structures or cellular components in more detail. Examples of special stains used in histopathology laboratories are Masson Fontana (staining melanin), elastic von Gieson (staining elastin) and Alcian Blue (staining acid mucins).

Cross reference
More detail on the use of stains in pathology may be found in the *Histopathology*, *Cytopathology* and *Haematology* volumes in this Fundamentals of Biomedical Science series.

In addition to dyes, more advanced methods of tissue investigation are possible. **Histochemistry** involves the use of biochemical reactions to form a coloured product that, ideally, will be localized to the area of tissue in which it is produced. An example of a histochemical technique is the Perls' Prussian blue method, used to identify iron deposits. In this method, tissue treated with hydrochloric acid releases ferric (Fe^{3+}) ions, producing an insoluble blue compound called Prussian blue following a reaction with the reagent potassium ferrocyanide. The blue deposit accumulates at the site of the reaction and is used to identify iron stores, which is particularly valuable when investigating iron overload in haemochromatosis.

Cross reference
More detail about the applications of fluorescence-labelled monoclonal antibodies can be found in the flow cytometry section.

Another method used in histopathology is **immunocytochemistry** (ICC), also known as immunohistochemistry (IHC). Utilizing **monoclonal antibodies** bound to either a fluorochrome—generating **immunofluorescence**—or an enzyme, the monoclonal antibody will bind to a specific antigenic target in a tissue and produce either a coloured product, in the case of enzyme conjugation, or fluorescence, allowing specific targets to be identified.

Monoclonal antibodies
These are immunoglobulins that all have the same specific target protein (epitope).

Clusters of differentiation

Clusters of differentiation (CD) antigen is the standardized term used to describe a range of molecules associated with the differentiation and maturation of cells.

Cross reference

More detail about CD nomenclature and diagnosis of haematological cancers can be found in the *Haematology* and *Histopathology* volumes in this Fundamentals of Biomedical Science series.

Of particular relevance in this field is the identification of **clusters of differentiation** (CD) markers. The term clusters of differentiation was coined by the Human Cell Differentiation Molecules' (HCDM) Human Leucocyte Differentiation Antigen (HLDA) workshops, the first of which was convened in Paris in 1982. The responsibilities of the HCDM include the validation of monoclonal antibodies and the characterization and naming of CD markers. These markers are expressed on the surface of cells and have important roles in, for example, cell signalling and cell adhesion, and can be used for cell identification. The CD markers are particularly useful in the diagnosis of haematological cancers, where both ICC and flow cytometry techniques can be used for the visualization and quantitation of antigens.

Following this consideration of the morphological examination of cells and tissues using microscopy, methods of cell counting and automated analysis will now be examined.

2.4 **Introduction to cytometrics**

Cytometrics is the process of measuring the physical and chemical characteristics of cells. In this section we will begin by examining the processes of flow cytometry, impedance, and light-scattering technology for the purposes of cell counting and identification, before moving into the world of immunophenotyping, which involves the application of monoclonal antibodies and the use of fluorescence-activated cell sorting.

2.4.1 **Impedance technology**

First developed by Wallace Coulter in the 1940s, impedance technology is used in various scientific instruments primarily for measuring the number and volume of blood cells in the haematology laboratory. Coulter, a graduate of the Georgia Institute of Technology, revolutionized the way in which blood cells could be counted. Originally, manual counting techniques using haemocytometers were employed, but these often lacked accuracy and precision. The introduction of automated counters improved accuracy and precision while significantly reducing the time taken for blood sample analysis. Today, using technology incorporating Coulter's technology of impedance measurements, a single instrument can easily analyse 1000 full blood count samples per day. Analysers using this technology are called Coulter counters. As demonstrated in Figure 2.15, in their most basic form, they measure impedance using a conductive buffer, or electrolyte solution, to which cells of interest (e.g. red blood cells) are added. The cell solution will be placed in a system, much like a beaker containing a tube with

FIGURE 2.15

The Coulter principle. Displacement of the conductive buffer by cells of interest within the sensing zone leads to a change in impedance. Each peak represents a single cell and the amplitude represents volume.

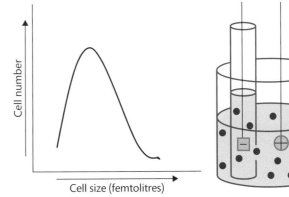

a small aperture in the side, which is submerged. In order for impedance to be detected, an electric field must be applied and two electrodes added, one to the main beaker and the other through the aperture within the tube. The region of this system, centred on the aperture, is called the 'sensing zone' where physical measurements are made. As cells pass through the aperture in single file, the buffer in the tube is displaced, leading to a change in impedance and a subsequent change in voltage or current. The data can be recorded as peaks and troughs; each peak represents a single event (i.e. an individual cell) while the amplitude of each peak represents cell volume. By controlling the volume of fluid that passes through the aperture for each analysis, the concentration of cells can be determined. In the case of red blood cells, this is often reported in units $\times 10^{12}$/L. The volume of red cells is averaged and reported as the mean cell volume (MCV), which is measured in femtolitres (fL).

2.4.2 Light-scattering technology

Commonly used for the identification of blood cell species, light-scattering technology utilizes the structural properties of cells in order to facilitate their identification. By aligning cells in single file, either by passing through a narrow channel or by **hydrodynamic focusing**, cells pass in front of a light source. The light source, whether a laser or a tungsten-halogen lamp, will generate light from a single point to interact with the cell, as demonstrated in Figure 2.16. The cell partially absorbs the light and the remainder is scattered by the cell. Photodetectors aligned within this system will detect forward- and side-scattered light and convert this to electronic signals. Depending on the size and volume of the cell, light will be scattered in a predictable way to allow identification. Light scattered at 180° to the source, called forward scatter, measures light absorbance representing cell size. Light reflected, called side scatter, represents internal complexity (Figure 2.16).

Hydrodynamic focusing
This is a technique used to deliver cells in single file to a set point within an enumeration analyser. The cells of interest are injected into the middle of a stream of fast-moving sheath flow. The sheath flow forms a wall, or barrier, to the injected cells and the different velocities of the fluids ensure they do not mix.

SELF-CHECK 2.6

What does forward scatter and side scatter represent in terms of cell structure and identity?

SELF-CHECK 2.7

Explain how hydrodynamic focusing can facilitate cell enumeration.

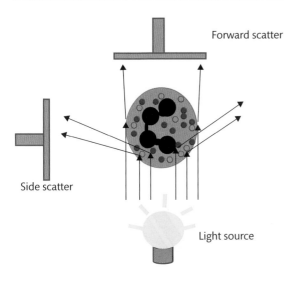

FIGURE 2.16
Light scattering technology. Light emitted from a source is scattered by cells maintained in single file. Forward scatter represents cell size, while side scatter represents internal complexity.

2.4.3 Flow cytometry

Flow cytometry uses many of the processes we have already considered in relation to light-scattering technology; however, it also utilizes fluorescence to aid cell identification or the investigation of a range of cellular processes including cell cycling and apoptosis. Cells can be identified according to their light-scattering properties. Light scattered in a forward direction (forward scatter) equates to cell size, whereas light scattered and measured at 90° to the light is called side scatter, representing internal complexity. To add value to the measurement and identification of these cells, fluorescence markers can be introduced.

Flow cytometry utilizes the work of Kohler and Milstein—who developed **hybridomas** to produce large quantities of identical immunoglobulins, called monoclonal antibodies (MAbs). These MAbs enable the identification of cells according to specific proteins that they express. These MAbs can be attached (conjugated) to fluorochromes, as demonstrated in Figure 2.17, with very specific excitation and emission spectra. When target cells are treated with fluorochrome-conjugated antibodies and pass in single file through a laser beam of the correct wavelength, the fluorochrome will be excited and emit energy in the form of light of a lower wavelength. This light is detected and confirms the presence of a particular antigen on, or within, the cell. Monoclonal antibodies are raised protein targets or CD antigens, which, when expressed, may be found within the cytoplasm of a cell or on its membrane. Evaluation of these antigens is called **immunophenotyping**. Depending on the technique used, we can target intracellular or extracellular antigens to enable the identification of the cell, or population of cells, of interest. Figure 2.18 demonstrates how flow cytometry can be applied to the identification of B cells. This is of particular importance when investigating cancers of the blood (i.e. leukaemia or lymphoma) to assist in determining a patient's prognosis or to evaluate the efficacy of an anticancer treatment.

Use of separate channels to detect light emitted from stimulated fluorochromes makes it possible to measure a number of different fluorochromes simultaneously. Dichroic filters are used to separate fluorescent light at different wavelengths. The filters are placed at a 45° angle within the system on the side-scatter path. These filters allow light of a specified wavelength to pass through, while other light is reflected towards a photodetector. Through the use of dichroic filters of different wavelength specificities organized in serial format, each fluorochrome can be detected in a multichannel system. Light of a specific wavelength generates a small change of current, which is then amplified and represented graphically for interpretation.

Hybridoma

This is a term used to describe the fusion of a B lymphocyte that has been stimulated to produce specific antibodies with a cancerous B-lineage cell called a myeloma cell. The fusion of these cells ensures continual production of identical immunoglobulins, and the fusion of the cancer cell ensures survival of the hybridoma under culture conditions.

Cross reference

More detail on flow cytometry and immunophenotyping can be found in the *Haematology*, *Cytopathology* and *Immunology* volumes in this Fundamentals of Biomedical Science series.

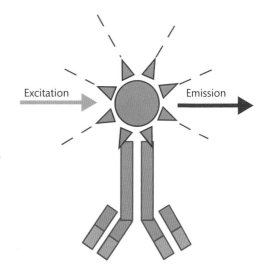

FIGURE 2.17

Illustration of a monoclonal antibody conjugated to a fluorochrome. The variable region of the antibody, which recognizes a specific epitope, is shown in gold. Following binding, exposure of the labelled cells to light at a particular wavelength causes excitation of the fluorochrome and emission of light in the form of fluorescence. Fluorescence can be detected to demonstrate the presence of a particular antigen.

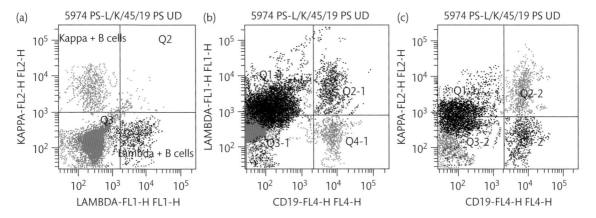

FIGURE 2.18

Scattergram illustrating a B-cell immunophenotype. (a) B-cell receptors express either kappa or lambda light chains. In this case, both are visible, suggesting a polyclonal B-cell population. Kappa is shown on the y-axis, and lambda on the x-axis, summarizing the data presented in (b) and (c). The B-cell marker CD19 shows co-expression with lambda light chains (b; purple) and kappa light chains (c; orange).

Image courtesy of Adnan Mani, Southampton General Hospital.

SELF-CHECK 2.8

Describe the work of Kohler and Milstein, and explain how this is applied to flow cytometry.

2.4.4 Fluorescence-activated cell sorting

Fluorescence-activated cell sorting (FACS) is an extension of flow cytometry and facilitates the isolation of cells that possess particular characteristics from a heterogeneous population and permits further analysis. Particular selection criteria are outlined by the user; for example, the expression of a particular cell marker (e.g. CD34). When cells conforming to these set criteria are identified by flow cytometry, as outlined above, the cell is marked for separation. As the cell of interest leaves the fluidics system, an electrostatic charge is administered to the cell and its associated sheath fluid, causing the sheath fluid containing the cell to form a droplet. This droplet separates from the heterogeneous population of cells and is collected in a vessel with other cells sharing the same characteristics. The result of FACS is the collection of a homogeneous population of cells for further downstream analysis.

Technologies available to enumerate and collect cells have been outlined here; however, consideration must also be given to further downstream analytical processes. We will begin with the examination of cytogenetic techniques—the visualization and study of chromosomes.

2.5 Cytogenetic analysis

The term cytogenetics is formed from a combination of cytology, the study of cells, and genetics, the study of genes. Cytogenetics involves the assessment of genetic material at the cellular level and can be used to investigate **heritable, congenital** and **acquired** diseases.

Cytogenetic analysis involves studying tissue at the chromosomal level or below. In order to analyse a patient's chromosomes, appropriate tissue harvest is essential. For example, patients with leukaemia—a cancer of the blood and bone marrow—will require collection of peripheral blood or bone marrow, whereas investigation of a suspected case of Down's syndrome *in utero* will require a biopsy of the placenta. Adequate tissue collection is essential to enable processing and treatment of the sample prior to analysis. In addition, the clinician collecting the specimen must be familiar with the specimen requirements of the technique. For example, live cells are required in order to determine chromosomal banding patterns. These cells are cultured to increase their population and to encourage their entry into the cell cycle. A cell-cycle inhibitor is then added, arresting the cycle at the point at which metaphase chromosomes have accumulated. Conversely, a commonly used technique called fluorescence *in situ* hybridization (FISH) requires cells with preserved, intact nuclei. In the majority of cases, samples received for cytogenetic analysis will be heterogeneous and a variety of different cell types, both normal and abnormal, will be apparent. For example, if a biopsy of bone marrow is taken for cytogenetic analysis of suspected leukaemia, normal and cancerous cells will be collected. While this does not present a major problem for cytogeneticists, one way of acquiring a pure population of cells is to harvest cells using laser capture microdissection, as outlined previously.

Key points

Often, we may use the terms **heritable, congenital** and **acquired**. However, students and lay persons often fail to distinguish between the terms heritable and congenital. It is important to be able to make this distinction in practice to prevent misleading an audience, in particular patients or other healthcare professionals. The term **heritable** means a genetic disorder that has been passed from one or both parents to their offspring. An example of this is the bleeding disorder haemophilia, where a tendency for uncontrollable bleeding is passed from parents to offspring. Females tend to be carriers, and males will be sufferers. Conversely, **congenital** means present from birth, but this does not mean that this disorder is genetic or heritable. For example, in certain countries, children may be born with congenital malaria. The malarial parasites cross the placenta and enter the child's circulation, and at birth the neonate shows signs of malaria. **Acquired** is a term used to define the development of a condition such as cancer during a patient's life. Even though some cancers are related to family history, acquired mutations ultimately will lead to the development of the disease.

The first and most commonly recognized investigation in cytogenetics involves the establishment of a patient's **karyotype**.

Karyotype

The karyotype is the chromosomal complement of a cell or population of cells, and it is important to determine whether or not a patient has the correct number of chromosomes within their cells. For example, Down's syndrome is associated with three copies of chromosome 21, termed **trisomy** 21. In a normal situation, an individual possesses 46 chromosomes in each somatic cell. The 46 chromosomes comprise 22 pairs of autosomes and two sex chromosomes. One chromosome in each pair is maternal in origin, while the other chromosome is paternal in origin. The patient's karyotype is established initially by producing a **karyogram**. A karyogram

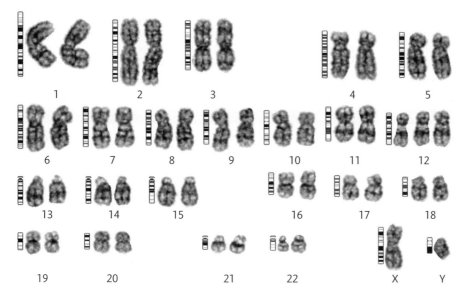

FIGURE 2.19

The karyogram, using G-banding, of an individual diagnosed with chronic lymphocytic leukaemia, a type of blood cancer. This patient has three copies of chromosome 12 (trisomy 12) and a deletion of the long arm of chromosome 13 (del13q).

Image courtesy of Anne Gardiner, Royal Bournemouth Hospital.

involves visualizing the patient's chromosomes during metaphase under the light microscope. An image is taken of the chromosomes and, using a computer, they are organized in descending order of size, banding pattern and centromere position. The karyogram for a patient with chronic lymphocytic leukaemia (CLL) is shown in Figure 2.19. Notice the three copies of chromosome 12, which has prognostic implications for this disease.

Cross reference

More detail on karyotypes can be found in the *Haematology* volume in this Fundamentals of Biomedical Science series.

SELF-CHECK 2.9

How many autosomes are in a somatic cell?

Banding patterns

The banding pattern of chromosomes is produced by applying various dyes to metaphase chromosomes, which are condensed and are the product of genomic replication prior to cell division. The most widely used stain is Giemsa, and the result is termed **G-banding**, which provides dark and light bands, as can be seen in Figure 2.19. The dark bands are formed of transcriptionally inactive heterochromatin, while the light bands, called euchromatin, are transcriptionally active. The opposite staining pattern can be obtained by **R-banding**, which incorporates heating prior to staining with Giemsa. **Q-banding** is the original technique used to visualize metaphase chromosomes, which uses the fluorochrome quinacrine and requires a fluorescence microscope to facilitate visualization of the banding pattern.

Although the generation of banding patterns is very helpful in chromosomal studies, the use of light microscopy for cytogenetic analysis provides only relatively poor resolution. Chromosomes can only be visualized by light microscopy in their metaphase state, and light microscopy has a resolution of approximately 5 Mb (**megabases**). While large structural alterations and numerical abnormalities of the chromosomes are clear, it may provide

Cross reference

Megabase and associated terms are defined in the key point entitled 'The definition of base pairs' later in this section.

Cryptic translocations

These involve the exchange of genetic material between chromosomes, where the morphology of the chromosomal structure and banding pattern are retained. Cryptic translocations are missed when using banding techniques.

insufficient resolution to visualize smaller mutations that retain the original banding pattern—**cryptic translocations**. In such cases, we need to utilize techniques that offer much greater resolution, and can therefore allow the scientist to observe much smaller chromosomal abnormalities. To facilitate this, we turn to *in situ* hybridization, of which a number of variants are available.

SELF-CHECK 2.10

Explain the difference between heterochromatin and euchromatin, and how these are demonstrated in G- and R-banding.

2.5.1 *In situ* hybridization

In situ hybridization is much more powerful than traditional cytogenetic methods, providing a resolution of between 50 kb for interphase fluorescence *in situ* hybridization (FISH) and 5 Mb for metaphase FISH. This improved resolution is attributable to the more open chromatin configuration seen in interphase nuclei compared to that in metaphase chromosomes. Whereas metaphase chromosomes are required for metaphase FISH, interphase FISH utilizes whole nuclei and allows three-dimensional (3D) analysis of the genome. *In situ* hybridization involves the use of DNA sequence-specific probes. These are usually labelled with fluorochromes, although chromogenic substrates may also be used (chromogenic *in situ* hybridization). The patient's double-stranded DNA is denatured, either during metaphase (in metaphase FISH) or during interphase (in interphase FISH). Using fluorochromes that have different excitation and emission wavelengths, a number of probes can be used simultaneously to identify a patient's chromosomal structure. It is therefore possible to visualize small mutations (e.g. chromosomal translocations) by assessing the relative position of the probes. Probes are designed for a specific DNA sequence and can also be raised against centromeres and telomeres. In Figure 2.20 it is possible to see that a single gene (*ATM*) has been lost in the majority of cells.

Key points

As the name would suggest, a base pair describes a pair of bases, either G-C or A-T present on opposing DNA strands. The term kilobase (kb) pairs describes 1000 base pairs and megabase (Mb) pairs represents one million base pairs. For illustration, a resolution of 1 kb is far greater than a resolution of 1 Mb.

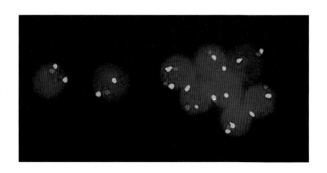

FIGURE 2.20

Fluorescence *in situ* hybridization of cancerous lymphocytes in chronic lymphocytic leukaemia. The tumour suppressor gene *ATM* is labelled with a red probe and the centromere is labelled with a green probe. Each cell should contain an equal number of red and green spots. Of the nine cells shown, only one contains two copies of the *ATM* gene.
Image courtesy of Anne Gardiner, Royal Bournemouth Hospital.

One application of *in situ* hybridization is multicolour karyotyping, which offers a resolution of approximately 1 Mb and uses various probes to give each chromosome a particular spectral signature. As a number of probes are used simultaneously, this technique is called multiplex FISH (M-FISH). The fluorescent signature permits identification of the chromosomes and can help the scientist identify regions where material has been exchanged between chromosomes.

2.5.2 Comparative genomic hybridization

Comparative genomic hybridization (CGH) is a relatively fast screening technique used to identify changes in chromosomal copy number. When investigating cancers, tumour DNA is extracted, conjugated with a green fluorescent probe and then mixed with reference (normal) DNA that has been conjugated with a red probe. This labelled DNA is then mixed with pre-prepared human metaphase chromosomes and the ratio of the red to green signal is measured, as shown in Figure 2.21. A ratio greater than one indicates chromosomal gain, whereas a ratio less than one indicates chromosomal loss. While CGH is a powerful technique, it does have certain limitations. Comparative genomic hybridization is unable to detect balanced chromosomal translocations, where genetic material is rearranged between chromosomes, although not lost. It is also insensitive to inversions, where regions of a chromosome have rotated while being retained by the original chromosome. Furthermore, CGH cannot detect cases of polyploidy, where more than one complete set of chromosomes is present. The use of CGH with metaphase chromosomes provides a resolution no greater than other metaphase-dependent technologies, where a resolution of approximately 5 Mb is possible. The resolution of metaphase CGH can be improved by embedding DNA fragments in a matrix on a glass slide, thus providing a resolution of approximately 50 kb. This variation in technique, called array CGH, can be applied to the analysis of the entire genome. Using these sensitive methods, cloned DNA is attached to a glass slide in small spots and combined with suitably labelled normal and tumour probes. Whole genome CGH can use one clone per Mb to provide very high levels of resolution.

The development of *in situ* hybridization has required the necessary comprehensive understanding of molecular biology. The next section will look at a number of commonly used molecular techniques used in modern-day research and diagnostics.

Cross reference

More detail on traditional and modern cytogenetic techniques can be found in the *Haematology*, *Histopathology* and *Cytopathology* volumes in this Fundamentals of Biomedical Science series.

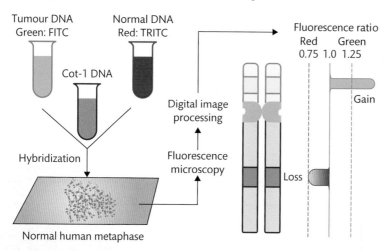

FIGURE 2.21

Principles of comparative genomic hybridization.
Adapted by permission from BMJ Publishing Group Limited. Weiss MM, *et al.* Comparative genomic hybridisation. *Molecular Pathology* 1999; **52**: 243–51.

2.6 **Molecular biology**

Our understanding of molecular biology and associated techniques has increased exponentially since Watson and Crick identified the structure of DNA in 1953. Currently, in some pathological arenas, the use of molecular biology and associated technologies is preferable to traditional techniques of diagnosis and monitoring on the basis that our genes affect our proteins and our proteins affect our cells. While this is an incredibly simplistic notion, understanding the molecular basis of disease has become a significant research direction in the twenty-first century. The fundamentals of transcription, translation and heritability are beyond the scope of this textbook; however, it is important to be familiar with these processes to enable understanding of the application of molecular techniques to cellular disease. We will begin by examining the process of the polymerase chain reaction (PCR).

2.6.1 **Polymerase chain reaction**

The polymerase chain reaction is a fundamental technique to master in any molecular biology laboratory, as it has important applications both in the diagnostic and research arenas. The underlying premise of PCR is the ability to amplify a very small DNA fragment exponentially, therefore allowing sufficient DNA to be available for downstream analysis. In order for amplification of target DNA sequences to occur, the nucleotide sequence either side of the target must be determined. Once this sequence is known, complementary primers can be developed to act as initiators for the replication of the target sequence. The first time the sequence completes, a single copy is made of each strand of DNA. Subsequent cycles copy the original template as well as the copies made from the previous cycle. Therefore, with every completed cycle of PCR, the number of copies doubles. This process of PCR provides a very effective method of producing large quantities of a chosen nucleotide sequence, enabling us to visualize the product using gel electrophoresis.

Consideration should also be given to the technical requirements behind the process of DNA amplification, as outlined in Figure 2.22. The first requirement of PCR is the need to be able to access the DNA sequence (i.e. the order of nucleotides making up the DNA code). Nucleotides are stabilized by phosphodiester bonds in a 5′→3′ direction. The complementary strands of DNA combining to form a double helix are maintained by hydrogen bonds. However, to enable complementary pairing of DNA strands, the strands must run in opposite directions (i.e. one strand runs in a 5′→3′ direction, and its complementary strand runs in a 3′→5′ direction.

Key points

Adenine (A) pairs with thymine (T) using two hydrogen bonds, whereas guanine (G) combines with cytosine (C) with three hydrogen bonds.

In PCR, access to the DNA sequence is essential. To enable sequence reading, the two opposing strands of DNA must be separated. This can be achieved by heating to 90°C. During this heating process, DNA melts—the hydrogen bonds joining the complementary strands are disrupted, while the phosphodiester bonds retaining the sequence are maintained. Separation of the two strands of DNA permits access to the nucleotide sequence.

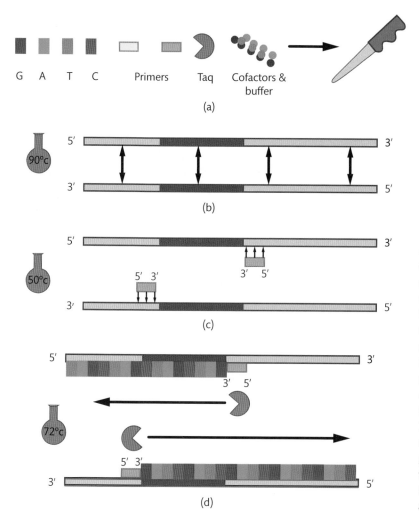

G A T C   Primers   Taq   Cofactors & buffer

(a)

5′   3′
90°c

3′   5′

(b)

5′   3′
50°c
                    5′ 3′   3′ 5′

3′   5′

(c)

5′   3′
                                   3′ 5′

72°c

                    5′ 3′

3′   5′

(d)

FIGURE 2.22

Polymerase chain reaction. (a) All components necessary for the reaction to proceed, including the DNA template, are added to a 'master mix'. (b) The temperature of the reaction vessel is increased to approximately 90°C, melting the double-stranded DNA. (c) The reaction vessel cools to approximately 50°C to allow the forward and reverse primers to bind to the flanking sequences of the target area (brown). (d) The temperature increases to approximately 72°C, allowing *Taq* polymerase to extend the primers in a 5′ to 3′ direction, generating a DNA strand complementary to the template. The entire process begins again, providing exponential amplification of the target sequence.

The next task in amplification involves the addition of deoxynucleotides (dNTPs) to the target sequence. If nucleotides were added alone, the rate of reaction between complementary nucleotides (i.e. dNTPs and target sequence) would be so slow that suitable products would not be produced. The use of enzymes increases the rate of this reaction. However, due to the enzymes' protein structure, the rise in temperature to 90°C denatures the majority of enzymes. Fortunately, a polymerase enzyme obtained from *Thermophilus aquaticus (Taq)*, a bacterium found in hot springs, is suitable as it is thermostable (heat resistant). The region of DNA to be amplified must also be controlled, as there may be an entire genome of DNA within the reaction vessel. The use of primers—sequence-specific oligonucleotides—will identify the starting points required for DNA synthesis. These oligonucleotides will not anneal at the high temperatures required for DNA melting, so a second phase of the cycle is required, at a temperature of approximately 50°C. The addition of the oligonucleotides to the template—a process called annealing—will then allow the third phase of the cycle to begin. An increase in the temperature of the vessel to approximately 72°C facilitates the addition of dNTPs by *Taq* polymerase and the replication of the target DNA sequence. Cooling of the reaction vessel and the DNA allows complementary strands to re-anneal, and the process can begin again. The exponential increase in DNA concentration following each cycle means that 20 cycles are often ample to provide a sufficient volume of DNA for further investigation.

Cofactors such as magnesium chloride must also be included to facilitate enzyme activity.

All of these reactions can be programmed in a thermal cycler, enabling a number of specimens, each within a designated reaction vessel, to be processed simultaneously. This is of importance in any application, and is particularly important where there is a high demand for results, such as in a diagnostic laboratory.

Cross reference

More detail on PCR can be found in the *Biomedical Science Practice, Histopathology, Haematology, Cytopathology* and *Medical Microbiology* volumes in this Fundamentals of Biomedical Science series.

Key points

The design of the primer is critical in ensuring the appropriate DNA sequence is targeted using PCR. Very short primers are unlikely to provide the specificity needed to recognize a single sequence within total human DNA, so is it better to design long primers? The answer is not necessarily. While longer primers are more likely to be specific for the flanking regions of the target DNA selected, they are far less efficient at annealing, thereby increasing the time taken for successful PCR to occur. Therefore, each primer should be designed based on the specific requirements of each PCR reaction and each target, rather than generically.

SELF-CHECK 2.11

Explain the role of *Thermophilus aquaticus* in PCR.

2.6.2 Following amplification of DNA, what's next?

There are various applications downstream of PCR, although this chapter will focus on two:

- Analysis of restriction fragment length polymorphisms (RFLP)
- DNA sequencing.

These applications will now be considered in turn.

Restriction fragment length polymorphisms

The term restriction fragment length polymorphisms is used to describe the genetic heterogeneity seen between individuals in the form of alternate alleles. To establish whether RFLPs differ between, or indeed within, individuals, restriction enzymes are used. Restriction enzymes recognize particular sequences (called recognition sequences) within the nucleotide sequence. Where a specific recognition sequence exists, DNA will be cut, producing restriction fragments. Before a clinical application of RFLP is considered, the role of restriction enzymes in molecular diagnostics must be examined.

Restriction enzymes

Restriction enzymes recognize specific sequences within DNA and cleave bonds between adjacent nucleotides. There are two types of restriction enzyme: endonucleases, which recognize regions within a DNA sequence, and exonucleases, which recognize terminal sequences. Where cleavage occurs, fragments are formed of a predetermined length. These fragments can be examined using gel electrophoresis.

The name of the restriction enzyme is derived from the Latin name of the organism from which it was obtained. Three letters are used; the first represents the genus and the following two, the species. A number is also allocated, representing the 'number' of the enzyme in relation to the variety produced by that organism. For example, the enzyme EcoR1 is derived from *Escherichia coli*, R-strain, and was the first to be identified in this bacterium. The name provides no clue to the recognition sequence, and this must be determined prior to use.

Restriction endonucleases cut both strands of DNA and can produce, depending upon the recognition sequence, blunt or overhanging ends. The sequence recognized by EcoR1 is shown below:

5′ - - - - G↓A A T C - - - - 3′
3′ - - - - C T A A↑G - - - - 5′

The products of this endonuclease reaction are as follows:

G + AATC and CTAA + G

Conversely, HaeIII, derived from *Haemophilus aegyptius*, endonuclease activity produces blunt ends, as illustrated below:

5′ - - - - G G↓C C - - - - 3′
3′ - - - - C C↑G G - - - - 5′

The products of this digestion are:

GG + CC and CC + GG.

An example of RFLP analysis and its clinical applications can be found in the clinical correlations box entitled 'Factor V Leiden'.

Cross reference

More detail on restriction enzymes can be found in the *Biomedical Science Practice* volume in this Fundamentals of Biomedical Science series.

SELF-CHECK 2.12

Define restriction fragment length polymorphism.

CLINICAL CORRELATION

Factor V Leiden

The identification of a heritable DNA mutation predisposing an individual to thrombosis (blood clots) can be achieved using restriction enzymes. Factor V Leiden represents the mutation G1691A. In this condition, there is a substitution of nucleotides guanine to adenine at nucleotide number 1691, producing an abnormal protein—R506Q. This is formed from the substitution of arginine (R) at amino acid position 506 for glutamine (Q), leading to the accumulation of the activated clotting factor Va, rather than its appropriate degradation *in vivo*.

The change in DNA sequence leading to G1691A introduces a recognition sequence into the gene encoding clotting factor V for the restriction enzyme HindIII (pronounced Hin dee three). HindIII is an endonuclease that recognizes the sequence AAGCTT and cleaves between the AA nucleotides.

When visualized using gel electrophoresis, as illustrated in Figure 2.23, and ethidium bromide (as outlined in the following section), the following patterns can be compared to the DNA ladder:

For individuals homozygous for G1691 (i.e. those without the mutation) there is no HindIII recognition sequence and therefore the product is undigested and a single band of 241 bp is demonstrated.

For those homozygous for the G1691A mutation, a HindIII recognition site is created, producing two fragments, one at 209bp and one at 32bp.

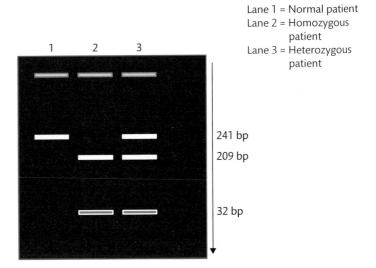

Lane 1 = Normal patient
Lane 2 = Homozygous patient
Lane 3 = Heterozygous patient

241 bp
209 bp
32 bp

FIGURE 2.23

Three specimens investigated for RFLP using the restriction enzyme HindIII to identify the factor V Leiden mutation. Each specimen was loaded onto the gel at the top. Lane 1 contains a sample from a patient with a normal factor V gene. There is no recognition sequence, therefore, only a single band of 241 bp is formed. Lane 2 is from a patient homozygous for the factor V Leiden mutation. A single HindIII recognition site generates two fragments (209 bp and 32 bp). Lane 3 is a sample from a heterozygous patient. The 241-bp fragment is generated from the normal allele and the 209 bp and 32 bp bands originate from the mutant allele.

In heterozygous individuals (i.e. those possessing both G1691 and 1691A), electrophoresis produces three fragments—one 241bp undigested fragment, one 209bp fragment plus a 32bp fragment.

By examining the pattern of RFLPs, it is possible to determine whether a mutation has occurred within a particular gene, potentially leading to significant pathology. It is far cheaper and easier to perform RFLP analysis of patients rather than to sequence particular regions of their genome, as outlined below.

DNA sequencing

Once a region of DNA has been amplified using PCR, it is possible to determine the sequence of the DNA within the target region. This can be achieved in one of two ways, a manual or an automated method. In order to understand the automated, high-throughput method, it is useful to understand the manual method.

The most widely accepted method of sequencing DNA is the Sanger method, first devised in 1977. This method, also called the dideoxy method, requires four single-stranded templates obtained from the PCR reaction divided into separate aliquots. To each of these aliquots, a radioactive-labelled primer is added (similar to the PCR reaction), and this primer provides a 3'-OH group upon which phosphodiester bonds can be formed, allowing the DNA sequence to be assembled. In addition to the primer, four deoxynucleotide triphosphates (dNTPS)— dATP, dTTP, dCTP, dGTP—are added to excess, and to each of the four tubes, one of a dideoxynucleoside triphosphate (ddNTP) terminator—ddATP, ddTTP, ddCTP, ddGTP. These ddNTP

Cross reference

More detail on DNA sequencing can be found in the *Biomedical Science Practice* and *Haematology* volumes in this Fundamentals of Biomedical Science series.

terminators have the 3'-OH missing, thereby preventing the formation of the phosphodiester bond. The ddNTPs are incorporated randomly into the sequence, although at the appropriate location for the specific complementary nucleotide in question, producing fragments of different lengths. Over the course of the analysis, each position within each aliquot for each dNTP will have been occupied by the ddNTP, producing sequences of variable length. Each aliquot is then loaded onto a polyacrylamide gel in columns A, C, G and T. The sequence of the newly synthesized fragments can then be read using **autoradiography**.

Instead of using radioactively-labelled probes, the ddNTP terminators are fluorescently labelled. Each ddNTP species—ddATP, ddTTP, ddCTP, ddGTP—is labelled with a different fluorophore, and, rather than using four aliquots, all can be combined within a single reaction vessel. Although over-simplified, the sequence of the DNA can be determined according to the order of fluorescence 'read' by the sequencer.

Autoradiography

This involves exposing an X-ray film to the decay emissions of a radioactive substance—in this case, the radioactively-labelled DNA probes used in Sanger sequencing—producing an image for interpretation.

Visualizing the products of PCR

In order to identify, purify and analyse DNA following PCR, electrophoresis is used. Electrophoresis is an important technique to master in the laboratory, and it has a wide range of applications. Electrophoresis separates molecules according to size and charge across an electric field using a gel immersed in buffer. The choice of gel depends on the degree of resolution required for the product. Agarose gel, derived from seaweed, is used in a slab of 0.5–2.0 per cent and can separate DNA sizes between 100 bp and 50 000 bp, while the complex process of using polyacrylamide gel electrophoresis can resolve fragments up to 500 bp. Specimens of interest are placed at a defined point, called the origin, and are accompanied by a loading dye to aid visualization of the loaded DNA on the gel. The electric field is switched on, creating a differential charge across the gel. On one side there is a positive charge (called the anode), and on the other a negative charge (the cathode). Nucleic acids possess a negative charge by virtue of the phosphate backbone, and will migrate towards the anode. Conversely, proteins may have a negative or a positive charge according to their amino acid constituents, and therefore will migrate between the two.

It is important that, following electrophoresis, visualization of bands is possible. In order to do this, a DNA ladder (each fragment forming a 'rung' separated according to DNA fragment size) is run alongside the samples. The dye most commonly used to visualize the bands is ethidium bromide (EtBr). This intercalates between bases and, following electrophoresis, will fluoresce orange when exposed to ultraviolet light. A photograph of the gel can then be taken for interpretation.

Cross reference

More detail on electrophoresis can be found in the *Biomedical Science Practice* volume in this Fundamentals of Biomedical Science series.

Key points

Ethidium bromide (3,8 diamino-5-ethyl-6-phenyl phenanthridinium bromide) intercalates between DNA bases, not only within a specimen but also in DNA to which it is exposed—including scientist DNA. As EtBr can bind to DNA, it can induce mutations and therefore is considered carcinogenic (cancer causing). It is essential that all scientists using EtBr use appropriate personal protective equipment (PPE), including protective gloves.

Evolving PCR

One of the limitations of standard PCR is that it is impossible to check that the amplification of DNA has been successful until the products of PCR are visualized on a gel. Real-time PCR

(RT-PCR) significantly improves on standard PCR, allowing the quantification of PCR products at the end of each cycle. Thus, RT-PCR can also be referred to as quantitative PCR (qPCR).

Real-time PCR

In addition to forward and reverse primers, RT-PCR utilizes reporter probes. These have a complementary design to the DNA template, binding a different region to the primer. Reporter probes contain two important elements. Situated at the 5′ end is a fluorescent tag, and at the 3′ end is a **quenching** tag. Separation of the two tags enables fluorescence to occur. When in close proximity (i.e. when retained by the reporter probe), the quenching tag suppresses the fluorescence from the 5′ end tag. During PCR, the forward primer is extended by DNA polymerase. When the reporter probe is reached, it is degraded by the polymerase. In the first instance, the 5′ fluorescent tag is released from the reporter probe. Separating the fluorescent tag from the quenching tag ensures fluorescence is generated, allowing the amplification process to be monitored using an *in situ* fluorimeter. Sufficient concentration of reporter probe is added to ensure that each amplification step can be monitored.

Quenching
This is a term used to describe the process of reducing the fluorescence generated by a given substance.

Alternatively, rather than using a reporter probe containing a specific binding sequence, a non-specific dye can be used. Although there are a number of dyes suitable for this application, SYBR green I will be discussed here. SYBR green I is a dye that absorbs light at 494 nm and emits light at 521 nm, within the green region of the spectrum. SYBR green I binds specifically to double-stranded DNA and, when bound, fluoresces intensely. This fluorescence intensity can be detected using a fluorimeter. As the double-stranded DNA is denatured, SYBR green I dissociates from the DNA, reducing fluorescence intensity. Each time double-stranded DNA forms, fluorescence intensity increases, allowing the PCR process to be monitored. The primary limitation of using dyes rather than reporter probes is that they will bind to any double-stranded DNA. Therefore, if the PCR mix is contaminated, non-target DNA will also be measured.

SELF-CHECK 2.13

Differentiate between reporter probes and SYBR green I.

An important figure when quantifying the amount of PCR product at the end of each RT-PCR cycle is the Ct value. The Ct (cross threshold) value is the cycle number at which the fluorescence generated reaches the threshold level. The lower the Ct value the greater the concentration of target DNA in the sample. A higher Ct value indicates a lower target DNA concentration.

RT-PCR is widely available both for diagnostic and research applications, and a number of kits are available for the rapid evaluation of a range of pathogens and diseases, including hepatitis, bloodborne parasites, sexually transmitted infections, markers of genetic diseases, and for the diagnosis and monitoring of a range of cancers.

Multiplex PCR

So far, the focus of this section has been on the amplification of a single gene, or region of a gene. However, providing sufficient materials are available (e.g. *Taq* polymerase, dNTPs etc.), it is possible to amplify more than one gene at a time. This forms the basis of **multiplexing**. This is also true for qPCR methods (RT-PCR). The addition of alternative primers ensures the amplification of additional templates, and consideration of the primer design can allow the scientist to control the timing of the denaturation of different double-stranded DNA. For example, including more GC regions in one of the primer pairs increases the melting

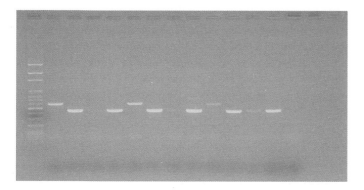

FIGURE 2.24
Output of multiplex PCR run alongside a DNA ladder (left) to detect three extended-spectrum β-lactamase (ESBL) enzymes expressed by bacteria isolated from clinical specimens. The three genes encoding SHV, CTX and TEM produce bands at 747 bp, 593 bp and 445 bp, respectively. Of the 13 specimens processed, three express CTX and seven express TEM. (courtesy of Sarah Fouch, University of Portsmouth).

temperature required for denaturation, and different products can be separated as fluorescence will cease at different melting temperatures when using SYBR green I. This is important for RT-PCR because, until relatively recently, it was believed that this technique could not be multiplexed. If using gel electrophoresis to evaluate the products following PCR, the products should be generated at different sizes to allow appropriate separation by the gel. An example of multiplex PCR using traditional PCR techniques and visualization by gel electrophoresis is shown in Figure 2.24.

Multiplex PCR has a number of applications both in diagnosis and research, all of which allow the differentiation of PCR products, leading to the identification of a particular molecular target. Multiplex PCR can be applied clinically to diagnose various pathogens, such as those causing hospital-acquired infections, sexually transmitted infections and tuberculosis, and can also be used to aid the understanding and elucidation of molecular pathways in genetic diseases and cancers.

Reverse transcriptase PCR

Genes encode proteins and the process of decoding a gene is called transcription. Transcription involves the production of an intermediate called messenger RNA (mRNA). This is processed in such a way that only the regions of the gene responsible for encoding the final polypeptide, called exons, are retained within the mature mRNA molecule. All intervening sequences, called introns, are removed.

It is possible to investigate DNA when only mRNA is present, and this can be accomplished using a modified PCR method called reverse transcriptase PCR. The problem with mRNA is that it is rather fragile and liable to degrade, whereas DNA is much more stable. It is possible, by isolating mRNA, to synthesize single-stranded DNA using the enzyme reverse transcriptase. Biologically, reverse transcriptase is used by RNA viruses, such as retroviruses, to generate DNA from their RNA genomes for incorporation into the host's genome.

Once target mRNA has been isolated, perhaps using laser capture microdissection, the mRNA can be converted into DNA using reverse transcriptase PCR. It is essential to know the sequence of the target mRNA as reverse primers are used to provide a starting point for generating the

DNA. Once a single strand of DNA has been synthesized from the mRNA template, a forward primer will bind to the newly synthesized single-stranded DNA template to produce double-stranded DNA. Traditional PCR techniques can then be used to amplify the double-stranded DNA to provide sufficient product for analysis or other downstream applications. The DNA synthesized through this process is called complementary DNA (cDNA) as it only contains the elements of the gene encoding the protein (i.e. the exons). As the content of cDNA is restricted, it is ideally suited to the synthesis of gene probes or for cloning genes for their incorporation into cells not usually expressing the protein product.

Key points

Real-time PCR and reverse transcriptase PCR both can be abbreviated to RT-PCR. It is important when reading around these types of PCR, or when designing and duplicating laboratory protocols, that the two techniques are not confused.

Utilizing PCR technology and the generation of cDNA, it has now become possible to evaluate the activity of genes within a cell or population of cells. Gene expression studies and microarrays are discussed next.

Gene microarrays

Gene microarrays are used to compare gene expression within at least two separate samples. This technology, introduced in the 1990s, has become increasingly useful for the detection and classification of disease. Although gene microarrays can be used to investigate a wide variety of disorders, they are used primarily for research purposes—particularly in cancer research. It has been demonstrated that a wide variety of cancers can be diagnosed according to their gene expression profile instead of utilizing the standard approaches of cell morphology, immunophenotype and cytogenetic profile.

All healthy nucleated cells share the same chromosomal complement, and therefore the genome of every healthy nucleated cell is the same. However, proteins expressed within different cell types vary according to cell identity, demonstrating that gene expression within cells and tissue is specific. Therefore, it stands to reason that genes are expressed according to cellular requirements, and this is a dynamic process.

Gene expression can be measured according to the proteins expressed within a cell. However, this can be problematic as protein isolation can be complex, time-consuming and costly— three unattractive facets within the diagnostic and research arenas. However, the product of gene transcription, mRNA, can be measured relatively easily using a standardized technique, and all mRNAs can be treated identically as they all have the same biological properties. Unfortunately, mRNA is rather unstable, making isolating and utilizing mRNA for gene expression profiling problematic, as mRNA may degrade prematurely, leading to the generation of misrepresentative data. Usually, the mRNA is converted to cDNA using reverse transcriptase PCR, as outlined above. The cDNA is more stable and easier to use than mRNA. The specimens to be investigated are fluorescently labelled, usually green, and compared with a control specimen, labelled red. These specimens are added to the microarray, composed of thousands of spots, with each spot containing probes for a specific gene. The probes for each gene, covalently bound within a particular spot, may number in the hundreds of thousands or millions. These probes may be composed of either cDNA or oligonucleotides. Each probe set represents a particular gene, the identity and location of which is recorded for later analysis. Each probe binds to the cDNA derived from the original specimen in a complementary fashion. The relative intensity of each fluorochrome is established and analyzed statistically. Ultimately, the data generated will demonstrate normal, low or high gene expression. This information can provide vital clues about how the cell machinery works, and, in the case of cancer, provides important prognostic markers or therapeutic targets for scientists and clinicians.

Cross reference

More detail on DNA microarrays can be found in the *Biomedical Science Practice, Haematology* and *Cytopathology* volumes in this Fundamentals of Biomedical Science series.

SELF-CHECK 2.14

What is isolated from target cells to measure gene expression?

CLINICAL CORRELATION

Chronic Lymphocytic Leukaemia and ZAP-70

The use of gene microarrays was essential in identifying a prognostic marker in chronic lymphocytic leukaemia (CLL), a relatively common haematological cancer caused by transformation of B lymphocytes. Variability in survival data between cases suggested CLL may be subdivided into different biological entities. It was shown that cases could be differentiated into *IgVH*-mutated and *IgVH*-unmutated cases. *IgVH* genes encode the heavy variable region of immunoglobulins, and mutation of these genes is a completely normal response to antigenic stimulation. *IgVH*-mutated cases have a better prognosis than unmutated cases, and identifying the status of *IgVH* for the patient adds value to clinical consultation. The technology used to differentiate between the two types—gene sequencing—and the technological expertise required is both expensive and demanding. Gene microarrays were used to compare both types and demonstrated that a protein called ZAP-70 was expressed in unmutated cases and absent in mutated cases, as demonstrated in Figure 2.25. This discovery led to the development of a flow cytometry-based application,

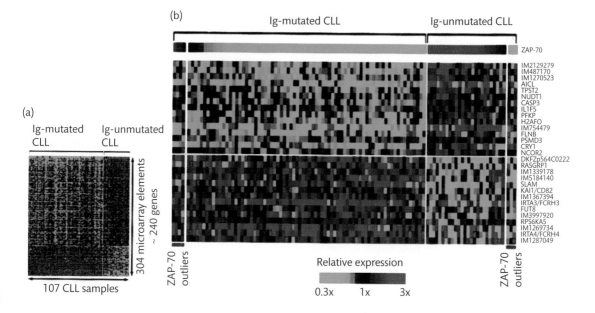

FIGURE 2.25

(a) Gene expression analysis of 107 patients diagnosed with CLL. (b) Gene expression analysis identifying ZAP-70 as a marker of unmutated *IgVH* in chronic lymphocytic leukaemia (top). In the majority of cases, *IgVH*-mutated CLL over-expresses ZAP-70 (red) compared to limited expression (green) in mutated cases.

From Weistner *et al.* (2003). ZAP-70 expression identifies a chronic lymphocytic leukemia subtype with unmutated immunoglobulin genes, inferior clinical outcome, and distinct gene expression profile. *Blood* **101(12)**: 4944–51.

using monoclonal antibodies, to evaluate ZAP-70 expression in CLL—a much cheaper and technologically available technique than sequencing *IgVH* genes. Nowadays, ZAP-70 is used clinically as a surrogate marker of *IgVH* mutation.

Chapter summary

In this chapter we have:

● introduced and discussed different types of commonly used microscope.

● explored fluorescence and its applications in the investigation of the cell.

● examined cell enumeration and identification through the use of electrical impedance and light-scattering technologies.

● discussed the use of monoclonal antibodies and demonstrated how these can be exploited to identify cells according to CD marker expression.

● considered various cytogenetic methods for the investigation of chromosomal abnormalities.

● outlined the basic molecular methods used in the laboratory and how these can be utilized to investigate genes, mRNA and protein expression and to identify mutations within particular genes.

Further reading

● Crocker J, Murray PG eds. *Molecular biology in cell pathology*. Chichester: Wiley, 2003.

● Dickersin GR. *Diagnostic electron microscopy. A text/atlas* 2nd edn. London: Springer, 2000.

● Glencross H, Ahmed N, Wang Q eds. *Biomedical science in practice*. Oxford: Oxford University Press, 2011.

● Hall A, Yates C eds. *Immunology*. Oxford: Oxford University Press, 2010.

● Lacey AJ ed. *Light microscopy in biology: a practical approach*. Oxford: Oxford University Press, 1999.

● Moore G, Knight G, Blann A. *Haematology*. Oxford: Oxford University Press, 2010.

● Orchard GE, Nation BR eds. *Histopathology*. Oxford: Oxford University Press, 2012.

● Pinkel D, Albertson DG. Array comparative genomic hybridization and its applications in cancer. *Nat Genet* 2005; **37**: S11-17.

● Shambayati B ed. *Cytopathology*. Oxford: Oxford University Press, 2011.

● Speicher MR, Carter NP. The new cytogenetics: blurring the boundaries with molecular biology. *Nature* 2005; **6**: 782-92.

● Tubbs RR, Stoler MH. *Cell and tissue-based molecular pathology*. Foundations in Diagnostic Pathology Series. Edinburgh: Churchill Livingstone, 2008.

Discussion questions

2.1   Our understanding of cell structure and function is derived from technological innova-
tion. Using appropriate examples, discuss this statement.

2.2   Compare and contrast the different forms of microscopy and provide appropriate
clinically-based applications for each.

2.3   Investigating cells often requires considerable sample preparation. Choose three methods
outlined in this chapter, investigate and write notes on the different processing requirements of each technique.

Answers to SELF-CHECK questions and discussion questions, hints and tips are provided
on the book's Online Resource Centre.

  Visit www.oxfordtextbooks.co.uk/orc/.

3

Introduction to anatomy and embryology

Joanne Murray and Ian Locke

Learning objectives

By the end of this chapter you should be able to:

■ Discuss the gross features of the human body using anatomically correct descriptors and understand the difference between fine and gross anatomy.

■ Discuss basic cell structure and be aware of the specialized features of the four principal cell families.

■ Explain the differences and relationships between tissues, organs and organ systems.

■ Describe the gross anatomy and roles of the 11 human organ systems.

■ Describe fertilization and implantation.

■ Discuss the early development of the human embryo and understand the roles and significance of the ectoderm, mesoderm and endoderm as sources of the different tissue types.

■ Understand the origins and early development of the major organ systems.

■ Discuss the causes and origins of congenital malformations/disorders, giving examples to illustrate the different types.

■ Describe diagnostic techniques that may be employed to screen for congenital malformations/disorders *in utero*.

■ Discuss the continued development of the ageing human through childhood, adolescence, adulthood and into old age.

The mature, fully developed human body is composed of various specialized organs and tissues which carry out discrete, but often interrelated, functions and combine to form a complex, adaptable and, above all, successful organism. In many ways the modern human represents perhaps the ideal 'compromise' organism with an anatomy and physiology that allows us to withstand various environments and perform a wide

range of activities within those environments, albeit with assistance in the form of tools and inventions devised as a result of our evolved abilities.

It is probably worthwhile at this point to consider what we mean by anatomy and physiology. The two terms are closely linked, so much so that it is unusual to find one discussed without the other; hence, books or learning modules in this area are generally titled 'anatomy and physiology'. If we wish to separate the two disciplines, however, anatomy refers primarily to the structure of the whole body and the spatial/structural relationships between the various parts and systems, whereas physiology is generally concerned with the function of the whole body and systems within that body. Anatomy may also be qualified as gross or fine anatomy, gross being that described above and fine being the detailed microscopical structure of individual organs or tissues, analogous to the biomedical science disciplines of histology and cytology. Between these two extremes, the terms regional and systemic anatomy are often used to describe the study of specific parts or organ systems within the whole organism. The fine anatomy of the various organs will be covered in more detail in subsequent chapters of this book, whereas this chapter will concentrate primarily on gross anatomy and the 'building blocks' of organ systems.

Key points

Fine anatomy refers to the detailed microscopic structure of individual organs or tissues.

There are many other descriptive terms used in anatomical discussion to describe, for example, the location of an organ within the body, the relationship of one organ to another, and the viewpoint from which one is examining a particular feature. In some cases, organs may also be described by the body cavity in which they reside, either the dorsal cavity (further subdivided into the cranial/vertebral cavities) or the ventral cavity (further divided into the thoracic and the abdominopelvic cavities). To help you follow the descriptions given in this and other chapters of the book, this 'language' of anatomy is summarized in Boxes 3.1 and 3.2, with the major directional terms illustrated in Figure 3.1.

BOX 3.1 The language of human anatomy: Directional terminology

1) **Superior/Cephalic**: Towards the head or above.

2) **Inferior/Caudal**: Away from the head or below (N.B. while these two terms are often used synonymously, Caudal, strictly speaking, means 'towards the tail' [i.e. the base of the spine in humans]).

3) **Anterior/Ventral**: Towards the front of the body or 'in front of'.

4) **Posterior/Dorsal**: Towards the back of the body or behind.

5) **Medial**: Towards the midline, an imaginary vertical line dividing the body into two equal halves, left and right.

6) **Lateral**: Away from the midline, on either side.

7) **Ipsilateral**: On the same side of the midline.

8) **Contralateral**: On the opposite side of the midline.

9) **Proximal**: Nearer to the point at which the limb attaches to the trunk.

10) **Distal**: Further from the point at which the limb attaches to the trunk.

11) **Superficial/External**: Towards the surface of the body.

12) **Deep/Internal**: Away from the surface of the body.

BOX 3.2 The language of human anatomy: Regional terminology

Anterior descriptors:

1) Cephalic (head). Subdivided into Cranial (skull) and Facial regions. Additional anterior cephalic descriptors include: Frontal (forehead), Temporal (temple), Orbital (eye), Nasal (nose), Buccal (cheek) and Oral (mouth).

2) Cervical (neck).

3) Thoracic (chest). Additional thoracic descriptors include: Sternal (sternum/breast bone) and Axillary (armpit).

4) Abdominal (abdomen). Additional abdominal descriptors include: Umbilical (navel).

5) Pelvic (pelvis). Additional pelvic descriptors include: Inguinal (groin) and Coxal (hip).

6) Pubic (genitals).

7) Upper limb. Anterior upper limb descriptors include: Acromial (joining point of shoulder), Brachial (arm), Antecubital (front of elbow), Antebrachial (forearm), Carpal (wrist), Palmar (palm), Digital (fingers).

8) Lower limb. Anterior lower limb descriptors include: Coxal (hip), Femoral (thigh), Patellar (front of knee), Crural (leg), Tarsal (ankle), Digital (toes).

Posterior descriptors:

1) Cephalic (head). Additional posterior cephalic descriptors include: Occipital (base of the skull).

2) Cervical (neck).

3) Dorsal (back). Additional dorsal descriptors include: Scapular (shoulder blade), Vertebral (spine), Lumbar (lower back/spine) and Sacral (area between hips).

4) Upper limb. Posterior upper limb descriptors include: Acromial (joining point of shoulder), Brachial (arm), Olecranal (back of elbow), Antebrachial (forearm), Digital (fingers).

5) Lower limb. Posterior lower limb descriptors include: Femoral (thigh), Popliteal (back of knee), Crural (leg), Sural (calf), Tarsal (ankle), Calcaneal (heel).

Inferior descriptors:

1) Plantar (sole of the foot).

It should also be noted that these terms are used on the assumption that the body being discussed is presented in a certain way and this is referred to as the 'anatomical position'—one in which the subject is standing, facing the observer with the feet parallel and the arms hanging by the sides of the body, palms forward. Other terms relating to the way in which a subject is observed are those applied to the view presented to the observer. These are referred to as sections or planes, of which there are three basic planes. In humans and other bipeds, the sagittal plane cuts the body vertically resulting in left and right parts; if these parts are equal then it is known as a mid-sagittal plane. The frontal or coronal plane also cuts the body vertically, but at 90° to the sagittal plane, creating front and back (anterior and posterior) parts. The transverse plane, also known as a cross-sectional plane, cuts the body horizontally at 90° both to the sagittal and frontal planes, resulting in upper and lower (superior and inferior) parts. The three basic planes are shown in Figure 3.2.

The complexity of the mature human body is in stark contrast to the origins of human life, where the joining of two haploid gametes results in the formation of a single diploid cell. Within the space of a few hours, this single cell, carrying the full complement of human genetic material, will commence a programme of cell division and differentiation that, all being well, will culminate in the formation of a complete, functional human being approximately nine months later. Of course, development does not end with the delivery of the new infant and will continue steadily over the next 15–20 years until the mature adult organism emerges. Even then, there is scope for further

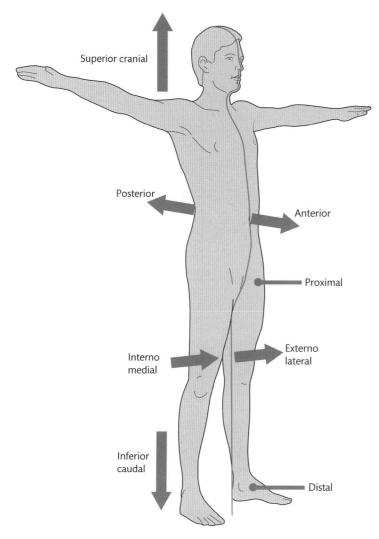

FIGURE 3.1
The major directional anatomical terms.
YassineMrabet/CC-BY-SA

development and adaptation in response to changing environmental stimuli and/or physiological status. As we progress through this chapter we will consider the development of the human body at all stages, from conception, through embryonic and fetal development to the juvenile, adult and aged individual, but in the meantime we will commence with a discussion of the 'basic' units of anatomy and physiology—cells, tissues, organs and organ systems.

SELF-CHECK 3.1

Using correct anatomical terminology, how would you describe the position of your navel/umbilicus:

a) In isolation?

b) Relative to your nose?

c) Relative to the small of your back?

d) Relative to your foot?

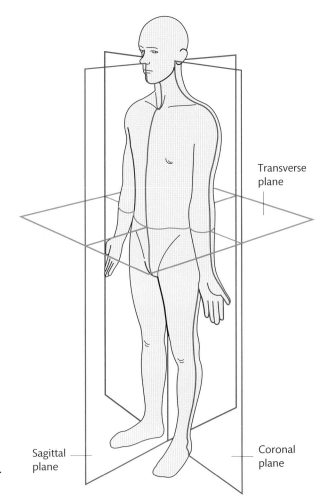

Transverse plane

Sagittal plane

Coronal plane

FIGURE 3.2

The three basic anatomical planes.
G. Pocock and C. Richards 2009.

SELF-CHECK 3.2

Using correct terminology, how would you describe the position of the right lung:

a) In isolation?
b) Relative to the left lung?
c) Relative to the left kidney?
d) Relative to the right clavicle/collarbone?

3.1 Fine and gross anatomy, cells, tissues, organs and systems

As discussed in Chapter 1, cells represent the basic unit of structure and function, which, either on their own or combined with others, produce an organism that satisfies the definition of 'life'. While some discussion remains, the general consensus is that there are seven characteristics of 'life':

● *Organization*: The organism is structurally composed of one or more cells.

- *Homeostasis*: The organism demonstrates the ability to regulate its internal environment to maintain a relatively stable state.
- *Metabolism*: The organism demonstrates the ability to utilize energy.
- *Growth:* The organism demonstrates during its life cycle the ability to increase in size or complexity through cell division, differentiation or an increase in cell size.
- *Response to stimuli*: The organism demonstrates specific responses to external stimuli, with such responses typically involving some manner of change or movement.
- *Adaptation*: The organism demonstrates the ability to change over a period of time in response to the environment.
- *Reproduction*: The organism is able to form new individual organisms, either by asexual mechanisms involving only one parent organism, or by a sexual mechanism involving two parent organisms.

There are somewhere in the region of 200 distinct cell types recognized within the mature human body, but these may be broadly grouped into four families based on the functions they perform: nerve cells, connective tissue cells, muscle cells and epithelial cells. Regardless of their family or the distinct, differentiated cell type, all cells share the same basic biology and structure, as described in the first two chapters of this book.

- Each cell type constitutes a discreet entity contained within a plasma membrane encompassing a cytoplasm, with a variety of functional organelles, and, in most cases, a nucleus. It is perhaps also worth noting, however, that some cells are multinucleate (e.g. skeletal muscle cells) and others are anucleate (e.g. erythrocytes).
- The basic structure of the cell membrane varies little between the various cell types, as all cells require a selective barrier to control substance entry to, or exit from, the cell. The cell membrane does, however, carry a number of receptor and ligand molecules involved in signalling between cells, and these will vary greatly depending on the cell type.
- The cytosol, the fluid component of the cytoplasm, is also very similar between cell types, but the nature and number of organelles suspended within depend on cell type.
- The nucleus contains the cell's genetic material organized, in humans, into 46 chromosomes (22 pairs of autosomes and a pair of sex chromosomes [XX in the female, XY in the male]). All cells of the body will carry the same complement of chromosomes within their nucleus, the exception being the gametes (ova and spermatozoa), which are haploid (see Box 3.3). Gametogenesis and the nature of the gametes will be discussed in greater detail in Chapter 13.

While the four cell families share many common structures and cell processes, each family exhibits a number of distinct features that enable them to carry out their specialized functions.

- Nerve cells are able to initiate and conduct electrical impulses, either as a result of a detected stimulus in the case of sensory nerve cells or in response to a signal relayed from another nerve cell. In this case, the signal is generally passed between neurons via chemical transmission at synapses (although some electrical synapses do exist, particularly in the central nervous system [CNS]) utilizing various neurotransmitters. When the electrical impulse has been passed to motor neurons they will then, in turn, elicit a response by stimulating, for example, muscular contraction or glandular secretion.
- Connective tissue cells include various cell types which appear, at first, to have little in common. Cells in this group include fibroblasts, chondrocytes, osteoclasts and osteoblasts, adipocytes and blood cells. Despite their lack of similarity, these cell types provide a supportive or connective role for the various structures of the body, including the production of extracellular matrices.

BOX 3.3 Discussion point: We describe gametes as being haploid but is the oocyte ever actually haploid?

In this chapter, as well as in Chapters 1 and 13, we describe the moment of conception as the joining of two haploid cells. At the end of spermatogenesis the spermatozoa are haploid (n); that is, each spermatozoon only contains 22 autosomes and either an X or Y sex chromosome. Until ovulation, the oocyte has been held in dictyate (meiotic arrest) at the diplotene stage of meiosis I. This means that each of the 23 homologous pairs of each chromosome have been duplicated, resulting in 46 bivalent chromosomes; that is, there are 92 chromatids in total. Recombination has also occurred but the first reductional division of meiosis has not occurred. The luteinizing hormone (LH) surge that causes ovulation also causes, within the ovulated oocyte, the completion of meiosis I. The result is that the ovulated oocyte is one of two cells contained within the glycoprotein structure called the zona pellucida. Most of the cytoplasm and cellular organelles have disproportionately moved to the oocyte while the second cell contains mostly the rejected nuclear material. This little cell is called the polar body and will soon degenerate. The now secondary oocyte again is arrested before completing meiosis II and still has 44 autosomes and two

X chromosomes (the bivalent chromosomes [sister chromatids] have separated to maintain ploidy at 2n). The final stages of meiosis are only completed in the secondary oocyte if a haploid spermatozoon is successful in penetrating the zona pellucida and the plasma membrane of the oocyte, resulting in the initiation of fertilization. Hence, at the initiation of fertilization, the secondary oocyte actually contains its 44 autosomes and two X chromosomes as well as the spermatozoon's 22 autosomes and either an X or Y sex chromosome (at this point the oocyte is 3n). Within a couple of hours of the spermatozoon entering the oocyte, the oocyte and the first polar body complete meiosis II and the unwanted sets of chromosomes are extruded into a second and third polar body—again with minimal loss of cytoplasm (therefore, completion of meiosis in the primary oocyte results in three haploid daughter cells, the polar bodies, as well as the oocyte). The haploid set of chromosomes from the oocyte and the spermatozoon are then free to complete syngamy, and fertilization is completed. Hence, at no point does an oocyte contain only 22 autosomes and one X chromosome; thus, strictly speaking, the oocyte is never truly haploid.

- Muscle cells are able to: contract in response to nervous stimuli, resulting in the production of forces that allow independent movement of one bone in relation to another (skeletal muscle cells); control and accomplish changes in organ activity and blood supply (smooth muscle), and ensure the continuous supply of oxygen and other nutrients around the body (cardiac muscle).

- Epithelial cells are adapted for selective adsorption and secretion and are generally located on the outer layers of organs to allow this function to be utilized effectively. Epithelial cell membranes contain various transporter or 'pump' proteins that enable the selective adsorption/secretion of ions and organic substances, in many cases through 'both sides' of the membrane; that is, first into the cytosol of the epithelial cell from the internal environment and then from the cytosol of the epithelial cell to the external environment, and *vice versa*.

Key points

There are four basic cell families: nerve cells, connective tissue cells, muscle cells and epithelial cells.

If the cell represents the basic functional unit of life, the next stage in the development and organization of a multicellular organism is the formation of tissues. Tissues represent a collection of cells with similar origins/functions, often with the addition of an extracellular matrix, that form the building blocks for the organs and organ systems of the mature human body. Analogous to the four cell families discussed above, there are also four basic tissue types: nervous tissue, connective tissue, muscle tissue and epithelial tissue. Each tissue type originates from one or more of the three primary germ layers observed in the early embryo: the ectoderm (the outer layer), the mesoderm (the middle layer) and the endoderm (the inner layer). The three germ layers and embryonic development will be discussed in more detail later in this chapter, but, essentially, nervous tissue is derived from the ectoderm, connective and muscle tissues are derived principally from the mesoderm, and epithelial tissues are derived from all three primary germ layers.

- Nervous tissue forms one of the major control mechanisms of the human body and comprises both the CNS and peripheral nervous system (PNS) with sensory and motor neurons.

- Connective tissue is probably the most functionally diverse of the four types and includes adipose tissue, bone, blood cells and the 'true' connective tissues in which cells such as fibroblasts and chondrocytes are responsible for the production of the extracellular matrix containing collagen, other proteins and polysaccharides. The extracellular matrix (ECM) is found in various locations throughout the body where it serves to anchor and support the components of the body, literally 'connecting' organs and tissues.

- Muscle tissue includes cardiac, striated and smooth muscle and is therefore central to the transport of nutrients and oxygen through the bloodstream, the movement of the whole organism, and the correct functioning of a number of body systems.

- Epithelial tissue is generally found at the boundaries between the organism and its environment, or it serves to compartmentalize the body by forming the outer structure of organs. As well as forming barriers, epithelial tissue also plays a distinct role in the selective absorption and secretion of ions, nutrients and endogenously produced secretions.

Key points

There are four basic tissue types: nervous tissue, connective tissue, muscle tissue and epithelial tissue.

Tissues may be considered as the intermediate level of organization between individual cells and organs, which are generally composed of more than one tissue type. However, some caution is needed here as the term 'tissue' is often used quite ambiguously, particularly in the world of biomedical science where it may be used to describe part of an organ; for example, skin tissue, lung tissue etc. Strictly, 'tissue' refers to a construct containing a distinct cell type, but when used in relation to part of an organ it will refer to a construct containing several, and quite possibly all four, of the basic tissue types.

Organs are the functional units of the body and represent a number of cells and tissue types combined to form a structure capable of performing a specific role in the biological processes of the whole organism. Individual organs, however, rarely function in isolation, and several organs usually combine to form a system. There are 11 distinct organ systems described in

SELF-CHECK 3.3

How many basic tissue types are there? Name them.

the human body: the cardiovascular system, the respiratory system, the digestive system, the urinary system, the nervous system, the endocrine system, the reproductive systems, the lymphatic system (often including the immune system), the integumentary system, the muscular system and the skeletal system—the last two are often described as a combined entity, the musculoskeletal system. The fine anatomy of the individual systems will be covered in greater detail in later chapters of this book, but the basic functions and gross anatomy of each of the systems is summarized below.

- The cardiovascular system is responsible for the transport of nutrients, oxygen, endocrine mediators and other substances around the body. It also plays a crucial role in the supply and transport of immune cells to areas of infection and inflammation. The principal components of the cardiovascular system are the blood vessels (i.e. arteries, arterioles, capillaries, venules, veins) and the heart (see Chapters 4 and 9 for more details).

- The respiratory system maintains effective gaseous exchange, ensuring that the body is constantly supplied with oxygen for cellular respiration and that the gaseous product of that respiration, carbon dioxide, is removed. The principal components of the respiratory system include the alveoli of the lung, the air passages (i.e. trachea, bronchi and bronchioles) and associated structures such as the larynx, pharynx and nose (see Chapter 6 for more details).

- The digestive system effects the breakdown of ingested foodstuffs, the absorption of nutrients and other materials of use to the body, and the elimination of solid waste products. The principal components of the digestive system are the mouth, oesophagus, stomach, small and large intestine, liver, gall bladder and exocrine pancreas (see Chapters 7, 8 and 11 for more details).

- The urinary system eliminates nitrogenous waste from the body and regulates water, electrolyte and acid–base balance within the body. The principal components of the urinary system are the kidneys, bladder, ureters and urethra (see Chapter 12 for more details).

- The nervous system is one of the two main control systems of the body and can be divided into the CNS, consisting of the brain and spinal cord, and the PNS, which connects the central nervous system to the rest of the body. The nervous system effects a rapid response to internal and external stimuli via a network of sensory/afferent neurons and motor/efferent neurons connected to the CNS by a third class of neurons, interneurons, which function to integrate the signals from the PNS into the CNS (see Chapter 5 for more details).

- The endocrine system represents the other of the two main control systems, but, in contrast to the nervous system, the endocrine system is generally concerned with slower, longer-term responses to stimuli. The principal components of the endocrine system are the endocrine glands that release chemical messengers known as hormones into the bloodstream where the cardiovascular system transports them to their target, which may be at considerable distances from the source of the hormone. The major endocrine glands include the pituitary, thyroid and adrenal glands, the endocrine pancreas and the gonads (see Chapter 14 for more details).

- The reproductive systems are gender-specific, and while both the male and female reproductive systems contribute to the same basic function of procreation, their contributions to that process dictate the nature of the organs involved in the two systems. The roles of the female reproductive system are to produce ova and provide a suitable site for the fertilization of those ova, the growth and development of the resulting embryo/fetus, and the birth of the formed infant. The principal components of the female reproductive system which achieve these functions are the ovaries, the oviducts, the uterus and the vagina. Also associated with the female reproductive system are the organs allowing nurturing of the offspring once born, the mammary glands. The roles of the male reproductive system essentially are to generate the spermatozoa and deliver them to the site of fertilization within the female. The principal components of the male reproductive system are the testes, the ducts and glands associated with the production and transport of seminal fluid, and the penis. Alongside their roles in producing ova and spermatozoa, the ovaries and testes also have a further role, as noted above, in the production of the female and male sex hormones, respectively (see Chapter 13 for more details).

- The lymphatic system has two main roles; first to return excess tissue fluid to the bloodstream, and second to provide an environment in which the cells of the immune system may interact with antigens present in the tissues draining into the lymphatic vessels. The components of the lymphatic system that perform the first function are the lymphatic vessels running from the periphery to join the larger right lymphatic and thoracic ducts, which empty into the right and left subclavian veins, respectively. Along the course of the lymphatic vessels are clusters of kidney-shaped structures, the lymph nodes, and it is these that provide the ideal environment for the lymphocytes and other cells of the immune system to interact and initiate an immune response. Along with the lymph nodes there are a number of other lymphoid tissues associated with the lymphatic system including the spleen, thymus and a number of sites of mucosa-associated lymphoid tissues (MALT), which, along with the cells of the immune system, provide the resources to fight invading organisms capable of causing disease (see Chapter 9 for more details).

- The integumentary system is composed of the skin and its associated structures such as sweat and sebaceous glands, hair and nails. The function of the integumentary system essentially is to provide a protective and insulating layer to maintain the internal status of the body, keep out potentially harmful microorganisms and chemicals, and provide a degree of cushioning from mechanical damage to the internal tissues (see Chapter 15 for more details).

- The skeletal system provides a rigid framework for the support of the body and protection of its delicate internal organs while allowing movement through various specialized joints between bones. The skeleton also serves in a storage capacity for many minerals and also fats in the form of yellow bone marrow in the internal cavities of the bone. The cavities of certain bones also contain red marrow and these provide the site of manufacture of the various blood cells. The adult human skeleton comprises 206 bones ranging from some no more than a few millimetres in length (bones of the inner ear) to some which may be half a metre long (femur) (see Chapter 10 for more details).

- The muscular system provides the force for bodily movement that, in conjunction with the joints of the skeletal system, permits the movement of the various skeletal components relative to each other. The functions of the muscular system are achieved by the contractions of striated muscle and are considered as a distinct group, different from cardiac muscle powering the heart and the smooth muscle involved in the functioning of various internal organs (see Chapter 10 for more details).

SELF-CHECK 3.4

Name the 11 major organ systems of the human body.

SELF-CHECK 3.5

What are the functions of the integumentary system?

> ## Key points
> The 11 organ systems are the functional units of the fully formed human body.

The organ systems described here represent the fully formed and developed body systems present in the mature adult organism. While these organ systems are present from birth, and essential for the young human to survive as a free-living organism, their level of development varies considerably. Some systems in the full-term fetus (e.g. integumentary system) are anatomically and physiologically very similar to that of the adult, albeit on a smaller scale, while others undergo rapid development and maturation following the transition from *in utero* to free-living life (e.g. respiratory and cardiovascular systems). Some systems will show maturation at distinct time points in an individual's life (e.g. the transition to reproductive maturity during puberty) while others will continue to develop throughout the greater part of an individual's lifetime (e.g. the nervous and immune systems). Regardless of the point at which these systems may be considered to have reached maturity, by far the greatest level of development occurs during the first nine months of *in utero* life (i.e. the embryonic and fetal periods), the majority of which takes place during the first two months.

3.2 Embryology and fetal development

Fertilization occurs in the ampulla of the oviduct. The completion of the second division of meiosis of the secondary oocyte is triggered by the penetration of the plasma membrane by the successful spermatozoon. Soon after this is completed, the male and female pronuclei fuse, a process referred to as syngamy, resulting in the formation of the diploid single-cell zygote. This cell continues to be surrounded by the zona pellucida, a glycoprotein shell produced by the primary oocyte. Syngamy, and therefore the formation of the zygote, is regarded as the beginning of embryonic development and is referred to as Day 1 of development. The stages of human embryogenesis may be seen graphically in Figure 3.3 and are described in detail below but can be difficult to follow from reading text alone. Hence we recommend viewing the excellent YouTube animations referred to at the end of the chapter in conjunction with the descriptions.

The first cleavage of the zygote occurs about half a day later to form a two-cell embryo. From Day 1½ to Day 3 the cells of the embryo continue to divide and with each division the cells become smaller. These daughter cells are called blastomeres. The blastomeres continue to be confined by the zona pellucida, which does not expand, hence the diameter of the embryo does not increase. The cell divisions become increasingly asynchronous and from the eight-cell stage the blastomeres begin to change shape. There are currently four different hypotheses to explain how the apparently homogenous cells of an eight-cell embryo diverge into two distinct populations as described below. These hypotheses have been reviewed recently by Wennekamp and Hiiragi (2012). While the 32-cell embryo, known as a morula, looks like a tight cluster of disorganized cells resembling a mulberry (from the Latin *morum*), the blastomeres have already started to organize themselves into a distinct arrangement. Flattened cells lining the zona pellucida form a shell, while the cells that are towards the centre of the embryo are forming the inner cell mass. The cells of the inner cell mass will eventually give rise to the embryo (hence, it is also referred to as the embryoblast), while the cells forming the shell, described as the outer cell mass, will become the placenta (trophoblast).

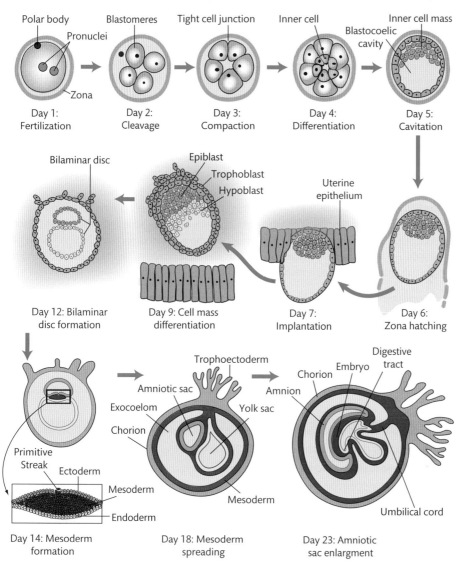

FIGURE 3.3
Human embryogenesis

Key points

The blastocyst contains two cell types: the inner cell mass which will become the embryo (also called the embryoblast); and the outer cell mass which will become the placenta (also called the trophoblast).

At about the time of morula formation, fluid starts to accumulate within the embryo and a fluid-filled cavity, the blastocyst cavity, forms. The cells of the outer cell mass become a thin single epithelial cell layer completely lining the zona pellucida. The inner cell mass is displaced and attached to one side, and the entire structure is referred to as the blastocyst. On about

Day 4 the blastocyst reaches the uterus and on Day 5, just before implanting in the endometrium, the blastocyst hatches. Hatching involves the embryo producing enzymes that act on the zona pellucida to produce a hole through which the blastocyst passes. Immediately, the embryo increases in size once released from the zona pellucida, demonstrating how constrained it had been. The embryo implants such that the side of the blastocyst with the inner cell mass attached enters the endometrium first. The cells of the trophoblast rapidly divide and differentiate into two cell types: the outer syncytiotrophoblast and the inner cytotrophoblast. As they divide and differentiate they are also invading the endometrium, seeking out the maternal blood vessels. The syncytiotrophoblast is formed from proliferating trophoblastic cells that lose their cell membranes and fuse to become a single unit, a synctium. The cells of the cytotrophoblast also proliferate and have already begun to synthesize and secrete human chorionic gonadotrophin (hCG). If a pregnancy is to be established then the cytotrophoblast must produce sufficient hCG to prevent the corpus luteum in the maternal ovary from undergoing luteolysis (i.e. ceasing to produce progesterone and degenerating). This is the hormonal basis of the maternal recognition of pregnancy. If the embryo cannot produce sufficient hCG then the corpus luteum will undergo luteolysis and the embryo will be lost when the endometrium is sloughed at the next menses.

Key points

Pregnancy is not established until the embryo has successfully signalled its presence to the maternal system (via the synthesis and release of hCG) and implanted in the endometrium.

SELF-CHECK 3.6

a) At what age does the embryo implant into the uterine wall?

b) Name the hormone secreted by the cytotrophoblast that maintains the corpus luteum and hence the pregnancy?

c) What is the principal hormone produced by the corpus luteum?

d) Describe the function of that hormone.

Before implantation, the inner cell mass/embryoblast has already begun to differentiate and two distinct layers of cells are identifiable by Day 8. The layer of columnar cells closest to the cytotrophoblast is called the epiblast or primary ectoderm, while the cells closest to the blastocystic cavity are cuboidal and called the hypoblast or primary endoderm. An extracellular basement membrane separates the epiblast and hypoblast: the entire structure is called the bilaminar germ disc. On Day 8 fluid starts to accumulate between the epiblast and the cytotrophoblast cells. Some cells of the epiblast specialize to surround this fluid and eventually form a very thin membrane, the amnion, around the fluid-filled cavity, the amniotic cavity. By the eighth week of development, this cavity and its membrane completely surround the developing embryo.

Also starting on Day 8, and taking about four days to complete, some cells of the hypoblast migrate and cover the cytotrophoblast cells lining the blastocystic cavity to form a membrane of extraembryonic endoderm, known as the extracoelomic membrane (also known as Heuser's membrane). The blastocystic cavity is now referred to as the primary yolk sac or extracoelomic cavity. On Days 10 to 11, between the extracoelomic membrane and the cytotrophoblast, acellular extraembryonic reticulum collects. On Days 11 and 12, extraembryonic mesoderm surrounds the extraembryonic reticulum. The origin of these cells remains uncertain: it is not known if they are from the epiblast, hypoblast or cytotrophoblast. Within the extraembryonic reticulum, lacunae start to appear on Day 12 as the acellular material is broken down and replaced with fluid. These spaces coalesce to form the chorionic cavity on Days 12 and 13. On Day 12 the secondary yolk sac endoderm is produced by cuboidal cells from the hypoblast multiplying and moving to line the inner surface of the extraembryonic mesoderm. The primary yolk sac is pushed towards the opposing side and soon degenerates. The former blastocystic cavity and extracoelomic cavity is now the secondary or definitive yolk sac. The extraembryonic mesoderm lining the definitive yolk sac is a major site for haematopoiesis. Tanaka and colleagues (2012) have used a conditional mouse model to demonstrate that the extraembryonic mesoderm is the site of origin of adult-like haematopoietic stem cells and have concluded that it is unlikely that other sites, although previously proposed, are involved.

Cross reference

Blood cell genesis is covered in Chapter 4

The primitive streak consisting of the primitive pit, node and groove appears caudally on the longitudinal midline on the surface of the epiblast of the bilaminar germ disc on Day 15 and 16. Gastrulation occurs on Day 16 and involves epiblast cells multiplying and migrating through the primitive streak into the space between the epiblast and the hypoblast. The hypoblast is replaced by some epiblast cells to become the definitive endoderm. The definitive endoderm will give rise to the lining of the future gastrointestinal tract and its derivatives (see Chapter 8). Other epiblast cells migrating through the primitive streak colonize the space between the epiblast and the newly formed definitive endoderm to create the intraembryonic mesoderm. The epiblast cells that enter via the primitive node migrate cranially to form the prechordal plate and the notochordal process, while the cells that enter via the primitive groove move either side of the midline to form a sheet. When the two new layers have formed, the epiblast is renamed the ectoderm. The now trilaminar germ disc is therefore composed of the ectoderm, the mesoderm and the definitive endoderm, and these can be clearly seen in Figure 3.4.

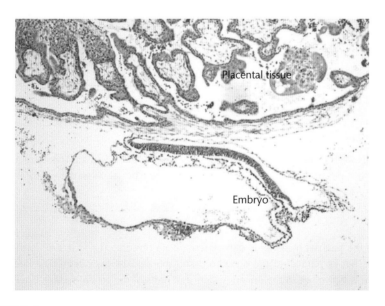

FIGURE 3.4

Early human embryo approximately two and a half weeks post-fertilization (original magnification ×1000).

Ed Uthman/CC-BY.

Key points

Gastrulation is the process by which the three germ cell layers (the ectoderm, the mesoderm and the endoderm) are formed during the third week of development.

From Day 16 to Day 26 the primitive streak slowly regresses until it disappears. Before this, at the level of the primitive node, epiblast cells enter and become a mesodermal structure known as the notochordal process. Over Days 16 to 22 the notochordal process is transformed from the hollow tube-like notochordal process to a plate within the endoderm and finally a solid cylinder of cells, known as the notochord, that lies longitudinally along the midline of the germ disc between the ectoderm and definitive endoderm.

The trilaminar germ disc is not completely trilaminar: at both the cranial and caudal ends of the disc there are two small depressions that only consist of the ectoderm and the definitive endoderm. The two depressions are the buccopharyngeal and the cloacal membranes. The buccopharyngeal membrane degenerates around Day 24 to form the cranial opening of the gut. The cloacal membrane does not degenerate until about Day 47 to form the openings of the urogenital tract and the anus.

The mesoderm that arises from epiblast cells that entered the primitive groove differentiates over weeks three and four. The mesoderm immediately lateral to the notochord, the paraxial mesoderm, initially forms rod-like structures that lie parallel to the notochord before condensing into sphere-like structures called somitomeres on Days 18 to 19. Most of the somitomeres will later develop into somites: the most cranial seven pairs do not and these will eventually give rise to the striated muscles of the face, jaw and throat.

SELF-CHECK 3.7

a) What is the key embryonic process that results in the formation of the three germ cell layers called?

b) Name the three layers.

c) Which tissues derive from each layer?

Key points

Somitogenesis is the process that results in a segmented body plan. There are 40 pairs of somites, mesodermal tissue, which have three functional parts: the sclerotome, the dermatome and the myotome.

Lateral to the paraxial mesoderm is the intermediate mesoderm and it, too, forms rod-like structures that lie parallel to the notochord and the paraxial mesoderm. The intermediate mesoderm will give rise to the urinary system and parts of the genital system. Extending further laterally from the intermediate mesoderm is sheet-like mesoderm that is divided into two layers. The layer in contact with the ectoderm is called the somatopleuric mesoderm and will give rise to the inner lining of the body wall, parts of the limbs and most of the dermis. The other layer, the one in contact with the definitive endoderm, is called the splanchnopleuric mesoderm and will give rise to the mesothelial lining of the viscera.

The somitomeres begin to differentiate into somites from Day 20 and do so in craniocaudal sequence. Pairs of somites form either side of the notochord, three to four each day until there are just over 40. However, several caudal pairs will disappear leaving in total about 37 pairs. The somites undergo further differentiation to become myotomes (which will initiate muscle formation), sclerotomes (which will initiate the formation of the vertebrae) and dermatomes (which differentiate into the dermis of the neck and trunk in conjunction with cells from the lateral mesoderm).

Cross reference

For the development of muscle, bone, and skin cells see Chapters 10 and 15.

On Day 18 the prechordal and cranial portion of the notochordal plates of the axial mesoderm induces the formation of the neural plate in the overlying ectoderm. The cells of the neural plate differentiate into columnar, pseudostratified neuroepithelial cells (the neuroectoderm). Over Days 18 to 20 the neural plate expands in a craniocaudal direction and is widest at the cranial end, narrowing towards the caudal end. The cranial end will become the brain while the narrow caudal end that overlies the notochord will become the spinal cord. Around Day 22 the fore-, mid- and hindbrain become identifiable. The hindbrain already appears segmented and the segments are called neuromeres (rhombomeres). Neurulation, the conversion of the neural plate into the neural tube, commences on Day 22. The process starts at the most cranial five somites and involves the lateral edges of the neural plate folding dorsally until they meet to form a tube. Fusion of the two lateral edges also involves surface ectoderm. As fusion occurs, the tube pulls away from the surface ectoderm and embeds in the posterior body wall along the midline. The tube initially is open at each end and the openings are called neuropores. The cranial neuropore closes on Day 24 and the caudal neuropore on Day 26. At the caudal end, the neural tube has arrested at about somite 31 within the sacral region. Secondary neurulation quickly follows, resulting in the extension of the neural tube from the initial point of closure caudally to end in the coccygeal region. The cells that give rise to this extension of the neural tube arise from pluripotent cells found within the mesodermal caudal eminence.

During neurulation, as the lateral edges of the neural plate begin to fold, some cells along the edges, referred to as the neural crest, detach from the neural plate and migrate to a wide range of locations within the body. The wave of detachment and migration starts at the cranial end of the neural plate and continues towards the caudal end over a few days. Cranial neural crest cells give rise to connective tissue surrounding the eye, papillary and ciliary muscles, dermal bones of the skull, the truncoconal septum, pharyngeal arch cartilages, odontoblasts, the dermis and hypodermis of the face and neck and some cranial nerve ganglia. Cranial and spinal neural crest cells give rise to the glial cells in peripheral ganglia, Schwann cells, arachnoid and pia mater, enteric ganglia and melanocytes. Spinal neural crest cells give rise to preaortic ganglia, the adrenal medulla and both dorsal root and chain ganglia. Starting on Day 24 the neuroepithelial cells that immediately surround the neural tube will undergo three consecutive waves of cytodifferentiation, first to form neuroblasts (origin of neurons), then glioblasts (origin of astrocytes and oligodendrites) and finally ependymal cells (nervous tissue is discussed in more detail in Chapter 5).

Key points

Neurulation is the first major event in organogenesis: ectoderm differentiates to form the brain and spinal cord.

Around Day 28 each of the somites starts further differentiation. Within each somite the ventromedial and core cells become the sclerotome and start to surround both the notochord and the neural plate. The remaining tissue within the somites form dermomyotomes, which

Cross reference

For more details on muscle
tissue, see Chapter 10

in turn soon differentiate into the dermatomes and myotomes. The myotomes divide again to
form the dorsal epimere and the ventral hypomere. The deep muscles of the back are formed
from the epimere while the hypomere results in the lateral and ventral muscles of the thoracic
and abdominal walls. The tendons and connective tissue of the body wall muscles are derived
from the lateral mesoderm. Migration of some myotome cells to the limb buds provides the
precursor cells for the limb musculature.

CLINICAL CORRELATION

Cell migration

Throughout development of the embryo/fetus, some groups of cells migrate from their point of
origin to their final destination. What is fascinating about these cells is how they find their way.
A classic example is the gonadotrophin-releasing hormone (GnRH) cells. These neuroendocrine
cells are first identified in the nasal placode outside the central nervous system. Early in gestation
they begin to move from the nasal placode through the cribiform plate into the olfactory bulb,
and then move through the developing forebrain following olfactory-derived vomeronasal and
terminal nerve axons until they reach the mediobasal hypothalamus. The cells appear to follow
the right axons by interacting with specific molecules produced on their axonal pathway: an example of **neurophilic/axophilic** migration. Along the axonal pathway there are several different
regions, each of which is characterized by a different cohort of specific signalling molecules. By
week 16 of embryonic development, all the GnRH cells (1000–2000 in total) will have reached
their final destination. These cells are instrumental in the control of the reproductive axis, hence
any failure of this migration can result in the affected individual being infertile.

In Kallmann's syndrome the GnRH cell bodies fail to migrate and as a result these individuals
are hypogonadotrophic (i.e. they have no circulating LH and follicle stimulating hormone [FSH])
and hypogonadal (i.e. the gonads fail to develop and mature; therefore, no gonadal steroids are
synthesized and released and no gametogenesis occurs). Initially, the GnRH cells are accompanied by some of the olfactory neurons, hence individuals with the monogenic X-linked form of
Kallmann's syndrome (the *KAL1* gene gives rise to the gene product anosmin 1, an extracellular
glycoprotein) are also anosmia or hyposmia; that is, they have no (or very little) sense of smell
due to the failure of the olfactory neurons to migrate correctly. The genetics of Kallmann's syndrome are not well understood because mutations in any one of the genes that encode for one
of the many signalling molecules could result in a Kallmann's syndrome phenotype; more recently, individuals with mutations in more than one of the genes involved have been identified.
The complexity of the phenotypes loosely grouped as Kallmann's syndrome is only now being
appreciated.

Dr Susan Wray recently published an excellent review that is worth reading (see Further reading);
however, more recently other genes not described in Wray's review have been associated with
this migration.

Starting at about Day 22, in the fourth week of development, the trilaminar germ disc begins to
fold to produce a three-dimensional embryo with an intraembryonic coelomic cavity, enclosed
within the amniotic cavity. The folding of the lateral edges of the germinal disc continues until
the two edges meet and fuse, thereby forming concentric circles of outer investing ectoderm,
the mesodermal connective tissue and muscles of the body wall and the inner endodermal gut
tube. Simultaneously, there is folding at both the cephalic and caudal ends resulting in flexing
of the embryo towards the connecting stalk and yolk sac. From about Day 42 the amnion will
form a sheath around the connecting stalk and the yolk sac, and this structure will become
the umbilical cord. Caudal to the cephalic fold in the region that will become the thorax, an
area of tissue known as the cardiogenic area will, through a series of folding and remodelling

events, result in the heart. The development of the vasculature starts on Day 17 when the first blood vessels begin to appear in the mesoderm of the yolk sac—a process called 'blood island formation'—and is in association with the appearance of blood cells (as discussed earlier in this section). These blood vessels will vascularize the chorionic villi of the placenta. Within a day, blood vessels begin to appear within the embryo but these vessels are not initially in association with blood cells. Mesodermal angioblastic cysts (differentiated cells of splanchnopleuric mesoderm origin) fuse and form cords that will give rise to the arterial, venous and lymphatic systems in a process referred to as vasculogenesis.

The origin of haematopoietic stem cells is somewhat controversial. The most accepted model in the human is that haematopoietic stem cells in the extraembryonic mesoderm of the yolk sac, which first appear on about Day 17, will colonize the fetal liver and give rise to myelo-erythroid lineages. Haematopoietic stem cells that arise in the region of the aorta-gonad-mesonephros (AGM) of the embryo first localize to the ventral surface of the aorta before beginning to migrate to, and expand in, the fetal liver on about Day 27. These haematopoietic stem cells will give rise to the myeloerythroid and lymphopoietic lineages (see Chapter 4 for more details). However, the work of Tanaka and colleagues (2012) in the mouse (referred to earlier) does challenge this accepted model.

Cross reference

For more on the vascular system, see Chapter 9

On Day 26 lung development commences. Two primary lung buds appear on the ventral surface of the anterior foregut. At about the same time, the single tube of the foregut begins to divide into two to form the dorsal oesophagus and the ventral trachea. The ventral trachea joins with the lung buds. The buds then undergo a process of branching morphogenesis, which involves a carefully coordinated interaction between the endoderm-derived epithelium and the mesoderm-derived mesenchyme, to form the bronchi, bronchioles and terminal airspaces. Branching morphogenesis occurs in the pseudoglandular stage (five to 17 weeks) of lung development and once ended allows for the development of the alveoli in the canalicular (16 to 24 weeks) and the saccular (24 weeks to term) stages. At 19 weeks the terminal airspaces are short, simple tubular endings. During the canalicular stage, vascularization of the lungs commences with the formation of an extensive capillary network in the parenchyma (canalization, hence the canalicular stage). By 30 weeks the terminal airspaces will have developed into short, shallow saccules. The terminal airspaces continue to develop for several months in the neonate before adult-like alveoli are found throughout the lung, and their size and number continue to increase up to eight years of age. The epithelial cells in the tips of the branches are believed to be multipotent progenitor cells and, depending on where they are located, will differentiate into one of the specialized cells of the lung (for a comprehensive review on the cellular and genetic mechanisms involved in lung formation, see Morrisey and Hogan, 2010). Skeletal muscle cells are not one of the cell types found in the respiratory tract; however, the respiratory muscles have an important role in the later stages of fetal lung development. In the later stages of gestation, fetal breathing-like movements, controlled by the respiratory muscles, trigger lung cell differentiation and stimulate lung growth to prepare the lungs for the external environment.

Also commencing in the fourth week of gestation is the formation of the liver. A pair of liver buds forms on the ventral surface of the foregut endoderm and extend into the mesenchyme of the septum transversum. The anterior liver bud gives rise to the liver and the intrahepatic biliary tree, while the posterior bud results in the gall bladder and extrahepatic biliary tree. As with the lung described above, the development of the liver involves the carefully coordinated interaction of endoderm- and mesoderm-derived tissues. The bulk of the cells in the adult liver are hepatocytes and these are derived from the hepatic endodermal cells, along with the biliary epithelial cells, which migrate into the septum transversum to populate the liver. The other cell types found in the liver are derived from the mesenchyme (for more details see Chapter 11 and Zorn, 2008).

Cross reference

Cells of the respiratory system are covered in more detail in Chapter 6

SELF-CHECK 3.8

a) At what age do the lungs start to develop?

b) At what age is lung development complete?

About the same time as the liver buds are forming, another ventral and a dorsal bud form on the endoderm of the foregut. This ventral bud will give rise to the head and uncinate process of the pancreas, and the dorsal bud to the remainder. The main duct of the pancreas is created when the two tissues eventually fuse. Again, epithelial endoderm-derived cells are important in populating the tissue and again these cells interact closely with the mesenchymal cells. The progenitor epithelial cells that form the exocrine cells of the pancreas undergo three phases of development: predifferentiated, protodifferentiated and differentiated. These three phases are largely defined based on their enzyme activity. The endocrine cells are a subset of the protodifferentiated cells and, depending on the pattern of transcription factor expression, differentiate to become either α and γ cells or β and δ cells (for more details see Chapter 14 and a review, albeit of mouse pancreas organogenesis, by Benitez et al., 2012).

Cross reference

For further information on the organogenesis of other endocrine organs see Chapter 14.

The pronephros and mesonephros are transitory organs that precede the metanephros, which will give rise to the permanent functional kidney. The metanephros is derived from the intermediate mesoderm. The Wolffian duct, an epithelial cell-lined tube and an important component of the male reproductive tract, arises within the pronephros. From this duct a bud forms, the ureteric bud, and epithelial cells from this bud are responsible, in conjunction with differentiation of the metanephric mesenchyme, for the formation of collecting ducts and nephron (for more details see Chapter 12, as well as a review by Hendry et al., 2011).

During the second week of development the primordial germ cells, which will give rise to either the male or female gametes (i.e. the germ line) depending on the genotype and gonadal phenotype of the embryo, originate in the primary ectoderm. These cells detach from the ectoderm and migrate to an extra-embryonic structure, the yolk sac. From Day 24 to about Day 37 the primordial germ cells migrate from the yolk sac to some loose mesenchymal tissue just deep to the epithelial lining of the coelomic cavity on the dorsal body wall. The primordial germ cells will induce this tissue, in which they have come to rest, to differentiate and form the genital ridge. As they migrate from the yolk sac to the putative genital ridge, the primordial germ cells begin to proliferate and mitosis continues after they have reached their destination.

Cross reference

Refer to Chapter 13 for more details on the sexual differentiation of germ cells and the gonadal cells.

Key points

Organogenesis, the formation of the organs, is an early event and by eight weeks of development all the systems are present in at least rudimentary form.

CLINICAL CORRELATION

Chromosomal/genetic basis of congenital malformations and disorders

Genetic factors include chromosomal and single gene defects and account for a wide variety of congenital malformations. Chromosomal abnormalities may be subgrouped into numerical chromosomal abnormalities and structural chromosomal abnormalities. Numerical chromosomal abnormalities (aneuploidy) can include an extra copy of a chromosome (trisomy) or the loss of one copy of a chromosome (monosomy) and can occur both in autosomes and the sex chromosomes. There are a number of autosomal and sex chromosome trisomies associated with congenital malformations seen in live births, and these include Down's syndrome (47 XX/XY +21), Edwards syndrome (47 XX/XY +18) and Klinefelter's syndrome (47 XXY), each resulting in a distinctive phenotype. However, autosomal monosomy is usually lethal, with the embryo being aborted shortly after fertilization. The only documented monosomy found in live births is Turner's syndrome, an X chromosome monosomy (45 XO). The term 'partial monosomy' is often used to describe structural chromosomal abnormalities in which chromosomal damage (often induced by environmental factors) is followed by deletion or inappropriate repair/recombination; for example, Cri du Chat syndrome in which there is a deletion of the short arm of chromosome 5.

Single, or indeed multiple, gene defects cause a diverse range of diseases, and while they may arise spontaneously, the majority are hereditary. Some show dominant inheritance, which requires only one copy of the mutated form to result in the disease phenotype; some are recessive and require two copies of the abnormal allele to result in overt disease. Some of these disorders result almost exclusively from a single mutation (e.g. achondroplasia, an autosomal dominant condition with 98 per cent of cases resulting from a single point mutation [G-A at nucleotide 1138] in the gene encoding fibroblast growth factor receptor 3 [*FGFR3*]), whereas others can result from a large number of different mutations (e.g. phenylketonuria, an autosomal recessive condition that may result from any one of over 400 different mutations).

As this section has shown, the development of the embryo is a series of complex and highly structured events. It is perhaps not surprising, therefore, that many embryos/fetuses do not develop to term and are aborted spontaneously as a result of abnormalities in the developmental process that are not compatible with life. However, some of these abnormalities do not result in spontaneous abortion/miscarriage and the fetus is carried to term (or near term) and delivered. When this occurs, the resulting structural and/or functional aberrations present at birth are referred to as congenital malformations/disorders. Such malformations can range from non-life-threatening problems, some of which can be rectified by medical intervention, to more severe abnormalities that result in stillbirth or perinatal death. The incidence of congenital malformation/disorder is in the region of 3 per cent of total births (live births and stillbirths) worldwide (WHO, 2012). While the occurrence of many congenital malformations is almost certainly multifactorial, there are three basic groups of abnormality caused by genetic, environmental and mechanical factors (explored in more detail in the Clinical Correlation on congenital malformations and disorders). Some of these defects may be diagnosed during routine ultrasound scans conducted during pregnancy. Where these abnormalities have a genetic basis, and where there is an increased risk (e.g. a family history of the condition or perhaps increased maternal age—a contributing factor in the case of Down's syndrome), more invasive techniques may be used to permit genetic screening of the fetus. These techniques are discussed in the Clinical Correlation on testing for genetic defects. Even in the absence of congenital defects, the growth, development and health of the adult is a function of the environment experienced *in utero* (see Box 3.4).

CLINICAL CORRELATION

Environmental factors and congenital malformations and disorders

Environmental factors can have profound effects on development, either as a direct causal agent or as a contributing factor in multifactorial induction of congenital malformations. Environmental factors can include the maternal nutritional status where deficiency or excess of dietary components result in fetal abnormalities; for example, deficiency of vitamin B_{12} and folate correlate with an increased risk of neural tube defects/spina bifida, whereas increased levels of vitamins A and D can also result in various abnormalities. Congenital abnormalities may also arise from maternal, and therefore fetal, exposure to teratogenic drugs and chemicals. These can include prescribed drugs (e.g. the characteristic deformities resulting from the use of the anti-emetic drug thalidomide), 'social' drugs (e.g. fetal alcohol syndrome resulting from the consumption of alcohol during pregnancy), environmental pollutants (e.g. the microcephaly and cerebral palsy associated

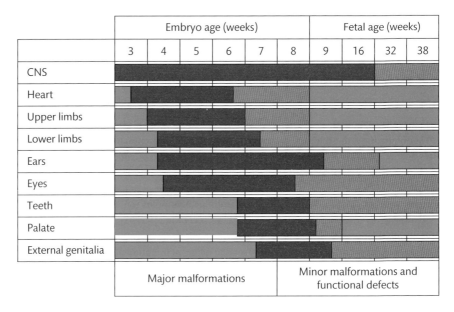

	Embryo age (weeks)						Fetal age (weeks)			
	3	4	5	6	7	8	9	16	32	38
CNS										
Heart										
Upper limbs										
Lower limbs										
Ears										
Eyes										
Teeth										
Palate										
External genitalia										
	Major malformations						Minor malformations and functional defects			

FIGURE 3.5

Organ susceptibility to environment teratogens. Red zones: time points when the developing organs are most sensitive to teratogenic influences; Amber zones: periods when the organs are less sensitive but exposure may still result in abnormality; Green zones: time points when teratogenic effects are unlikely to occur.

with methylmercury ingestion) and exposure to ionizing radiation (e.g. growth retardation and abnormalities *in utero* following maternal exposure to diagnostic/therapeutic irradiation). The other main group of environmental factors to cause severe congenital malformations are micro-organisms, where infection by certain organisms (e.g. the rubella virus and *Treponema pallidum*, the causal agent of syphilis) can result in severe congenital malformations affecting multiple systems. It should also be noted that there are distinct 'windows' during fetal development in which the developing organs and body parts are most susceptible to teratogens. In general, exposure to teratogens in the first four to eight weeks results in major malformations, while exposure after seven to eight weeks often results in less-severe malformations and functional disorders. These timeframes are, however, dependent on the organs concerned—the upper limbs, for example, are most susceptible to teratogenic effects between four and six weeks, whereas the palate is most susceptible between six and eight weeks. More examples are shown in Figure 3.5

CLINICAL CORRELATION

Mechanical factors and congenital malformations and disorders

Mechanical factors result in a group of abnormalities known as congenital postural deformations, and these are usually a result of abnormal mechanical forces *in utero* resulting from a decreased uterine space and/or restricted fetal movement. This may be caused by various factors but commonly include uterine/pelvic abnormalities in the mother and reduced amniotic fluid volume as a result of placental insufficiency or premature rupture/leakage of the amniotic membranes. The position of the fetus can also contribute to congenital postural deformations (e.g. fetuses in the breech position often have restricted opportunity to move, and congenital dislocation of the hip is common in infants carried in this position).

CLINICAL CORRELATION

Testing for genetic defects

When there is a high risk that an unborn child might have a genetic defect, perhaps because of familial history or the age of the parents, the following procedures can be performed to acquire cells for genetic screening.

Preimplantation genetic diagnosis (PGD) for embryos created using assisted reproduction technology (ART): Most usually, one or two blastomeres are removed from the eight-cell embryo prior to being implanted although cells can be removed at any stage of early development. The diagnostic test needs to be rapid because, unless the embryo is to be frozen, it must be placed in the uterus of the woman within a specific and very short time.

Chorionic villus sampling (CVS): At 10–13 weeks of pregnancy, using ultrasound guidance, cells from the chorionic villus (the placenta) are removed. These cells are derived from the trophoblast of the embryo and therefore have the same genetic composition as the fetus.

Amniocentesis: A sample of amniotic fluid removed under ultrasound guidance, usually around 15 to 20 weeks into the pregnancy. The fluid will contain cells sloughed from amnion derived from the trophoblast and therefore have the same genetic composition as the fetus. The amniotic fluid can also be assayed for α-fetoprotein, which, if high, can indicate that the fetus has a neural tube defect (e.g. spina bifida) or, if low, can indicate Down's syndrome.

In all three procedures, the cell(s) can be used to determine if there are chromosomal abnormalities. For a specific inherited monogenic disorder (e.g. mutations in the cystic fibrosis transmembrane conductance regulator [*CFTR*] gene are associated with cystic fibrosis, an autosomal recessive disorder) a polymerase chain reaction (PCR)-based test would be used, while for determining if there is a chromosomal abnormality, such as trisomy (e.g. Down's syndrome [trisomy 21; see Chapter 1] and Edwards syndrome [trisomy 18]), or a translocation/deletion, then fluorescence *in situ* hybridization (FISH) is used. If the embryo/fetus is a mosaic then the results of the test may be misleading.

Chorionic villus sampling and amniocentesis are highly accurate, but are considered to be 'invasive' techniques and carry a small, but significant, risk of miscarriage. In recent years there have been significant advances in non-invasive prenatal testing (NIPT) including the use of ultrasound based Nuchal scans for Down's syndrome and more recently the use of fetal cells or cell-free DNA present in the maternal blood to identify fetal genetic abnormalities or assist in risk assessment by identifying fetal sex. Currently, NIPT techniques are best seen as initial screening options and whilst they are useful in confirming low-risk pregnancies, follow-up CVS/ amniocentesis is recommended following positive results.

BOX 3.4 Discussion point. The Barker hypothesis: A developmental model for the origins of chronic disease?

In 1995, David Barker from the University of Southampton reported a relationship between low birth weight and an increased risk of coronary heart disease (Barker, 1995). Based on this observation, he proposed the hypothesis that the *in utero* environment (e.g. macro- and micro-nutrient availability) has long-term effects on an individual's health. This led to a huge interest in exploring the effects of *in utero* health on long-term health. In animal models and in man, it has been shown that the *in utero* environment can influence an individual's propensity to develop not just coronary heart disease but hypertension, stroke, type 2 diabetes, and other late-onset adult diseases. Barker and colleagues recently published a review of resource allocation *in utero* and health in later life (see Further reading).

3.3 **Growth and development**

3.3.1 Infancy, childhood, adolescence and puberty

The transition from an *in utero* environment to the external environment requires some significant adaptations that include rapid anatomical changes. A classic example is that of the fetal circulatory system. Fetal circulation needs to include the placenta and mostly bypasses the non-functional lungs and liver. At birth, the neonate's lungs have to become functional suddenly and the circulatory system alters to allow for the loss of the placenta. During *in utero* growth and development, blood bypasses the non-functioning lungs via two cardiac shunts: the foramen ovale (an opening between the two atria) and the ductus arteriosus (a vessel between the pulmonary artery and the aortic arch). At birth, pulmonary pressure decreases because the inflation of the lungs allows blood flow into its capillary bed. The pressure in the right atrium decreases while the pressure in the left increases, resulting in the functional closure of the foramen ovale. The tissue of the foramen ovale fuses with the septum over the ensuing weeks. Likewise, the ductus arteriosus collapses and is no longer patent following the first inflation of the lungs: this is thought to be due to the contractile action of bradykinin, released by the lungs in response to mechanical stimulation of the first breaths, on the smooth muscle of the ductus. It remains as the non-functional ligamentum arteriosum following birth.

Throughout childhood, all the body systems continue to grow and develop towards maturation, albeit at very different rates. The differing rates are very important to remember, especially for pharmacologists—a child is not simply a down-sized adult. The liver of a small child is disproportionately much larger than in older children and adults when expressed as a proportion of total body weight. This may therefore affect the clearance rate of some drugs. The size of the liver alone does not fully explain the differences seen in pharmacokinetics and pharmacodynamics in children of different ages. In hepatocytes there appear to be differences in the age of onset of expression of the different enzymes involved in drug metabolism. The renal system also undergoes a maturation process, hence it is difficult to estimate the ability of the kidney to eliminate some metabolites in the young child.

Another body system that continues to grow and develop throughout childhood and adolescence is the central nervous system. There is extensive published research reporting sexually dimorphic morphological changes in different areas of the brain that occur during puberty; for example, the volume of the amygdala in the male increases with age in the adolescent compared to the female, while the volume of the hippocampus increases significantly in the female. Several psychopathologies appear to manifest during the pubertal period, hence there has been much recent research activity using advances in non-invasive technologies to map the anatomical changes that occur in the brains of children and adolescents (Asato *et al.*, 2010; Lenroot and Giedd, 2010).

Although puberty is associated with profound psychological changes, we are all more familiar with the significant physical and physiological changes. In both the male and female, puberty of the gonads (gonadarche) is stimulated by the reactivation of the GnRH neuroendocrine neurons of the hypothalamus. Some pulses of GnRH act on the gonadotrophs of the anterior pituitary, resulting in small pulses of luteinizing hormone (LH) and follicle stimulating hormone (FSH). The LH and FSH then stimulate steroidogenesis in the gonads. The first outward sign of puberty in the male is an increase in testicular size. The increase is due to FSH stimulating spermatogenesis in the seminiferous tubules and LH stimulating testosterone production by the Leydig cells. Testosterone not only acts on the testes and the rest of the male reproductive tract (as well as the hypothalamic-pituitary axis via negative feedback) but also acts on many other tissues, resulting in the body changes that we associate with a male phenotype (e.g. testosterone acts on the larynx, resulting in lowering of the voice). Similarly, in the female the production of oestrogens stimulates the development of the secondary sexual characteristics, and the first observed change is the growth and development of the breasts, although growth and development is not completed during puberty. The hormonal changes induced by pregnancy are required to complete breast growth and development in anticipation of lactation. Some of the changes induced at puberty are irreversible, while others depend on continued exposure to steroid. This is illustrated by the clinical observation that testosterone production in many men declines with age and that the decline in circulating testosterone concentration is associated with loss of muscle bulk and strength.

3.3.2 Adulthood

The organ systems present in the adult human represent, in most cases, the fully developed systems that will function cooperatively to maintain a constant internal environment for the lifetime of the individual unless affected by stimuli which require adaptation, disease processes, or deterioration in function as a result of the ageing process.

The process of maintaining a constant internal environment, or, more accurately, the process of maintaining a dynamic equilibrium where the internal environment is in constant flux but within very narrow limits, is known as homeostasis. While the term 'internal environment' can encompass many factors, the principal concern is the composition of the extracellular fluid (ECF), as this is the cells' source of oxygen, nutrients, ions and trace elements. It is also the route for the excretion of waste products from cells. If the composition of the ECF varies significantly from normal for a particular tissue then the cells within will not function properly. Therefore, homeostasis is concerned with the supply of oxygen and nutrients, the removal of cellular waste products, and the maintenance of the ECF in terms of, for example, volume, pH and ion content.

While all the organ systems are involved to some extent in the homeostatic process, two systems in particular—the nervous system and the endocrine system—are responsible for the control and coordination of the process. Homeostasis is achieved essentially through feedback systems, and the majority of these are negative feedback; that is, a homeostatic imbalance triggers sensory receptors that activate, via a control centre, an effector system to produce an effect that counters the imbalance and removes or blocks the original trigger. Positive feedback loops also exist, although in this case the homeostatic imbalance results in the activation of an effector mechanism which, rather than returning the system to a state of equilibrium, will push it even further out of balance until such point as the loop is broken. Positive feedback loops are often found in situations where a rapid amplification of the response is required and a classic example would be the clotting response following an injury. This positive feedback drives the local clotting response a long way from the homeostatic process, resulting in repair of the wound. While this local repair mechanism is a long way from the homeostatic balance of the clotting system, it does restore homeostasis on the systemic level as blood loss is stopped, returning the circulatory system to something approaching its normal state and removing the stimulus for the positive feedback loop.

Throughout life, many tissues have the capacity to respond to different stimuli thereby demonstrating a degree of plasticity. An excellent example is the anterior pituitary gland's response to pregnancy. As pregnancy advances, the amount of prolactin produced by the lactotrophs increases dramatically. The prolactin is biologically inactive until placental steroid is removed from the maternal system at the time of parturition. Prolactin has a crucial role in the maintenance of milk synthesis. In late gestation and throughout lactation the volume of the anterior

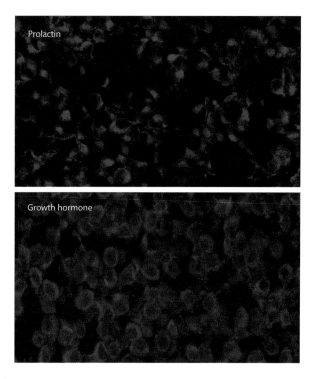

FIGURE 3.6

Immunohistochemical staining of sections of mouse pituitary for prolactin (top panel) and growth hormone (bottom panel) show a distinct organization of the two cell types into networks. Whilst prolactin cells are organized in a honeycomb-like structure, growth hormone cells are typically found in clusters, which are linked by strings of cells. In different physiological states when there are increases in physiological demand, the network organization is more pronounced and likely mediates the increased hormone output.
Images kindly provided by Leonard Cheung (National Institute for Medical Research, London) and Paul Le Tissier (Institute of Child Health, London).

pituitary increases. It remains unclear what are the relative contributions of lactotroph hypertrophy and hyperplasia to this increase in gland volume, but it is clear that there is an increase in lactotroph tissue to meet the physiological demand. As weaning commences, the volume of total lactotroph tissue decreases. In mice it has been shown that the lactotrophs move relative to each other to ensure a pattern of contact with other lactotrophs that uniquely occurs in lactation (i.e. cells can rearrange their positions within structures to facilitate certain functions; Figure 3.6, Hodson *et al.*, 2012).

SELF-CHECK 3.9

What does 'tissue plasticity' mean?

Some tissues in the body undergo constant remodelling throughout life (e.g. connective tissue, bone). From birth until death, the skeleton undergoes remodelling. Osteoblasts are responsible for bone deposition. These cells produce a collagen matrix and then regulate the mineralization of the matrix. The activities of the osteoblasts are balanced by those of the osteoclasts, which are responsible for bone resorption. The ability to remodel is important for reshaping and/or replacing bone in the event of trauma, as well as repairing the 'micro' breaks that occur due to wear and tear. Bones also change density in response to changes in mechanical loading. The rate at which bone remodelling occurs is regulated mostly by the parathyroid glands, the kidney, the intestine and the nervous system: an excellent illustration of how multiple tissues interact in concert to maintain homeostasis.

SELF-CHECK 3.10

Name the two main organ systems responsible for the control and coordination of homeostasis?

Key points

While the adult human is composed of fully functional organ systems, some of these systems continue to develop or remodel throughout the life of the individual.

3.3.3 Ageing

The ageing process has a widespread effect on virtually all the systems of the body, resulting in gradual deterioration and reduced function. At the cellular level, ageing generally results in a reduction as a result of decreased cell division and increased cell death and/or senescence (a change or reduction in the cell's functional phenotype). As organ function is essentially a product of the functions of the various cell types within that organ, then as the cells fail so will the organ. However, it is important to make the distinction between functional deterioration resulting from ageing *per se* and that from functional deterioration resulting from disease processes associated with the ageing individual.

To illustrate this, let us consider dementia. The various forms of dementia are diseases predominantly of the older individual, affecting approximately 7 per cent of those over 65 years of age and less than 0.05 per cent of those under 65 in the UK population (Alzheimer's Society, 2012). While age is clearly a risk factor for dementia, the fact that 93 per cent of those aged over 65 show little sign of serious cognitive impairment, and something in the region of 20 000 'younger' people in the UK are also afflicted, would indicate that dementia is not a result of ageing. The vast majority of older individuals do in fact have good cognitive function, and where minor memory losses, for example, are reported that could be attributed to the ageing process, there is also evidence to suggest that this may be a lifestyle factor as those who remain active and engaged are less likely to report such issues. These complications aside, there are a number of well-documented, age-associated anatomical/physiological changes that occur in virtually all tissues. Some examples include:

- *Decreased rates of protein synthesis*: contributes to age-related problems in virtually all tissues but especially connective tissues.
- *Decreases in tissue flexibility and elasticity*: contributes to reductions in cardiovascular and respiratory system performance.
- *Decreases in muscular tone, strength and adaptability*: contributes to cardiovascular problems, digestive system problems and skeletal muscle weakness.

Age-related defects in certain organ systems have a knock-on effect on others:

- Decreased cardiovascular output can affect the nervous system, resulting in the death of brain tissue due to a reduced supply of nutrients and oxygen. It can affect the urinary system where the reduced blood flow results in less-efficient kidney function.

- Age-related changes in the integumentary system, in particular changes in collagen and elastin content and regenerative capacity, mean that elderly skin is more easily damaged, and when injury does occur it takes longer to repair. This reduction in skin integrity is also thought to contribute to the reduced efficiency of the immune system in older individuals, as the contribution of the skin as a physical barrier to infection is reduced.

Key points

While some systems continue to develop throughout life, all will eventually demonstrate a reduction in performance as the individual ages.

Chapter summary

- Anatomy defines the spatial/structural relationships between the various components and organ systems of the body. Physiology is the study of the function of the body and the systems within. Fine anatomy is the detailed, microscopical structure of individual organs or tissues.

- The human body is composed of 11 major organ systems: the cardiovascular, respiratory, digestive, urinary, nervous, endocrine, reproductive, lymphatic, integumentary, skeletal and muscular systems.

- Each of these systems is composed of various cell and tissue types, all of which can be classified as one of four basic types: nervous, connective, muscle and epithelial tissues.

- The fully formed adult human body arises from a single-celled zygote following the fusion of a single, haploid female gamete (ovum) with a single, haploid male gamete (spermatozoon).

- The zygote undergoes a process of rapid cell division and migration until the blastocyst is formed with two distinct cell types: an inner cell mass known as the embryoblast, which will eventually become the embryo, and the outer cell mass known as the trophoblast, which will eventually become the placenta.

- Gastrulation occurs on Day 16 of human embryogenesis and results in the formation of the three primary germ layers (ectoderm, mesoderm and endoderm) from which all cells and tissues develop.

- At the start of the fourth week of development, the trilaminar germ disc of ectoderm, mesoderm and endoderm begins to fold to produce the three-dimensional embryo with an internal coelomic cavity, and organogenesis commences. This is a time of rapid development and the next couple of weeks will see the development of the brain and spinal cord, the heart will begin to beat, and basic facial features will start to appear, as will limb buds.

- Errors may occur in the development process, many of which are incompatible with life while others result in congenital malformations or disorders. Where these disorders are genetic in origin, and there are predisposing factors, screening procedures are available.

- Following the growth and development that occurs *in utero*, a number of important adaptations occur at birth to allow the infant to adapt to free living, particularly in the cardiovascular system. Following these rapid transitions, the infant will slowly develop over a period of years until the next significant developmental milestone of puberty where the body undergoes yet another series of dramatic changes to form the mature adult body.

- While the adult body represents the mature, fully developed organism, there is still scope for change under the influence of a range of external and internal stimuli and thus some tissues remodel continuously throughout life. The ageing process affects all body systems, with most showing marked deterioration with age; however, this can be influenced by lifestyle choices.

 # Further reading

- Alzheimer's Society. *What is Dementia?* Factsheet No. 400, 2012.

- Asato MR, Terwilliger R, Woo J, Luna B. White matter development in adolescence: a DTI study. *Cereb Cortex* 2010; **20** (9): 2122–31.

- Barker DJ. Fetal origins of coronary heart disease. *BMJ* 1995; **311** (6998):171–4.

- Barker DJ, Lampl M, Roseboom T, Winder N. Resource allocation *in utero* and health in later life. *Placenta* 2012; **33** (Suppl 2): e30–4.

- Benitez CM, Goodyer WR, Kim SK. Deconstructing pancreas developmental biology. *Cold Spring Harb Perspect Biol* 2012; **4** (6): Pii: a012401.

- Hendry C, Rumballe B, Moritz K, Little MH. Defining and redefining the nephron progenitor population. *Pediatr Nephrol* 2011; **26** (9): 1395–406.

- Hodson DJ, Schaeffer M, Romano N *et al*. Existence of long-lasting experience-dependent plasticity in endocrine cell networks. *Nat Commun* 2012; **3**: 605.

- Lenroot RK, Giedd JN. Sex differences in the adolescent brain. *Brain Cogn* 2010; **72** (1): 46–55.

- Morrisey EE, Hogan BL. Preparing for the first breath: genetic and cellular mechanisms in lung development. *Dev Cell* 2010; **18** (1): 8–23.

- Tanaka Y, Hayashi M, Kubota Y *et al*. Early ontogenic origin of the hematopoietic stem cell lineage. *Proc Natl Acad Sci USA* 2012; **109** (12): 4515–20.

- Wennekamp S, Hiiragi T. Stochastic processes in the development of pluripotency *in vivo*. *Biotechnol J* 2012; **7** (6): 737–44.

- World Health Organization. *Congenital anomalies*. Fact sheet No. 370. Geneva: WHO, 2012 (www.who.int/mediacentre/factsheets/fs370/en/index.html).

- Wray S. From nose to brain: development of gonadotrophin-releasing hormone-1 neurones. *J Neuroendocrinol* 2010; **22** (7): 743–53.

- Zorn AM. Liver development. In: *StemBook*. Harvard: HSCI, 2008 (www.stembook.org/node/512).

Useful websites

- UNSW Embryology, http://php.med.unsw.edu.au/embryology Dr Mark Hill has created a very useful resource on embryology with contributions from an international panel of scientists. The site has many images and movies that will aid your understanding of the key events

- The Multi-Dimensional Human Embryo. http://embryo.soad.umich.edu/ The multi-dimensional human embryo project contains images and animations of human embryos at various stages of development. The images were gathered using magnetic resonance microscopy (MRM) at the Centre for In-vivo Microscopy at Duke University with image processing and data management performed at the School of Art and Design, University of Michigan.

The following videos are excellent animated visualizations of embryo folding and gastrulation in embryogenesis.

- http://www.youtube.com/watch?v=qMnpxP6EeIY

- http://www.youtube.com/watch?v=sUuX-4fEF3A

- http://www.youtube.com/watch?v=x-p_ZkhqZ0M

 Discussion questions

3.1 The four basic tissue types combine to form organ systems which perform the many diverse functions that occur in the human body, but what does each tissue type bring to the functioning organ/organism?

3.2 Discuss why gastrulation may be considered the most important event in embryogenesis.

3.3 Which of the 11 major organ systems are involved in homeostasis and what is their contribution to the maintenance of stable internal environment?

Answers to self-check questions and discussion questions, hints and tips are provided on the book's Online Resource Centre.

 Visit www.oxfordtextbooks.co.uk/orc/orchard_csf/.

4

Blood cell genesis: red cell, white cell and platelet families

Gavin Knight

Learning objectives

After studying this chapter you should confidently be able to:

- Describe the main types of blood cell found within the peripheral blood.
- Describe the architecture of the main organs responsible for blood cell production.
- Explain the mechanisms controlling blood cell production rate.
- Describe the role of stem cells in blood cell production.
- Describe the role of effector blood cells in maintaining the health of the individual.

In order to facilitate oxygen delivery to metabolizing tissues, fight infection, aid wound healing, and to prevent blood loss, we need a ready and continuous supply of red blood cells, white blood cells and platelets. This chapter describes the processes by which blood cells are produced to enable these functions to occur.

4.1 Cellular components of the blood

Blood is composed of two main elements: plasma and blood cells. Plasma has a straw-coloured appearance and carries a wide range of chemical substances including proteins, nutrients, metabolic by-products, clotting factors and hormones, all of which are suspended in water. Blood cells contain a heterogeneous population of cells derived from the bone marrow or thymus gland. This chapter will focus on the cellular element of blood. Figure 4.1 demonstrates the components of whole blood following centrifugation.

FIGURE 4.1

A peripheral blood sample following centrifugation. At the bottom, red cells have sedimented, while above can be seen straw-coloured plasma. Separating the two is a thin white buffy coat, comprising white blood cells.

Key points

Blood volume is largely dictated by plasma volume. In cases of acute blood loss, plasma volume expands to maintain blood pressure. In cases of dehydration, blood volume falls and blood cells become concentrated, a process called haemoconcentration. Redistribution of tissue fluid to maintain blood volume in extreme cases, following acute haemorrhage for example, leads to dilution of the cellular components of the blood.

The red colouration of our blood is derived from the **red blood cells**, also called erythrocytes, and in particular the intracellular respiratory pigment **haemoglobin**. As haemoglobin binds oxygen, forming oxyhaemoglobin, a bright red colour is imparted to the blood. As oxygen is released and red cells enter the venous circulation, blood loses its vibrant red colouration, becoming dark red. Red blood cells have a biconcave disc shape, affording them a great deal of flexibility to deform in order to pass through the narrow-lumen blood vessels that supply distant tissues with oxygen.

Our capacity to fight infection comes from heterogeneous populations of **white blood cells**, also called leucocytes, each population having a different function. Under the umbrella term of white blood cells are three types of cell characterized by the types of granule within their cytoplasm. These cells are broadly called **granulocytes**, and more specifically are **neutrophils, eosinophils** and **basophils** (Figures 4.2, 4.3, 4.4, and 4.5). These cells may also be called **polymorphonuclear** cells (polymorphs), due to the lobulated structure of their nuclei.

Neutrophils form the largest proportion of polymorphs and are responsible for providing the body's first defence, once the main barriers to disease (e.g. skin) have been breached by microbes. Neutrophils are effective phagocytic cells capable of removing foreign bodies

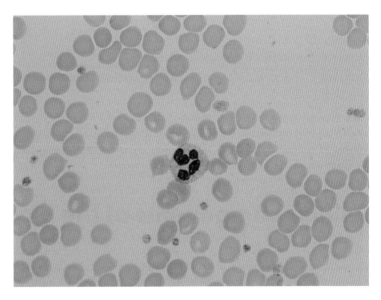

FIGURE 4.2

Typical neutrophil morphology (centre). Note the five nuclear lobes and cytoplasmic granules (Romanowsky stain, original magnification ×600).

(courtesy Haematology Laboratory, Queen Alexandra Hospital, Cosham).

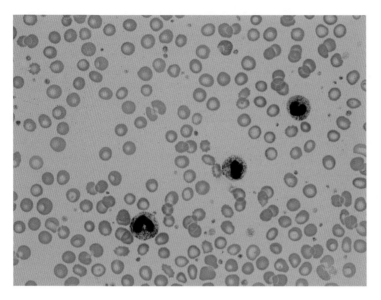

FIGURE 4.3

Reactive neutrophils. The morphology seen in this image is called left shifted with toxic granulation. The nuclei are horseshoe-shaped as these cells have been released from the bone marrow early to fight infection. The intense cytoplasmic granulation is present to combat infection (Romanowsky stain, original magnification ×600).

(courtesy Haematology Laboratory, Queen Alexandra Hospital, Cosham).

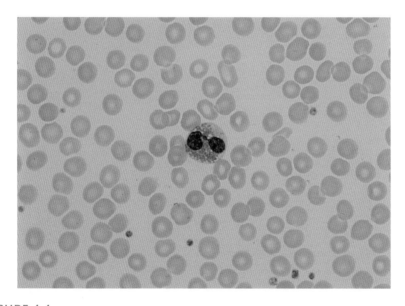

FIGURE 4.4

Typical eosinophil morphology. Note the bright orange cytoplasmic granules and bilobed nucleus. (Romanowsky stain, original magnification ×600).

(courtesy Haematology Laboratory, Queen Alexandra Hospital, Cosham).

(e.g. bacteria) and are invaluable in removing damaged tissue. Frequently, neutrophils and monocytes work in conjunction, and their migration to peripheral tissues follows inflammation and vasodilation. Neutrophils have two effective mechanisms for controlling bacterial infection: oxidative mechanisms induce the generation of hydrogen peroxide (H_2O_2) superoxides, hydroxyl radicals and nitric oxide; and non-oxidative mechanisms, which tend to be bacteriostatic. The morphology of neutrophils in health is shown in Figure 4.2, and during bacterial infection in Figure 4.3. Notice the differences in nuclear structure and granulation between the two states.

Eosinophils possess large, bilobed nuclei and bright orange granules within their cytoplasm (hamburger granules). These granules contain a wide range of chemical mediators that help to execute their function. The main roles of eosinophils include mediation of the allergic reaction by opposing basophil and mast cell actions, and also anti-**helminth** activity. Figure 4.4 shows the typical morphology of eosinophils. Notice the vibrant orange colour of the granules compared to those in neutrophils and basophils.

Helminth

This is a parasitic worm commonly found within the intestine.

Basophils possess an indented nucleus and its outline is often obscured by numerous blue-black granules. These granules are important as they help to initiate inflammation by releasing histamine and inducing increased vascular permeability, leading to angioedema (commonly called welts). Inflammation is an important process for inducing healing and repair. Basophil morphology is shown in Figure 4.5. Blue-black granules form the hallmark of this species of white blood cell.

Mononuclear cells are also present in the peripheral blood and contain a single, often uniformly shaped nucleus of either the **monocyte** (Figure 4.6) or **lymphocyte** (Figure 4.7) lineages. Monocytes possess an indented nucleus and are important cells. They transit from the bone marrow into the peripheral blood and then enter the tissues where they become

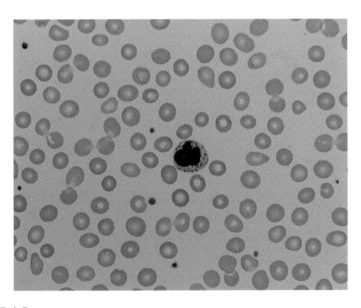

FIGURE 4.5

Typical basophil morphology. In this case, the nuclear outline is apparent. Note the blue-black cytoplasmic granules that help the identification of this species and allow the basophil to effect its function (Romanowsky stain, original magnification ×600).

(courtesy Haematology Laboratory, Queen Alexandra Hospital, Cosham).

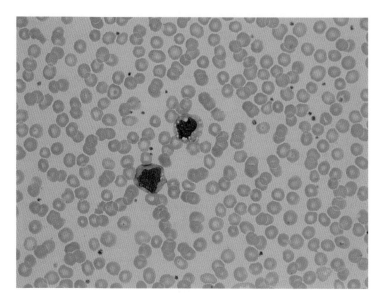

FIGURE 4.6

Two monocytes. The upper, central monocyte contains cytoplasmic vacuoles (Romanowsky stain, original magnification ×600).

(courtesy Haematology Laboratory, Queen Alexandra Hospital, Cosham).

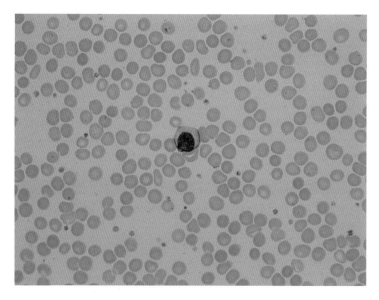

FIGURE 4.7

Lymphocyte. In this cell line, morphology is often highly variable and can be considered reactive, with deep blue (basophilic) cytoplasm, or atypical in appearance—often with scalloped edges. This lymphocyte is considered normal in structure and is the most common lymphocyte morphology seen in the peripheral blood (Romanowsky stain, original magnification ×600).

(courtesy Haematology Laboratory, Queen Alexandra Hospital, Cosham).

tissue macrophages. Different tissues possess different types of macrophages, although all are phagocytic in nature. Small numbers of granules present within the cytoplasm of monocytes are important to aid in the degradation of phagocytosed material

Lymphocytes can be further subdivided into **B cells, T cells** and **large granular lympho-cytes.** B cells are so called because they are derived from the Bursa of Fabricias in birds, although in humans the B-cell maturation process occurs in the bone marrow. Alternatively, T-cell development is initiated in the bone marrow, although maturation concludes in the thymus gland, and this provides the 'T' prefix in the name. Morphologically, B cells and T cells cannot be distinguished, although large granular lymphocytes are often rather irregu-larly shaped and include some granules within their cytoplasm. Large granular lymphocytes, or natural killer cells, comprise approximately 10 per cent of the lymphocyte population and provide cytotoxic activity against cells displaying cancerous changes and cells infected by some viruses.

Immunoglobulin

Abbreviated to Ig, is a type of protein secreted into bodily fluids that binds specifically to a particular antigen. Different classes of Ig include IgG, IgM, IgE, IgA and IgD. Each has a different structure and role within the immune response. You will read more about this in the other books in this series, most notably in the *Immunology* and *Histopathology* volumes.

Naïve B cell

This is a cell of B lineage that has yet to encounter an antigen.

Cross reference

More details about white blood cells and their structure and function can be found in the *Haematology* volume in this Fundamentals of Biomedical Science series.

As B cells mature, the level of expression of **immunoglobulin** on their surface increases. These membrane-bound immunoglobulins, also called B-cell receptors, enable **naïve B cells** to interact with antigens expressed on a particular bacterium or other foreign body. Complementary binding between the B-cell receptor and antigen initiates a proliferative drive, leading to the expansion of antigen-specific B cells. This population of B cells then differentiates into either **plasma cells**, actively secreting immunoglobulin with the same specificity as the B-cell receptor, or **memory B cells**. Memory B cells are long-lived and can be reactivated following a second encounter with an antigen, enabling a rapid immunologi-cal response.

There are three important populations of T cells, including T helper cells (T_H), expressing CD4 on their surface, cytotoxic T cells (T_C) that show surface expression of CD8, while the final T cells are regulatory T cells (T_{regs}). There are approximately twice as many CD4-expressing cells (termed CD4-positive T cells) as those expressing CD8 (CD8-positive T cells). Helper T cells are so called because, following interaction with antigen, T_H cells aid the immune response by facilitating the activation of T_C cells, B cells and macrophages. Following stimulation, for example by tumour or viral antigens, T_C cells proliferate and differentiate into cytotoxic T lymph-ocytes, which can induce the death of cells expressing these antigens. Regulatory T cells oppose the actions of T_H cells, inhibiting the immune response. In addition to controlling a prompt immune response, T_C, T_H cells and T_{regs} can also form memory T cells and, in a similar fashion to memory B cells, lead to a rapid immune response following exposure to a previously encountered antigen.

SELF-CHECK 4.1

Name the different types of cell found in peripheral blood and provide a brief explanation of the functions of each.

Key points

Cluster of differentiation (CD) markers are discrete antigens expressed by cells. In excess of 350 antigens are currently recognized, and their expression is associated with cell dif-ferentiation and maturation. Some CD antigens may be expressed in a lineage-restricted manner (i.e. a single CD antigen is expressed by only one cell lineage, while others may be expressed in a multilineage fashion.

4.2 Introduction to blood cell production

In order to survive, we need a continuous supply of blood cells. In the adult, blood cells are produced in the bone marrow of the long bones of the body, the skull, pelvis and ribs (Figure 4.8), although at an early stage (i.e. during early development) blood cell production occurs throughout the body. Of particular importance are the liver, spleen and yolk sac. During the third week of gestation, within developing yolk sac-derived blood-filled vessels, both endothelial and haemopoietic cells are produced. Following this, haemopoietic stem cells emerge. These stem cells are initially associated with the dorsal aorta and, during the fourth week of gestation, with the ventral wall of the dorsal aorta and vitelline artery. At this stage, intraembryonic **haemopoiesis** begins. The position and morphology of the extra-embryonic yolk sac is demonstrated in Figure 4.9. During embryonic development, the haemopoietic system is one of the first major organs to form, primarily to facilitate the delivery of oxygen to the developing embryo. Throughout embryonic development, changes in the location of blood cell production and the type and structure of the oxygen carrying pigment, haemoglobin, occurs. These changes improve oxygen delivery and aid tissue remodelling within the embryonic microenvironment.

Haemopoiesis, the term used to describe blood cell production, is derived from the Greek haima, meaning blood, and poiesis, meaning production.

SELF-CHECK 4.2

Name the main haemopoietic sites associated with the embryonic stage of development.

The term haemopoiesis can be subdivided according to the different types of blood cells produced. It is useful in medical and biomedical practice to be able to make the distinction between different levels of blood cell production, as this can aid the description of

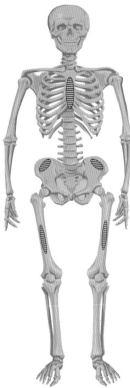

= major sites of physiological haemopoiesis (sternum, femurs, pelvis)

FIGURE 4.8

The skeleton showing major sites of haemopoiesis. The major sites of haemopoiesis in the adult are the sternum, femur and pelvis. Minor sites include the ribs.

© Gary Moore, Gavin Knight, and Andrew Blann, 2010

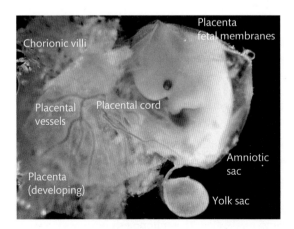

Placenta
fetal membranes
Chorionic villi
Placental vessels
Placental cord
Placenta (developing)
Amniotic sac
Yolk sac

FIGURE 4.9

Embryo. The embryo is situated within the uterus. Attached is a placenta and below the embryo, the yolk sac. The yolk sac is the primary site for haemopoiesis during embryological development.

© MA Hill, 2004.

Cross reference

Stem cells are discussed in section 4.4

Haemopoiesis

The production of blood cells.

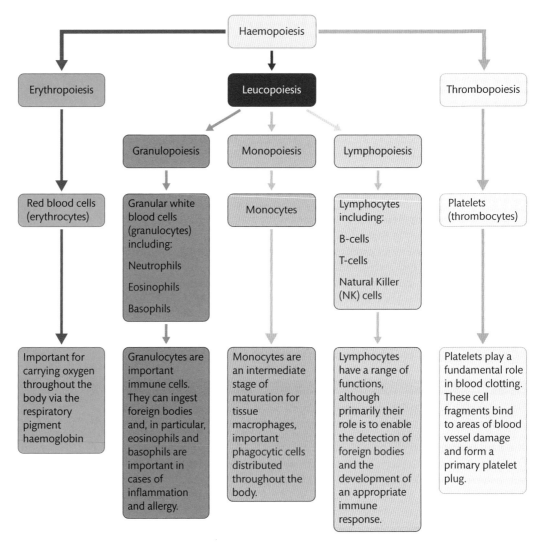

FIGURE 4.10
An illustration of the subdivisions within the haemopoietic pathway.

haematological disease and its manifestations. Figure 4.10 demonstrates the subdivision of the processes leading to the production of different types of blood cell.

Haemopoiesis can be subdivided into **erythropoiesis**, which describes the production of red blood cells. Disorders of red blood cells or their production normally can be seen in the erythropoietic pathway. **Leucopoiesis** describes the production of the immune part of the haemopoietic system, the leucocytes. Leucopoiesis can be further divided in to **granulopoiesis, lymphopoiesis**, describing the production of lymphocytes, and **monopoiesis**, the process through which monocytes develop. **Thrombopoiesis** describes the process by which platelets develop. Together, the red cells, white cells and platelets make up the cellular component of our blood, and these components, along with their developmental sites, are of particular interest to haematologists. One way of measuring these elements is through the use of the full blood count (FBC), and results for a healthy patient are presented in Figure 4.11. Notice the broad range of parameters that can be measured from a single sample.

TEST	RESULT	ABN	NORMALS			UNITS	
WBC	9.6		(4	-	10)	x10.e9	/L
RBC	4.55		(3.8	-	4.8)	x10.e12	/L
HGB	132		(120	-	150)	g/L	
HCT	0.42		(0.37	-	0.46)	L/L	
MCV	93.0		(83	-	101)	f/L	
MCH	29.1		(27	-	32)	pg	
MCHC		314	(315	-	345)	g/L	
CHCM	319		(	-	)	g/L	
PLT	335		(150	-	400)	x10.e9	/L
#NEUT		7.5	(2	-	7)	x10.e9	/L
#LYMPH	1.3		(1	-	3)	x10.e9	/L
#MONO	0.5		(0.0	-	1)	x10.e9	/L
#EOS	0.2		(0	-	0.5)	x10.e9	/L
#BASO	0.1		(0	-	0.2)	x10.e9	/L
#LUC	0.1		(0	-	0.5)	x10.e9	/L
MPXI	-3.5		(-10		10)		

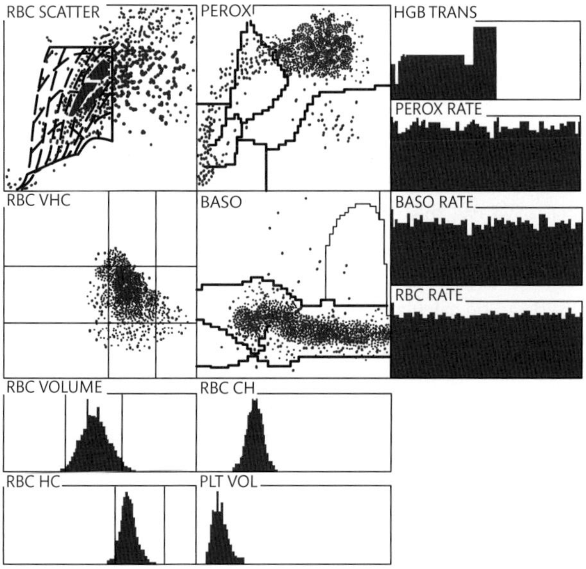

FIGURE 4.11

Typical full blood count results generated from an automated counter. Each of the elements listed in the test column relate to a particular cell type. Of relevance to this chapter are WBC = white blood cells, RBC = red blood cells, HGB = haemoglobin, PLT = platelets, Neut = neutrophils, Lymph = lymphocytes, Mono = monocytes, Eos = eosinophils, Baso = basophils. The # prefix demonstrates these are absolute values, rather than percentages. Expected values (reference range) are depicted as 'normals' and the units for reporting are shown in the right-hand column. Although the neutrophil count is slightly raised, this patient is considered normal haematologically (Romanowsky stain, original magnification ×600). (courtesy Haematology Laboratory, Royal Bournemouth Hospital).

In a healthy adult, the bone marrow will produce approximately 3.5×10^{11} cells per day, sufficient to allow the peripheral blood to complete all appropriate physiological functions. Table 4.1 shows the expected concentrations of blood cells within the peripheral blood of an adult.

This level of production ensures that the number of peripheral blood cells is maintained in a steady state. Should the demand for blood cells increase, due, for example, to infection or **anaemia**, then blood cell production will increase.

Anaemia

This describes a haemoglobin concentration below the reference range for a particular gender and a particular age.

TABLE 4.1 Reference ranges for the different peripheral blood cell species.

Blood cell species	Peripheral blood concentration	Units
Red blood cells	4.3–5.7 (M) 3.9–5.0 (F)	$\times10^{12}$/L
Platelets	143–400	$\times10^9$/L
White blood cells	4.0–10.0	$\times10^9$/L
Neutrophils	2.0–7.0	$\times10^9$/L
Lymphocytes	1.0–3.0	$\times10^9$/L
Monocytes	0.2–1.0	$\times10^9$/L
Eosinophils	0.02–0.5	$\times10^9$/L
Basophils	0.02–0.1	$\times10^9$/L

Haematocrit

This describes the proportion of whole blood occupied by red blood cells. This proportion is reported as a decimal fraction. For example, if red blood cells occupy 450 mL of 1000 mL blood, the haematocrit would be 45%, or 0.45 L/L.

A normal adult body contains approximately 5000 mL blood. Of this, approximately 2250 mL consists of red blood cells. We call this proportion of red blood cells to whole blood the **haematocrit**. Knowing that the lifespan of a red blood cell is 120 days, the minimum volume of blood to be produced each day in order to maintain the concentration of red blood cells within the peripheral blood can be calculated, as follows:

Production (P) = red cell mass (M) / red cell life span (L) Therefore, 2250 / 120 = 18.75 mL per day

So, 18.75 mL red cells are required every day to maintain the red cell component of the peripheral blood. If the bone marrow is unable to produce this volume of blood each day, perhaps due to poor nutrition, anaemia develops.

The **maximum functional capacity** of the bone marrow describes the degree to which the bone marrow can compensate for blood loss. Provided that all the appropriate micronutrients (e.g. iron, vitamins B_9 [folic acid] and B_{12}) are present, the bone marrow should be able to upregulate erythropoiesis five- to tenfold.

SELF-CHECK 4.3

Using the equation P = M/L outlined earlier, state the volume of red cells that would need to be produced each day, if the red cell life span reduced to 75 days.

4.3 Structure of the bone marrow

Beyond the embryonic stage, the main haemopoietic tissue is found within the bone marrow and accounts for approximately 5 per cent of a human's body weight. The bone marrow is composed of two distinct compartments. The red marrow, or haemopoietic marrow, forms discrete islands within the bone marrow environment and is responsible for the production of blood cells (Figure 4.12). The yellow marrow, composed of fatty (adipose) tissue, is largely inert. The red and yellow marrow are surrounded by vascular sinuses. These sinuses

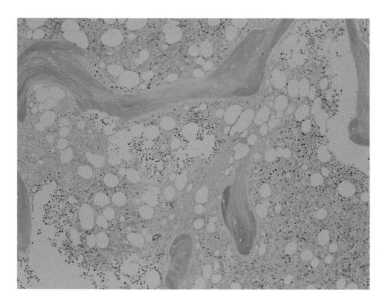

FIGURE 4.12

Bone marrow trephine biopsy showing normal marrow architecture. The cellular component is haemopoietic marrow, comprising largely blood cell precursors. Clear spaces are adipose tissue, or fatty (yellow) marrow. Pink swathes running through the marrow are bone trabeculae (haematoxylin and eosin [H&E], original magnification ×100).
(courtesy Haematology Laboratory, Queen Alexandra Hospital, Cosham).

are responsible for providing micronutrients and hormones to the marrow environment and provide a mechanism for mature cells to exit the bone marrow and enter the circulatory system.

A trabecular bone meshwork extends throughout the bone marrow, providing structural support and sites facilitating haemopoiesis. We can describe the relationship between the components of the marrow as the bone marrow architecture. The architecture of the bone marrow can be examined using a trephine biopsy, as shown in Figure 4.12. Red and yellow marrow can be clearly differentiated.

When we examine a bone marrow biopsy under a light microscope we can assess cellularity, which describes the proportion of haemopoietic cells within the biopsy. Up to the age of three months an individual has 100 per cent bone marrow cellularity, meaning that their entire bone marrow is haemopoietic (red marrow). As we age, the proportion of red marrow reduces and the proportion of yellow marrow increases. Up to ten years of age, bone marrow cellularity reduces to approximately 80 per cent, and a 30-year-old adult can expect 50 per cent marrow cellularity. By the time an individual reaches 70 years of age, bone marrow cellularity is reduced to 30 per cent. Importantly, there is a dynamic relationship between red and yellow marrow; should the demand for blood cell production increase, it is not uncommon for yellow marrow to be partially replaced by red marrow, increasing haemopoiesis. An increase in bone marrow cellularity is also evident in various disease states, including **leukaemia** and **haemolytic anaemia**.

Leukaemia

Describes the presence of cancerous haemopoietic cells in the bone marrow and peripheral blood.

Haemolytic anaemia

Describes the process whereby an individual develops a low red blood cell count and haemoglobin concentration as a consequence of reduced red blood cell survival.

SELF-CHECK 4.4

Provide a brief comparison between red and yellow marrow.

SELF-CHECK 4.5

Define cellularity and explain the consequences of haemolysis on bone marrow cellularity in terms of red and yellow marrow.

Cross reference

More detail regarding leukaemia and haemolytic anaemia can be found in the *Haematology* volume in this Fundamentals of Biomedical Science series.

Cross reference

Erythroblastic islands are discussed in section 4.7

The structure of the bone marrow microenvironment is essential to support haemopoiesis. The microenvironment should facilitate the retention of haemopoietic stem cells (discussed below) while providing the appropriate micronutrients and growth factors to promote maturation and differentiation into the appropriate blood cell lineages. Thin-walled blood vessels must also be present to facilitate the release of blood cells into the peripheral blood.

The organization of cells in haemopoietic tissue is tightly regulated to encourage haemopoiesis to occur. Red blood cell precursors associate with iron-laden macrophages to produce **erythroblastic islands**, while megakaryocytes develop near vascular sinuses to allow the release of platelets directly into the peripheral blood. Sites of granulopoiesis are less discernable and occur at sites throughout the red marrow.

4.4 Stem cells as the ultimate haemopoietic precursor

The major stages of differentiation and maturation of blood cells occur within the bone marrow, enabling multiple haemopoietic lineages to be produced. However, before the detail of differentiation and maturation is considered, the identity of the cell ultimately responsible for the production of the full range of effector cells must be examined. Haemopoiesis begins with **haemopoietic stem cells** (HSC), which are present in very low numbers in the bone marrow, with between 11 000 and 22 000 per individual. From these HSCs, red blood cells, white blood cells and platelets are produced. Stem cells have various characteristic features of critical importance in supporting life. These features include:

- self renewal
- differentiation into a range of cell lineages
- slow replication.

Self renewal is an important characteristic of stem cells and ensures that the stem cell population is not consumed in the process of cell division and maturation. Self renewal describes the production of additional stem cells from the parent, while enabling maturation to occur. Two models are used to describe stem cell division: asymmetric cell division is used to describe the process whereby, following cell division, one daughter cell becomes lineage restricted and matures to form an effector cell, while the other daughter cell retains its stem cell characteristics and maintains the stem cell pool.

Cross reference

The process of differentiation will be examined in section 4.6

Conversely, in symmetric cell division, two identical daughter cells, either effector cells or stem cells, are produced depending on the body's requirements at that time. These models of stem cell division are illustrated in Figure 4.13. At present it is unknown whether the identity of the daughter cells is selected prior to mitosis or through exposure to environmental factors (e.g. growth factors) following cell division.

Within the bone marrow, stem cells can be found in one of two 'factions'. Stem cells directly involved in the haemopoietic process, with actively cycling and dividing cells, may be found

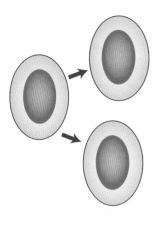 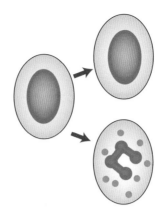

Symmetric division   Asymmetric division

FIGURE 4.13

Stem cell division. Left, symmetrical division leading to the production of two identical daughter cells, in this case stem cells. Right, asymmetric division. One daughter retains stem cell characteristics whilst the other follows the neutrophil lineage of maturation.

within the active pool, whereas inactive, or quiescent, stem cells are found within the reserve compartment. Although, in the steady state, the size of the active pool is retained, during periods of increased bone marrow activity, HSCs from the reserve compartment may begin cycling and enter the active pool.

Key points

All cells, at some stage in their lives, will pass through the cell cycle. This is important because it allows daughter cells to grow and develop into tissues, and to allow these tissues to be maintained. The cell cycle begins with the exposure of a cell to a mitogen, a chemical that induces mitosis. Following this exposure, four stages are completed. Each stage is designated as follows: G1, S, G2 and M. G1 and G2 are gap stages, while S is the stage at which synthesis of DNA and cell cycling machinery occurs. The cell then organises itself and separates into two different cellular compartments during mitosis. Finally, the cytoplasm divides - a process called cytokinesis - producing two daughter cells.

At some point during their lives, cells enter a G0 stage, called the quiescent stage, where cycling ceases temporarily. Stem cells are retained in this quiescent state for much of their lives. Each of the stages contained within the cell cycle are shown in figure 4.14.

SELF-CHECK 4.6

Explain the differences between the two different forms of cell division associated with stem cells.

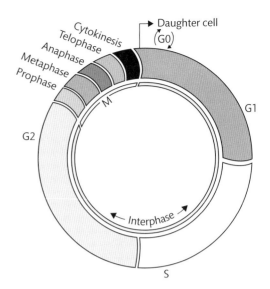

FIGURE 4.14

The cell cycle. Gap 1 (G1), synthesis (S) and gap 2 (G2) phases comprise interphase. Interphase is the entire process of the cell cycle leading to the preparation of cell division (mitosis).

4.5 **Control of haemopoiesis**

In order to ensure that the correct number of blood cells is produced every day, and that our bodies are capable of responding to increased need, control mechanisms are necessary. Haemopoiesis is largely controlled through the synthesis and secretion of growth factors and interleukins, which are produced in response to physiological stimulation.

Erythropoiesis is largely controlled by the exposure of red blood cell precursor cells, called progenitors, to the growth hormone **erythropoietin** (EPO). Erythropoietin, a 34 kDa glyco-protein produced primarily by the peritubular cells of the kidney, consists of 165 amino acids and acts as a hormone. Erythropoietin is also produced in smaller quantities by the liver. The production of EPO is stimulated by tissue hypoxia, low oxygen tension in the blood, in order to improve oxygen delivery to metabolizing tissues. Erythropoietin functions synergistically with a range of other growth factors to ensure cell viability is maintained by avoiding **apoptosis** and increasing proliferation and maturation in red cell precursors.

Erythropoietin must bind to specific cells that express EPO receptors (EPOR) in order to initiate cell survival, proliferation and maturation. When the receptor is bound to its complementary binding partner (**ligand**), an intracellular signalling cascade is activated resulting in the tran-scription and translation of genes encoding peptides responsible for cell survival, proliferation and maturation. Erythropoietin receptors are expressed primarily on CFU-E and pronormo-blasts (see section 4.6); as these cells mature then the expression of EPOR reduces.

Erythropoietin is synthesized and secreted by peritubular cells in response to the stimula-tion of oxygen tension receptors in the kidney. When low oxygen tension is detected, EPO is secreted into the blood where it will circulate freely until it encounters EPO receptors expressed on the surface membrane of red cell precursors. The interaction of EPO and its receptor stimulates a signalling cascade within the cell to initiate the production of a range of proteins. These proteins increase the rate of cell division and prevent programmed cell death (apoptosis).

Erythropoietin-mediated signalling ensures that red cell production is intrinsically linked to tissue oxygen demands or to the oxygen carrying capacity of the blood. We can clearly see

Apoptosis

This is an inflammation-independent mechanism through which controlled cell death is initiated.

Cross reference

Different stages of erythroid maturation, including that of CFU-E and pronormoblasts, are discussed in section 4.8

this mechanism of erythropoiesis in place when we examine the blood of individuals who live at, or visit for a significant period of time, high altitudes (above 4000 metres). This mechanism of erythropoiesis can be exploited, for example, by athletes who train at high altitude prior to athletic events. The low partial pressure of atmospheric oxygen at high altitude induces an EPO-dependent response leading to an elevated red blood cell concentration, called **erythrocytosis**. When these trained athletes then return to sea level, the raised red cell concentration can improve cardiovascular performance as a consequence of the improved oxygen carrying capacity of the blood. By the same token, some athletes have been known to inject exogenous hormones, such as EPO or its analogues. Exogenous erythropoietin is now considered a banned substance by sporting regulatory bodies.

During thrombopoiesis, platelet production is controlled by the hormone **thrombopoietin** (TPO). This is a 332 amino acid glycoprotein produced primarily in the liver, although it is also synthesized by the kidney and bone marrow stroma. Thrombopoietin receptors, called Mpl, are expressed on the surface of platelet progenitors called **megakaryocytes** When TPO binds to Mpl, this receptor-ligand interaction induces proliferation and budding of megakaryocyte membranes, leading to platelet production. In contrast to EPOR expression, TPO receptors are expressed on the surface of mature platelets, and, as a consequence, the number of platelets in the peripheral blood can negatively regulate the thrombopoietic drive by binding TPO in the peripheral blood. As platelet number begins to drop, less TPO is sequestered by the circulating platelets, ensuring more TPO is bioavailable to enhance thrombopoiesis. This dynamic process, leading to the self-regulation of platelet numbers, is an example of homeostasis in action.

> **Megakaryocyte**
> This describes the platelet progenitor cell and is derived as follows: mega = large, karyo = chromosomal complement, cyte = cell.

The effects of TPO and EPO binding to their receptors are shown in Figure 4.15. Notice the role phosphorylation of JAK2 plays in mediating signalling to the cell nucleus.

SELF-CHECK 4.7

Outline the role of EPO in red cell production.

SELF-CHECK 4.8

Explain how low platelet count can facilitate thrombopoiesis.

In addition to EPO and TPO (Figure 4.15a and b), a range of cytokines are used to facilitate differentiation and maturation of haemopoietic cell lineages. These cytokines will be discussed in context in the next section.

4.6 First stages of haemopoiesis

Haemopoiesis begins with the haemopoietic stem cell. This cell differentiates into a more restricted type of stem cell, called the **colony forming unit** (CFU) granulocyte, erythrocyte, monocyte, megakaryocyte (**CFU-GEMM**), which has the potential to form any cells of **myeloid** lineage, or the CFU-L, the lymphoid equivalent of the CFU-GEMM.

> **Myeloid**
> This describes all cells of the bone marrow, with the exception of lymphocytes.

From this point, differentiation becomes increasingly lineage restricted. The next CFU formed from CFU-GEMM is restricted to either:

- CFU-EMk—colony forming unit erythrocyte and megakaryocyte
- CFU-GMo—colony forming unit granulocyte and monocyte.

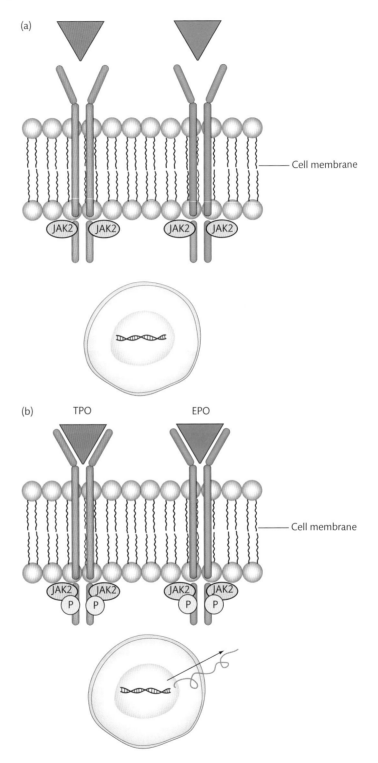

FIGURE 4.15

(a) Thrombopoietin (TPO)–blue and erythropoietin (EPO)–green signalling. Both growth factors bind to their receptors causing a structural change in the receptor. This leads to phosphorylation of JAK2 (b) activating a series of chemical mediators in a signal transduction pathway. The end result is DNA transcription and translation, leading to platelet (TPO) and red cell (EPO) production respectively.

Beyond the **CFU-EMk** and **CFU-GMo** stages, lineage specificity is achieved. Burst forming unit (BFU) erythrocyte (**BFU-E**) cells derived from the CFU-EMk mature into **CFU-E** and **CFU-Mk**, whereas CFU-GMo become restricted to **CFU-N**, **CFU-Eo**, **CFU-Baso** and **CFU-Mono** progenitors, representing neutrophil, eosinophil, basophil and monocyte lineages, respectively. These colony-forming units then become cells designated as 'blast' cells, which have a high proliferative capacity and are able to generate considerable numbers of effector cells of discrete lineage. The blasts generated are of the following specificities:

- Pronormoblast
- Megakaryoblast
- Myeloblast
- Lymphoblast
- Monoblast.

Under the light microscope, these blasts all look very similar, and it is very difficult to differentiate between these lineages using standard **Romanowsky** staining techniques.

Importantly, there are typical features of blasts of which one needs to be aware, so that, even if the lineage of these cells initially cannot be established confidently (leading to more specific tests to elucidate their lineage), the presence of blasts can be confirmed. This is particularly important in suspected cases of acute leukaemia. See the next Key points box.

Blast cells, as shown in Figure 4.16, invariably have a high **nuclear:cytoplasmic ratio**; that is, the nucleus of the cell is often almost as large as the cell itself, and only a thin rim of cytoplasmic material is visible. In addition, the nucleus appears open and lacy. The nucleus of all eukaryotic cells is composed of **chromatin**, which is a composite of DNA and histone proteins wound together, forming nucleosomes. In mature cells, chromatin is condensed and appears clumped, or chunky, whereas in immature cells, such as blasts, chromatin has an open

> **Romanowsky**
>
> This is a family of stains utilized for the visualization of blood cells using light microscopy. These stains contain a mixture of dyes, commonly including azure B and Eosin Y.

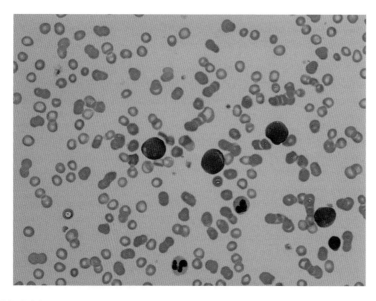

FIGURE 4.16

Blast cells. Four blast cells, the larger of the seven cells shown demonstrating a high nuclear:cytoplasmic ratio. The chromatin is open, facilitating transcription and translation. Prominent nucleoli are apparent. Other white cells in the image include two neutrophils and a lymphocyte (lower right).

(courtesy Haematology Laboratory, Queen Alexandra Hospital, Cosham).

configuration. The physiological reason for this change in chromatin structure relates to the need to transcribe DNA into RNA, and the processing machinery—polymerases, requiring access to the DNA template ultimately to produce specific peptides encoded by the DNA sequence. In mature cells, only limited access to DNA is required and therefore chromatin condenses. In addition to the open chromatin configuration typical of blasts, one or more prominent **nucleoli** are present. Nucleoli contain genes that encode ribosomal RNA and facilitate the translation of transcribed proteins.

Key points

The ability to recognize and determine the percentage of blast cells in a peripheral blood sample or bone marrow specimen is important when considering a diagnosis of leukaemia. In a normal situation, the blast percentage should be very low—less than 4 per cent in the bone marrow—and no blasts should be apparent in peripheral blood. If a patient has more than 20 per cent blast cells in the bone marrow then a diagnosis of acute leukaemia is made.

SELF-CHECK 4.9

List the main characteristics of blast cells and provide a brief explanation of each.

The chemical control of haemopoiesis is largely achieved through the actions of a range of cytokines and growth factors. Growth factors are largely produced by the supportive cells of the bone marrow, called **stromal cells**. We have already discussed the EPO and TPO members of the cytokine family. Cytokines can function in either a lineage-restricted manner or can regulate the genesis of multiple lineages. In some situations, multiple cytokines are required to function in an additive fashion in order to promote proliferation. Interleukins (chemical mediators released by white blood cells that act on cells of all lineages) are also inherently involved in the control of haemopoiesis.

Haemopoiesis begins at the stem cell level. Ultimate control of stem cell activity is regulated by the soluble factor, stem cell factor (SCF), which can induce stem cell proliferation by binding to the SCF receptor (c-kit). This interaction ensures that there is sufficient stem cell activity to support haemopoiesis.

Interleukin 3 (IL-3) has the broadest range of haemopoietic-related functions, interacting with cells from the stem cell level down to and including the later stages of CFUs— CFU-E, CFU-Mono, CFU-N and CFU-Eo. Granulocyte-macrophage colony stimulating factor (GM-CSF) acts synergistically with IL-3 to, in the main part, stimulate multipotent progenitors. CFU-GEMMs are sensitive to GM-CSF, and the activity of this growth factor encompasses BFU-E, CFU-EMK, CFU-GMo and CFU-E. Beyond this stage of maturation, specific targeted growth factors control the final stages of differentiation and maturation. Erythropoiesis is controlled by erythropoietin. Thrombopoietin controls platelet production, macrophage-colony stimulating factor (M-CSF) enables monocytes to form, granulocyte-colony stimulating factor (G-CSF) controls neutrophil maturation and IL-5 controls eosinophil development.

The synergistic activity of all these growth factors enables blood cell genesis to occur as a continuum, and also to be able to react to the body's specific requirements at any point in time. It is important to recognize that inflammation and inflammatory cytokines can influence haemopoiesis and enhance the production of neutrophils, monocytes and platelets, ensuring there are sufficient numbers of cells to fight infection, remove tissue debris and facilitate wound healing.

We shall now examine erythropoiesis in more detail.

4.7 **Erythropoiesis**

Erythropoiesis is the term used to describe the production of red blood cells. Depending on the age or gender of the individual, the red cell concentration in the peripheral blood should be maintained within the range 3.9–5.7×10^{12}/L. Maintaining the concentration of red blood cells in the peripheral blood is important. The primary role of red blood cells is to transport oxygen to metabolizing tissues and metabolic by-products carbon dioxide and hydrogen ions to the lungs. This transportation utilizes the intracellular respiratory pigment, haemoglobin. As we see with all biological systems, erythropoiesis is in a state of balance between cell death and cell proliferation. Retaining the balance between these two elements ensures the maintenance of a steady state (or equilibrium).

Arguably, the most important structural element in erythropoiesis is the erythroblastic island. Erythroblastic islands form in the bone marrow and are composed of a central macrophage, surrounded by erythroid precursors. The number of erythroblastic islands in an individual's marrow varies, although there is a relationship between the number of islands and the body's requirement for red cells. An eloquent study examining the number of erythoblastic islands in hypertransfused rats showed that, as the red cell concentration increased, so the number of erythroblastic islands decreased.

Immature red cells require a nucleus and other organelles to enable the important enzymes required for red cell metabolism, antioxidants, and membrane-associated proteins to be produced. Mature red blood cells need to be anucleate to facilitate their role in oxygen delivery, in which they need to be sufficiently flexible to squeeze through narrow-lumen blood vessels to supply tissue with oxygen. Red blood cells are approximately 7 μm in diameter, and the lumen of the smallest of blood vessels is approximately 3 μm. The red blood cells' biconcave disc shape allows them to traverse these small blood vessels. Red cells are thickest around their circumferential periphery, and become thinner towards the centre. The presence of a nucleus would impede the flexibility of red blood cells, thereby compromising oxygen delivery. Typical red blood cell morphology is shown in Figure 4.17. The area of central pallor represents the thinnest part of the biconcave disc.

The respiratory pigment haemoglobin is responsible for the carriage of oxygen to metabolizing tissues. An individual red blood cell contains approximately 640 million molecules of haemoglobin, each of which is capable of carrying four oxygen molecules. Haemoglobin is composed of two important components: four globin chains, each of which has a haem moiety attached. Within each haem moiety, a central ferrous iron (Fe^{2+}) molecule is located. This ferrous iron facilitates the carriage of oxygen from the lungs to tissue.

The central macrophage in the erythroblastic island is important for the **phagocytosis** of the extruded red cell nucleus, prior to the release of red cells into the peripheral blood. Additionally, the various cell-to-cell and cell-to-extracellular matrix interactions are important for retaining developing red cells within appropriate erythroid niches in the bone marrow.

Although the mechanisms responsible for iron delivery to red cell precursors has yet to be fully elucidated, it is believed that iron, retained within macrophages by the storage protein **ferritin**, is released into the erythroid niche, and is endocytosed by these red cell

Phagocytosis

This describes the process whereby certain cells engulf and consume foreign material such as a bacterium or damaged tissue.

Ferritin

This is an iron storage molecule containing the protein apoferritin combined with ferric iron.

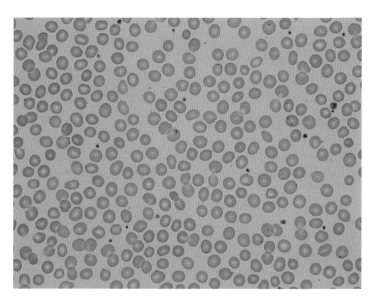

FIGURE 4.17

Normal red blood cells. Note the red cells are roughly the same size and shape. The cells, if examined in profile, are of a biconcave disc shape. Red cells are thickest around the periphery and thinnest at the centre. Haemoglobin accumulates around the periphery compared to the central area—the area of central pallor (Romanowsky stain, original magnification × 600).

(courtesy Haematology Laboratory, Queen Alexandra Hospital, Cosham).

precursors. Iron is then thought to be released by acidification and proteolysis and incorporated into haem within mitochondria in preparation for haemoglobin synthesis in the cytoplasm. However, it could be the case that iron is delivered to red cell precursors independently of erythroblastic island-associated macrophages, as is the case with other iron-requiring cells.

Intriguingly, erythroblastic islands are mobile. Initially, these islands are dispersed throughout the red marrow; however, as the red cells associated with these islands mature, the islands move towards vascular sinuses. It is believed that this is in preparation for red cell release into the peripheral blood to effect red cell function. The architecture of an erythroblastic island is shown in Figure 4.18. The central macrophage is clearly evident.

SELF-CHECK 4.10

Provide a description of the role of erythroblastic islands, and state how the central macrophage may be intrinsically involved in iron control.

4.8 Hierarchy of erythropoiesis

Beyond the CFU-E, the pronormoblast forms. From each pronormoblast, a total of 16 mature red cells can be produced. This red cell production is achieved through successive cell divisions of maturing cells. Red blood cell precursors proliferate, becoming progressively smaller. When

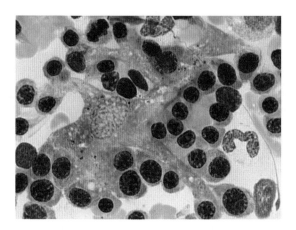

FIGURE 4.18
Erythroblastic island. A central macrophage (light blue) is surrounded by developing red cells. Red cells are nucleated until extrusion occurs within the bone marrow.
© 2010 Blackwell Publishing Ltd, *British Journal of Haematology*, **150**, 499.

examined by light microscopy, their cytoplasm changes from deeply basophilic (dark blue) as a consequence of high concentrations of mRNA, to pink/red, as cytoplasmic mRNA is replaced by mature proteins such as haemoglobin.

As the erythroid lineage shows cytoplasmic maturation, nuclear changes also become apparent (Figure 4.19). As transcription slows within the cell, the nuclear chromatin condenses and the nucleus shrinks. We call this nuclear structure **pyknotic**.

Pronormoblasts measure 14–19 µm in diameter, with the next stage of development, the erythroblast (also called normoblast), measuring 10–15 µm. The final stage restricted to the bone marrow is the nucleated red blood cell (nRBC). At this stage, approximately 60–70% of the cell's haemoglobin has been synthesized, although all the RNA required to ensure the

> **Pyknosis**
> This describes a condensation or shrinking and thickening of a cell or nucleus.

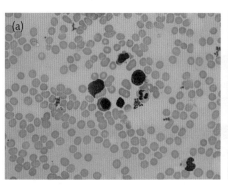
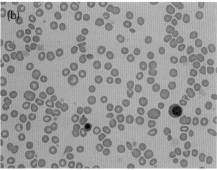

FIGURE 4.19
Different stages in the erythropoietic pathway. (a) Two different stages of normoblast are present, with a neutrophil, two lymphocytes, a smeared cell nucleus (centre) and two platelet clumps. The pronormoblast (right) and slightly more mature intermediate normoblast (lower left) both demonstrate basophilic cytoplasm as a consequence of high RNA content. (b) As cells mature, the cytoplasm contains more haemoglobin, staining pink, and the nucleus becomes pyknotic (centre).
(courtesy Haematology Laboratory, Queen Alexandra Hospital, Cosham).

synthesis of the remaining haemoglobin has been transcribed. The nucleus is now pyknotic, and is expelled (or extruded) from the cell into the erythroid niche. Within the eythroblastic island, the central macrophage will phagocytose the nucleus, largely in response to high levels of phosphatidylserine expressed on the expelled nuclear membrane, acting as a recognition site for macrophages.

With the extrusion of the nucleus, the nRBC becomes a reticulocyte, so called because these cells have a striated (reticulated) pattern running throughout their cytoplasm which can be visualized using light microscopy and a **supravital stain**. This pattern is a consequence of the presence of high numbers of residual ribosomes within these cells.

Reticulocytes measure approximately 7–10 μm in diameter and are released into the venous sinuses from the erythroblastic islands, and enter the peripheral blood. Reticulocytes are larger than red blood cells and lack the characteristic biconcave disc shape seen in mature red blood cells. As reticulocytes mature over a period of one to three days in the peripheral blood, they undergo a series of important changes, including:

* completion of haemoglobin synthesis

* becoming smaller, homogenous and biconcave

* degradation of internal organelles

* adoption of mature red blood cell staining characteristics.

Examination of a Romanowsky-stained peripheral blood film in which a high proportion of reticulocytes is present demonstrates a bluish tinge called **polychromasia**. This colouration is due to the high levels of RNA retained within these cells; as they mature, the blue colouration is replaced by the typical pink/red colour seen in mature red blood cells. The expected number of reticulocytes in a healthy individual's blood is approximately $25–125 \times 10^9$/L, and in cases where individuals have lost blood due to trauma, for example, or are receiving treatment for anaemia, an increase in reticulocytes, termed **reticulocytosis**, becomes apparent. In these types of anaemia, the reticulocyte count can be used as an indicator of a bone marrow response, called compensation.

Erythropoiesis is a dynamic process during which a gradual transition from immature to mature cells is evident. As a consequence, the polychromasia seen in peripheral blood films represents only the least mature reticulocytes rather than the whole population of these immature cells.

SELF-CHECK 4.11

Compare and contrast reticulocytes with mature red blood cells.

SELF-CHECK 4.12

Explain the value of measuring the reticulocyte count in the evaluation of bone marrow function.

Supravital stains

Such as new methylene blue, are used to stain live, unfixed cells. Usually, stains are applied to fixed cells, and as such are only used when the cells are no longer alive.

Endomitosis

This involves the replication of chromosomes without the division of the cell nucleus. This leads to cells with, potentially, very high chromosome numbers.

4.9 Thrombopoiesis

Following the formation of CFU-EMk, megakaryoblasts develop. These highly proliferative cells mature into megakaryocytes. The formation of platelets from megakaryocytes follows, in the first instance by a process called **endomitosis**. As megakaryocytes progress through their cell

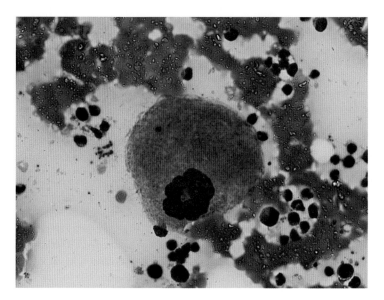

FIGURE 4.20

Megakaryocyte. This is a large bone marrow-derived cell that produces platelets. The nucleus forms multiple lobes from successive phases of endomitosis, while the cytoplasm becomes increasingly granular. Pro-platelets bud from the cell membrane, each containing a number of granules from the parent megakaryocyte. From these pro-platelets, platelets form.

(courtesy Haematology Laboratory, Queen Alexandra Hospital, Cosham).

cycle, rather than dividing and producing daughter cells, the chromosomes replicate. The cell membrane and cytoplasm remain intact. This process of endomitosis leads to the formation of sequentially larger megakaryocytes, with increasingly complicated nuclei (Figure 4.20). Rather than possessing the usual 46 chromosomes (called 2n, where n = 23), a diploid chromosome complement, many megakaryocytes may possess up to 368 chromosomes (16n). Endomitosis ensures that sufficient organelles are available and cytoplasmic volume and maturation occurs to support platelet production.

As megakaryocytes increase in size, internal rearrangement of the cytoskeleton and organelles occurs. A fine microtubular network is deposited around the periphery of the cell, close to the cell membrane, and this is associated with a secondary, internal demarcation membrane, used for the generation of platelet precursors called pro-platelets. From the surface of the megakaryocyte, slender pseudopodia form. This, now dense, microtubular network extends within the pseudopodia, enabling organelles and granules to move into the pseudopodia, facilitating the development of pro-platelets. Throughout this process, the cytoplasm of the megakaryocyte is consumed through the formation of pro-platelets and these are then released into the plasma. Pro-platelets elongate, forming a dumb-bell shape, concentrating organelles and granules at opposing ends. The bridging shaft between each end seals and pinches off, releasing platelets rich in organelles and granules, ready to initiate haemostasis.

SELF-CHECK 4.13

Explain the role of endomitosis in thrombopoiesis.

4.10 Haemostasis and the role of platelets

Haemostasis is the process whereby blood is maintained in a fluid state within blood vessels. In order to function, blood must be in a fluid state, facilitating the delivery of oxygen, nutrients and hormones to metabolizing tissue, and to remove metabolic by-products for excretion. Equally, blood must be retained within the blood vessels, or circulation, because it is functionally useless outside this environment, and when deposited within extravascular tissues it is called a bruise.

Cross reference

More detail regarding haemostasis can be found in the *Haematology* volume in this Fundamentals of Biomedical Science series.

As summarized in Figure 4.21, haemostasis is divided into two main components: primary and secondary. Primary haemostasis involves cellular interactions between the vascular subendothelium and platelets, often mediated by a large protein complex called von Willebrand factor, while secondary haemostasis involves biochemical interactions between a network of soluble clotting factors, terminating in the formation of an insoluble fibrin meshwork. Primary and secondary haemostasis function concurrently to stop bleeding, first through the production of a primary platelet plug, sealing the blood vessel, and second through stabilization of the platelet plug with a fibrin meshwork, which acts as a supporting scaffold.

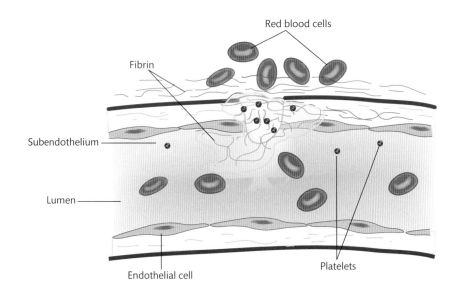

FIGURE 4.21

An overview of haemostasis. Blood cells travel through the lumen of blood vessels. When vascular injury occurs, the surface below the endothelial cells (the subendothelium) is exposed. The exposed subendothelium acts as a procoagulant surface, allowing platelets (blue) to bind, forming a primary platelet plug. This plug is strengthened through the formation and deposition of fibrin over the platelet plug (blue strands). The formation of the plug forms a barrier to infection, prevents further blood loss and facilitates wound healing.

4.11 **Granulopoiesis**

Following the formation of CFU-GM from the CFU-GEMM stem cell, further lineage restriction occurs leading to the formation of CFU-N (a neutrophil precursor), CFU-Eo (an eosinophil precursor), CFU-Baso (a basophil precursor) and CFU-Mono (the monocyte precursor). More detail on monopoiesis can be found in Section 4.12. With the exception of the CFU-Mono, these lineage-restricted colony forming units then mature into myeloblasts.

Myeloblasts account for up to 4 per cent of the nucleated cells in the bone marrow and are highly proliferative progenitor cells. Measuring 12–20 μm in diameter, myeloblasts have the capacity to form large numbers of granulocytes, although at this stage the specific identity of their daughter cells cannot be determined morphologically. Myeloblasts have an open chromatin configuration with prominent nucleoli and a high nucleocytoplasmic ratio.

As myeloblasts mature they form promyelocytes, which are larger than myeloblasts (15–25 μm in diameter) and contain primary granules in their cytoplasm. Promyelocytic cytoplasm shows a much deeper blue colouration than myeloblasts as a consequence of increased RNA transcription. Prominent nucleoli are retained within their indented nucleus. As promyelocytes undergo mitosis and mature, myelocytes form.

Myelocytes are smaller than promyelocytes, and are approximately the same size as myeloblasts (10–15 μm in diameter). The cytoplasm is much paler than that found in promyelocytes, and discrete secondary granules can be seen throughout the cytoplasm. Secondary granules are lineage-specific and so myelocytes are the first stage at which neutrophil, eosinophil or basophil lineage can be definitively established. Myelocytes represent the final mitotic stage and following division and maturation, metamyelocytes form.

Metamyelocytes possess primary and secondary granules, and a kidney bean-shaped nucleus. A smooth transition in cell structure and maturity occurs from the metamyelocyte stage through an intermediate band-form stage, where the cell nucleus resembles a horseshoe shape, and then further maturation occurs leading to the mature neutrophil, eosinophil and basophil stages. The entire production of effector cells takes approximately five to six days to complete.

Each of these stages is represented in Figure 4.22. Notice how each of the mitotic stages, outlined above, are retained within the bone marrow microenvironment.

SELF-CHECK 4.14

List the key maturation stages in the development of granulocytes. For each stage, state whether or not mitosis occurs, and confirm the presence or absence of granules.

4.12 **Monopoiesis**

Monoblasts, which form from CFU-mono, typically have the same morphological characteristics as myeloblasts, although they have become specialized to produce monocytes. Following the monoblast stage, promonocytes form (Figure 4.23) that are larger than monoblasts and monocytes, and possess a large C-shaped nucleus. From this promonocyte stage, monocytes form and migrate from the bone marrow into the peripheral blood. At this point, monocytes

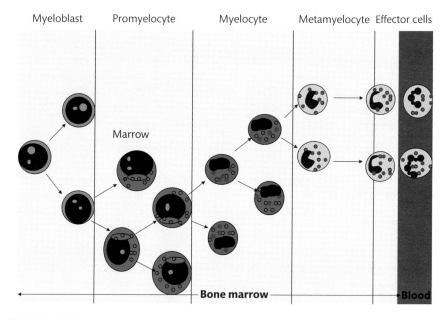

FIGURE 4.22

Granulocyte maturation in the bone marrow. Myeloblasts divide into larger promyelocytes, and successive cell divisions produce myelocytes and metamyelocytes. Metamyelocytes do not undergo mitosis, but mature to form effector cells. Species-specific granules can be seen from the myelocyte stage.

© Gary Moore, Gavin Knight, and Andrew Blann, 2010.

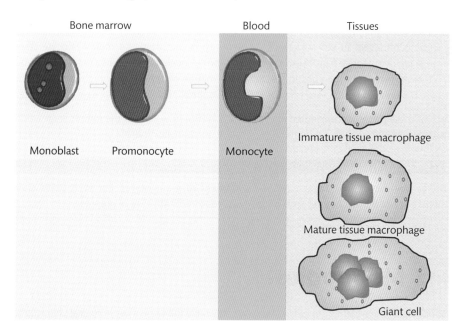

FIGURE 4.23

Monopoiesis. This begins with monoblasts present in the bone marrow. Larger promonocytes are produced following mitosis of monoblasts. Promonocytes mature into monocytes, which can be visualized in the peripheral blood. Following exit from the blood vessel, monocytes mature into tissue macrophages.

© Gary Moore, Gavin Knight, and Andrew Blann, 2010.

circulate for two to three days before entering the tissues to become tissue macrophages. In the tissues these macrophages will recommence cell cycling until they become mature macrophages. Mature macrophages may then embark on the process of endomitosis, as with megakaryocytes, leading to the formation of giant cells.

4.13 Abnormal haemopoiesis

So far, we have considered the normal process of haemopoiesis. Every day of our lives, our bone marrow endeavours to provide sufficient cells for the peripheral blood to complete its necessary functions. However, there is a wide range of disorders that can alter the bone marrow's capacity to produce blood cells, or the function or survival of blood cells once released into the peripheral blood. In this final section, we consider some of the broad effects and consequences of abnormal haemopoiesis.

Cross reference

More detail of the causes and consequences of abnormal haemopoiesis can be found in the *Haematology* volume of the Fundamentals of Biomedical Science series.

4.13.1 Red blood cells

The most common disorder of blood is anaemia. Although there is a very strict definition of anaemia, it can be defined as reduced oxygen-carrying capacity of the blood as a consequence of reduced haemoglobin concentration. This leads to clinical symptoms such as increased heart rate (tachycardia), breathlessness, pallor and lethargy. The most common cause of anaemia is iron deficiency, although there are many others including blood loss, bone marrow suppression, vitamin B_{12} or folate deficiency, haemolysis, inherited abnormal haemoglobin production and autoimmunity.

Cross reference

More detail on the precise definition, causes and consequences of anaemia can be found in the *Haematology* volume in this Fundamentals of Biomedical Science series.

SELF-CHECK 4.15

List the common causes of anaemia.

In contrast to anaemia, we may see an increase in red blood cell production. This is due most commonly to an increased demand for red cells (e.g. due to compensation). Compensation was discussed earlier in the context of living at high altitude. Another example of where we may find an increased red cell count is in patients with chronic carbon monoxide poisoning. Haemoglobin has a high affinity for carbon monoxide, which is some 200 times greater than that of its affinity for oxygen. When carbon monoxide is bound to haemoglobin it forms carboxyhaemoglobin. The formation of carboxyhaemoglobin is irreversible, and so the body produces more red blood cells in order to compensate for the reduced oxygen carrying capacity of the blood. A reactive erythropoietic response to a lower oxygen carrying capacity results in **erythrocytosis**. However, patients sometimes develop a cancer of the red cells, a consequence of erythropoietin-independent proliferation of red cell precursors. This inappropriate and enhanced red cell production leads to increased numbers of red blood cells in the peripheral blood, and is called **polycythaemia**. It is important to note that, historically, the terms erythrocytosis and polycythaemia have been used interchangeably, although this is now no longer the case.

SELF-CHECK 4.16

Briefly compare and contrast erythrocytosis with polycythaemia.

4.13.2 Platelets

When a patient's platelet count falls below the reference range of 150–400 $\times$ 10^9/L, the individual is at increased risk of bruising and bleeding. Platelet counts less than 20 $\times$ 10^9/L result in a significantly higher risk of spontaneous bleeding, and levels lower than this may require the infusion of donor platelets. Platelets are necessary for the formation of the primary platelet plug in response to vascular injury. Bleeding may result if there are insufficient platelets to fulfil this role. A reduced platelet count is called **thrombocytopenia**. Causes of thrombocytopenia can be broadly subdivided into impaired platelet production, increased platelet destruction or consumption, or increased pooling of platelets in the spleen. Platelets are retained in the spleen in a reserve capacity as a normal physiological process; however, in patients with an enlarged spleen, called **splenomegaly**, or if the patient has antibodies directed against their own platelets, more platelets than normal will be retained by the spleen and perhaps be destroyed.

Cross reference

More detail regarding the role of platelets in haemostasis can be found in the *Haematology* volume in this Fundamentals of Biomedical Science series.

We may see an increase in platelet count as a consequence of cancerous changes to megakaryoblasts. We call an accumulation of cancerous platelets in the blood **essential thrombocythaemia** (ET), in which case the platelet concentration is TPO-independent.

Together, polycythaemia and essential thrombocythaemia can be described as **myeloproliferative neoplasms**.

4.13.3 White blood cells

Cross reference

More detail on lymphomas can be found in other books in this Fundamentals of Biomedical Science series, most notably in the *Histopathology* and *Haematology* volumes.

Large numbers of white blood cells in the peripheral blood is called **leucocytosis**. If leucocytosis is associated with cancerous changes in the patient's DNA, we call this **leukaemia**. An increased white blood cell count does not always signal leukaemia. For example, white blood cells may also be increased in response to infection, inflammation and allergy.

Leukaemias can be associated with all types of white blood cell and may be apparent at different stages of white blood cell maturation (Figure 4.24). An accumulation of 'blast' cells of

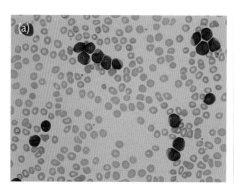
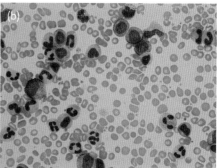

FIGURE 4.24

Two forms of leukaemia. (a) acute lymphoblastic leukaemia in a child. Note the monomorphic nature of the lymphoblasts (i.e. they all look similar). Cells have a very high nuclear:cytoplasmic ratio with open chromatin and prominent nucleoli. Mature white blood cells have been replaced by these cancerous cells. (b) Chronic myeloid leukaemia. A very high white blood cell count can be seen with all stages of maturation, except myeloblasts, present. These cells are granular and can clearly be identified as myeloid in origin. They are heterogeneous and retain many of the morphological characteristics of normal myeloid cells. (courtesy Haematology Laboratory, Queen Alexandra Hospital, Cosham).

either myeloid or lymphoid origin (i.e. myeloblasts or lymphoblasts) is called acute leukaemia, whereas an accumulation of more mature cells is called chronic leukaemia.

A mass of cancerous lymphoid cells (B cells, T cells or their precursors) restricted to lymphoid tissues, including the spleen, liver, bone marrow, lymph nodes or central nervous system, is called **lymphoma**.

The disorders listed here are a small number of examples of disease associated with abnormal haemopoiesis and it is important to recognize that, in each case, patients must receive appropriate investigation and treatment to ensure they have the best possible opportunity to recover.

Cross reference
More detail regarding cancers of the haemopoietic system can be found in the *Haematology* volume in this Fundamentals of Biomedical Science series.

 ## Chapter summary

In this chapter we have:

- established the role of stem cells in the generation of blood cells.
- discussed the key features of stem cells and their ability for self-renewal and differentiation.
- discussed the role of a variety of growth factors in stimulating haemopoiesis.
- considered the differential effects of erythropoietin and thrombopoietin in producing red cells and platelets, respectively.
- considered each of the blood cell lines, the processes by which they develop, and their ultimate function.
- defined the terminology associated with blood cell genesis.
- introduced a range of disorders of haemopoiesis.

 ## Further reading

- Kaushansky K. Thrombopoietin: a tool for understanding thrombopoiesis. *J Thromb Haemost* 2003; **1** (7): 1587–92.
- Manwani D, Bieker JJ. The erythroblastic island. *Curr Top Dev Biol* 2008; **82**: 23–53.
- Ottersbach K, Smith A, Wood A, Gottgens B. Ontogeny of haematopoiesis: recent advances and open questions. *Br J Haematol* 2010; **148** (3): 343–55.
- Travlos GS. Histopathology of bone marrow. *Toxicol Pathol* 2006; **34** (5): 566–98.

 ## Discussion questions

4.1 Discuss the application of haemopoietic growth factors in the treatment of named haematological diseases.

4.2 Based on your understanding of the genesis of blood cells, provide an overview of the role of a diagnostic haematology laboratory.

4.3 Two important clinical procedures used to determine bone marrow function include the bone marrow aspirate and bone marrow trephine biopsy. Investigate each of these procedures and write brief notes on their individual merits.

Answers to self-check questions and discussion questions, hints and tips are provided on the book's Online Resource Centre.

 Visit www.oxfordtextbooks.co.uk/orc/orchard_csf/>.

Nerves: the cells of the central and peripheral nervous systems

Rosalind King and Richard Mathias

Learning objectives

After studying this chapter you should confidently be able to:

- Describe the cells that make up the nervous system and their functions.
- Explain the mechanisms of neuronal communication.
- Outline the arrangement and integration of the central and peripheral nervous systems.
- Summarize the basic functions of the nervous system.
- Describe some of the major disease processes that affect the nervous system.

The nervous system is an essential component of the body, without which we could not survive. Nerve cells form a connected network, extending throughout every part of the body. By means of signalling, from cell to cell throughout this network, information is received and processed, and a response is generated. Thus, the nervous system provides a means of coordination and helps to maintain homeostasis.

The nervous system is often thought of as having two main components. The central nervous system (CNS) comprises the brain and spinal cord. From the CNS, nerve fibres project to the receptors and effectors in every other organ and tissue throughout the body, forming the peripheral nervous system (PNS). However, while this distinction may be convenient in structural terms, it is important to remember that the CNS and PNS are continuous and largely function together.

In this chapter we will examine the cells that make up the structures of the nervous system, and how they act together to facilitate all the functions of human life. End organs such as muscle and sensory endings will be dealt with in Chapters 10 and 15.

5.1 **Neurons**

The functional unit of the nervous system is the **neuron**. Neurons are excitable; that is they are able to receive a stimulus, and in some way generate a response to that stimulus. They can then transmit a stimulus to another cell. Neurons exist in a complex network providing multiple connections between themselves, receptor cells and effector cells. Receptor cells are neurons that are able to receive an external stimulus. These neurons are often referred to as sensory neurons because they detect a stimulus or change in environment (e.g. touch receptors in skin or light receptors in the retina). These neurons transmit information into the CNS and hence may also be called **afferent** neurons. Effector cells are neurons that induce an external response (e.g. contraction of muscle fibres), in which case they are motor neurons (Figure 5.1). They are also described as **efferent** neurons because they transmit a signal away from the CNS to cause an action (e.g. muscle contraction). Interneurons form the multitude of connections between other neurons. These functional distinctions between different types of neuron are important in understanding the nervous system.

The number of neurons in the CNS is contentious, but several estimates put the figure to be in the order of 100 billion. They are highly diverse both in structure and function, and neurons found in one anatomical area are often similar to each other, but different from those found in other areas. This reflects the way in which many functions are localized to specific anatomical areas. The cell body, or soma, can be up to 130 μm in diameter, e.g. spinal cord motor neurons (Figure 5.2), whereas others are much smaller (e.g. cerebellar granular neurons, which may be only around 5 μm). Neurons are highly metabolically active, and thus require a constant supply of nutrients, particularly glucose and oxygen, for the generation of energy. Large amounts of energy are needed to maintain the membrane potential, synthesize neurotransmitters and transport materials over long distances. This is reflected by large numbers of mitochondria, and the abundance of rough endoplasmic reticulum, termed Nissl substance, which is characteristically seen in the cell body. There are numerous lysosomes, reflecting the high turnover of cell components, including the maintenance of a large surface area of membrane. Another feature commonly present, especially with ageing, is **lipofuscin**. This is residual lysosomal material. As in all cells, neurons have a cytoskeleton comprised of microfilaments and intermediate filaments. This is important for intracellular transport mechanisms, as well as maintaining cell integrity. The neuronal cytoskeleton contains intermediate filaments, constructed from three types of neurofilament protein (light [NFL], medium [NFM], and heavy [NFH]). These are unique to neurons and therefore can be used as neuronal markers. In addition, there are microtubules, the function of which is to transport organelles (more on this later).

The fundamental function of the neuron is communication. Neurons usually have processes called **axons** and **dendrites** to make connections with other cells via junctions called **synapses**. Dendrites are relatively short, branching processes that communicate with adjacent cells. Axons can be long and convey signals to cells that may be some distance away (e.g.

Lipofuscin
A wear and tear pigment, common to ageing cells, including those of the CNS.

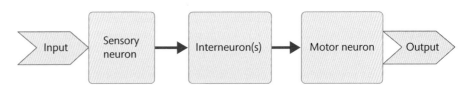

FIGURE 5.1
Schematic diagram of a nervous system pathway. In some instances, the interneuron step is omitted and a sensory neuron communicates directly with a motor neuron.

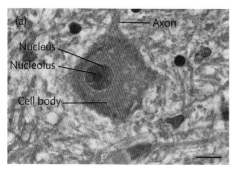
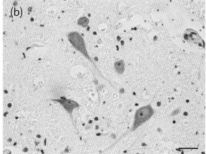

FIGURE 5.2
(a) Section of a large neuron. Note the large nucleus with prominent nucleolus and the abundance of granular endoplasmic reticulum (haematoxylin and eosin [H&E] stain; Scale bar: 20 µm). (b) Neurons in the spinal cord (Cresyl fast violet stain; Scale bar: 50 µm).

consider a motor neuron in the spinal cord whose axon extends to a muscle in the foot). Generally, dendrites are the means by which a neuron receives a signal, and axons are the means by which it transmits a signal. Axons may be coated in a sheath of **myelin**, a lipid-based substance that aids faster nerve impulse transmission. The myelin sheath is laid down in segments by adjacent support cells (more on this later). In between adjacent segments of myelin there is a very short length of the axon without myelin. This is called the **node of Ranvier** (Figure 5.11), and there will be many of these along the length of a myelinated axon. Towards the terminus, the axon branches into numerous terminals, each of which broadens slightly at its end into a **terminal bouton**, which forms part of the synaptic apparatus.

Axons and dendrites establish a complex network between receptors, effectors and other neurons. A single neuron may have thousands of connections with other neurons. Most, if not all, of the non-myelinated region of a neuronal cell membrane in the CNS, including the nodes of Ranvier, will be in direct contact either with other neurons, via synapses, or supporting astrocyte processes.

Key points

Neurons and their processes form an interconnected network to provide communication throughout the entire nervous system and between organs.

5.1.1 Structural classification of neurons

While we mentioned earlier that neurons can be classified by their function (i.e. sensory or motor), neurons are also commonly classified structurally (Figure 5.3) by their **polarity** (i.e. the number of processes extending from the cell body). **Pseudounipolar** neurons have a single process, which then diverges at a 'T' junction into an axon. One end of the axon branches into dendrites, which contact other neurons in the CNS, and the other process contacts sensory endings in the PNS. Hence, these are sensory neurons. True unipolar neurons with a single, non-separating process are not seen in the human body. **Bipolar** neurons have two separate cell body processes, one axonic and one dendritic. Again, these are usually sensory in function (e.g. light-sensitive cells in the retina). **Multipolar** neurons have numerous processes,

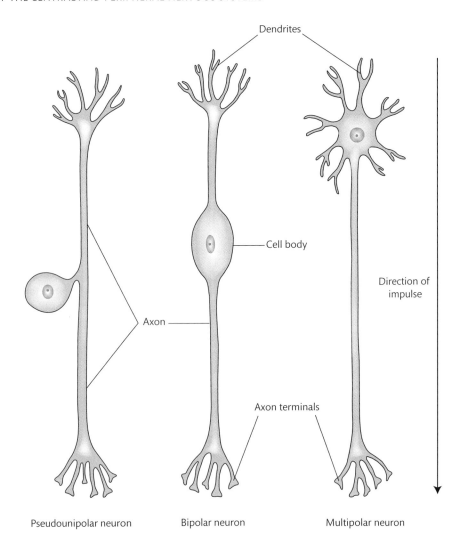

FIGURE 5.3
Schematic representation of different neuronal structures.

usually many dendrites with a single axon. These are motor and interneurons, and are the most abundant structural type. Neurons have a clearly defined direction of transmission, away from the dendritic terminus and towards the axonal terminus. It is also worth noting that some specialized neurons, such as amacrine neurons in the retina, have no true axons.

5.1.2 Intracellular transport

It is essential, especially given that an axon may be up to a metre in length, that there is an efficient mechanism for intracellular transport. Organelles and structural components need to be moved throughout the cell. As well as microfilaments and intermediate filaments, the cytoskeleton comprises a system of microtubules. These run throughout the cell body and its processes, and provide a mechanism for the movement of organelles and substances throughout the cell.

There are two systems of transport. In **fast axonal transport**, motor proteins link particulate objects such as mitochondria and vesicles to the microtubule and move them along the microtubule. Transport away from the cell body (e.g. vesicles of neurotransmitter to a synapse) is termed **anterograde** transport, and is mediated by the motor protein kinesin. Transport in the opposite direction is termed **retrograde**, and is mediated by dynein. This has a speed of up to 400 mm/day. Soluble proteins such as neurofilament components and tubulin travel much more slowly, at a speed of less than 10 mm/day, in what is termed **slow axonal transport**.

Disruption of intracellular transport is commonly one of the effects of neuronal injury.

5.1.3 Resting and action potentials

As we mentioned earlier, neurons are excitable. Under appropriate conditions, the neuron receives a stimulus, either from a receptor or another neuron, and can then transmit its own stimulus to another cell. This is an electrochemical process that relies on the movement of charged ions to generate changes in electrical potential.

The neuronal cell membrane contains numerous ion pumps, which transport ions across the membrane by active (energy requiring) processes, and ion channels, which allow ion passage by passive diffusion. In the resting state, ion pumps in the membrane carry sodium ions out and carry potassium ions in. Although both ions are also able to diffuse in the opposite direction down the resulting concentration gradients, the membrane is in fact much less permeable to sodium ions than potassium ions. This means that the concentration of extracellular sodium ions is much higher than the intracellular concentration. Of course, anions are also present, but these are much less able to pass through the cell membrane to restore the ionic balance. This unbalanced distribution of charged particles creates an ionic gradient, and thus a potential difference, across the cell membrane. This is the **resting membrane potential**. The resting membrane potential is around −70 mV. Maintenance of this potential by active processes is one of the reasons why neurons have such high energy demands.

When the neuron receives an appropriate stimulus, sodium ion channels in the membrane open, allowing an influx of sodium ions by diffusion. This causes an increase in the membrane potential to around +30 mV. However, this increase is short-lived and almost immediately the sodium channels reclose, whereupon potassium channels open, allowing the diffusion of potassium ions out of the cell to restore the ionic balance. However, if the increase in membrane potential reaches a critical threshold level, the so-called **action potential**, sodium channels in adjacent regions also open, and then in regions adjacent to them, and so on, and thus the depolarization is transmitted along the membrane. In this manner, the action potential can be transmitted along an entire axon. The action potential is described as an 'all or nothing' event. This means that either the threshold is reached and the action potential is conducted along the axon, or the threshold is not reached and nothing happens. There is no partial transmission.

Axons that are myelinated allow much faster conduction than do non-myelinated axons. This is because myelinated regions of the axon have no ion channels; only the nodes of Ranvier have these. Instead of moving along the axon in a wave, the action potential 'jumps' from node to node. This is called saltatory conduction.

Key points
Transmission of nerve impulses requires the generation of an action potential across the cell membrane.

Key points

Myelination of axons allows for faster conduction throughout.

After the action potential has passed, potassium ions passively entering the cell allow the ionic balance to be restored, and then the membrane once again pumps sodium ions out and potassium ions in to restore the depolarized resting membrane potential. The entire process, from stimulus to return to resting potential, takes around two milliseconds.

5.1.4 Synaptic communication

Neurons communicate with other cells by junctions called synapses, by which the signal is passed from cell to cell. The receiving cell may be an effector cell such as muscle (**the neuro-muscular junction**) or secretory, or may be another neuron. The synaptic apparatus of the cell conveying the signal is called the presynaptic component; that of the cell receiving the signal is called the post-synaptic component. The presynaptic component is most commonly at a terminus of the axon, but the post-synaptic component may be anywhere on the cell body (where the synapse is described as axo-somatic), on a dendrite (axo-dendritic), or on a non-myelinated region of the axon (axo-axonic). A single neuron may have thousands of synapses. There are two types of synapse: electrical and chemical.

Chemical synapses

At chemical synapses, a substance, the neurotransmitter, is released from the presynaptic cell and passes to, and exerts an effect on, the post-synaptic cell.

The axon terminus broadens slightly into a structure called the terminal bouton, which contains membrane-bound vesicles of neurotransmitter substance. There is a gap of around 20 nm, the synaptic cleft, between the membrane of the terminal bouton and the post-synaptic cell membrane. The neurotransmitter vesicles are anchored by membrane protein complexes to proteins in the presynaptic membrane. While the neuron is at rest, the neurotransmitter vesicles are prevented from fusing with the presynaptic membrane by a protein, synaptotag-min, present in the vesicle membrane. When the action potential reaches the presynaptic terminal, voltage-gated calcium channels in the region open, allowing an influx of calcium ions to the terminal bouton. The calcium ions induce a change in the synaptotagnin molecules, allowing fusion of the neurotransmitter vesicle with the presynaptic membrane. Thus, the neurotransmitter substance is released from the presynaptic cell into the synaptic cleft, where it binds to receptors on the post-synaptic cell membrane (Figure 5.4).

Receptor binding affects the post-synaptic cell in some way (i.e. stimulatory, inhibitory, or modulatory). This is achieved by the modification of ion channels such that the passage of ions in or out of the cell is either allowed or prevented. There are two receptor mechanisms by which the post-synaptic cell responds to neurotransmitter binding. Receptors can be incorporated into **ligand gated** ion channels. The ion channel is then modified directly by the binding of neurotransmitter to its receptor. Other receptors have no direct connection with an ion channel, instead being coupled to G-proteins. On binding the neurotransmit-ter, the G-protein is activated and dissociates from the receptor to initiate a cascade of intracellular messengers that modify ion channels separate from the receptor. Such signal

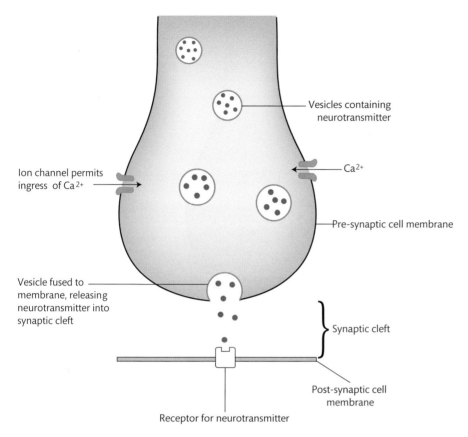

Vesicles containing neurotransmitter

Ca^{2+}

Ion channel permits ingress of Ca^{2+}

Pre-synaptic cell membrane

Vesicle fused to membrane, releasing neurotransmitter into synaptic cleft

Synaptic cleft

Post-synaptic cell membrane

Receptor for neurotransmitter

FIGURE 5.4
Schematic representation of a chemical synapse.

transduction pathways may have a slower onset but the effect may last for hours. There may also be an amplifying effect whereby one bound receptor may lead to the opening of many ion channels.

Following transmission, it is important that the neurotransmitter is quickly removed, or else inappropriate prolonged stimulation may occur. This is achieved by several means. Some are broken down enzymatically (e.g. acetylcholinesterase), there may be diffusion away from the synapse, including perhaps uptake by astrocytes, or there may be reuptake by the presynaptic neuron where the neurotransmitter substance can be recycled.

Electrical synapses

At electrical synapses there is no neurotransmitter substance and the signal is conveyed by ions passing directly from the presynaptic to the post-synaptic cell.

These are structurally different to chemical synapses and comprise a gap junction that contains ion channels. The synaptic cleft is only 3.5 nm. The ion channels allow the passage of ions from one cell to the other. Thus, depolarization or hyperpolarization of the presynaptic cell is conveyed to the post-synaptic cell (Figure 5.5). Unlike in chemical synapses, the post-synaptic effect is always the same as the presynaptic signal.

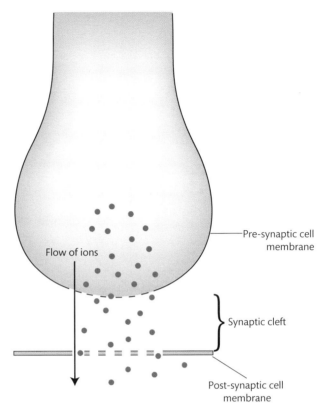

FIGURE 5.5
Schematic representation of an electrical synapse.

Electrical synapses are few in number compared to their chemical equivalent but they permit much faster transmission, hence they occur in situations where a rapid response is vital (e.g. flight mechanisms).

Following transmission, the effect is to generate a **post-synaptic potential** (PSP) in the post-synaptic neuron. This is a change in the potential of the post-synaptic neuron due to the opening or closing of ion channels in the membrane. These PSPs are either excitatory or inhibitory, depending on the polarity and direction of the altered ion passage. An excitatory PSP alters the potential towards threshold, and an inhibitory PSP away from threshold. As the post-synaptic neuron bears many synapses, its potential will change according to the net combined effect of all the PSPs it creates: a phenomenon called **summation**. If an excitatory PSP received at a single synapse is followed rapidly by a second, before the cell has returned to its resting potential, the excitatory PSPs will combine to create a summated potential greater than a single excitatory PSP. This is called **temporal** summation. Similarly, if simultaneous (or near-simultaneous) excitatory PSPs are received at different synapses, then they will combine by what is termed **spatial** summation. The net effect of all the excitatory PSPs and inhibitory PSPs manifests at the axonal hillock. This is the area of the cell body from which the axon extends. If, here, the threshold is exceeded then an action potential is generated in the post-synaptic cell and is then transmitted along the axon to the next receiving neurons. The axonal hillock serves as a type of gateway between the signals received on the dendrites and cell body, and the axon along which an action potential is transmitted.

Key points

Transmission of nerve signals involves two processes: conduction of electrical impulses throughout the neuron, and communication between adjacent cells via synaptic transmission.

SELF-CHECK 5.1

Which key properties of neurons allows them to combine to form a nervous system?

SELF-CHECK 5.2

What two main types of synapse are there and how do they differ?

5.1.5 Neurotransmitters

Neurotransmitters are synthesized from their precursors in the cell body and transported to synapses in membrane-bound vesicles via the microtubule axonal transport systems. A range of neurotransmitters exists (Table 5.1); many of them are relatively simple molecules (e.g. amino acids and biogenic amines), but there are also polypeptides. The most abundant within the brain are the amino acids glutamate and gamma amino butyric acid (GABA), but many

TABLE 5.1 Examples of some important neurotransmitters, showing their distribution within the nervous system.

Neurotransmitter	Distribution
Acetylcholine	Neuromuscular junction pre- and post-ganglionic parasympathetic autonomic system, preganglionic sympathetic
Catecholamines:	
Noradrenalin	Post-ganglionic sympathetic
Dopamine	Hypothalamus and substantia nigra
Serotonin	Raphe nuclei of brainstem
Amino acids:	
Glycine	Spinal cord
Glutamate	Widespread in CNS
GABA	Widespread in CNS
Peptides:	
β-endorphin	Hypothalamus

CLINICAL CORRELATION

Myasthenia gravis

This is an autoimmune disorder characterized by muscle weakness. It is caused by autoantibodies to acetylcholine receptors on the post-synaptic neuromuscular junction. These antibodies block the receptors and thus muscle function is impaired.

---

others, including acetylcholine, 5-HT (serotonin), dopamine and its derivatives, nitrous oxide, somatostatin and a range of amino acids, are present. Many neurotransmitters are predominantly or exclusively specific to discrete systems, including systems within the PNS.

Neurotransmitters are often described as inhibitory or excitatory, but strictly speaking the effect does not depend on the neurotransmitter but on its receptor. For example, receptors for glutamate, the most abundant neurotransmitter in the CNS, always generate an excitatory effect upon binding, whereas acetylcholine transmission commonly gives rise both to excitatory and inhibitory effects, depending on the type of receptor to which it binds.

5.1.6 Response of neurons to injury

Neurons have limited metabolic resources and high energy requirements, thus they are extremely vulnerable to any disruption to blood flow, or to low levels of oxygen or glucose in the blood, as may occur with respiratory disease or hypoglycaemia. Even a brief disturbance can lead to neuronal injury, while adjacent glial cells and vascular cells may survive. The first effect of hypoxic or ischaemic injury is failure of the membrane ion pumps, as the energy required to power them is no longer provided. Initially there is swelling of membrane-bound organelles (e.g. mitochondria and endoplasmic reticulum), and changes at this stage may be reversible. Later there is shrinkage of the nucleus with condensation of the chromatin before the nucleus disappears, and the cytoplasm becomes more intensely eosinophilic. Neuronal death occurs through necrosis and apoptosis. In the adult, neurons are unable to divide to replace dead cells.

CLINICAL CORRELATION

Neurodegenerative disease

The term neurodegenerative disease is used to describe a wide range of diseases in which there is a progressive and irreversible loss of neuronal function. Such diseases are ultimately fatal. In many cases there is a genetic basis. A common feature is disruption of normal protein pathways with accumulation of proteins, sometimes abnormal, perhaps through an inability to remove them by the usual means. Although apparent clinically, definitive classification of neurodegenerative disease usually only occurs after post-mortem neuropathological examination. Some of the more prominent pathological features are described below.

Alzheimer's disease A progressive degenerative disease characterized by dementia. Tau, a normally soluble microtubule-binding protein, is rendered insoluble by hyperphosphorylation and forms aggregates known as neurofibrillary tangles within neuronal cell bodies. These are demonstrated histologically as characteristically flame-shaped inclusions occupying the entire cell body, except for the nucleus. Accumulation of extracellular amyloid plaques is also a characteristic feature, which can be demonstrated histologically using Congo red stain.

Pick's disease This is a severe, progressive disease, clinically similar to Alzheimer's, but much rarer. There is severe atrophy of the frontal and temporal lobes. There is disorganization of several proteins, including tau, in the neurons, which can be demonstrated histologically as Pick bodies by staining with antibodies to tau or by silver techniques. Unlike neurofibrillary tangles, Pick bodies are spheroids and cause distension of the cell body.

Lewy body disease Lewy bodies are spherical, eosinophilic, intracytoplasmic inclusions comprising abnormal NFP, α-synuclein and ubiquitin. Thus, Lewy bodies can be demonstrated by immunohistochemistry (IHC) using antibodies to any of these proteins, and also by special staining techniques. One type is *Parkinson's disease*, where a region of the brain stem is affected, and hence movement disorders are seen.

Huntington's disease Huntington's disease is characterized by movement disorder with progression to dementia. There is severe loss of neurons in parts of the cerebrum, with astrocytosis. It has an autosomal dominant genetic inheritance. Affected individuals have an increased number of CAG trinucleotide repeats in the gene for the protein huntingtin. The disease exhibits the phenomenon of anticipation, whereby the number of repeats increases from generation to generation, with a corresponding earlier age of onset. The function of the huntingtin protein is unknown.

Creutzfeldt-Jakob disease This is one of a number of so-called transmissible spongiform encephalopathies that can affect humans. The causative agent is actually a structural isoform of a normally occurring cell-surface protein, the prion protein. This abnormal form can induce conformational change of the normal form. The disease is characterized pathologically by vacuolation of the cortex and the presence of amyloid plaques. The abnormal protein can be demonstrated by IHC.

Motor neuron disease Also known as amyotrophic lateral sclerosis, the disease shows loss of motor neurons, especially anterior horn cells of the spinal cord. This leads to muscle atrophy and there is sclerosis and astrocytosis of the lateral columns of the spinal cord. Most cases are sporadic, but around ten per cent are familial. Surviving neurons show ubiquitin-rich inclusion bodies.

5.2 Glial cells

Besides neurons, nervous tissue comprises many other cells that provide structural and physiological support. Glial (from the Greek for glue) cells provide a supporting function to the neurons in the CNS. They are derived embryologically from the same neural stem cells as neurons. There are around ten times as many glia as neurons in the CNS. Despite the collective name, glia comprise several quite distinct cell types that have different structures and functions.

Key points
Neurons are supported by specialised glial cells that protect from infection and toxins by modification of their blood supply.

5.2.1 Astrocytes

Astrocytes are characterized by their stellate shape as a result of the numerous processes radiating from the cell body. They are rich in the intermediate filament glial fibrillary acidic

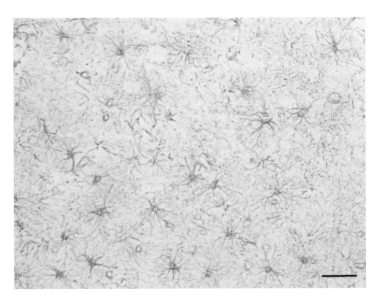

FIGURE 5.6

Brain tissue showing astrocytes. Note the numerous processes. (GFAP immunostaining; Scale bar: 100 µm).

Grey and white matter

These are commonly used terms based on macroscopic appearance. Grey matter is neuron rich. White matter lacks neurons, but has an abundance of myelin coated axons giving it the whiter colour.

Cross reference

Further information on demonstration techniques may be found in the *Histopathology* volume in this Fundamentals of Biomedical Science series.

protein (GFAP), which means they can be demonstrated easily by IHC (Figure 5.6). They provide numerous functions, including biochemical and structural support to the neurons, and can be divided into two major types: protoplasmic and fibrillary. Protoplasmic astrocytes are found in the **grey matter** where their processes help to enclose synapses. Fibrillary astrocytes are found in **white matter**, where their processes help to enclose the nodes of Ranvier.

Astrocytes provide metabolites for neurons and are able to take up excess neurotransmitters for breakdown. They play a role in the modulation of myelination by oligodendrocytes. Astrocytes communicate with one another via cytokines and carry receptors for a number of neurotransmitters that mediate their response to stimuli. Astrocytes surrounding blood vessels contribute to the **blood-brain barrier** that protects the CNS from bloodborne infections and toxins.

Both types of astrocyte play an important pathological role. Injury to neural tissue (e.g. an infarct) results in neuronal cell death. The necrotic tissue is removed by phagocytosis, but neurons are unable to regenerate to replace the lost tissue. Instead, astrocytes are stimulated to proliferate, increasing the number of processes and production of intermediate filament GFAP. The reactive astrocytes form a region of 'gliotic' scar tissue in the remaining space. This helps to restore the blood-brain barrier around the site of injury and promote revascularization of the tissue. However, it does prevent axonal regrowth from surrounding uninjured neurons and the formation of new connections across the injured area.

5.2.2 Oligodendrocytes

Oligodendrocytes are distinguished from astrocytes by their smaller size, with smaller, more condensed nuclei. They also lack the intermediate filament GFAP. Like astrocytes though, they bear numerous processes. Their major function is to provide a sheath of myelin around the neuronal axons. Processes from the oligodendrocyte cell body wrap around portions of the neuronal axons to form the myelin sheath. It is the presence of myelin that gives the white

BOX 5.1 *Blood-brain barrier*

Unlike much of the vascular system, the endothelial cells of the capillaries in the brain are not fenestrated and are joined by occluding (tight) junctions that prevent the passage of blood constituents out of the capillary. The capillaries have relatively few pinocytotic vesicles and are surrounded by a basement membrane upon which lie astrocytic foot processes that join to form a continuous layer around the vessel. Together, these features form what is known as the blood-brain barrier. The endothelial cells have systems for the active transport of basic metabolites such as glucose and oxygen, but passage of many other substances between the vascular system and the intercellular spaces of nervous tissue is greatly restricted. Not all of the brain lies 'behind' the blood-brain barrier, as the anterior pituitary, for example, has a conventional vasculature. Furthermore, the blood-brain barrier has important implications, for example, in the administration of therapeutic reagents to the brain.

matter of the brain its characteristic appearance, distinct from the grey matter. Hence they occur mostly in the white matter. See Figure 5.7.

In the PNS, Schwann cells have an analogous function. We will examine the process of myelination of axons later.

SELF-CHECK 5.3

What is the role of myelin in the nervous system?

SELF-CHECK 5.4

How does the vascular endothelium in the CNS differ from that of most other tissues?

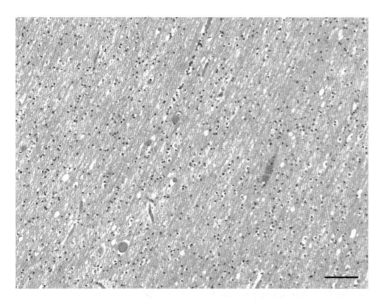

FIGURE 5.7

Cerebral white matter. There are no neurons in this region. There are abundant oligodendrocytes exhibiting small, darkly stained nuclei (H&E stain; Scale bar: 200 μm).

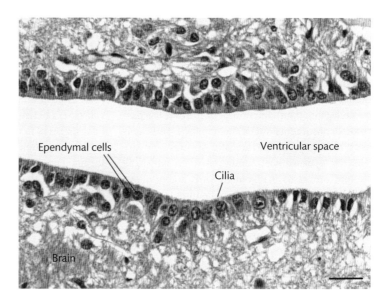

Ependymal cells

Ventricular space

Cilia

Brain

FIGURE 5.8

Ependymal cells lining part of the ventricular system. Note the basal processes extending into the neuropil, and the numerous apical cilia (H&E stain; Scale bar: 50 μm).

5.2.3 Ependyma

Ependymal cells line the central lumen of the spinal cord and the greater part of the ventricles in the brain. They are short columnar cells and form a tissue structurally similar to epithelium seen in many organs outside the CNS, except that there is no basement membrane (Figure 5.8). Ependymal processes extend deep into the nervous tissue on which they sit. The apical surface is ciliated and the action of the cilia assists the circulation of cerebrospinal fluid (CSF; Box 5.2) through the ventricular system.The ependyma provides a selective barrier between nervous tissue and CSF.

BOX 5.2 *Cerebrospinal fluid*

Cerebrospinal fluid (CSF) is normally a clear, colourless fluid containing glucose, electrolytes and a small amount of protein. The brain effectively 'floats' in a bath of CSF and consequently this reduces its effective weight and provides protection from trauma in the event of sudden movement. The rate of CSF production is normally around 500 mL/day and its volume is turned over several times a day. It drains back into the bloodstream via sinuses in the subarachnoid space. This facilitates the removal of metabolites from the CNS.

Laboratory analysis of CSF plays an important role in the investigation of neurological disorders, as a sample can be obtained relatively easily compared with neurosurgical biopsy. Blood breakdown products can be detected as an indication of subarachnoid haemorrhage. Microbiological analysis is used to confirm a diagnosis of meningitis. Non-infective inflammatory disorders can manifest microscopically as an increase in inflammatory cells. Cells from CNS tumours may sometimes metastasize into the CSF.

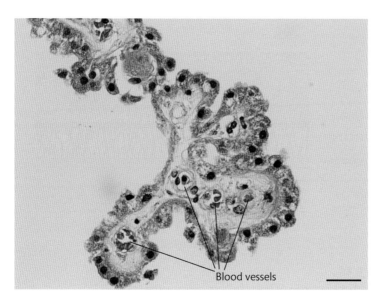

Blood vessels

FIGURE 5.9

A portion of choroid plexus. Note the abundance of blood vessels in the underlying stroma (H&E stain; Scale bar: 50 μm).

5.2.4 Choroid plexus

Some regions of the ventricles contain structures called choroid plexi (singular: plexus), and these are the sites of CSF production. They comprise modified ependymal cells that are more cuboidal in shape. Choroid plexi are highly folded structures with numerous microvilli on the apical surface to create a large surface area. Beneath the folds are networks of capillaries (Figure 5.9). Cerebrospinal fluid is produced in, and secreted from, the cells of the choroid plexi, which use their rich vascular supply as a source. The vascular endothelium of a choroid plexus is normally fenestrated (unlike most of the CNS; see blood-brain barrier, Box 5.1), hence there is availability of plasma in the extracellular space for uptake by the choroid plexus cells in the production of CSF.

5.2.5 Microglia

Microglia are immune cells which do not have a neural origin. They are in fact part of the monocyte/macrophage system and are formed in the bone marrow from haematopoietic stem cells. After migration to the CNS, they differentiate into microglia that have elongated nuclei and numerous processes. Injury to the CNS tissue causes activation and proliferation of microglia, whereupon they assume a phagocytic role and act as antigen presenting cells to T lymphocytes. Additionally, they have an important role in neurodegenerative disease.

Cross reference

See more about immunity in Chapter 4.

5.2.6 Meninges

The brain and spinal cord are surrounded by three membranous layers of non-nervous tissue. Outermost is the dura, which is the thickest layer, and is composed of tough fibrocollagenous

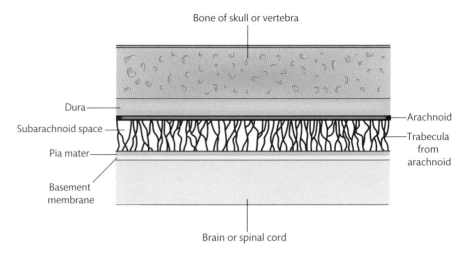

FIGURE 5.10
Schematic representation of the meninges covering the CNS.

tissue. It adheres tightly to the skull, but less so to the vertebrae. The internal aspect is covered by simple flattened cells. Beneath the dura is the arachnoid, which is a thinner layer of fibrocollagenous tissue. Beneath the arachnoid is a gap, the subarachnoid space, which is filled with circulating CSF and carries blood vessels. Projections (trabeculae) from the arachnoid extend deep across the subarachnoid space to connect to the third and thinnest layer, the pia. The pia follows the contours of the brain closely and sits on a basement membrane that is abutted by the astrocytic foot processes beneath (Figure 5.10).

Numerous venous sinuses are present within the dura. Outgrowths of the arachnoid, known as granulations, project into these sinuses, providing a route for drainage of CSF into the venous circulation.

5.3 Cells and structures of the peripheral nervous system

The peripheral nervous system (PNS) is the name given to the connections between the brain and spinal cord and the muscles and sensory endings.

5.3.1 Mature nerve trunk

Ganglia
Collections of neurons and support cells lying outside the CNS.

Nerve trunks of the PNS consist of several fascicles (Figure 5.11). Each contains bundles of axons that are extensions of neurons in the CNS or dorsal root **ganglia** outside the PNS. The axons are ensheathed by Schwann cells and surrounded by a tube of modified fibroblasts, the perineurium, that isolate them from surrounding tissues. Fascicles are supported and connected by the epineurium. This is composed of fibrous collagen and contains the blood vessels supplying the nerves and lymph vessels. A nerve may be recognized with the naked eye by its white, glistening appearance crossed by diagonal dark bands.

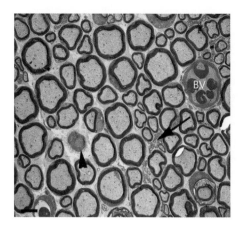

FIGURE 5.11

Part of a peripheral nerve fascicle showing myelinated fibres, unmyelinated axons (arrow) and a small blood vessel (BV). One of the myelinated fibres is sectioned through the node of Ranvier (arrowhead). Scale bar: 5 µm.

Peripheral nerves connect the sensory and motor endings with the CNS. Sensory neurons in the dorsal root ganglia are ensheathed by capsule cells that are modified Schwann cells. Rarely, these cells make a thin myelin sheath around the neuronal cell body (e.g. ciliary ganglia in mice [not man]).

5.3.2 Schwann cells

Schwann cells are the glial cells of the peripheral nerves, and all axons have a closely associated Schwann cell. In myelinated fibres (see section 5.3.6) the Schwann cell cytoplasm is not evenly distributed along the internode. A large part of it occurs in the nuclear region in the middle of the internode, which is the site of the Golgi apparatus and endoplasmic reticulum. Schwann cell mitochondria lie mainly in pockets of cytoplasm adjacent to the node. Connecting the nuclear and paranodal cytoplasmic regions are narrow ribbons or canaliculi that run along the outer surface of the myelin sheath. Between them the myelin is only separated from extracellular space by Schwann cell basal lamina. The outer surface of all Schwann cells has a basal laminal coating composed of collagen V and VI, laminin and glycosaminoglycans. The cytoplasm also contains lysosomes, which, in non-myelinating Schwann cells, are small, dense bodies. In myelinating Schwann cells they have a lamellar component and are known as Reich granules. In abnormal situations there may be other types of lipid deposit.

5.3.3 Fibroblasts

Small numbers of fibroblasts lie in the endoneurium and produce the fibrous collagen that provides mechanical support to the nerve fibres. Often there are small patches of basal laminal-like material adjacent to the cell membrane; these are useful for discriminating fibroblasts from lymphocytes and macrophages.

5.3.4 Perineurial cells

Perineurial cells are derived from fibroblasts but are closely connected and form a tubular sheath consisting of several laminae of cells (Figure 5.12), bounded by basal laminal coating and separated by fibrillar collagen, around the endoneurium. The basal lamina differs from

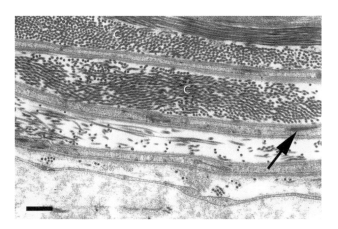

FIGURE 5.12
Several layers of the perineurium. The denser regions on the cell membrane (arrow) indicate attachment plaques of bundles of actin fibrils that give the perineurium its contractile properties. Collagen fibrils (C) separate the cells. Scale bar: 0.5 μm.

that around Schwann cells in being thicker and lacking a lamina lucida, except in the innermost and outermost layers. The function of the perineurium is to isolate the endoneurium from the epineurium. Transport across the perineurium is by means of pinocytotic vesicles. This is a non-specific process by which invaginations of the cell membrane engulf material or particles in the extracellular space. These then close and become vesicles lying within the cytoplasm. The contents may be ingested into the cell or transported across the cell.

At the distal end of the nerve, the perineurium terminates before reaching the skin, leaving a short length of nerve unprotected. At the proximal end, the perineurium merges with the meninges, resulting in a continuous pathway subperineurially from the subarachnoid space to the nerve ending. In capsulated endings, the perineurium forms the outer layers of the capsule. The motor nerves end as specialized structures, called **motor end plates**, on the muscle fibres. The perineurium terminates before the nerve contacts the muscle, and the final, unprotected Schwann cell has slightly different properties from the majority of the Schwann cells in the nerve trunk.

5.3.5 Blood vessels

The blood vessels in the endoneurium are small and have a sleeve of perineurial cells where they enter the perineurium; this does not continue far into the endoneurium. Pericytes surrounding endothelial cells have the morphological characteristics of smooth muscle.

5.3.6 Myelinated fibres

Naked-eye examination of fresh peripheral nerve shows a white, shiny structure with darker cross-hatching. Myelin is almost runny and is very easily damaged mechanically. In the PNS, axons with a diameter greater than 2–3 μm are normally myelinated (slightly larger than in the CNS). Although the presence of a myelin sheath increases conduction velocity, it also makes the fibre more vulnerable to injury and disease. Hence, small unmyelinated fibres function to facilitate touch and the sensation of pain and temperature, where conduction speed is less important and robustness more vital. Myelin thickness is directly related to axon diameter, and in a large myelinated axon with a diameter of 10 μm there may be as many as 120 lamellae in the myelin sheath. Fibre size (axon plus myelin) is related to conduction velocity and also to function: the largest fibres are sensory afferents to the muscle spindles, followed by motor fibres, with the smallest myelinated fibres and those that remain unmyelinated being sensory nerve fibres. The Schwann cell nucleus is at the midpoint of the myelin sheath on the outer surface of the sheath. The myelin sheath remains connected to the Schwann cell plasma membranes by extensions of these membranes called the inner and outer mesaxons (see box 5.3).

BOX 5.3 Myelination

We mentioned myelin earlier when considering the structure of neurons and the transmission of nerve impulses along axons. Axons both in the CNS and the PNS may or may not be myelinated, depending on their location and the function of the neuron. Myelination of axons in the CNS is performed by oligodendrocytes, and in the PNS by Schwann cells; the process is very similar in both cases. One major difference, however, is that a single oligodendrocyte is involved in the myelination of several axons, whereas a Schwann cell will only myelinate a single axon. This has implications for recovery after demyelination or axonal damage, as the simpler topographical relationship means that it is possible for Schwann cells to multiply and remyelinate demyelinated or regenerative axon sprouts, whereas this is not possible with oligodendroglia.

During development, as axons increase in diameter, oligodendroglia (in the CNS) or Schwann cells (in the PNS) divide (Figure 5.13). In the CNS, each oligodendocyte sends out

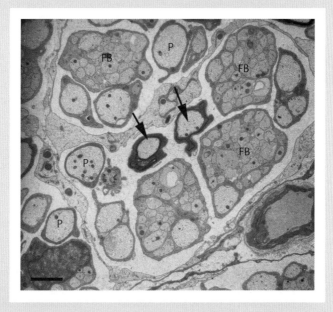

FIGURE 5.13
By a gestational age of 23 weeks, nerves are transforming from fetal bundles of axons (FB) via promyelinated fibres (P) to thinly myelinated fibres (arrows). Scale bar: 2 μm.

several processes. As each contacts an axon, it rotates about it to form numerous layers of cell processes. These compact, the cytoplasm is eliminated and additional proteins are incorporated to form the mature myelin sheath. In the PNS, individual axons about 2 μm in diameter come to lie individually in separate Schwann cells. Once this one-to-one relationship is established, the Schwann cell process rotates around the axons and myelin is formed by the compaction of these membranes, as in the CNS. Each Schwann cell myelinates about 200 μm of the axon (the internode). As the fetus grows and the axon diameter increases, the internodal length also increases so that it is directly related to the axon diameter. In the adult CNS the smallest axons to be myelinated are noticeably smaller than in the PNS; the largest nerve fibres have an axon diameter of about 10 μm and an internodal length of about 1500 nm.

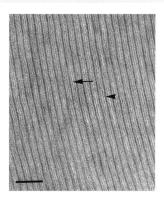

FIGURE 5.14
Peripheral nerve myelin; arrow marks the intermediate line and arrowhead the major dense line. Scale bar: 50 nm.

The resultant myelin sheath is composed of alternating layers of lipid and protein. In suitably fixed and stained nerves, this is seen as dense and less-dense (intermediate) lines, the dense lines being derived from compaction of the inner (cytoplasmic) surfaces of the plasma membrane, and the interperiod (less-dense) line from the outer surfaces (Figure 5.14). The periodicity in fresh nerves is about 17 nm in the PNS and about 14 nm in the CNS. The node of Ranvier is the region where two cells abut. In the CNS, this region is covered by astrocytic processes, while in the CNS there is a ring of villous-like Schwann cell processes covered by an extension of the Schwann cell basal laminal sheath. Loss or failure of myelination is an important cause of nervous system pathology.

5.3.7 Node of Ranvier

The node of Ranvier (Figure 5.15) is the electrically active region. Adjacent Schwann cells terminate in interdigitating villous-like processes that turn to end close to the bare axon. This region is covered by a basal laminal tube extending from each Schwann cell, forming a continuous sheath from the cell body or spinal cord, right to the fibre ending. The sodium channels are in the axolemma at the node.

Adjacent to the node is the region of myelin termination, the paranode, where each lamella opens up to enclose a small pocket of cytoplasm, before terminating adjacent to the axon. These terminations are attached to the axolemma by specific proteins. Further away from the node is the juxtaparanode, where the potassium channels are located and myelin finally becomes compacted; most of the Schwann cell mitochondria are in this region. The degree of compaction in the paranode varies in some pathological conditions and may fail to form at all (uncompacted myelin) or may only form the major dense line, with the outer surfaces of the Schwann cell being further apart than normal (widely spaced myelin). Myelin-associated glycoprotein, the protein associated with myelin compaction, is found in this region.

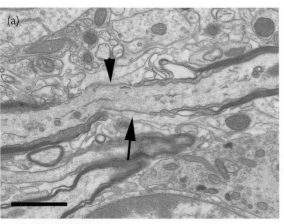
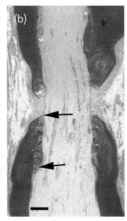
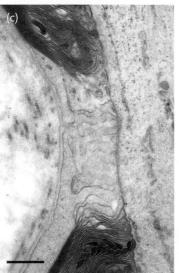
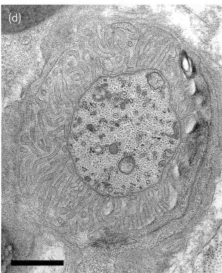

FIGURE 5.15

(a) A CNS node on a small myelinated fibre in the CNS (arrowhead). The nodal gap is covered by an astrocyte process (arrow). There is no connective tissue in the CNS. Scale bar: 1 μm. (b) In the PNS, the nodal gap is covered by Schwann cell nodal processes that extend from either side of the node and approach the nodal membrane very closely. In a small fibre, the myelin lamellae terminate consecutively in the paranode in a similar way as in Figure 5.14a. In larger fibres the terminations are more complex and the paranodes (between the arrows) relatively shorter. Collagen fibrils support and separate the nerve fibres. Scale bar: 2 μm. (c) The nodal gap is covered by Schwann cell microvillae and the basal laminal sheath. The actual gap is about 1 μm, regardless of axon diameter. The number of microvillae increases with the diameter of the fibre. Scale bar: 0.5 μm. (d) In cross-section the close apposition of SC microvillae to the nodal axolemma is clearly seen. The axoplasm is denser in this region due to the dephosphorylation of the neurofilaments that allows them to pack more closely together. Scale bar: 1 μm.

5.3.8 Schmidt–Lanterman incisures

Schmidt–Lanterman incisures (Figure 5.16) are conical segments in the myelin sheath where the lamellae are uncompacted and there is a helical cytoplasmic pathway running through the sheath. Its function is probably to provide metabolic support by connecting the abaxonal Schwann cell (outside the myelin) to the adaxonal (adjacent to the axon) Schwann cell.

5.3.9 Mature nerve trunk

Nerve trunks of the PNS consist of several fascicles and can be recognized by the naked eye by a white, glistening appearance crossed by diagonal dark bands. Each nerve fascicle contains bundles of axons that are extensions of neurons in the CNS or dorsal root ganglia outside the PNS. The axons are enwrapped by Schwann cells and surrounded by a tube of modified fibroblasts, the perineurium, that isolates them from surrounding tissues. Fascicles are supported and connected by the epineurium, which is composed of fibrous collagen and contains the blood vessels supplying the nerves. There are also small lymph vessels in the epineurium.

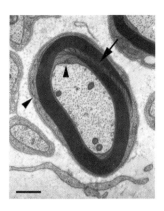

FIGURE 5.16
Transverse section of a small myelinated fibre where the compact myelin structure opens up to form a Schmidt–Lanterman incisure (arrow). The inner and outer mesaxons (respectively) that connect the myelin lamellae to the Schwann cell plasma membrane are indicated by arrowheads. Scale bar: 0.5 μm.

5.3.10 Unmyelinated fibres

Small axons with a diameter less than about 2 μm do not acquire a myelin sheath and do not have the same relationship with their supporting Schwann cells. Instead, between one and four axons associated with one Schwann cell split up and travel independently to other Schwann cells so there is not a chain of Schwann cells as with myelinated fibres but rather adjacent Schwann cells with axons weaving their way between them. It is convenient to call a Schwann cell and its associated unmyelinated axons a Remak fibre (Figure 5.17).

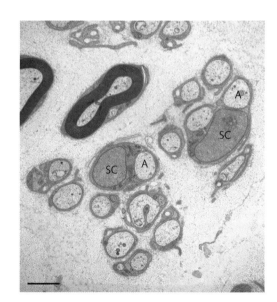

FIGURE 5.17
Normal Remak fibres. Small axons (less than 3μm in diameter) (A) are partially or wholly embedded in Schwann cell cytoplasm (SC). Scale bar: 2 μm.

5.4 Structural organization of the nervous system

Now that we have studied the major functional cell types in the nervous system, it is convenient to consider briefly the structure and organization within its main regions. The formation of the CNS begins around the third week of embryonic development. The ectoderm layer invaginates to form the neural tube. This will give rise to the brain and spinal cord. Cells

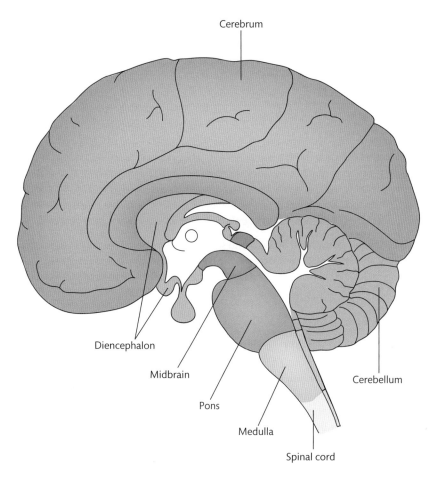

Cerebrum

Diencephalon

Midbrain

Pons

Medulla

Cerebellum

Spinal cord

FIGURE 5.18
Schematic representation of some of the major regions of the CNS.

peripheral to the neural tube, the neural crest, migrate laterally and give rise to the PNS, Schwann cells, ganglia and meninges. Three vesicles develop at the rostral (head) end of the neural tube, corresponding to the forebrain, midbrain and hindbrain. The forebrain develops into the **diencephalon** and the **cerebrum**, the hindbrain develops into the **pons**, the **medulla oblongata** (often just called the medulla) and the **cerebellum** (Figure 5.18). The midbrain, pons and medulla are collectively known as the **brainstem**. The spinal cord and brainstem comprise the CNS component of the autonomic nervous system. A system of interconnected ventricles runs deep within the CNS, from the cerebrum to the spinal canal, and is permeated by circulating CSF.

The evolution of brain development can be seen by the comparison of different organisms. The acquisition of spinal cord, hindbrain, midbrain and forebrain, respectively, represents an evolutionary pathway from the most primitive animals to the most advanced, and corresponds to the spectrum of functions from basic life support through to higher cognitive functions.

The spinal cord is involved in reflex actions and basic locomotion, as well as providing a conduit between the PNS and the brain. It is a feature common to all vertebrates. The midbrain is involved with vision, hearing and arousal. The forebrain is involved with voluntary movement and higher functions (e.g. memory, language, reason and perception), many of which are not seen in the lower animals. It must be noted, however, that this description is a great over-simplification, and most functions involve interaction between the different regions.

5.4.1 Cerebrum

In humans, the cerebrum accounts for the majority (around 80 per cent) of the brain mass. It is separated sagitally into two hemispheres, connected by a thick band of white matter called the corpus callosum. Each hemisphere comprises four lobes—frontal, temporal, parietal and occipital. In section, the cerebrum is divisible macroscopically into two well-demarcated regions—the outer grey matter (cortex) and the inner white matter. The white matter gains its appearance by an abundance of myelin, which is only minimally present in the grey matter. There are no neuronal cell bodies in the white matter, only their processes and associated glial cells.

The cortex is highly convoluted (folded) into sulci (peaks) and gyri (troughs) in order to maximize the available volume. Microscopically, the cortex can be seen to contain layers of neurons morphologically different from layer to layer. The meshwork of intertwining processes and the extracellular space in which they lay is termed the neuropil (Figure 5.19).

5.4.2 Diencephalon

The diencephalon is the region of the forebrain beneath the cerebrum that represents the earliest evolution of the forebrain. It includes the hypothalamus, posterior pituitary and pineal. Attached to the posterior pituitary is the anterior pituitary, which is an important endocrine gland. We will consider these structures and their function in more detail later in the chapter.

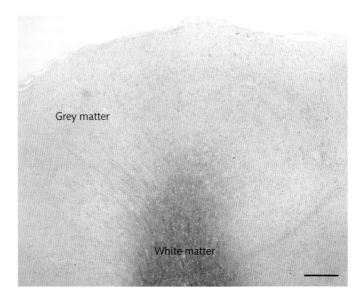

FIGURE 5.19
Photomicrograph showing section through the cerebral cortex, along with underlying white matter. Numerous neuronal cell bodies (stained red) can be seen in the cortex of the cerebral cortex. Myelin is stained dark blue and this permits demonstration of fibres within the white matter (Luxol fast blue/neutral red stain; Scale bar: 400 µm).

Brain stem death

Such is the importance of the brainstem to vital functions that UK law permits certification of death by demonstrating a state known as brain stem death. In comatosed patients, being kept alive only by artificial support, the diagnosis is made by rigorously establishing the absence of all brain stem reflexes.

5.4.3 Brainstem

The brainstem is the collective name for the midbrain, the pons and medulla, the last two being derived from the hindbrain. The brainstem is composed predominantly of tracts of white matter with occasional interspersed small groups of neurons, referred to as nuclei.

The brainstem is involved in the transfer of information between the PNS and the forebrain. All sensory information in, and motor information out, passes through the pons and medulla. The brainstem acts as a sensory gateway. One specific area, the reticular formation, acts as a sensory filter to block familiar or repetitive signals that might otherwise overload the forebrain. The medulla plays an important role in automatic life support and homeostatic functions (e.g. breathing, heart rate, swallowing, vomiting and digestion). The brainstem is also responsible for many reflex actions.

5.4.4 Cerebellum

Although the cerebellum (Latin: little brain) develops from the embryological hindbrain, it is not considered to be part of the brainstem. The cerebellum has a characteristic, easily recognizable structure. As in the cerebrum, there is an outer cortex of neuronal cell bodies lying over the inner white matter. However, the cerebellar cortex has a more distinct layering than that seen in the cerebrum. The outer **molecular** layer is grey matter that is rich in large neuronal cell bodies, with a row of very large, distinctive neurons called **Purkinje cells** lining the interface with the granular layer beneath. The **granular** layer is composed of very small, closely packed neurons with relatively little cytoplasm. Beneath this cortex lies white **matter** (Figure 5.20).

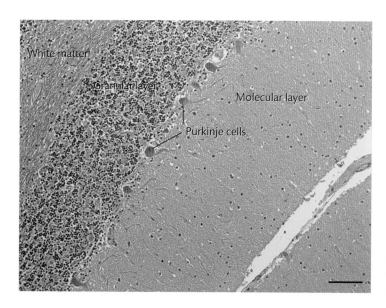

FIGURE 5.20
Cerebellum showing the different strata (H&E stain; Scale bar: 200 µm).

The cerebellum receives sensory information about the position of joints and the length of muscles. It assimilates this information to coordinate complex body movements that require multiple muscle actions at once or in strict sequence. It also receives and uses sensory visual and hearing information to facilitate hand-eye coordination.

5.4.5 Spinal cord and spinal nerves

The spinal cord (Figure 5.21) extends from the medulla in the brainstem down the vertebral column before tapering out in the sacral region. It has two clearly demarcated layers—white matter externally and grey matter internally surrounding the CSF-filled spinal canal. Note that this distribution is reversed compared to that found in the cerebrum. The grey matter approximates an H shape, forming a pair of anterior and a pair of posterior 'horns'.

The spinal nerves of the PNS exit the spinal cord in left/right pairs throughout its length. Efferent fibres arising from neurons in the anterior horn exit the cord via the ventral roots. Afferent fibres enter the cord via the dorsal roots, terminating in the posterior horn. Outside the vertebrae, the afferent and efferent fibres converge to form mixed spinal nerves. We will examine this arrangement in more detail shortly. The spinal cord is described anatomically by levels, in the same way as the corresponding vertebrae of the spine. Thus, there are eight pairs of cervical nerves, 12 thoracic, five lumbar and five sacral. The nerves serving the lower limbs arise from the lumbosacral plexus formed from numerous interconnected branchings of the distal spinal nerves.

The spinal cord carries two main pathways. First, it connects the PNS to the brain. This allows sensory information to reach the brain, and for the conduction of voluntary motor signals. Second, spinal cord neurons also communicate laterally with sensory and motor structures in the body to provide the mechanism for reflex actions that do not involve the brain. We will consider these pathways a little later, as part of the PNS.

Following development of CNS neurons and their organization into brain and spinal cord, motor axons in the ventral horn of the spinal cord become collected into bundles, leaving

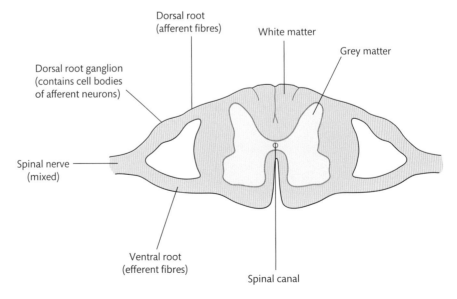

FIGURE 5.21
Diagram of the spinal cord showing arrangement of the nerve roots.

the cord proper at regular intervals between the vertebrae to form the ventral roots. Groups of sensory neurons come to lie adjacent to the cord and between the dura mater and the arachnoid.

The dorsal root ganglion cells are pseudounipolar neurons; the centrally directed axons enter the cord via the dorsal roots and the peripherally directed ones join the bundles of motor axons to form the peripheral nerves. Nerves in the thorax follow the ribs, while lumbar and sacral nerves follow the developing limbs.

SELF-CHECK 5.5

The ventral and dorsal spinal nerve roots carry fibres in which directions?

5.4.6 Cranial nerves

The cranial nerves are so called because they arise from the CNS within the cranium (skull). They arise as left-right pairs, but, in contrast to the spinal nerves, they do not have separate dorsal and ventral roots. They provide motor and sensory connections to the structures of the head and neck. All the cranial nerves (except I and II) provide pathways for the **somatic nervous system**; some also provide pathways for the **autonomic nervous system**. We will look at these systems in the next section.

Cranial nerves I and II are in fact extensions of the CNS and are composed of the same cells, hence they are not part of the PNS. However, it is convenient to consider them along with the other cranial nerves.

TABLE 5.2 Cranial nerves.

No.	Name	Arises from	Type	Functional connectivity
I	Olfactory	Cerebrum	Sensory	Olfactory (smell) receptors in nasal cavity
II	Optic	Diencephalon	Sensory	Retina
III	Occulomotor	Midbrain	Motor	Eye muscles (external and pupillary dilation)
IV	Trochlear	Midbrain	Motor	Eye muscles (external)
V	Trigeminal	Pons	Both	Facial sensation and movement
VI	Abducens	Pons	Motor	Eye muscles (external)
VII	Facial	Pons	Both	Facial muscles, salivation and lacrimation and taste buds
VIII	Vestibulocochlear	Pons	Sensory	Hearing and movement sense in ear
IX	Glossopharyngeal	Medulla	Both	Taste buds and salivation
X	Vagus	Medulla	Both	Larynx and pharynx muscles and parasympathetic innervation to thoracic and abdominal organs
XI	Accessory	Part from medulla, part from spinal cord	Motor	Neck and shoulder muscles
XII	Hypoglossal	Medulla	Motor	Tongue and throat muscles

5.5 **Functional divisions of the peripheral nervous system**

The PNS connects the CNS to all the organs and tissues throughout the body. It provides pathways for their motor control and sensory uptake. Functionally, different pathways exist, and it is convenient to consider these separately.

5.5.1 **Somatic nervous system**

The somatic nervous system allows voluntary contraction of skeletal muscle. Skeletal muscles are innervated by fibres that originate from motor neurons in the anterior horns of the spinal cord. These neurons have long myelinated axons that terminate at the neuromuscular junctions. The axons exit the spinal cord via the ventral roots. Motor neurons are large multipolar cells. Sensory (afferent) fibres enter the spinal cord via the dorsal roots. These are pseudounipolar neurons, whose cell bodies reside in the dorsal root ganglia. The axons terminate in the posterior horns, providing sensory information about the state of the muscle. In the simplest arrangement, an interneuron links the sensory and motor neurons, but there are also many vertical connections to the higher CNS.

Not all skeletal muscle contractions are voluntary, as many movements occur by reflex (e.g. withdrawing the hand from a hot surface). This pathway connects pain receptors in the hand with the appropriate motor neurons in the spinal cord. These sensory neurons have their cell bodies in ganglia on the dorsal roots of the spinal cord (Figure 5.22). This reflex reaction occurs before a conscious awareness of the pain is felt, and happens entirely independently of the brain.

There are many other known reflexes (e.g. **knee jerk reflex**). The pathways are not all schematically identical; for example, in some cases the spinal interneuron step is omitted, whereby the sensory axon synapses directly with the motor neuron.

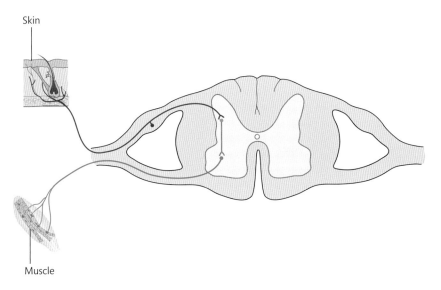

Skin

Muscle

FIGURE 5.22

Schematic representation of a spinal reflex pathway.

> **Key points**
>
> The somatic nervous system provides both voluntary and reflex control over skeletal muscle contraction.

5.5.2 Autonomic nervous system

The **autonomic nervous system** (ANS) is an important subsystem of the PNS that provides involuntary control of visceral and vascular smooth muscle, cardiac muscle and glandular secretions. Some examples of the effects of the ANS are given in Table 5.3. These actions are all involuntary and occur normally in response to the appropriate sensory input. It is this involuntary nature that gave rise to the name autonomic; however, it is important to recognize that the ANS does not function independently of the CNS. The autonomic **nuclei** are connected to, and are largely under the control of, the CNS.

There are two divisions to the ANS—**sympathetic** and **parasympathetic**. Where organs are innervated by both systems, the effects are generally opposing; for example, the sympathetic system raises heart rate, whereas the parasympathetic system decreases heart rate. The sympathetic system is generally regarded as functioning to prepare the body for activity, whereas the parasympathetic promotes routine physiological activities (Figure 5.23).

Anatomically, the system comprises sensory and motor fibres running between the target organ and nuclei in the brainstem or spinal cord via ganglia. The efferent fibres of the sympathetic division exit the CNS via the ventral roots of the thoracic and some of the lumbar spinal nerves. These are the preganglionic fibres and they run to large ganglia in the neck or abdomen (part of the sympathetic chain), where they synapse with the post-ganglionic fibres that run to the target organ. The preganglionic fibres are cholinergic (i.e. the neurotransmitter is acetylcholine). The post-ganglionic fibres are unmyelinated, and the neurotransmitter is noradrenalin. The sympathetic chain lies adjacent to the spine, permitting vertical communication, and houses some of the sympathetic ganglia.

The parasympathetic fibres exit the CNS via the III, VII, IX and X cranial nerves, and some of the sacral spinal nerves. The ganglia are situated close to, or within, the target organ. Hence, compared to the sympathetic division, the preganglionic fibres are generally longer and the postganglionic fibres shorter. Both pre- and post-ganglionic fibres are cholinergic.

> **Nuclei**
>
> In the context of the CNS, the term nuclei is used to describe discrete functional groups of neurons, rather than the chromosome-containing structures of cells.

TABLE 5.3 Examples of some of the actions of sympathetic and parasympathetic systems. It can be seen that the two systems often have opposing effects.

Target tissue	Sympathetic	Parasympathetic
Gastrointestinal tract	Inhibits peristalsis and contracts sphincter muscles	Promotes peristalsis and relaxes sphincter muscles, promotes digestive glandular secretion
Respiratory system	Predominantly relaxes bronchiolar smooth muscle	Contraction of bronchiolar smooth muscle
Heart	Increases cardiac output and heart rate	Decreases cardiac output and heart rate
Urinary system	Relaxes bladder muscle and contracts urethral sphincter	Contracts bladder muscle and relaxes urethral sphincter

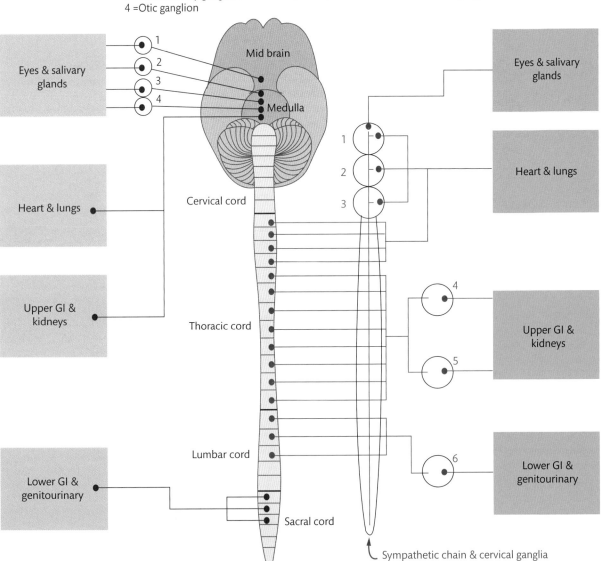

Parasympathetic

1= Ciliary ganglion
2= Sphenopalatine ganglion
3 = Submaxillary ganglion
4 =Otic ganglion

Sympathetic

1, 2, 3 = Superior, middle & inferior cervical ganglia
4 = Coeliac ganglion
5, 6 = Superior & inferior mesenteric ganglia

FIGURE 5.23
Simplified representation of the pathways of the autonomic nervous system. In reality, the arrangement of each is bilateral. The sympathetic chains lay adjacent to the vertebrae. Note that the parasympathetic ganglia are close to or within the target organ; those of the sympathetic are distant from the target organ.

The majority of autonomic fibres are small and unmyelinated; approximately 10 per cent are small, myelinated fibres.

Key points

The autonomic nervous system provides involuntary control over visceral smooth muscle and glandular secretions.

5.6 Sensory reception in the peripheral nervous system

The sensory fibres entering the brain and spinal cord carry impulses generated by the activation of receptors at their distal endings. These provide the CNS with information about the environment outside the nervous system, which is then processed and can be responded to with an appropriate motor action. You will immediately think of the five 'classical' senses (taste, touch, vision, hearing and smell), but there are also various other sensory receptors such as baroreceptors, which respond to changes in blood pressure, chemoreceptors, which respond to the levels of blood gases, proprioceptors, which respond to the stretch of muscles and tendons, and of course pain receptors. These specialized cells help the body to maintain its homeostatic environment, respond to insults, and generate sensations such as pleasure and hunger. We will look at some of the more important sensory receptor systems in the following paragraphs.

5.6.1 Olfactory reception

Olfactory receptor cells are bipolar sensory neurons that sit among the epithelial cells lining the roof of the nasal cavity. From the cell body, a single dendrite extends from which non-motile cilia project into the covering layer of mucus. Basally, the axon extends through the cranium and terminates within the **olfactory bulb**. The olfactory bulb is the distal portion of the olfactory nerve (cranial nerve I). Within the olfactory bulb, the axons make synaptic contact with the dendrites of **mitral cells**, from which axons extend to several regions of the forebrain (Figure 5.24).

The cilia of the olfactory neurons bear many receptors with affinity for odorant molecules that dissolve in the mucus. These receptors are coupled to a G-protein signal transduction system, and binding can lead to the opening of membrane cation channels and depolarization of the cell. If sufficient binding takes place, an action potential is generated. It is noteworthy that an olfactory neuron has only one type of receptor and thus can bind only one type of odorant molecule. However, the variety and number of olfactory neurons allows the detection of thousands of different odours.

5.6.2 Gustatory (taste) reception

Taste receptor cells are situated in taste buds, which are bulb-shaped structures recessed in the epithelium lining the oropharyngeal cavity. Taste buds occur throughout the oropharynx but are most abundant on the tongue. The taste bud narrows towards the apical surface, where

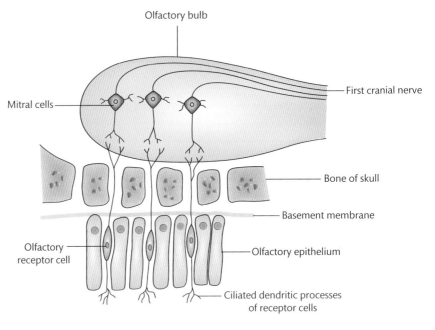

FIGURE 5.24

Diagram showing the arrangement of the olfactory receptor structures.

microvilli project from the taste receptor cells. The receptors themselves are located on these exposed microvilli. There are numerous receptor cells in a single taste bud. Also present are basal cells, which provide a regenerative potential, and the termini of axons projecting through the basement membrane, which synapse to the taste receptor cells.

There are five basic types of taste receptor, responding to sweet, sour, salty, bitter, and savoury, respectively, although there is a range of molecules that can be detected by each type of cell. For example, various simple carbohydrates will stimulate sweet receptors, as well as substances such as aspartame and saccharin. Ingested molecules make contact with the receptors on the microvilli of the taste receptor cells. Signal transduction occurs, resulting in depolarization of the receptor cell, and the release of neurotransmitter into the synaptic cleft. Innervation of taste buds is via the VII, IX and X cranial nerves.

5.6.3 Auditory reception

The auditory receptor cells are located in a region of the inner ear called the organ of Corti. These so-called **hair cells** are arranged in an epithelium and are tall columnar cells with numerous stereocilia projecting from the apical surface. The fluid covering this epithelial surface into which the stereocilia project is called endolymph. Basally, the hair cells synapse with nerve fibres of the VIII (vestibulocochlear) cranial nerve.

Sound waves from the middle ear are transferred through the endolymph and exert a mechanical force on the stereocilia. As the stereocilia move, ion channels open at their tips, causing an influx of potassium that depolarizes the hair cell. This results in an influx of calcium via voltage-gated channels that causes neurotransmitter release into the synaptic cleft.

5.6.4 Visual reception

Light reacts with the retina of the eye to produce a signal that travels along the optic nerve to the brain. There are eight layers in the retina (Figure 5.25) which is a very narrow strip covering the back of the eyeball. The light-sensitive region is at the back of the retina furthest from the pupil where light enters the eye. The optic nerve passes through the retina at the blind spot. Here, most of the myelin sheaths are lost and the naked axons continue to the outer edge of the retina where they turn to run centripetally for the most part, although some run centrifugally and end in the inner plexiform and inner nuclear layers. Under the **optic fibre layer** lies the **ganglion cell layer** consisting of a single layer of large ganglion cells. These send an axon into the optic fibre layer above, and numerous dendrites into the **inner plexiform layer** below. These ganglion cells vary in size; the dendrites of the smaller ones terminate almost immediately on the **inner plexiform layer**, whereas the larger ones send dendrites through almost to the inner nuclear layer. The **inner plexiform layer** is a complex of interlacing dendrites from the ganglion cells and the inner nuclear layer. The inner nuclear layer contains three types of neuron—bipolar, horizontal and amacrine. The amacrine neurons are unusual in not having an axon; they connect together the neurons of the ganglion cell layer. The inner processes of the inner nuclear layer bipolar cells contact bipolar neurons in the **outer nuclear layer**. These neurons are also connected by the horizontal cells that are multipolar neurons. The outer parts of the bipolar neurons lay in the **rod and cone layer**. These neurons are the receptor cells that contain specialized light-detecting apparatus—the rods and cones. Between the inner nuclear layer and the outer nuclear layer is the **outer plexiform layer** where processes from inner nuclear layer neurons terminate on the numerous fine ramifications of the outer nuclear cell layer neurons. The rods and cones contain the visual pigment. This is a protein called opsin and a vitamin A-derived chromophore called retinal. The combination, rhodopsin, reacts to a photon of light by isomerizing from the 11-*cis* form to a *trans* form, causing a conformational change resulting in bleaching. Visual pigment is in a cone-shaped, lamellated structure in the cone and a thinner rod-like structure in the rods. There are three different types of cone sensitive to different wavelengths of light and hence responsible for colour perception.

External to the rod and cone layer is the **pigment cell layer** that sends processes between the rods and cones. This layer absorbs light hitting it after passing through all the other layers. It lacks pigment in albinos and stray light interferes with the functions of the layers above.

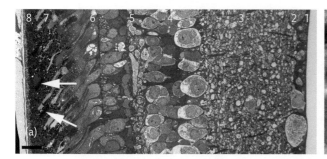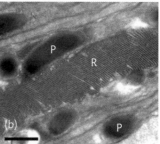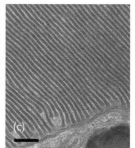

FIGURE 5.25

(a) Low-power view of the retina showing optic nerve fibre layer (1), ganglion cell layer (2), inner plexiform layer (3), inner nuclear layer (4), outer plexiform layer (5), outer nuclear layer (6), rod and cone layer (7) and pigment cell layer (8); arrows indicate rods. Scale bar: 10 µm. (b) Higher-power view of a rod (R) showing lamellated structure. Pigment granules are also shown (P). Scale bar: 0.5 µm. (c) Very high power of lamellations in rod. Scale bar: 0.2 µm.

5.6.5 Nerve endings

These endings permit the CNS to receive signals from the surface or interior of the body, or information about its position (proprioception). Endings may be either **free nerve endings** or **encapsulated endings**. **Motor end plates** are the means by which the CNS sends signals to the muscles.

5.6.6 Free nerve endings

Most sensory endings in the skin are of this type. They are derived from small unmyelinated axons. These lose their Schwann cell and basal laminal ensheathment on reaching the epidermis, and naked sprouts penetrate the layers of epidermal cells. These axons are specialized to respond to different stimuli (e.g. heat, cold or pain). There is a complex arrangement of sensory endings around hair follicles which could therefore be counted as a form of specialized nerve ending.

5.6.7 Encapsulated nerve endings

Pacinian corpuscles (Figure 5.26) are mechanoreceptors that detect movement, especially high-speed vibration. They are found in connective tissue in many parts of the body, including nerve trunks and muscles. There are large numbers in the finger tips and in the sole of the foot. The ovoid corpuscles consist of 30–50 concentric layers of cells separated by fibrillar collagen and centred on a single sensory myelinated axon, which may be as much as 1 mm long. Schwann cells form the inner layers of the capsule, and the outer layers are derived from fibroblasts. In the inner core, the myelin sheath is lost and the axon terminates in fine spines that

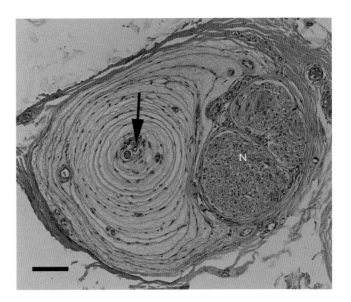

FIGURE 5.26

Pacinian corpuscle. The central axon is unmyelinated (arrow) and is surrounded by numerous concentric layers of cytoplasmic processes. There are three closely apposed sural nerve fascicles (N) (H&E stain; Scale bar: 50 μm).

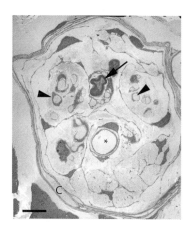

FIGURE 5.27
Golgi tendon organ showing capsule (C) derived from perineurium surrounding unstained fibrillar collagen in which are embedded Schwann cells that encircle collagen bundles (arrowheads). The myelinated fibre innervating the organ is shown (arrow). There is one small blood vessel (asterisk). Scale bar: 5 µm.

protrude between radial clefts of the capsule. In some sites (e.g. cat mesentery) the corpuscle is additionally innervated by unmyelinated axons, probably sympathetic ones.

Meissner's corpuscles are smaller encapsulated sensory endings 80–150 µm in length and 20–40 µm in diameter found in the dermal papillae of glabrous skin and in some mucous membranes. The sensory axon supplying the corpuscle loses its myelin sheath and forms several branches before entering the corpuscle, where they form a complex of interdigitating processes within a supporting Schwann cell. These terminate within this cell. There are 10–24/mm^2 of skin surface area. They respond to slower vibrations than do Pacinian corpuscles and have a smaller receptor field.

Golgi tendon organs (Figure 5.27) are spindle-shaped receptors about 1 mm long consisting of a thin capsule covering the complex terminal branchings of a sensory nerve. They are found most often at myotendinous junctions. The capsule contains numerous bundles of fibrous collagen that are continuous with the adjacent tendon, and also some extrafusal muscle fibres inserted between the collagen bundles. The nerve supply is via a single large myelinated fibre that loses its sheath on entering the tendon, and divides into small unmyelinated axon branches that are entwined with the collagen bundles. The endings are activated either by stretching or contraction of the associated muscle. Their function is as proprioceptors, giving information about the position of the body.

Muscle spindles (Figure 5.28) are also proprioceptors supplied by large myelinated fibres. These are encapsulated endings found in varying numbers in all striated muscles in man, with the greatest numbers found where fine control of the muscle is needed. They are ovoid and about 1.5 mm

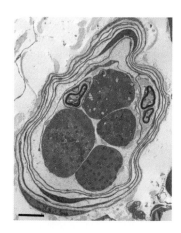

FIGURE 5.28
Muscle spindle showing capsule, three small myelinated nerve fibres and four intrafusal muscle fibres. Scale bar: 5 µm.

long and 0.5 mm wide. The capsule consists of several laminae of flattened fibroblasts connected by desmosomes, and contains several small muscle fibres running along its longer axis. These intrafusal muscle fibres (as opposed to the extrafusal fibres in the main part of the muscle) are shorter in length and narrower in diameter but run parallel to the extrafusal fibres. The intrafusal muscle fibres are of two types—nuclear bag fibres and nuclear chain fibres. The nuclear bag fibres are larger and have several nuclei in the central region, whereas nuclear chain fibres are smaller and have a longitudinal chain of peripherally placed nuclei. There are no microtubules in the equatorial region of either fibre type The innervation of muscle spindles is somewhat complex; primarily there is a large myelinated fibre that loses its sheath and gives rise to several unmyelinated branches that spiral around the intrafusal fibres, and are deeply invaginated into them without an intervening basal lamina. This is the primary sensory ending. In addition, there are smaller sensory fibres that terminate in spirals on the nuclear chain fibres. These are the secondary sensory endings. There are also small motor fibres that terminate on the polar regions of the intrafusal fibres. Finally, some large motor fibres innervating the adjacent extrafusal muscle fibres occasionally also give rise to a branch that terminates on an intrafusal fibre.

Muscle spindles provide a feedback loop enabling the CNS to control the contraction of the muscle.

Ruffini corpuscles are similar to Golgi tendon organs. They are found in joint capsules and the dermis of glabrous and hairy skin. The capsules have the fewest lamellae of the encapsulated receptors and in some the capsule is incomplete or missing. An elongated fusiform connective tissue capsule is penetrated by several bundles of fibrous collagen that are continuous with the collagen in the dermis. The supplying nerve is a single myelinated fibre that loses its myelin on entering the capsule and divides into several unmyelinated branches. These intertwine with the collagen bundles. These receptors respond to sustained pressure.

Merkel cells are specialized cells in the epidermis and hair follicles of mammalian skin that act as touch receptors in association with sensory endings. They are not encapsulated and can be distinguished from keratinocytes by their rounded shape and lack of tonofibrils. The associated sensory nerves lose their myelin sheath just before they contact the Merkel cell and terminate in numerous small unmyelinated branches, each of which innervates a cell by contacting it at a flattened ovoid end-bulb. The specialization of the ending resembles that of a synapse, and the Merkel cell contains numerous dense-cored vesicles.

5.6.8 Motor end plates

When motor nerves approach their muscle target, they lose their myelin sheath and split into several finer, unmyelinated extensions. These end in flattened, oval plates that lie close to the muscle, which is also altered in this region. The sarcoplasmic surface forms several deep indentations that have electron-dense undercoating. The axon terminals show numerous small vesicles that contain acetylcholine. When an impulse arrives at the ending, these vesicles are ejected and travel the small distance to the muscle surface. Here they merge with specialized receptors and instigate the process of muscle contraction (Figure 5.29).

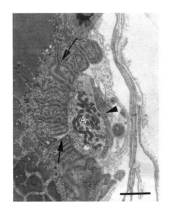

FIGURE 5.29

A motor axon (A) contacting a muscle at a motor end plate. The sarcolemma forms extensive folds (arrows). The Schwann cell (arrowhead) covering is very tenuous in this region. Scale bar: 2 μm.

Key points

The nervous system incorporates a wide range of sensory receptors to provide information about changes in the physiological environment in order that an appropriate response can be made.

5.7 Endocrine function and control in the central nervous system

The diencephalon region of the brain contains important endocrine structures.

5.7.1 Pituitary

The pituitary (also known as the hypophysis) is sited in a small recess in the base of the skull, the pituitary fossa. It is a bean-shaped structure, around 1 cm in length, and is enveloped by a meningeal capsule. It is divided into two distinct parts—the posterior and the anterior. The posterior (neurohypophysis) is connected directly to the brain by the pituitary stalk and is composed of nervous tissue. The anterior pituitary is separated from the posterior by a thin sheet of connective tissue. The anterior pituitary is not of neural origin but derived from epithelial tissue arising in the embryonic nasopharynx, and is composed of endocrine glandular tissue. However, it is convenient to consider it here as part of the nervous system. The anterior pituitary secretes a range of important hormones into the circulation. Control of this endocrine activity is mediated by the hypothalamus.

Cross reference

See Chapter 14 for more information on the endocrine system.

Anterior pituitary

This contains glandular tissue that is composed of cords of epithelial cells interwoven with a fenestrated capillary network. The hormonal secretions are released into these capillaries and subsequently enter the circulation. Secretory cells of the anterior pituitary are characterized by abundant cytoplasmic granules that contain the hormones. The anterior pituitary hormones

TABLE 5.4 Anterior pituitary hormones and their major effects.

Cell type	Hormone	Major effects
Acidophilic (secrete polypeptide hormones)		
Somatotroph	Growth hormone	Promotes bone and soft tissue growth among other anabolic functions
Lactotroph	Prolactin	Promotes milk secretion by mammary tissue
Basophilic (secrete glycoprotein hormones)		
Corticotroph	Adrenocorticotrophic hormone	Promotes release of corticosteroids by the adrenal cortex
Gonadotroph	Luteinizing hormone	Promotes maturation of ovarian follicles (females), and production of testosterone by Leydig cells (males)
Thyrotroph	Thyroid-stimulating hormone	Promotes release of thyroxin by the thyroid
Gonadotroph	Follicle-stimulating hormone	Promotes release of ovarian follicles (females) and spermatogenesis (males)

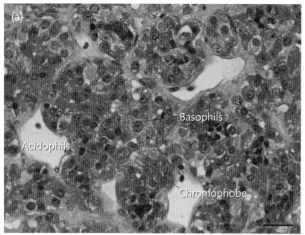
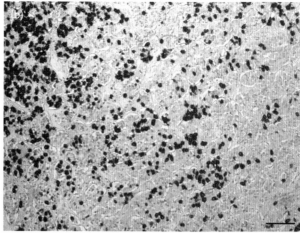

FIGURE 5.30

(a) Section of anterior pituitary showing basophils, acidophils and chromophobes. (b) Section of anterior pituitary stained with an antibody to growth hormone. A subset of cells shows intense cytoplasmic reactivity. Scale bar: 200 µm.

are of two broad types, glycoproteins and polypeptides; thus, routine staining techniques permit the classification of secretory cells as either basophilic or acidophilic, based on the nature of the granule contents (Figure 5.30a). The granules containing polypeptide hormones stain with acid dyes, while those containing the glycoprotein hormones stain with basic dyes. There is a third type of cell, the chromophobe, which stains with neither, and this may be a degranulated or undifferentiated cell. Secretory cells of the anterior pituitary can be classified further according to the specific hormone produced.

Immunohistochemistry using antibodies against each of these hormones can be applied to tissue sections to identify each type of cell. Some cell types show a characteristic location within the pituitary, while others are more diffusely situated (Figure 5.30b).

Posterior pituitary

Unlike the anterior pituitary, the posterior pituitary is composed of neural tissue continuous with the rest of the brain. The main components are the axons extending from hypothalamic neurons, but there are also specialized glial cells called pituicytes. The hormones antidiuretic hormone (ADH) and oxytocin are secreted from the posterior pituitary, but they are actually produced in the cell bodies of the supraoptic and paraventricular hypothalamic neurons and transported via the long axons to the posterior pituitary to be released into the circulation. The ends of axons bulging with neurosecretions abutting capillaries are known

Cross reference

See Chapters 12 and 14 for further information.

CLINICAL CORRELATION

Pituitary adenoma

Adenomas (benign tumours) can arise in the anterior pituitary. They often manifest with symptoms of hypersecretion of one of the pituitary hormones; for example, an adenoma comprised of growth hormone-producing cells can give rise to acromegaly, which is characterized by increased soft tissue and enlargement of the skull, or an adenoma of corticotrophs, in which hypersecretion of adrenocorticotrophic hormone (ACTH) leads to Cushing's syndrome.

as Herring bodies, with which are 100–200 nm membrane-bound neurosecretory granules. Antidiuretic hormone acts on the collecting ducts of the kidney to promote the reabsorption of water from urine. Oxytocin stimulates contraction of uterine smooth muscle and mammary myoepithelial cells.

5.7.2 Hypothalamus

The hypothalamus is a specialized region of the diencephalon, sited above the pituitary. It consists of small groups of neurons (nuclei) involved in many neural pathways. In addition to producing ADH and oxytocin secreted from the posterior pituitary, the hypothalamus directly controls anterior pituitary hormone secretion via a range of hormones that act upon the anterior pituitary. These hormones are produced in the dorsal medial, ventral medial and infundibular hypothalamic nuclei and travel to the anterior pituitary via the pituitary portal vessels (Figure 5.31).

Key points

The hypothalamus and pituitary provide vital endocrine control over numerous body systems.

SELF-CHECK 5.6

Which hormones are produced in the posterior pituitary?

5.7.3 Pineal

The pineal is another region of the brain with a neuroendocrine function. It is a structure of around 5 mm diameter projecting rearwards from the roof of the third ventricle covered by

TABLE 5.5 Hypothalamic hormones, with their target and effects.

Hormone	Target	Effect
Antidiuretic hormone (ADH)	Kidney	Water reabsorption
Oxytocin	Uterus and mammary tissue	Smooth muscle contraction
Thyrotrophin-releasing hormone (TRH)	Anterior pituitary	Release of TSH
Gonadotrophin-releasing hormone (GnRH)	Anterior pituitary	Release of LH and FSH
Growth hormone-releasing hormone (GHRH)	Anterior pituitary	Release of GH
Corticotrophin-releasing hormone (CRH)	Anterior pituitary	Release of ACTH
Prolactin inhibiting hormone (PIH)	Anterior pituitary	Inhibition of prolactin release
Somatostatin	Anterior pituitary	Inhibits release of GH and TSH

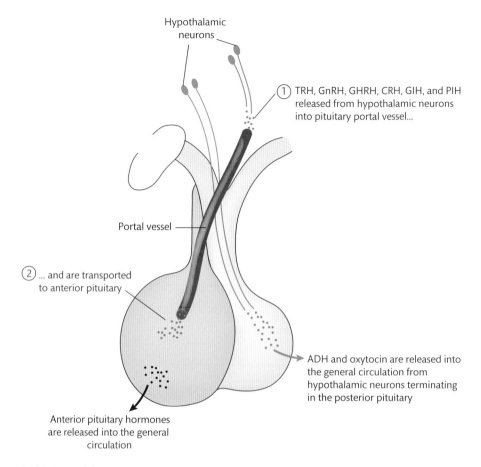

Hypothalamic neurons

① TRH, GnRH, GHRH, CRH, GIH, and PIH released from hypothalamic neurons into pituitary portal vessel...

Portal vessel

② ... and are transported to anterior pituitary

ADH and oxytocin are released into the general circulation from hypothalamic neurons terminating in the posterior pituitary

Anterior pituitary hormones are released into the general circulation

FIGURE 5.31
Schematic representation of the neurosecretory pathways between the hypothalamus and the pituitary gland.

meninges. Histologically, it is composed of astrocytes and modified neurons called pinealocytes. These cells secrete melatonin into the circulation and this has important modulating effects on the hypothalamus. Melatonin is important in the regulation of biological rhythms (e.g. the sleep/wake cycle). The pineal has neural connections to the retina, and secretion of melatonin is reduced by light and increases in darkness.

5.8 Tumours of the central nervous system

Tumours of the CNS are relatively uncommon, representing 2 per cent of all cancers in the UK, and a similar number of benign tumours. However, brain tumours in children are the second most common form of cancer (after leukaemia). The discreet definitions of benign and malignant used to describe tumours elsewhere in the body are less useful in the case of brain tumours, which are graded (I–IV) according to histopathological features. Grade I tumours are relatively benign and exhibit a generally slow course, whereas grade IV tumours are highly malignant and progress rapidly. It is probable that the grades seen represent in many cases a progression of disease over time. Little is known about the aetiology of brain tumours, but many genetic alterations are now recognized.

5.8.1 Glial cell tumours

Tumours can arise from any of the different types of glial cell and collectively these are the most common types of neoplasm seen in the CNS.

Astrocytomas arise from astrocytes and occur over the full spectrum of grades. Grade IV tumours are called glioblastoma multiforme (GBM) and have an extremely poor prognosis.

Oligodendrogliomas arise from oligodendrocytes and are graded as II or III. There has been some success using chemotherapy directed at these tumours. Ependymomas arise from ependymal cells lining the ventricles and spinal canal. These tumours commonly obstruct the flow of CSF. Tumours of the choroid plexus also occur.

5.8.2 Neuronal tumours

Tumours exhibiting neuronal differentiation are much less common and generally are relatively benign. Their origin is uncertain because mature neurons do not undergo cell division and therefore are thought not to be able to give rise to tumours.

5.8.3 Embryonal tumours

Embryonal tumours are most common in childhood, presumed to arise from remnant embryonal cells, and often exhibit features of neuronal differentiation. Medulloblastoma is a highly malignant paediatric tumour arising from the cerebellum. Primitive neuroectodermal tumour (PNET) is similar and appears to arise from the forebrain.

5.8.4 Meningiomas

Meningiomas are usually benign in behaviour and amenable to complete surgical excision due to their cohesive nature. Invasive forms arise very occasionally. They can arise anywhere on the surface of brain and spinal cord. Histologically, they usually show characteristic whorls of meningothelial cells.

5.8.5 Secondary tumours

The brain is a common site to which carcinomas (cancers of epithelial tissues) metastasize from other sites, especially the bowel, prostate and lung. Melanomas and lymphomas may also spread into the CNS.

Cross reference

See the *Histopathology* volume in this Fundamentals of Biomedical Science series for an explanation of this technique

BOX 5.4 Laboratory investigation of brain tumours

Clinical management of brain tumours requires accurate diagnosis and grading by histopathology. Often, the surgeon will require a diagnosis to be confirmed during an operation. This can be performed rapidly by means of frozen section or more commonly, due to the fluidity of brain tissue, by smear preparation. The slides are prepared from a sample of tissue, stained and examined microscopically, and the findings communicated directly to the surgical team. The diagnostic information may then be used to influence surgical progress (e.g. whether or not to attempt a resection).

Alternatively, confirming that the lesion has been reached and sampled is important if surgery has been undertaken simply for the purpose of a biopsy. Owing to time constraints, intraoperative diagnoses can often only provide limited information, and any remaining and subsequent tissue specimens will be subjected to routine processing to permit full histopathological interpretation.

Grading tumours is also important for patient management, especially for oncology. It is achieved by thorough histopathological assessment of the tumour and takes account of features such as the degree of dedifferentiation, mitotic activity, vascular proliferation, nuclear **pleomorphism** and whether or not necrosis is present. Tumour cells may more or less mimic the cell of origin. If the tumour cells bear little resemblance to the cell of origin (i.e. dedifferentiated) then this is termed anaplasia and is often a feature of higher grade tumours and a poorer prognosis. Immunohistochemistry is an important tool, and can be used to confirm cell type; for example, antibodies to GFAP indicate astrocytic origin, or neuron-specific markers such as NFP or CD56 can identify neuronal differentiation. The rate of proliferation is an important factor in grading tumours, and antibodies to Ki67 can be used to identify nuclei of non-resting cells. Immunohistochemistry is also used to identify potential primary sites of metastatic tumours, which may have been undetected prior to the appearance of neurological symptoms.

Patient management requires close liaison between neuropathologists, surgeons, radiologists and oncologists.

Molecular techniques are also now being used increasingly. For example, deletions of 1p and/or 19q occur in a subset of oligodendrogliomas that are more amenable to chemotherapy. Fluorescence *in situ* hybridization is used to identify these deletions.

5.9 Pathological processes in the peripheral nervous system

5.9.1 Axonal degeneration

The reaction of peripheral axons and their Schwann cells is very different to those in the CNS. Although axons can produce regenerative sprouts (Figure 5.32b), the degree to which this can produce successful recovery depends on the amount of damage to the nerve. When there is physical separation of the nerve trunk the first reaction is multiplication of fibroblasts around the wound. These produce fibrillar collagen that hinders outgrowth of axon sprouts. When an axon degenerates (Figure 5.32a) or is damaged by some external trauma it breaks and fails to support

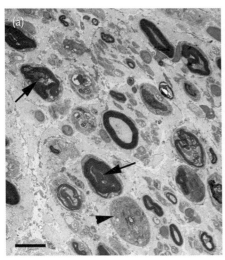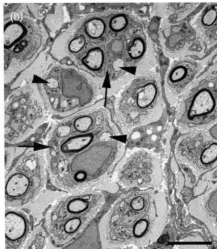

FIGURE 5.32
(a) Recent axonal degeneration showing loss of axons (arrows) and myelin breakdown
(arrowhead). Scale bar: 10 μm. (b) Regenerative axonal sprouts occur either singly
or in clusters of several smaller fibres close together (arrows). Some axon sprouts
are thinly myelinated and others have yet to acquire a myelin sheath (arrowheads).
Scale bar: 5 μm.

its myelin sheath; this then collapses inwards at the nodes and Schmidt–Lanterman incisures to
form large ovoids of myelin enclosing degenerating axonal components. These ovoids degenerate further into smaller bodies and finally into their component lipids and simple fats.

The largest myelinated fibres are most susceptible to damage from ischaemia or trauma. Axons
can recover if the initial cause of damage is removed, by producing from one to six smaller
sprouts from the last undamaged node of Ranvier. At the same time, Schwann cells multiply
and align along the sprouts in a recapitulation of normal development. Myelin can form in the
same way when the axon's diameter is large enough; however, the internodes remain at the
fundamental length of 200–300 μm. This increases with growth during development, but in an
adult there is no growth so internodal length stays at the basic length. Therefore, a regenerated fibre can be distinguished from normal and remyelinated fibres by the succession of short
internodes of equal length.

5.9.2 Segmental demyelination

This is the loss of the myelin sheath surrounding one or more internodes. The axon is undamaged initially but, as there is an intimate connection between axon and myelin, damage to any
part of the system may eventually lead to breakdown of the rest.

Primary demyelination is found where a disease primarily affects the Schwann cells and
they are unable to support the myelin sheath, resulting in its breakdown into the constituent
proteins and lipids. This is seen as myelin debris in the Schwann cell cytoplasm, never as thin
myelin.

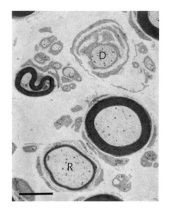

FIGURE 5.33
Recent demyelination (D) and remyelination, where the new myelin sheath is thinner than normal (R). Scale bar: 5 µm.

Myelin can adapt to some extent to alterations in axon diameter by slippage between the lamellae; however, when changes in axon diameter are too large, **secondary demyelination** results. Minor degrees of demyelination produce widened nodes of Ranvier.

Allergic demyelination is seen in autoimmune diseases where activated macrophages attack the myelin sheath and strip the lamellae from the axon.

Remyelination (Figure 5.33) of an axon can occur following demyelination from whatever cause. To achieve this the Schwann cells multiply and move onto bare lengths of axon. More Schwann cells may be generated than are required to remyelinate the demyelinated portion of the axon. If there are several episodes of demyelination, followed by remyelination, the excess Schwann cells move radially to lie alongside the fibre, forming circles around it that are known as onion bulbs. This Schwann cell reduplication, plus the excess collagen formed at the same time, can enlarge the nerve trunk to the extent that it can be felt through the skin.

5.9.3 Acquired neuropathies

Diabetic neuropathy is probably the most common neuropathy, particularly in the developed world. While several diseases are related to diabetes, in all of these the damage is primarily to the axons, but there may also be demyelination. All components of the nerve trunk are affected. The mechanism is not completely understood but it is related to glycation; however, normalizing blood sugar level does not necessarily result in improvement and may produce a painful exacerbation, possibly due to cross-talk between regenerating axon sprouts.

Leprosy, caused by infection with *Mycobacterium leprae*, is a common cause of neuropathy worldwide. The bacterium is the only one that primarily attacks nerves and is due to an attraction with the Schwann cell basal lamina. Details of the neuropathy are variable due to the degree of reaction of the host. In tuberculoid leprosy there is a strong reaction to the bacilli, which are mostly killed, but there is severe nerve damage. Axons and Schwann cells are killed, and there is replacement of most of the normal structure with fibrillar collagen. In lepromatous leprosy the bacilli multiply extensively in all the cells of the nerve, but particularly in Schwann cells, and there is granuloma formation with extensive invasion by macrophages and lymphocytes. The foam cells typical of lepromatous leprosy are macrophages that contain large clumps of bacilli. The common perineurial involvement distinguishes leprosy from most other neuropathies. There is a continuum of changes from lepromatous to tuberculoid leprosy.

Autoimmune demyelinating neuropathy results from direct attack on the myelin sheath by activated macrophages, and gives rise to two main diseases, Guillain-Barré syndrome (GBS) and chronic inflammatory demyelinating polyneuropathy (CIDP). Differences between the two manifest mainly in the clinical picture, with GBS being an acute disease, often following an infection such as influenza or gastrointestinal disease, whereas CIDP is a chronic disease that may persist in a fluctuating manner for years. In both there is invasion of the nerve by lymphocytes, and the initial demyelination may often be followed by secondary axonal degeneration (particularly in CIDP).

Systemic vasculitis is inflammation of the blood vessels and can occur in several diseases. If this affects the blood vessels supplying the nerve trunk then the resultant ischaemia leads to nerve degeneration. Very minor damage produces demyelination but the characteristic pathology accompanying vasculitis is degeneration of the myelinated fibres, the largest ones being most vulnerable to ischaemia.

Sarcoid neuropathy can be associated with a cranial nerve neuropathy (most commonly the VII nerve) but may also affect peripheral nerves; however, it is rare to find granulomas in the nerve. Damage may be a combination of axonal degeneration and demyelination.

SELF-CHECK 5.7

What are the organs that provide communication between skin and muscles?

SELF-CHECK 5.8

Both produce a myelin sheath but how do Schwann cells differ from oligodendroglial cells?

SELF-CHECK 5.9

How is a peripheral nerve fascicle protected from external influences?

5.9.4 Hereditary neuropathies

Charcot-Marie-Tooth disease (CMT): Type 1 are demyelinating diseases and type 2 are axonal diseases. Type 1A has the highest incidence, is the best understood and the first neuropathy to have its underlying genetic defect discovered. It is due to a duplication of the part of chromosome 11 coding for PMP-22, a small myelin protein that comprises just 10 per cent of the total. This genetic abnormality is seen about 20 times more often than any of the others. Repeated episodes of demyelination and remyelination may produce extensive onion bulb formation. Rarely, point mutations may be found in the same gene and these produce a much more severe neuropathy.

CMT type 1 are dominantly and type 4 recessively inherited. At present, about 40 genes have been identified as causing one of the CMT diseases. Although some of the genes are directly related to myelin or axonal structure, many are less obviously related to neuropathy.

Inherited disorders of metabolism mostly show onset in early childhood but there are rare variants characterized by adult onset. The pattern of pathology is mixed in most; there may be demyelination initially but this is often superseded by axonal damage and loss. There are characteristic Schwann cell lipid inclusions in metachromatic leucodystrophy, Krabbe globoid cell leucodystrophy, Tangier disease and Farber disease. In Fabry disease there are inclusions in endothelial cells of blood vessels and in perineurial cells. Refsum disease is due to phytanoyl-CoA hydroxylase deficiency.

5.9.5 Nerve sheath tumours

These are usually benign **Schwannomas**, arising from Schwann cells around the axons. The VIII cranial (vestibulocochlear) nerve is a common site for these tumours, but they can occur throughout the PNS.

Chapter summary

In this chapter we have:

- Described the different types of cells that make up the nervous system.

- Examined how neurons communicate throughout the central and peripheral nervous systems and the processes of neurotransmission.

- Considered the roles and functions of the supporting cells including the glia, Schwann cells and the meninges.

- Outlined the structure and organization of the nervous system including the brain and spinal cord and the interrelationships between the central and peripheral nervous systems.

- Examined the role of the nervous system in homeostasis and coordination and the organisation of the autonomic nervous system.

- Discussed the mechanisms of sensory reception and motor activation.

- Described from a laboratory perspective some of the important disease processes that occur in the nervous system.

Further reading

- **Orchard GE, Nation BR eds. *Histopathology*. Oxford: Oxford University Press, 2012.**

Discussion questions

5.1 What different types of glial cells are there in the CNS and what are their functions?

5.2 Why is it essential for neurons to have an efficient system of intracellular transport? What mechanisms exist for this?

5.3 Describe the structural arrangement of large nerve fibres in the PNS.

5.4 How does the autonomic nervous system contribute to normal physiology and homeostasis?

Answers to self-check questions and discussion questions, hints and tips are provided on the book's Online Resource Centre.

 Visit www.oxfordtextbooks.co.uk/orc/orchard_csf/.

Lungs: the cells of the respiratory system

Behdad Shambayati and Andrew Evered

Learning objectives

After studying this chapter you should be able to:

- Describe the anatomy of the respiratory tract.
- Describe the basic histology of different tissues that make up the respiratory tract.
- Relate histology and cytology to relevant physiology.
- Give examples of methods involved in the investigation of respiratory disease.

In this chapter you will learn about the tissues and cells involved in respiration. Respiration is the transport of oxygen into cells and tissues, and the removal of carbon dioxide. The respiratory system consists of two parts: a system of tubes and passages for conduction of the air, and a respiratory portion in which gaseous exchange occurs.

The conduction portion includes the nose, pharynx, trachea and bronchi. The respiratory portion includes respiratory bronchioles, alveolar ducts and alveoli. Alveoli are sac-like structures that make up the majority of lung tissue and are the sites where exchange of O_2 and CO_2 takes place. The very thin wall of the alveoli and their close proximity to a dense network of capillaries allows the diffusion of oxygen from inspired air into arterioles, and the removal of carbon dioxide from venules into the alveolar space for removal during expiration. (Figures 6.1 and 6.2)

6.1 **Respiratory epithelium**

Before we look at the anatomy and histology of the respiratory tract in detail, it is important to learn about the principal cell type that covers a large portion of the respiratory tract. **Respiratory epithelium** is classified as **ciliated pseudostratified columnar epithelium**. Pseudostratification describes the apparent multilayering (stratification) of cells exhibited by certain types of epithelia. However, each cell is in contact with the basement membrane so, strictly speaking, **pseudostratified epithelium** constitutes a single layer of cells. The nuclei, however, are not lined up in the same plane and this gives the epithelium the false impression

Respiratory epithelium
This is classified as ciliated pseudostratified columnar epithelium. The principal function of respiratory epithelium is to protect the respiratory tract.

Conducting Passages

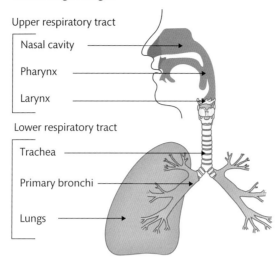

Upper respiratory tract

Nasal cavity

Pharynx

Larynx

Lower respiratory tract

Trachea

Primary bronchi

Lungs

FIGURE 6.1

The main components of the upper and lower respiratory system.

that there is more than one layer of cells (pseudo means false). The principal function of respiratory epithelium is to protect the respiratory tract. This will be explained in more detail below but briefly this is achieved by keeping the area moist with mucus secretions and removal of any trapped particles by movement of cilia.

SELF-CHECK 6.1

What does pseudostratification mean?

Neuroendocrine cells

These receive neural input and can release various messenger molecules into the blood. In the lung, they may act as receptors for oxygen levels.

The respiratory epithelium as seen by light microscopy consists of four main cell types: a) ciliated columnar cells, b) mucus-secreting **goblet cells**, c) **basal cells** and d) **neuroendocrine cells** (Figures 6.3a and 6.3b).

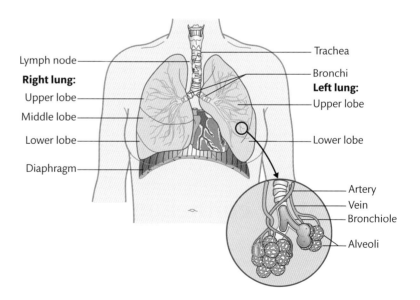

Lymph node

Right lung:

Upper lobe

Middle lobe

Lower lobe

Diaphragm

Trachea

Bronchi

Left lung:

Upper lobe

Lower lobe

Artery

Vein

Bronchiole

Alveoli

FIGURE 6.2

Components of the respiratory tract.

(a)
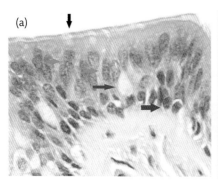
(b)
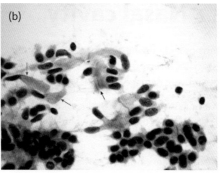

FIGURE 6.3

(a) Respiratory epithelium (haematoxylin and eosin [H&E] stain, high power). Black arrow: cilia; green arrow: ciliated columnar cells; red arrow: goblet cell; and blue arrow: basal cells. (b) Bronchial brushing sample (Papanicolaou stain, medium power). Red arrows: ciliated bronchial epithelial cells and occasional goblet cells; and black arrow: cilia attached at terminal plate.

Ciliated columnar cells make up the majority of cells and have numerous cilia on their upper surface. These cilia perform a rhythmic waving or beating motion and move mucus and particulate matter.

The next most abundant cells are mucus-secreting goblet cells, so named because of their resemblance in shape to that of a wine goblet. Goblet cells exude large quantities of mucus, helping to keep the area moist.

The basal cells are small, rounded, nearly cuboidal cells believed to have the ability to undergo mitosis and differentiate into other cell types found in the epithelium.

Cross reference

To find out more about the mechanism of cilia motility, see Chapter 1

SELF-CHECK 6.2

Name the four main cell types that can be seen with a light microscope?

The last cell type to consider can only be recognized using electron microscopy or by means of histochemical stains. These are the neuroendocrine cells located throughout the entire respiratory tract, from the trachea to the terminal airways. These cells, which are approximately the same size as basal cells, contain neuroendocrine granules and are specialist cells that release hormones as a result of neural stimuli; they are found throughout the body. Pulmonary neuroendocrine cells may be involved in regulation of oxygen levels. This is probably achieved by detecting decreased oxygen or increased carbon dioxide levels and sending chemical messages to help the lung adjust to these changes. It is noted that people living at high altitudes where oxygen levels are lower have higher numbers of neuroendocrine cells in their lungs.

SELF-CHECK 6.3

The function of mucus secreted by the goblet cells is to:

a) lubricate the bronchial mucosa

b) provide nutrients to the cells

c) carry inhaled particulate matter to the exterior

d) increase the surface tension on the cells

e) all of the above.

6.2 Nasal cavity

The nose admits the air for respiration and expiration. The main function of the nose and its associated structures is to prepare the air for passage to the lungs by warming and humidifying it to prevent damage to the delicate alveoli.

The mucus secreted by the nasal epithelium traps dust particles and particulate matter and moves them to the pharynx where they are swallowed. The nose is also involved in the sense of smell.

The most anterior (front) part is called the **nasal vestibule** and is surrounded by cartilage and lined by stratified squamous epithelium. Around the inner surface of the nostrils (nares) are numerous **sebaceous** and sweat glands, and the epithelium also bears a few short hairs called **vibrissae**. Further up the vestibule the epithelium changes into the typical respiratory epithelium that covers the majority of the respiratory tract (Figure 6.4).

Key points

The main function of the nose and its associated structures is to prepare the air for passage to the lungs by warming and humidifying it to prevent damage to the delicate alveoli.

The internal nasal cavity is referred to as the **internal nasal fossa**. This large air-filled cavity is divided into two areas: a respiratory portion and an area involved in **olfaction** (sense of smell). The surface area of the nasal fossa is greatly increased by three bony projections called the **nasal conchae**. The middle and lower conchae are covered by respiratory epithelium. The

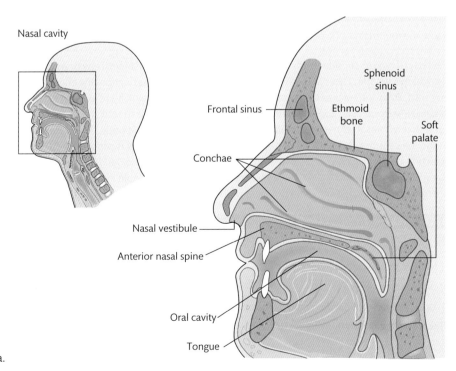

FIGURE 6.4
Main components of the nose.
© 2003 Encyclopaedia Britannica.

lamina propria in these two conchae is highly **vascularized**. The large **venous plexuses** in this area are known as **swell bodies**. These periodically engorge with blood to restrict airflow through each nasal fossa; this cyclical reduction in airflow allows the epithelium on the closed side to recover from the drying effect caused by airflow.

CLINICAL CORRELATION

Swell bodies

During inflammation caused by allergic reaction or the common cold, both swell bodies may engorge with blood and restrict airflow, resulting in a 'stuffy nose'.

The roof of the nasal cavity and the upper concha are lined by specialized pseudostratified columnar epithelium called **olfactory epithelium**, which contains receptors for the sense of smell (Figure 6.5). This specialized epithelium consists of three different cell types: **supporting cells**, basal cells and olfactory cells. The supporting cells are tall columnar cells with surface

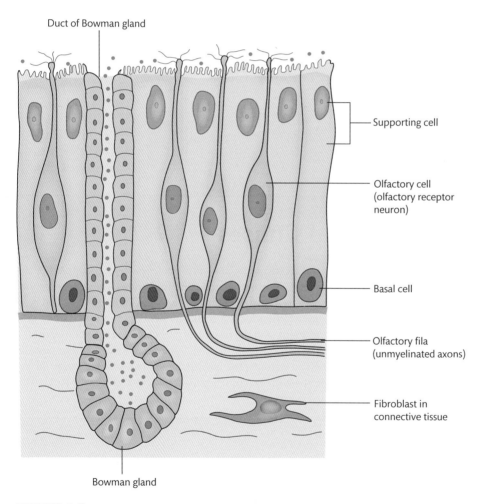

FIGURE 6.5
Drawing of the olfactory mucosa.
Modified from original image: Cui *et al.*: Atlas of Histology with Functional and Clinical Correlations © 2011 Lippincott Williams & Wilkins.

Junctional complexes
Complex attachment point between epithelial cells.

microvilli and nuclei that are situated towards the upper surface of the cells. The supporting cells are attached to adjacent olfactory epithelium by **junctional complexes** and thus physically and metabolically support the olfactory cells. The basal cells are small, almost spherical cells and form a single layer at the base of the epithelium. Basal cells are **stem cells** able to differentiate into either olfactory or supporting cells.

SELF-CHECK 6.4

Name the cells that make up the olfactory epithelium.

The olfactory cells are **bipolar neurons** (neurons with two extension). The uppermost layer is covered by very long non-motile cilia, and the afferent axons form the olfactory nerve which terminates in the olfactory bulb in the brain. The function of **non-motile cilia** is to increase the surface area to facilitate contact between odorous substances and the olfactory cells.

Serous secretions
Watery fluid produced by Bowman's gland.

The **Bowman's gland**, situated in the lamina propria, produces **serous secretions** that help to dissolve the incoming odours to facilitate access of odours to the sensory receptors on the cilia.

SELF-CHECK 6.5

Multiple choice questions.

The function of the lining of the nasal cavity includes:

a) sense of smell

b) warming of inspired air

c) trapping foreign matter

d) moistening of inspired air

e) all of the above.

6.3 **Paranasal sinuses**

The paranasal sinuses (paranasal means around the nose, sinus = cavity) are four pairs of air-filled spaces in the **frontal, maxillary, ethmoid** and **sphenoid bones**. The arrangement of these bones and associated spaces is shown in Figure 6.6. The sinuses are lined by a thin layer of ciliated columnar epithelium which includes a small number of goblet cells. Goblet cells secrete mucus to keep the area moist, and the beating of cilia drains the secretions from the sinuses into the nasal cavity. The purpose of the paranasal sinuses is not certain, but possible functions include humidifying and warming the inhaled air and reducing the weight of the skull. They may also play a role in speech by resonating the sound emitted from the vocal cords and by affecting the tone of the voice.

Key points

The paranasal sinuses (paranasal means around the nose, sinus= cavity) are four pairs of air-filled spaces in the frontal, maxillary, ethmoid and sphenoid bones.

SELF-CHECK 6.6

What are the supposed purposes of the paranasal sinuses?

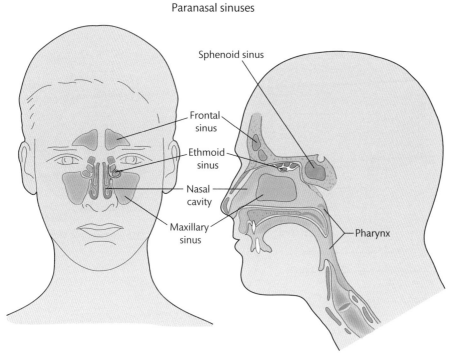

FIGURE 6.6
Drawing of the position of paranasal sinuses.
Modified from original image © 2012 Terese Winslow.

CLINICAL CORRELATION

Paranasal sinuses

Poor air flow and slow drainage that can occur when one is suffering from a cold often leads to sinusitis (infection).

Cancers arising in paranasal sinuses are rare in the UK, with fewer than 600 new cases diagnosed each year, but more common in Japan or South Africa. These cancers mostly arise in the maxillary sinus. The exact cause of paranasal malignancy is unknown, but people who work in the furniture industry and are exposed to wood dust are at higher risk of developing paranasal sinus cancer. Smokers are also at risk. This type of cancer is rare in people aged under 40 and the incidence in men is double that in women. The majority of these cancers are **squamous cell carcinoma**; **adenocarcinoma** is the second most common type, with occasional cases of **sarcoma** and **lymphoma**.

Cross reference

For further information see the *Cytopathology* volume in this Fundamentals of Biomedical Science series.

SELF-CHECK 6.7

Perceived functions of paranasal sinuses include:

a) humidifying air

b) warming inhaled air

c) affecting the tone of the voice

d) reducing the weight of the skull

e) all of above.

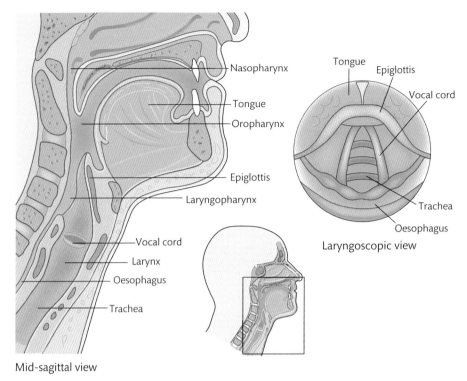

Mid-sagittal view

FIGURE 6.7
Drawing of the anatomy of the larynx.
Modified from original image © 2008 Trialsight Medical Media.

6.4 **Nasopharynx**

The nasopharynx is the first part of the **pharynx** (naso [from the Latin for nose], pharynx [from the Greek for mouth]) and lies at the back of the nose, above the roof of the mouth (**soft palate**). It connects the nose to the back of the mouth (**oropharynx**) and is lined with respiratory epithelium (Figure 6.7)

6.5 **Larynx**

The larynx is a tubular structure about 5 cm in length that links the **oropharynx** to the trachea. The larynx can be seen as a lump in the front of the neck, commonly known as the Adam's apple. It has two main functions: acting as a valve to close the airway during swallowing to prevent food from entering the trachea, and producing sound by means of the **vocal cords**. The wall of the larynx is supported by nine cartilages linked by ligaments and connective tissue. The larger cartilages are called **thyroid** and **cricoid** (Latin for ring-shaped) **cartilage**. The arrangement of these cartilages can be seen in Figure 6.8.

This type of cartilage contains **hyaline** material and is similar to that found on many joint surfaces. The **epiglottis** is a thin flap of **fibroelastic cartilage** situated at the entrance of the larynx and keeps food from entering the respiratory tract. During the swallowing process, the

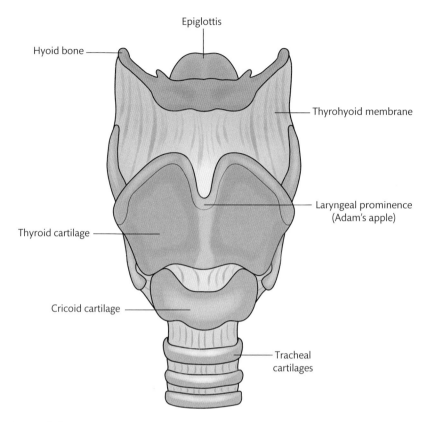

FIGURE 6.8
Drawing showing the position of cartilages supporting the larynx.

epiglottis folds down and temporarily blocks the opening to the trachea as food travels to the oesophagus. The epithelium covering the anterior surface of epiglottis (the surface in contact with food) is squamous epithelium; the posterior surface is also partially covered with squamous epithelium, but towards the base it gradually changes to ciliated pseudostratified epithelium.

SELF-CHECK 6.8

What is the function of epiglottis?

Below the epiglottis, two pairs of tissue folds overhang from the lining of larynx. The upper pair, lined by respiratory epithelium, called the **false vocal cords** protects the more delicate **true vocal cords** that lay beneath. Figure 6.9 demonstrates this physical relationship.

The true vocal cords are composed of elastic fibres covered with stratified squamous epithelium. Running parallel to the vocal cords is the **vocalis muscle** that can affect the tension in the cords and thus influence the frequency of the sound produced.

SELF-CHECK 6.9

The true vocal cords are:

a) made up of voluntary muscles

b) composed of elastic fibers covered with simple squamous epithelium

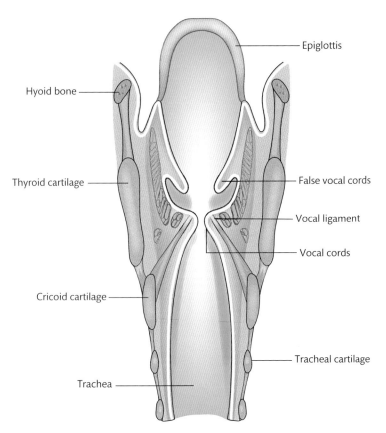

Epiglottis

Hyoid bone

Thyroid cartilage

False vocal cords

Vocal ligament

Vocal cords

Cricoid cartilage

Tracheal cartilage

Trachea

FIGURE 6.9
Relationship of false vocal cords to true vocal cords.

c) lined by pseudostratified columnar epithelium

d) linked to smooth muscle fibres

e) none of above.

CLINICAL CORRELATION

Cancer of larynx

(Laryngeal cancer) is considered a rare tumour, with an incidence of three per 100 000 in UK (2200 persons) and five per 100 000 in the USA (12 000 persons). Smoking is the major risk factor for laryngeal cancer, with heavy consumption of alcohol the second contributory factor. The risk increases in heavy smokers and drinkers. Other risk factors include a diet deficient in fruit and vegetables, immunosuppression and exposure to **human papillomavirus (HPV)**.

6.6 **Trachea**

From the larynx, the conducting portion of the respiratory tract continues to the trachea (Figure 6.10).

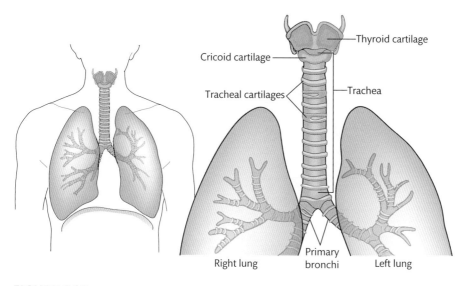

FIGURE 6.10
Drawing of the trachea.
Modified from original image © 2009 WebMD, LLC.

Colloquially known as the windpipe, it lies in front of the oesophagus, connecting the larynx to the lungs. The trachea is a flexible tube with a diameter of 2.5 cm and 11–15 cm in length. Its walls are strengthened by 16–20 C-shaped hyaline cartilages which are incomplete posteriorly (Figure 6.11).

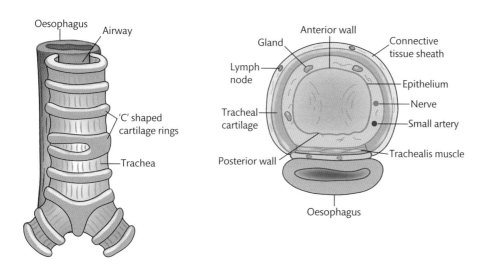

FIGURE 6.11
Relationship of tracheal C-shaped cartilages. The drawing on the right is a cross section of the same.
Cross section modified from Chung *et al.* (2011) CT of Diffuse Tracheal Diseases, *American Journal of Roentgenology* 196: W240–W246 © American Roentgen Ray Society, ARRS, All Rights Reserved.

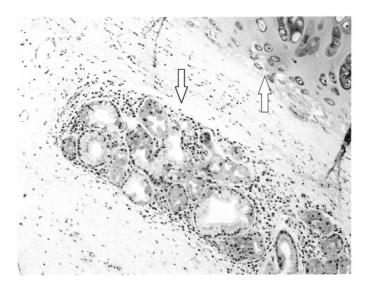

FIGURE 6.12
Medium power view of the trachea showing cartilage (blue arrow) and mucus glands (black arrow).

The open ends of these cartilages are joined to ligaments covered by smooth muscle that contracts during the cough reflex to decrease the diameter of the trachea; this increases the speed of airflow and helps to clear the air passage.

The trachea is lined by typical respiratory epithelium and beneath the lamina propria there are numerous glands that produce mucus and watery solutions. These secretions are delivered to the surface of the epithelium through ducts (Figure 6.12).

SELF-CHECK 6.10

Why does the trachea have cartilage rings?

CLINICAL CORRELATION

Cancer of the trachea

This is very rare, accounting for 0.1 per cent of all cancers. Adenoid cystic carcinoma and squamous cell carcinoma are the most common types. The exact causes of these cancers are unknown but squamous cell carcinoma is more common in smokers.

SELF-CHECK 6.11

The trachea:

a) has complete rings of cartilage

b) contains elastic cartilage

c) has pseudostratified columnar epithelium

d) has stratified squamous epithelium

e) has stratified columnar epithelium.

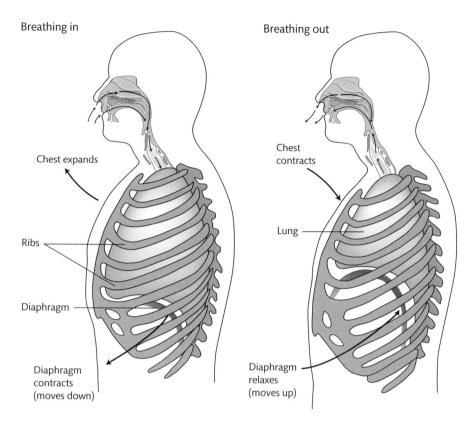

Breathing in

Breathing out

Chest expands

Chest contracts

Lung

Ribs

Diaphragm

Diaphragm contracts (moves down)

Diaphragm relaxes (moves up)

FIGURE 6.13
Drawing showing the mechanics of breathing.

6.7 **Lungs**

The lungs are situated in the left and right side of the thorax and are separated by the heart and the contents of **mediastinum**. The left lung has two lobes and the right lung three lobes. The shape of the lungs fits the shape of the chest cavity. During inhalation, the **diaphragm** and **intercostal muscles** (muscle between the ribs) contract. This causes an enlargement of the chest cavity which results in the air pressure being lower inside than that outside. This change in air pressure causes the air to flow in from the outside and expand the lungs. The reverse occurs during exhalation; the diaphragm and intercostal muscles relax causing the chest cavity to become smaller. This decreases the volume and increases the air pressure in the chest cavity compared to the outside, causing the air from the lungs to flow out. This cycle repeats itself at every breath. Figure 6.13 illustrates the mechanics of breathing.

> **Key points**
>
> The diaphragm has an important function in breathing; during inhalation the diaphragm and intercostal muscles contract, causing enlargement of the chest cavity which results in the air pressure within being lower than that outside. This change in air pressure causes the air to flow in from the outside and expand the lungs. The reverse occurs during exhalation; the diaphragm and intercostal muscles relax causing the chest cavity to become smaller. This decrease in volume and increase in air pressure in the chest cavity compared to the outside, causes the air to flow out from the lungs.

The lungs are separated from each other by:

a) pleural membrane

b) diaphragm

c) cartilage

d) intercostal muscles

e) mediastinum.

CLINICAL CORRELATION

Chest X-ray (radiograph)

A chest X-ray is a common radiology test generally used in conjunction with clinical data to diagnose many chest conditions. X-rays utilize electromagnetic radiation that penetrates different structures depending on the make-up of the tissue. For example, bones absorb most of the X-rays, whereas tissues filled with air absorb very little. Solid structures appear as white shadows on the chest X-ray and air-filled structures as a dark background. Chest X-rays can be used to diagnose many conditions including cancer, **pneumonia** and fluid around the lung.

6.8 **Bronchial tree**

In the mediastinum the trachea at the anatomical site called the **carina** (Figures 6.14 and 6.15) divides into the left and right primary bronchi and enters the lungs at an opening in the middle of each lung called the **hilum** (Figure 6.15). The hilum is also the site at which other structures such as the pulmonary artery, pulmonary veins and lymphatic vessels enter the lungs.

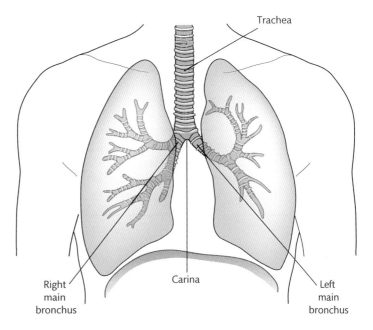

FIGURE 6.14
Anatomical position of the carina.

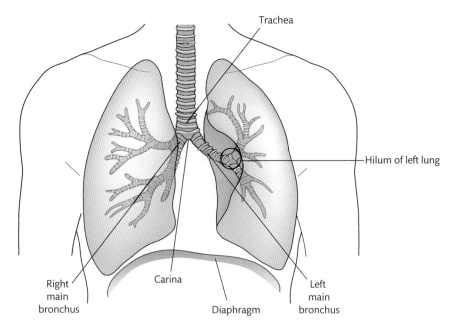

FIGURE 6.15
Anatomical position of the carina and hilum.

The large bronchi are lined by:

a) simple ciliated epithelium

b) squamous epithelium

c) stratified ciliated epithelium

d) pseudostratified ciliated epithelium

e) pseudostratified non-ciliated epithelium.

What structures enter/leave the lung at the hilum?

The structure of the primary bronchus is similar to that of the trachea, but gradually the amount of hyaline cartilage in the walls decreases and the cartilaginous rings are replaced by a small plate of cartilage. As the branching continues, the cartilage disappears and is replaced by smooth muscle. The mucous membrane also undergoes a gradual transition from ciliated pseudostratified columnar epithelium to simple cuboidal epithelium.

The right bronchus is slightly wider than the left bronchus. The right bronchus divides into three lobar bronchi, and the left bronchus divides into two. These divide into segmental bronchi (Figure 6.16) which supply the **bronchopulmonary segments**. There are ten bronchopulmonary segments in the right lung and 8–10 segments in the left. A bronchopulmonary segment is separated by connective tissue and has its own supply of blood vessels and **lymphatics**, making it a separate anatomical and functional unit that, if necessary, can be removed surgically by a thoracic surgeon without affecting the function of other segments.

Lymphatics
Lymphatics (or lymphatic vessels) are thin-walled structures that carry lymph.

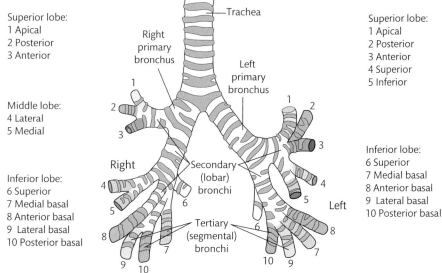

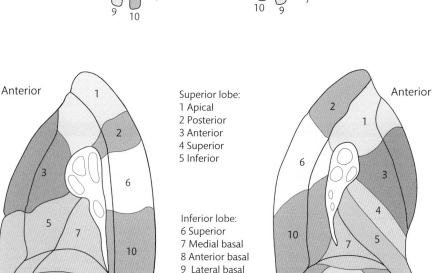

FIGURE 6.16
**Drawing of the
segmental bronchi and
bronchopulmonary segments.**
Modified from Tortora:
Principles of Human Anatomy
9th edn. © John Wiley & Sons.

Medial and basal views of right lung

Medial and basal views of left lung

Cross reference

For further information see the
Cytopathology volume in this
Fundamentals of Biomedical
Science series.

The segmental bronchi are divide further into **primary bronchioles**, and these further divide
into **terminal bronchioles**, which are the last part of the conducting portion of the respiratory tract.

Key points

Each lung is divided into ten bronchopulmonary segments. A bronchopulmonary segment
is separated by connective tissue and has its own supply of blood vessels and **lymphatics**,
making it a separate anatomical and functional unit that, if needed, can be removed surgically by a thoracic surgeon without affecting the function of other segments.

METHOD Bronchoscopy

Bronchoscopy is a procedure that allows a respiratory physician or surgeon to visualize the lower respiratory tract (up to the fourth division of the bronchial tree) through a bronchoscope. The bronchoscope is essentially a camera, attached to a light source, that can be inserted through the mouth or more commonly through the nose.

There are two type of bronchoscopes in use: a rigid and a flexible type.

The flexible bronchoscope (Figure 6.17) has a very small diameter and uses fibre optics to produce very high-quality images of the larynx, trachea, bronchi and bronchioles. The flexible design allows visualization of smaller bronchioles, which is not possible with a rigid bronchoscope. In addition, there are interior channels in the flexible bronchoscope that allow the passage of smaller instruments with which to take biopsies and cytology brush samples. This procedure is normally carried out using mild sedation and local anaesthesia.

The rigid bronchoscope is a straight metal instrument that is used mainly for visualizing the larger airways, or for certain therapeutic uses such as removing foreign objects from the airways or controlling bleeding.

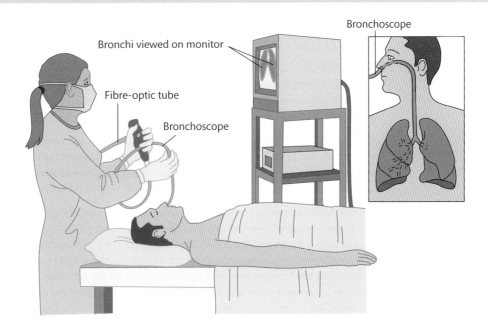

FIGURE 6.17

Diagrammatic representation of the flexible bronchoscopy procedure.

Modified from original image © 1995–2013 Healthwise Incorporated.

METHOD Endobronchial ultrasound (EBUS) transbronchial needle aspiration (TBNA) of mediastinal lymph nodes (EBUS TBNA)

This procedure is an emerging technique and a less-invasive alternative to mediastinoscopy for obtaining samples from the mediastinum. The procedure is normally carried out under mild sedation and local anesthesia, and has a lower rate of side effects compared to mediastinoscopy.

The technique combines a special bronchoscope that has a miniaturized ultrasound probe at its tip. This bronchoscope is slightly wider in diameter than the instrument used in normal bronchoscopy and hence passed through the mouth. The bronchoscope contains a channel for the fine, long sampling needle. Once the correct anatomical area is reached, the ultrasound probe provides images of lymph nodes and allows the physician to directly sample the area. In many hospitals use of EBUS has greatly reduced the need for mediastinoscopy.

EBUS TBNA is often used to diagnose lung cancer, stage lymph nodes in lung cancer, and diagnose other conditions affecting the mediastinum, including sarcoidosis or tuberculosis.

Cross reference

For further information see the *Cytopathology* volume in this Fundamentals of Biomedical Science series.

CLINICAL CORRELATION

Lung cancer

Lung cancer is the most common cancer in the world, with 1.3 million cases diagnosed every year. In the UK, it is the second most common cancer after prostate cancer in men, and in women it is the third most common after breast cancer and bowel cancer.

Lung cancer is the most common fatal malignancy both in men and women. The poor survival rate is mainly due to its late diagnosis, when the cancer is already at the stage where curative treatment is not possible. Also, many patients are elderly who have existing medical conditions that make them unfit for radical treatment. Most cases of lung cancer are caused by smoking.

Incidence: Most cases of lung cancer are diagnosed in people over the age of 60, and rarely diagnosed in people under 40 years of age. The highest incidence is in the 75–84 age group. The lung cancer incidence rate in the UK has decreased overall between 1997 and 2006, due to the reduction in smoking, mostly among men, where the rate has decreased by 21 per cent. The incidence rate in women has changed little over the same period. In the UK, there is wide geographical variation in lung cancer rates, with the highest incidence in Scotland and northern England, and lower incidences in Wales and southern England, mirroring smoking habits.

Mortality: Lung cancer is the most common cause of death due to cancer in both men and women. In the UK, variations in mortality follow the smoking habit, with the highest recorded mortality in Scotland: male cancer mortality has decreased steadily between 1982 and 2007, while in women during the same period the rates increased until the late 1990s, but then levelled off. The prediction for the future is that death rates will generally fall. However, in Scotland the death rate in women will continue to rise slightly up to 2010–2014. The different patterns of lung cancer death rates in men and women could be explained by previous smoking behaviour; men started smoking earlier than women, and smoked more heavily, and it is now apparent that more men than women gave up smoking.

Survival: Survival outcome in lung cancer is very low, as most cancers are discovered late, when the tumour has spread to other areas, or it is situated in an area that cannot be removed surgically. Radical treatment is also not feasible in some patients who may have an underlying medical condition. After diagnosis, 20 per cent of patients are alive after one year, falling to 5 per cent at five years. Survival rates for different subtypes of lung cancer are discussed later in this chapter.

Risk factors: Cigarette smoking is the main cause of lung cancer. Risk of developing lung cancer is related both to the number of cigarettes smoked and the duration of the smoking habit. A person smoking ten cigarettes a day has an eight times greater chance of dying from lung cancer–and a person smoking 25 cigarettes a day has 25 times the risk–compared to that of a non-smoker. Second-hand smoke or passive smoking (breathing the smoke of others) increases the risk of developing lung cancer. Lifelong non-smokers exposed to second-hand smoke at home or in their occupation have their risk of lung cancer raised by about a quarter, while heavy exposure at work doubles the risk. Tobacco smoke contains 55 carcinogens that have been evaluated by the International Agency for Research on Cancer (IARC), which showed that there is sufficient evidence for carcinogenicity in laboratory animals and humans.

Radon gas, a naturally occurring radioactive gas released during decay of uranium 238, is considered the second most important factor after cigarette smoking. Extensive epidemiological evidence from studies of miners and complementary animal data has documented that radon causes lung cancer in smokers and non-smokers. Radon is present in the air we breathe at very low levels, but at high concentration radon becomes a risk factor to the general population; this typically occurs in well-insulated, tightly sealed buildings. Ground floors or basements, due to their closeness to the ground, typically have the highest radon levels. In certain parts of the UK, particularly in Devon and Cornwall, high levels of radon have been recorded. Radon gas may be responsible for 9 per cent of lung cancers in Europe.

Other risk factors include occupational exposure to asbestos and polycyclic hydrocarbons.

Cross reference

For further information see the *Cytopathology* volume in this Fundamentals of Biomedical Science series.

CLINICAL CORRELATION

Types of lung cancer

Lung cancer is divided into two main types based on its histological appearance.

Non-small cell lung cancer (NSCLC) is often called non-small cell carcinoma, while **small cell lung cancer** (SCLC) is commonly called small cell carcinoma. Treatments for SCLC and NSCLC are very different, and therefore it is essential to have an unequivocal cytological or histological diagnosis.

Non-small cell carcinoma is the most common type, with approximately 80 per cent of lung cancers being of this type. Non-small cell carcinoma is subdivided into three main subtypes

Cross reference

For further information see the *Cytopathology* volume in this Fundamentals of Biomedical Science series.

METHOD *Positron emission tomography (PET) scan of the chest*

This method of imaging uses a radioactive substance called a tracer to look at areas of activity in the body. The most common tracer used is fluorine 18, also known as FDG-18, which is a radioactive version of glucose. The tracer, given through a vein (intravenous [iv]), travels through the blood and collects in biologically active tissues. A scanner detects positrons (positively charged electrons) emitted by a radio-nuclide in the organ or tissue being examined.

The PET scan is used to diagnose cancer, as tumour tissue is metabolically active and takes up more tracer than the surrounding normal tissues. This type of scanning is also used commonly to stage cancer and assess suitability for different treatment options.

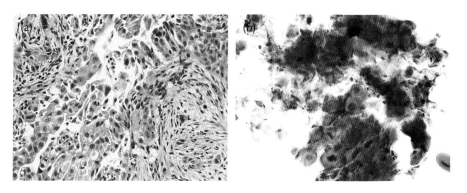

FIGURE 6.18

(a) Squamous carcinoma, bronchial biopsy (H&E stain, high power). (b) Squamous carcinoma, bronchial brush sample (Papanicolaou stain, medium power).

(squamous carcinoma, adenocarcinoma and large cell carcinoma) as approaches to diagnosis, staging, **prognosis** and treatment are similar.

Squamous carcinoma makes up 20–30 per cent of lung cancers, and is associated with smoking. Tumours most often arise in the hilum of the lung (Figure 6.18a and 6.18b).

Adenocarcinoma constitutes 30–40 per cent of lung cancers. It usually occurs at the periphery of the lung as a discrete mass arising from the proximal airways. Adenocarcinomas arise from glandular cells, such as the mucus-producing goblet cells, Clara cells and type 2 pneumocytes. The incidence of adenocarcinoma is rising and it is the most common lung cancer in ex-smokers (Figures 6.19a and 6.19b).

Large cell carcinoma (10–15 per cent of lung cancers) are aggressive, fast-growing tumours and by light microscopy show no squamous or glandular differentiation (Figures 6.20a and 6.20b).

Small cell carcinoma (15–20 per cent of lung cancers) was known previously as oat cell carcinoma as the small nuclei resemble oat grains (Figures 6.21a and 6.21b).

This tumour usually occurs centrally and metastasizes very early to the hilar lymph nodes, brain, and liver. Sometimes a patient presents with metastases and the primary may be too small and difficult to find. The majority of patients have a strong smoking history.

Cross reference

For further information see the *Cytopathology* volume in this Fundamentals of Biomedical Science series.

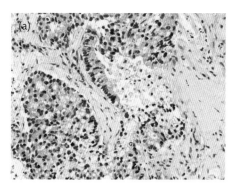
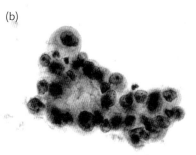

FIGURE 6.19

(a) Adenocarcinoma, bronchial biopsy (H&E stain, medium power). (b) Adenocarcinoma, bronchial brush sample (Papanicolaou stain, medium power).

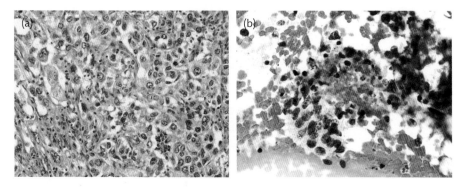

FIGURE 6.20
(a) Large cell carcinoma, bronchial biopsy (H&E stain, high power). (b) Large cell carcinoma, bronchial brush sample (Papanicolaou stain, medium power).

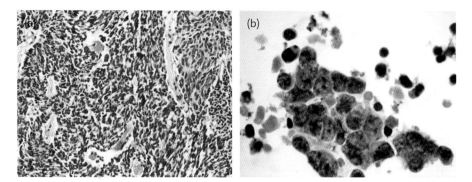

FIGURE 6.21
(a) Small cell carcinoma, bronchial biopsy (H&E stain, medium power). (b) Small cell carcinoma, bronchial brush sample (Papanicolaou stain, high power).

6.9 **Bronchioles**

The bronchioles have a diameter of 5 mm or less. Figure 6.22 shows the anatomical position of bronchioles in relation to the bronchi.

Unlike those of the the bronchi, the bronchiolar walls do not contain cartilage or glands. They do, however, contain a loose network of smooth muscles (Figures 6.23 and 6.24), some running circular and others at an angle. Their diameter can be regulated by a signal from the **autonomic nervous system**. Within this loose network there are **parasympathetic and sympathetic** nerve fibres; the parasympathetic fibres stimulate constriction of bronchioles while the sympathetic fibres have the opposite effect of relaxing the muscle and allowing the bronchioles to return to their normal diameter.

Autonomic nervous system
The autonomic nervous system is an involuntary division of the nervous system. It conducts impulses to cardiac muscle, smooth muscle and glands.

Key points

The walls of the bronchioles do not contain cartilage, but instead contain smooth muscle that can increase or decrease the diameter of the bronchioles in response to signals from the autonomic nervous system. Parasympathetic stimulation causes smooth muscle contraction (bronchoconstriction) while sympathetic stimulation causes relaxation of smooth muscle, leading to bronchodilation.

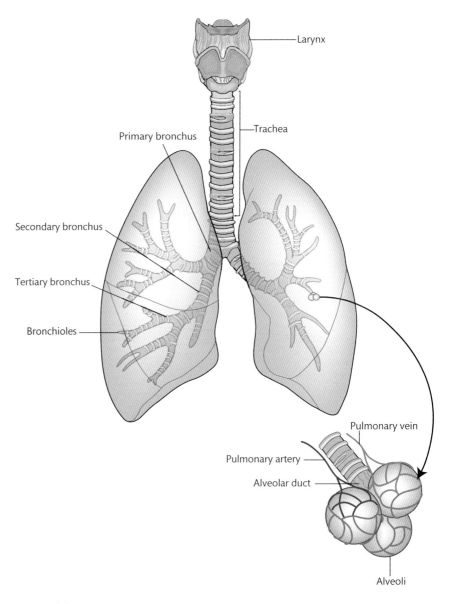

FIGURE 6.22
Various divisions of the bronchial tree.
Modified from original image © Dr Michelle Peckham, Faculty of Biology Sciences, University of Leeds.

CLINICAL CORRELATION

Bronchospasm

The sudden constriction of bronchioles is called bronchospasm.

This is caused by contraction of smooth muscles present in the wall of bronchioles due to chemicals released by **mast cells** and **basophils**. This causes difficulty in breathing, which can range from mild to very severe. In extreme cases, death by **asphyxiation** has been reported. Bronchospasm occurs in many conditions including the common inflammatory respiratory disease **asthma**. An

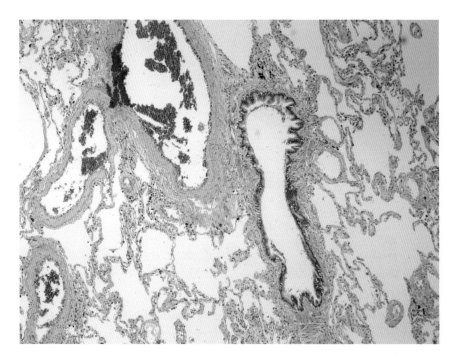

FIGURE 6.23
Respiratory bronchiole (H&E stain, medium power).

acute asthma attack can be treated by use of drugs such as inhaled short-acting **beta-2 agonist** (such as **salbutamol**). This drug acts by relaxing the smooth muscle resulting in dilation of the bronchial passages.

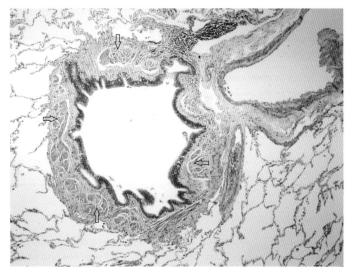

FIGURE 6.24
Respiratory bronchiole (H&E stain, high power). Arrows highlight smooth muscles.

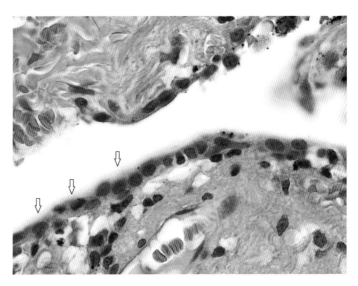

FIGURE 6.25

Clara cells (arrows) in the epithelium of a respiratory bronchiole (H&E stain, high power).

SELF-CHECK 6.15

What mechanism regulates the diameter of the bronchioles?

The epithelium of large bronchioles is ciliated pseudostratified columnar but gradually changes to become simple columnar or cuboidal in smaller bronchioles. The number of goblet cells is also reduced. The epithelium of terminal bronchioles contains non-ciliated cells called **Clara cells** (Figure 6.25).

Clara cells, originally described in 1881 by Kolliker, are named after Max Clara who in 1937 carried out a comparative study of human and rabbit bronchioles.

Clara cells are dome-shaped with numerous short microvilli on the surface. The function of Clara cells has been the subject of debate for many years, but it is generally agreed that their main function is to protect bronchiolar epithelium from pollutants and inflammation by producing a substance called **Clara cells secretory protein** (CCSP).

Clara cells may also act as **stems cells**, which means that they have the capacity to form new bronchiolar epithelium in the event of cell damage or death.

Each bronchiole undergoes branching to form about five smaller terminal bronchioles. The epithelium of terminal bronchioles changes from simple columnar to cuboidal.

Clara cells

These cells produce various proteins that protect the bronchiolar epithelium. They also have the potential to differentiate into ciliated cells.

METHOD *Computed tomography (CT) scan*

The CT scan is a diagnostic test that uses X-rays and computers to produce high-quality images of internal organs. Since its introduction in the 1970s, CT scanning has become an important diagnostic tool in patients with chest symptoms. The CT images can be enhanced by injection of a contrast medium (dye). A CT scan can also be used to target tissues accurately for the purpose of taking core biopsy or FNA samples.

SELF-CHECK 6.16

The epithelium of large bronchioles is lined by:

a) simple columnar

b) non-keratinizing squamous

c) pseudostratified columnar

d) ciliated pseudostratified columnar

e) simple cuboidal.

6.10 **Respiratory bronchioles**

The terminal bronchiole undergoes further branching to give rise to respiratory bronchioles. The respiratory bronchiole is the point at which the conducting portion of the respiratory tract ends and the respiratory portion begins (Figure 6.23).

The epithelium of respiratory bronchioles is identical to that of terminal bronchioles, but it also includes numerous sac-like alveoli where gaseous exchange occurs. As the branching of respiratory bronchioles continues, the number of alveoli increases. Smooth muscle and elastic connective tissue are present in the epithelium of respiratory epithelium.

6.11 **Alveolar ducts**

The terminal branches of the respiratory bronchioles give rise to alveolar ducts. Alveolar ducts are tiny ducts that connect the respiratory bronchioles to alveolar sacs, each of which contain a number of alveoli. The alveolar ducts and the alveoli are lined by squamous cells which facilitate air exchange. The alveolar ducts open into small spaces called atria, which connect to the alveolar sacs. The connective tissue supporting the alveolar sacs and alveoli consists of elastic and reticulin fibres.

Cross reference

For further information see the *Histopathology* volume in this Fundamentals of Biomedical Science series.

6.12 **Alveoli**

Alveoli are the terminal portion of the respiratory tract where uptake of oxygen and excretion of carbon dioxide by the process of gaseous diffusion takes place between blood and air (Figure 6.26)

These sac-like structures project from alveolar ducts or alveolar sacs and give the lungs their spongy appearance. Viewed under a microscope, they appear as round cavities similar to honeycombs of a beehive (Figure 6.27). It is estimated that there are up 700 million alveoli, giving a very large total surface area to facilitate gaseous exchange. The structure also enhances the process of gaseous diffusion; the epithelium lining the alveoli is simple squamous and two neighbouring alveoli are separated by a very thin interalveolar septum. The alveolar septum is made up of two squamous cells from adjoining alveoli, separated by a rich network of capillaries supported by connective tissue components of reticulum and elastic fibres.

There is no connective tissue material between the air in the alveoli and endothelial cells lining the capillaries, thus minimizing any resistance to the diffusion of oxygen and carbon dioxide.

Respiratory portion: Alveolar ducts and alveoli

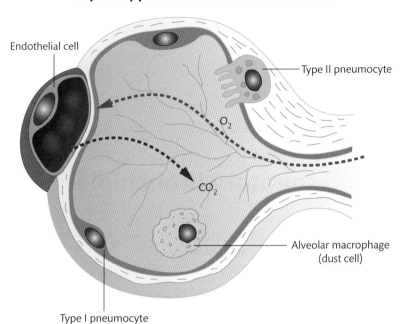

FIGURE 6.26

Drawing showing an alveolus and the air exchange.

Modified from original image: Cui *et al.*: *Atlas of Histology with Functional and Clinical Correlations* © 2011 Lippincott Williams & Wilkins.

This physical separation of air from blood is known as the **blood-air barrier**. The blood-air barrier exists to prevent air bubbles from forming in the blood, and from blood entering the alveoli. The barrier is permeable to oxygen, carbon dioxide, carbon monoxide and many other gases. The interalveolar septum includes occasional openings called alveolar pores that allow movement of air between adjacent alveoli.

Two major cell types make up the alveolar wall, **type 1 pneumocytes** and **type 2 pneumocytes** (Figure 6.28), which are also referred to as type 1 and type 2 alveolar cells.

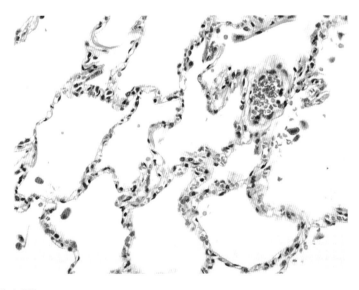

FIGURE 6.27

Honeycomb appearance of alveoli (H&E stain, high power).

FIGURE 6.28
Type 1 (large arrow) and type 2 (small arrow) pneumocytes (H&E stain, high power).

Type 1 pneumocytes are extremely thin squamous cells responsible for gaseous air exchange and occupy 95 per cent of total alveolar surface area. Type 2 pneumocytes are scattered among the type 1 cells (Figure 6.29). They have the potential to divide and differentiate into type 1 pneumocytes, which are unable to replicate. The other major function of type 2 pneumocytes is the production of pulmonary **surfactant**. This reduces the surface tension of, and the inspiratory force needed to expand the alveoli, and reduces the possibility of the alveoli collapsing during expiration. The surfactant is composed of phospholipids and includes proteins.

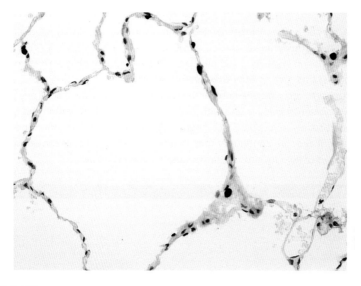

FIGURE 6.29
Type 2 pneumocytes stained brown (TTF1 antibody).

Key points

The blood-air barrier exists to prevent air bubbles forming in the blood, and to prevent blood entering the alveoli. The barrier is permeable to oxygen, carbon dioxide, carbon monoxide and many other gases.

SELF-CHECK 6.17

What do you understand by the blood-air barrier?

SELF-CHECK 6.18

The function of a type 1 pneumocyte is:

a) production of surfactant

b) removal of foreign matter

c) gaseous exchange

d) phagocytic activity

e) producing antibodies.

6.13 Macrophages

Alveolar macrophages (also known as **histiocytes**) are macrophages found in the pulmonary alveolus. They are derived from monocytes that originate in bone marrow. They form the first line of defence against infection. They are **phagocytes**; their role is engulfing, digesting and the removal of inhaled particles and pathogens. They also stimulate lymphocytes and other immune cells to respond to foreign matter by releasing various secretory products and interacting with other cells and molecules through the expression of several surface receptors.

Alveolar macrophages vary in appearance depending on their activity. In cytological preparations, the cytoplasm may be vacuolated and evidence of inhaled particles (usually carbon) may be seen as brown granules in Papanicolaou stain preparations (Figure 6.30) or black with **Romanowsky stains** such as **May Grunwald Giemsa** (MGG) and Diff-Quik. These carbon-laden macrophages are particularly common in smokers or city dwellers. They may have a single round or bean-shaped nucleus with granular chromatin. Binucleation and multinucleation occur, particularly in response to foreign matter or infection that cannot be eliminated. Alveolar macrophages are continually removed from the lungs, being carried upwards in a moving thin layer of fluid towards the larynx, and are then swallowed.

SELF-CHECK 6.19

Alveolar macrophages are derived from:

a) type 1 pneumocytes

b) type 2 pneumocytes

c) Clara cells

d) monocytes

e) neuroendocrine cells.

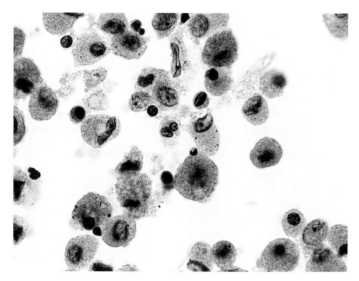

FIGURE 6.30
Alveolar macrophages. The cytoplasm is foamy and includes ingested carbon particles.

6.14 **Blood vessels**

The lungs have a twin blood supply; the **pulmonary circulation** for gas exchange with the alveoli, and the **bronchial circulation** to supply the tissue of the lung itself with blood and nutrients.

The pulmonary arteries carry deoxygenated blood to the lungs, where the carbon dioxide is released and oxygen is picked up. The oxygenated blood then leaves the lungs via the pulmonary veins, and returns to the left atrium and is distributed to the systemic circulation via the left ventricle.

The lung receives blood and nutrients from the bronchial arteries.

6.15 **Pleura**

The pleura is the serous covering of the lung, and is lined by a membrane called the **mesothelium**. The cells forming the mesothelium are called **mesothelial cells**. Immediately below these cells is a thin layer of fibrous connective tissue, which includes lymphatic vessels and capillaries. Mesothelium is composed of two layers that define the pleural cavity: one layer covers the wall of the cavity and the other lines the surface of the lung. The layer covering the wall of the cavity is called the **parietal layer**, and the layer covering the lung is referred to as the **visceral layer** (Figure 6.31). The mesothelium produces **serous fluid**, a lubricant that permits smooth movement of the lung during respiration. Under normal conditions there is only

Pulmonary circulation
The portion of the cardiovascular system that carries oxygen-depleted blood away from the heart, to the lungs, and returns oxygenated blood back to the heart.

Bronchial circulation
The bronchial circulation is the systemic vascular supply to the lung, and it supplies blood to conducting airways down to the level of the terminal bronchioles.

Pleura

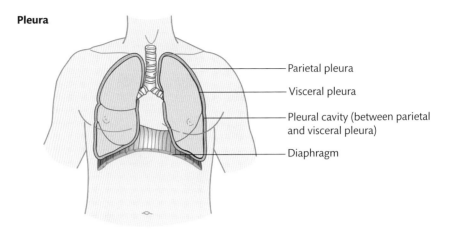

FIGURE 6.31
Schematic diagram of the pleural cavity.
© Oxford University Press 2011.

Hydrostatic pressure
In blood vessels it is due to
the height of the blood above,
which exerts a force on the
blood lower in the body.

a very small amount of pleural fluid present in the pleural cavity, but in certain pathological states excess fluid is produced. This excess fluid is called an **effusion**. Pleural fluid is formed by filtration of plasma through the capillary endothelium, and its rate of production is determined by several factors, including **hydrostatic pressure** within the capillary lumen, **colloid oncotic pressure** (osmotic pressure caused by proteins in plasma), rate of lymphatic drainage, and permeability of capillaries. Imbalance in these factors may result in an accumulation of fluid. Under normal circumstances the fluid contains a small number of mesothelial cells and lymphocytes.

SELF-CHECK 6.20

What is the function of serous fluid and how is it produced?

Key points

An effusion develops when the amount of fluid entering the cavity exceeds the amount that is removed. This may be due to a combination of factors including changes in hydrostatic pressure within the capillary lumen, colloid oncotic pressure, rate of lymphatic drainage, and permeability of capillaries.

SELF-CHECK 6.21

The pleura are:

a) lined by mesothelium

b) composed of a parietal and a visceral layer

c) produce serous fluid for lubrication

d) found around the lungs

e) all of the above.

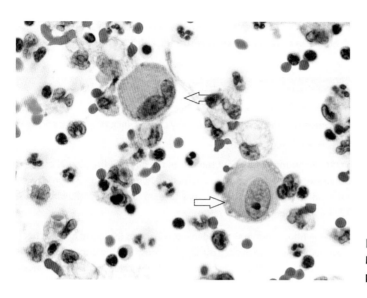

FIGURE 6.32
Mesothelial cells (arrowed) (Papanicolaou stain, high power).

6.15.1 Mesothelial cells

Cytologists are very interested in the microscopy of cells present in serous effusions, as many advanced cancers often metastasize to the pleural cavity. Recognizing the normal morphology of mesothelial cells is an important step in learning cytology and here we will briefly describe their appearances.

Mesothelial cells are usually present as discrete single cells, or in small clusters. Mesothelial cells generally have a single nucleus, but bi-nucleation and multinucleation can sometimes be seen, particularly in response to injury or inflammation. The nucleus is round, positioned centrally or just off centre, and the chromatin pattern is often described as being finely stippled (appearing as small dots), but this can vary (Figures 6.32 and 6.33).

Cross reference

For further information consult Chapter 9 of the *Cytopathology* volume in this textbook series.

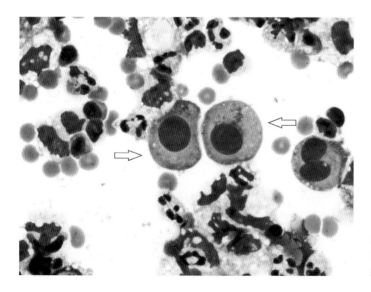

FIGURE 6.33
Mesothelial cells (arrowed) (May Grunwald Giemsa stain, high power).

The nuclei often have a single nucleolus. Mesothelial cells that are actively dividing in response to injury or other stimuli are often called **reactive mesothelial cells**.

CLINICAL CORRELATION

Malignant mesothelioma

The pleural cavity is often involved in advanced stages of many cancers, including those arising in the lung, breast and gastrointestinal tract.

Malignant mesothelioma, an uncommon malignant tumour of the lining of the pleural cavity, is caused by long-term asbestos exposure. Most people with malignant mesothelioma have worked or lived in places where they inhaled asbestos fibres. It may take up to 40 years following exposure to asbestos for malignant mesothelioma to develop.

Asbestos is a naturally occurring fibrous mineral mined in many parts of the world and has been in use for thousands of years. Asbestos use increased during the industrial revolution, when it was employed initially in heavy industries for insulating boilers and pipes. Later, however, its unique property of being resistant to heat, acid and alkali made it an ideal building material. Asbestos was also used in some common household goods such as oven gloves, mats and flooring. At the height of its use, it is thought that up to 3000 products contained asbestos. Use of asbestos continued into the twentieth century, finally being banned in the UK in 1999.

Mesothelioma is a rare disease: in 2005 in the UK there were 2164 new cases (2.7 per 100 000 [age-standardized]) comprising 1827 cases in men and 337 in women. Most of these cases were in people who had worked with, or had been exposed to, asbestos; however, a number of cases occurred in those with no history of asbestos exposure. Currently, there is no effective therapy for mesothelioma and hence mortality rates are very high (median survival 10 months); in 2006 there were 1996 deaths (2.3 per 100 000 [age-standardized]). The expected number of deaths among males is predicted to increase to a peak of 2038 around the year 2016.

Chapter summary

- The function of the respiratory tract is to remove oxygen from inspired air and eliminate carbon dioxide from the blood.

- The respiratory system is exposed to an array of particulate matter and microorganisms present in the inspired air, and there are several mechanisms in place to remove these.

- Large particles are retained in the nasal passages and smaller particles are trapped in the mucus secreted by the respiratory epithelium.

- Microorganisms and microscopic particles are removed by alveolar macrophages.

Further reading

- Ferlay J, Shin HR, Bray F, Forman D, Mathers C, Parkin DM eds. GLOBOCAN 2012: Cancer Incidence and Mortality Worldwide: IARC Cancer Base No. 10. Lyon: International Agency for Research on Cancer, 2014 (http://globocan.iarc.fr).

- Mescher A. *Junqueira's Basic Histology: Text and Atlas* 12th edn. New York: McGraw-Hill Medical, 2009.

- Office for National Statistics. Cancer Registration Statistics 2008 England. Series MB1, No. 39, 2011.

- Parkin DM. Tobacco-attributable cancer burden in the UK in 2010. *Br J Cancer* 2011; **105** (Suppl 2): S6–S13.

- Parkin DM. Cancers attributable to consumption of alcohol in the UK in 2010. *Br J Cancer* 2011; **105** (Suppl 2): S14–8).

- Shambayati B. *Cytopathology*. Oxford: Oxford University Press, 2011.

 # Discussion questions

6.1 Discuss the morphology of the various cell types lining the respiratory tract in relation to their function.

6.2 Describe the function of the nasal cavity, relating the structures and cells to the function.

6.3 What are the main functions of the larynx? Describe its structure and relate it to its function.

6.4 What do you understand by bronchopulmonary segments?

6.4 Compare and contrast type 1 and type 2 pneumocytes and relate their structure to their function.

6.6 What is the function of serous fluid in the pleura? How is serous fluid produced and how does it accumulate in disease?

Answers to self-check questions and discussion questions, hints and tips are provided on the book's Online Resource Centre.

 Visit www.oxfordtextbooks.co.uk/orc/orchard_csf/.

7

Digestive system

Tony Warford and Tony Madgwick

Learning objectives

After studying this chapter you should confidently be able to:

- Name and explain the functions of the major anatomical divisions of the gastrointestinal tract.
- Explain the relationship between histological structure and function in the alimentary tract.
- Recognize and explain the different patterns of epithelial organization within the alimentary tract.
- List and state the functions of the major cell types involved in the control and functional regulation of secretions and peristalsis in the alimentary tract.
- Describe the major immune surveillance components of the alimentary tract.

The digestive system is a complex association of different integrated anatomical components with the primary task of extracting nutrients, water and electrolytes from ingested food before collecting and excreting the indigestible waste. Thus, the system can be divided into a region for mechanical disruption and initial digestion of food (the oral pharynx), with delivery of this mass via the oesophagus to the stomach for further processing by extracellular digestion. Most nutrient uptake occurs in the small intestine (further divided into the duodenum, jejunum and ileum), while water and electrolyte absorption occurs mainly in the colon. Finally, the undigested remnants are collected in the distal colon and rectum before expulsion by defaecation. A secondary, but important, function is that of immune surveillance provided by aggregates of lymphoid tissue associated with the entire alimentary tract. Other important accessory organs, including the liver, pancreas and gall bladder, are also considered to be part of the digestive system, but these will be considered separately elsewhere in this textbook.

The core structure of the alimentary tract is essentially a tube of moist epithelium arising in the mouth and finishing at the anus. Although variable in length between individuals, the average intestine spans some nine metres from start to finish. Its function is the processing of food and fluids to provide the essential water and nutrients required by the cells of the body for growth, metabolism and repair. This is achieved by numerous and exquisite specializations of the epithelium and its supporting structures. The contents of the alimentary tract are propelled in one direction by regular waves of contraction of the outer layers of smooth muscle. The transit time for food from ingestion to voiding is on average 50 hours. However, the actual time taken varies with age, sex, type of food and health (Camilleri *et al.*, 1986; Metcalf *et al.*, 1987; Weaver and Steiner, 1984).

In this chapter we will explore these functional adaptations and seek to explain the relationship between tissue function and its structure.

7.1 **General structural arrangement**

Although all of its structures are found within the body, the internal surface of the alimentary tract is essentially an external covering as it is in contact with the outside world.

From the oral cavity, through the oesophagus to the rectum, the alimentary tract has a common structural organization (Figure 7.1). This basic plan consists of a tube of epithelium organised into a **mucosa** and bounded concentrically by thin **muscularis mucosae**, the **submucosa**,

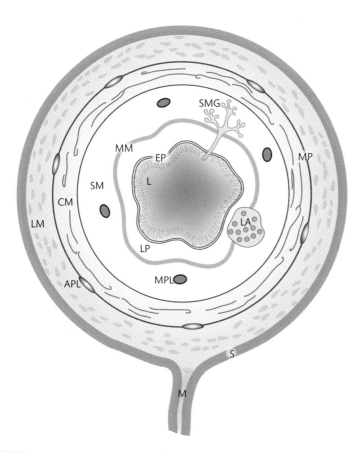

FIGURE 7.1

Diagram of the general arrangement of the alimentary tract. Lumen (L); mucosa consisting of epithelium (EP), lamina propria (LP) and muscularis mucosae (MM); submucosa (SM) containing submucosal glands (SMG), lymphoid aggregates (LA) and Meissner's nerve plexus (MPL); muscularis propria (MP) with longitudinal muscle (LM) and circular muscle coat (CM) and between them Auerbach's nerve plexus (APL); serosa (S) and mesentery (M).

and a substantial **muscularis propria**. The complete unit is bounded by a thin **serosa** and an **adventitia** or a mesentery containing the supporting blood vessels and lymphatic system.

The epithelium of the **mucosa** is specialized according to region but can be broadly divided into simple columnar or stratified squamous epithelium. The columnar epithelium has either an absorptive or secretory function (i.e. within the small intestine) while the stratified squamous epithelium has a protective function (i.e. within the oral mucosa). The epithelium of the intestine is mitotically very active and continually produces new cells to replace those lost by exfoliation into the lumen (Figure 7.2).

The epithelium is supported by a **lamina propria** generally formed of loose connective tissue that, with the exception of the oesophagus, harbours a lymphoid cell population. At the inner boundary of the mucosa, the **muscularis mucosae** is composed of a thin layer of smooth muscle and facilitates local movement of the mucosa, promoting, for example, the release of mucus from epithelial cells.

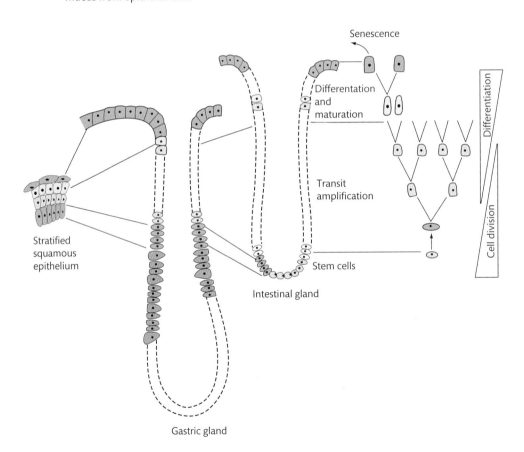

FIGURE 7.2

Maintaining mucosal integrity in the mucosal epithelium. This diagram illustrates the source of the cells that populate the intestinal mucosa. The cells populating the mucosa arise from the stem cell compartment. Asymmetric divisions here renew the multipotent stem cell and generate a symmetrically dividing transit cell. These immature cells go through several cycles of division (amplification), forcing the older cells to move up the structure (and also down in the case of gastric epithelium). Next, the cells differentiate and mature into the cells typical of their functional unit. Following a brief functional life, these cells become senescent and either undergo apoptosis or are shed from the mucosal surface.

External to the mucosa is the **submucosa**. This is composed of loose connective tissue and is the main delivery area for blood vessels and draining lymphatics. Adipocytes are often found in the submucosa and lymphoid aggregates that form gut-associated lymphoid tissue (GALT; see Box 7.1) are frequent and may protrude into the mucosa. The submucosa also contains a complex branching network of parasympathetic nerve tracts and ganglia (**Meissner's nerve plexi**) from which a finer network of nerves enters and controls the contractile activity of the muscularis mucosae and the secretory activity of the glandular epithelium (Wedel *et al.*, 1999).

BOX 7.1 Gut-associated lymphoid tissue (GALT)

Mucosa-associated lymphoid tissue (MALT) is the name given to the immune system associated with all mucosal tissues. The term usually refers to distinct tissue structures comprised of numerous white blood cells (mostly T cells and B cells) held together by a network of loose connective tissue in close association with the epithelium.

Gut-associated lymphoid tissue (GALT) is present throughout the alimentary tract, although its presence or absence will reflect the precise anatomical site or the health and age of an individual. It plays an important role in immune surveillance and innate immunity. Primary components are the tonsils and adenoids situated in the oral cavity, Peyer's patches of the small and large intestine, and extensive lymphoid aggregates in the appendix. In addition, GALT aggregates may be found in the stomach and these increase in incidence with age.

Situated in the submucosal regions, lymphoid aggregates can also penetrate the mucosa to provide more intimate sampling of the luminal contents. In these situations the epithelial covering is also modified. Accordingly, in the tonsil the basement membrane of crypt epithelium is discontinuous and lymphocytes are present within the stratified epithelium itself. Where mucosal penetration occurs in the small and large intestine as well as the appendix, epithelial glands are absent and a single layer of covering epithelium (termed follicle-associated epithelium [FAE]), containing microfold (M) cells, provides a functionally modified barrier between the lymphoid tissue and the luminal contents (Kraehenbuhl and Neutra, 2000; Corr *et al.*, 2008).

The M cells have a specialized apical surface complex ultrastructure with the characteristic microvilli of absorptive cells replaced by broader folds of the cell membrane. The overlying glycocalyx is also thinner facilitating the ingestion of antigens by endocytosis. Within the cells, antigens are transported to pockets on their basolateral surfaces and from there to lymphocytes and antigen-presenting cells. The 'fast track' transportation offered by M cells is predominantly beneficial, but organisms such as *Salmonella typhimurium* and individual proteins such as the prions of bovine spongiform encephalopathy can also gain entry to the body using these cells.

It should be remembered that in areas where lymphoid aggregations are absent there is still an abundant population of small lymphocytes and plasma cells present in the lamina propria, reflecting the continuous immune challenge to the alimentary tract, and the body's response. Information on the structure of GALT tissue is provided in the main text.

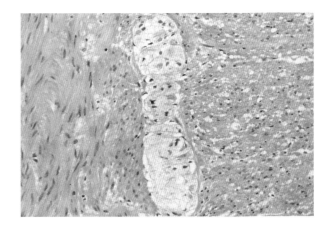

FIGURE 7.3
Auerbach's plexus (centre) between the muscle coats of the muscularis propria (haematoxylin and eosin [H&E] stain; high-power magnification).

The next layer is the **muscularis propria**, comprising inner circular and outer longitudinally orientated bundles of smooth muscle. These muscular components are responsible for the waves of peristaltic contraction that propel the luminal contents downwards along the alimentary tract. Both muscle layers are under parasympathetic nervous control and its nerve tracts and ganglia can be seen clearly between these muscle layers as the contrastingly pale-stained **myenteric** (or Auerbach's) **plexi** (Figure 7.3)

The outermost layer of the mobile, intraperitoneal segment of the alimentary tract is the **serosa**: a thinner version of the submucosa supporting a thin layer of **mesothelial cells** that produce a lubricating fluid. The serosa also provides attachment points to the mesentery. The fixed, extraperitoneal segments (including the oral mucosa, the oesophagus and parts of the distal colon) are covered with a loose connective tissue termed the **adventitia**.

The main divisions of the alimentary tract are outlined in Table 7.1.

The blood supply to the alimentary tract varies according to region. Supply to the oesophagus is via branches of the thoracic aorta, whilst to the divisions of the alimentary tract situated in the abdominal cavity three vessels arising from the abdominal aorta form an anastomozing system supplying the stomach, small and large intestines (Geboes *et al.*, 2001). Nutrients absorbed from the stomach, small and large intestines are transported to the liver via the hepatic portal vein.

TABLE 7.1 **Divisions and primary functions of the alimentary tract.**

Division	Primary function
Oral cavity	Mastication of food and initial digestion of carbohydrate
Oesophagus	Transport of masticated food to the stomach
Stomach	Acid and pepsin digestion to produce chyme
Small intestine	Absorption of nutrients and water
Large intestine	Absorption of water and electrolytes. Preparation of faeces for excretion
Rectum	Storage and expulsion of faeces.

7.2 **Oral cavity**

The primary function of the oral cavity (or pharynx) is the mastication and ingestion of foods and liquids. The oral cavity is bound by the jaws and their alveolar arches anteriorly and by the pharyngeal cavity posteriorly and is lined by a stratified epithelium moistened by saliva produced by a range of external and accessory salivary glands. Food taken into the mouth is mechanically disrupted (mastication or chewing) through the combined action of the jaws, teeth and tongue. During mastication food is mixed with a watery saliva to form a lubricated bolus ready for swallowing. While the act of chewing and swallowing of the bolus is under voluntary control, once swallowed its movement through the intestine is under a complex but predominantly autonomic control.

The jaws, comprising the dorsal maxilla and ventral mandible, form the bony matrix supporting the teeth. The maxilla is fixed, but the mandible is able to move in a dorso-ventral axis (for cutting and crushing) and laterally (for grinding).

7.2.1 **Teeth and gingiva**

Teeth are remarkably complex and different types of tooth structure are associated with their function. For example, the incisors have a wedged cutting surface for slicing, while the canine teeth are pointed and have deep roots and are used for holding and tearing food items. The molars have a flat grinding surface for mechanically reducing food in the mouth during chewing.

Teeth are the only body parts that can be shed and replaced, although this will happen only once. The deciduous (milk) teeth are the first to erupt from the bony matrix of the jaws. Beginning at the age of about six months, eruption is usually complete between the ages of six to eight years, by which time there are 20 teeth in the mouth. Between the ages of 10 and 12 years, these deciduous teeth are lost sequentially and replaced by the permanent dentition. Additional molars erupt so that by about 18 years of age a healthy mouth will contain 32 teeth (see Box 7.2).

The teeth are held in place within bony sockets in the upper and lower jaws. This is not a solid attachment and there is some movement of the teeth, reflecting the shock-absorbing function of the **periodontal ligament**, which is the connective tissue holding the tooth in place. The pulp cavity of the tooth contains blood vessels and nerves, which are important in maintaining the integrity of the tooth and also in mechano-proprioception. The neural component helps to protect the teeth against damage. Most of us will have experienced the reflex response where the jaws snap open if an unexpectedly hard foreign body is found in the mouth during chewing soft material. The force generated by the muscle of the jaw can be strong enough to break teeth if concentrated into a small area.

The **gingiva** (gum) is an important protective epithelium that helps to seal the gap between the tooth and the bony socket. If this becomes damaged or inflamed, it can pull away from the tooth and allow food and bacteria to build up into a thick biofilm. Microbial action in

BOX 7.2 *Teeth in forensic pathology*

Teeth can be of important service to forensic pathology. The patterns of tooth eruption can be used to age individuals up to about 20 years of age both in the living and the dead.

Teeth are tough and often survive when all else has been consumed, thus they can be important in helping to identify individual bodies from dental records after natural disasters.

BOX 7.3 Dental health

The oral mucosa has a rich microflora, but one rarely sees attachment to tissue except between the gingiva and the teeth, or in cases where there is damage to the integrity of the mucosa and its protective saliva (e.g. in cancer patients undergoing radiotherapy or chemotherapy). The dental profession is often keen to point out that the oral mucosa can reflect underlying disease. Regular visits to the dentist are more than just checks for damage caused by dental caries; they can reveal underlying systemic conditions such as vitamin deficiency, diabetes or even life-threatening local disease such as oral cancer.

this biofilm can produce a sufficient quantity of acids to break down the enamel of the tooth, resulting in dental caries (tooth decay, see Box 7.3).

SELF-CHECK 7.1

Why do we have two waves of tooth eruption?

7.2.2 Saliva and the salivary glands

Saliva is a serous secretion (containing water and proteins) but may also variously contain mucus and other protective molecules such as immunoglobulins, and, in some special secretions, enzymes. The role of saliva is to keep the oral mucosa moist and lubricated and to help protect the underlying epithelium from microbial attack. In addition to its role in lubrication, saliva serves an additional digestive function by containing salivary enzymes (amylases) that begin the primary digestive process by breaking down complex carbohydrates.

Saliva is produced by a number of specialized exocrine glands. There are three pairs of lobulated glands distributed around the oral cavity: the submandibular, sublingual and parotid glands. The major salivary glands are activated by the process of eating or in response to externally sensed stimuli (famously illustrated in the salivation response by Pavlov's dogs to a ringing bell [see Box 7.4]). These different mechanical events are served by a range of specializations within the salivary glands. The epithelium of the acinar glands are organized into quite distinct regions producing either a watery secretion containing enzymes (serous glands)

BOX 7.4 Pavlov's dogs and salivation

Ivan Pavlov was a Russian physiologist who is most remembered for his work on 'classical conditioning' in the early twentieth century. Dogs will begin to salivate when they come into contact with food (the unconditioned stimulus). By ringing a bell every time the dogs fed, the animals were able to associate the ringing bell (the conditioned stimulus) with food such that they would salivate in response to the bell, without smell or sight of food. However, it has become clear in recent times that this is not a straightforward physiological response (Rescorla, 1988). Many animals, including humans, may learn to link different conditional stimuli to food presentation very quickly and they may develop a cognitive response where there is prediction of the availability of food. Furthermore, the composition of saliva will change according to whether it is initiated by the sympathetic or parasympathetic system (Humphrey and Williamson, 2001). Interestingly, while the mechanics of chewing and taste of acid (think of the popular sugary confection Tangfastics) elicit a strong salivary flow, smelling food is not as effective.

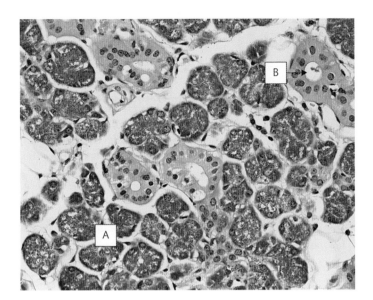

FIGURE 7.4
Submandibular gland showing serous glands (A) and ducts (B) (H&E stain; high-power magnification).

or a dilute mucus (mucous glands). Figures 7.4 and 7.5 show how these epithelial components are organized in the submandibular gland. The products of the secretory epithelium are transported to the oral cavity by small tubes (ducts) that collect the secretions into a common duct which opens into the oral cavity. These ducts are lined by a simple cuboidal epithelium.

Embedded in the submucosa throughout the oral cavity are hundreds of smaller accessory salivary glands, the role of which is to keep the oral mucosa constantly moistened and lubricated (see Box 7.5).

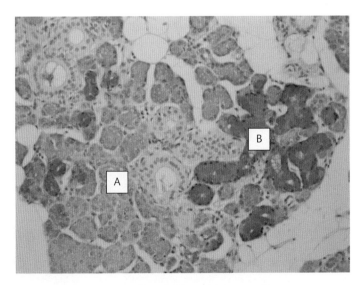

FIGURE 7.5
Submandibular gland showing the mixed distribution of serous (A) and mucus (B) producing glands (Short Maxwell staining method; high-power magnification).

BOX 7.5 Epidermal growth factor (EGF) in salivary secretions

Salivary gland secretions contain a number of hormones and other constituents that have a physiological role (Groschl, 2009). Epidermal growth factor is important in stimulating proliferation in basal epithelial cells and is an important initiator of regeneration and repair. It is quite noticeable that a cut in the mouth will heal much more quickly in the oral mucosa than in the keratinized squamous epithelium of the skin. Equally, the response to injury whereby a wound is licked is a common mammalian response.

SELF-CHECK 7.2

Explain why few people can eat more than 5–8 dry cheese biscuits (or crackers) without drinking.

7.2.3 Tongue

The tongue is an incredibly versatile and multifunctional organ. It is important for speech, manipulation of food in the mouth during chewing, aiding with the sensation of taste and with lipolysis. To a lesser extent, it is also important in maintaining clean teeth and gums.

The tongue is covered by keratinizing stratified squamous epithelium into which loose connective tissue papillae extend (Figure 7.6). Below this there are eight striated muscular elements organized into the intrinsic and the extrinsic muscles. The intrinsic muscles allow the tongue to fold and bend and change its shape, while the extrinsic muscles determine the position of the tongue—whether it is retracted, protruded or moving from side to side. The extrinsic muscles have their origins in the bony components of the oral cavity (e.g. the mandible, hyoid and styloid).

The upper surface of the tongue contains numerous small, raised bumps (papillae) containing gustatory cells within the taste buds. Taste buds are present not only on the dorsal aspect of

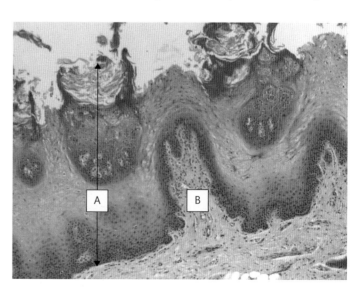

FIGURE 7.6
Tongue showing stratified keratinized squamous epithelium (A) and filiform papillae (B) (H&E stain; medium-power magnification).

BOX 7.6 The tongue and pathology

The complex surface structure of the tongue provides plenty of opportunity for fungal and microbiological infection to take hold in the nooks and crevices of the superficial epithelium. For example, *Candida albicans* is a frequent colonizer of the upper surface of the tongue, and where an individual becomes immunocompromised then the yeast will proliferate, resulting in a painful inflammatory response (oral thrush).

the tongue, but also on the soft palate, pharynx and the upper part of the oesophagus. Taste buds are composed of assemblages of 100 or so specialized cells that express a range of receptors belonging to three classes of cell type that detect the presence of tastants. These cells are important for detecting the main taste sensations of sweet, sour, salt, bitter and umami. Umami refers to the sensation of 'meatiness' and the prototypical tastant is monosodium glutamate (Roper, 2013). Activation of these cells elicits the sensations of salty (type I), sweet, bitter and umami (type II) and sour (type III). Specific areas across the tongue can be identified where these taste sensations are localized, reflecting the grouping of taste buds with specialized function.

Von Ebner's glands are found around the tongue papillae and these secrete a serous fluid containing lipases. These enzymes are important for the first stages of fat digestion and may also be important for feedback on the nutritional quality of ingested food (Kawai and Fushiki, 2003). Servicing all these functions, the tongue is richly supplied with blood vessels and nerves (see Box 7.6)

SELF-CHECK 7.3

The five main sensations do not explain the great variety of taste and flavours we experience. Why is that?

7.2.4 Oral mucosa

The mucosa of the oral cavity (along with the oesophagus and the anal canal) is quite different from the other mucosae of the alimentary tract in that it contains a stratified squamous epithelium. The function of the mucosa is the protection of the oral cavity from mechanical or chemical damage and to act as a barrier against the entry of microorganisms into the underlying structures of the oral pharynx. The oral mucosa is also different from other mucosae by not having a muscularis mucosae and in many areas not even having a clearly defined submucosa, instead being often attached to underlying bone or muscle by a loose fatty or glandular connective tissue (adventitia).

Within the oral cavity, the epithelium shows further specialization and can be divided into the masticatory mucosa, a lining mucosa and a specialized mucosa on the dorsal surface of the tongue. The masticatory mucosa lines the gingiva and the hard palate (roof of the mouth). Toughened through keratinization and firmly attached to underlying connective tissue, it provides protection against the mechanical and chemical stresses arising through the action of chewing. The lining mucosa covers the inner surface of the cheeks. It is a non-keratinized lining with a loose attachment to its underlying tissues, allowing flexibility of movement with the working of cheeks, lips and tongue during mastication, swallowing or speech.

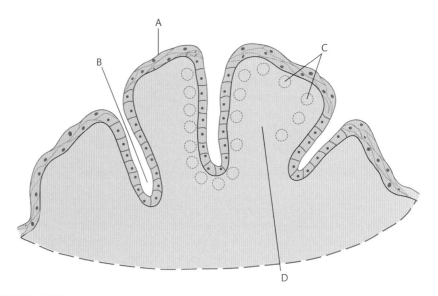

FIGURE 7.7

Diagram of the structure of the palatine tonsil. Stratified squamous epithelium (A), crypt with discontinuous epithelial lining (B), B-cell germinal centres and mantle zones (C), and interfollicular T-cell areas (D). Note there is no 'capsule' separating the lymphoid tissue from the underlying connective and muscular tissue (broken line).

7.2.5 Tonsils and adenoids

The tonsils and adenoids form a protective ring of lymphoid tissue in the oral cavity. The tonsils are aggregations of lymphoid tissue situated below a covering of stratified squamous epithelium. There are several types named according to their location; for example, sublingual (beneath the tongue) and palatine (within the soft palate). The tonsils are characterized by the presence of blind-ended crypts in their associated epithelium that 'trap' ingested substances, including potential pathogens, and stimulate immune responses, providing the body with a good first line of defence against new antigens (Figures 7.7 and 7.8, and Box 7.1). Adenoids are

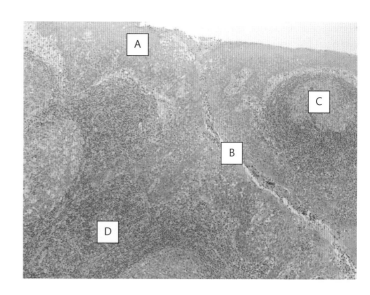

FIGURE 7.8

Survey of tonsil. Stratified squamous epithelium (A), crypt (B), germinal centre and mantle zone (C), and interfollicular area (D) (H&E stain; low-power magnification).

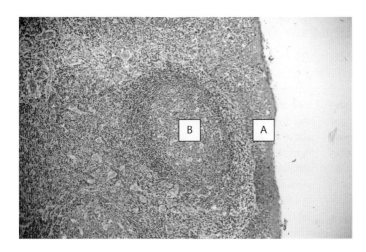

FIGURE 7.9
Adenoid. Note the presence of stratified epithelium (A) and lymphoid tissue below including a germinal centre (B), but an absence of crypts (H&E stain; low-power magnification).

also present and are similar to the tonsils in structure and function, but do not contain crypts (Figure 7.9).

7.3 **Oesophagus**

The oesophagus is situated at the back of the thoracic cavity and is about 25 cm in length. Its function is to provide a connection by which masticated food can pass from the pharynx to the stomach. Passage of food into the oesophagus is controlled by the voluntary action of swallowing and thus the upper third of the muscularis propria consists of striated muscle in place of smooth muscle. The middle third of the muscularis propria contains both muscle types, while the lower third is composed of smooth muscle only (Goyal and Chaudhury, 2008).

Reflecting its simple function and the 'coarse' state of the luminal contents, the oesophageal epithelium provides a protective role only. In humans, it is composed of stratified, non-keratinizing squamous epithelium (Figure 7.10) while in other animals, such as rodents, it

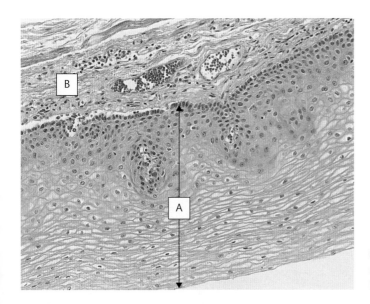

FIGURE 7.10
Stratified squamous epithelium of the oesophagus (A) and underlying connective tissue (B) (H&E stain; low-power magnification).

BOX 7.7 Epithelial metaplasia in the oesophagus and stomach

The epithelial mucosa of the oesophagus can be phenotypically dynamic in response to chronic stress; for example, the stratified squamous epithelium of the lower oesophagus will transform into glandular epithelium in response to constant exposure to stomach contents in gastric reflux—a condition known clinically as Barrett's oesophagus (Fitzgerald, 2005). These changes can be the consequence of obesity, smoking, high alcohol intake as well as weakening of the cardiac valve that normally seals the digestive chyme in the stomach. It remains controversial as to whether or not Barrett's oesophagus is a premalignant condition leading to oesophageal cancer. Chronic inflammation of the gastric mucosa may result in intestinal metaplasia, where glands of the stomach begin to resemble those of the bowel, losing chief and parietal cells to be replaced with acid mucus-secreting goblet cells.

is keratinized for enhanced protection due to the intake of food containing a much higher proportion of cellulose and other roughage. Changes in the structure of this epithelium are associated with a range of chronic conditions (see Box 7.7). The epithelium is occasionally punctuated by the openings of submucosal seromucous glands that aid lubrication of the food before it enters the stomach via the cardiac valve.

7.4 Stomach

The stomach is a relatively large and distensible component of the alimentary tract. Its function is the production of acid and enzymes and their mixing with food into a digestive gruel (**chyme**). Chyme is stored and churned in the stomach for up to five hours in the second phase of digestion. To achieve this, there is a specialized secretory and protective mucosa and a further specialized muscularis propria. The common structural organization of the gastric mucosa is a protective layer of surface mucus-secreting cells organized into **gastric pits** or **foveolae** (Figure 7.11). At the base of each of the pits are a number of simple secretory glands.

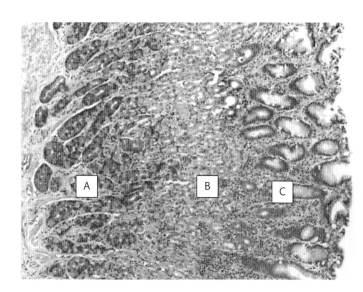

FIGURE 7.11
Survey of gastric mucosa. Basal region of glands (A), neck region of glands (B) and gastric pits (C) (H&E stain; low-power magnification).

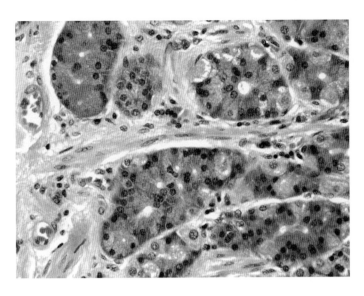

FIGURE 7.12
Basal region of gastric glands (H&E stain; high-power magnification).

The border area between the base of the pits and the glands is known as the neck region and is characterized by the presence of mucus-producing neck cells (Figure 7.13) and the proliferative component of the epithelium. It is from this region that new cells arise to repopulate the surface epithelium and the underlying glands.

Different regions of the stomach have specialized functions and this is reflected in the different types of cell seen in the glandular compartment. The largest functional mucosa is that of the body (corpus) of the stomach. Here, the glands are rich in parietal cells in the upper part of the gland, while chief cells predominate in the lower part of the gland. Basophilic chief cells can be found from the base to the middle half of the glands (Figure 7.12). These cells are rich in rough endoplasmic reticulum and secretory vesicles, reflecting their function to produce and secrete pepsinogen into the lumen of the glands. The parietal cells are large 'fried egg'-shaped cells found mostly in the upper half of the gland (Figure 7.13). These cells have a distinctive eosinophilic cytoplasm. Their role is to produce hydrogen and chloride ions and hydrogen carbonate

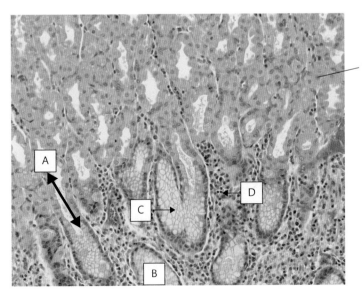

Parietal cells

FIGURE 7.13
Neck of gastric glands (A) and gastric pits (B). Note the abundant presence of mucus-producing cells in the pits (C) and lymphocytic infiltration of the intervening lamina propria (D) (H&E stain; medium-power magnification).

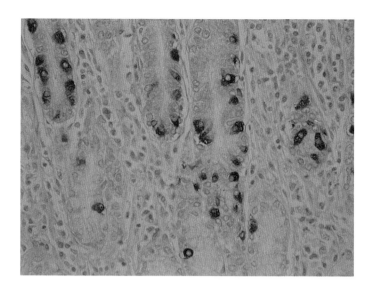

FIGURE 7.14

Gastrin (brown) in neuroendocrine cells situated in the basal and neck regions of gastric glands (Anti-gastrin immunohistochemistry [IHC]; medium-power magnification).

ions. The hydrogen and chloride ions are secreted via separate intracellular canaliculi. This maintains cellular integrity. The resultant hydrochloric acid lowers the pH of chyme and acts to convert pepsinogen into active pepsin. Furthermore, this acid helps to kill any potentially harmful bacteria that may have been ingested. The hydrogen carbonate ions are secreted into the lamina propria and taken up by the surface mucous cells where they are secreted into the mucus barrier to help neutralize any acid that may penetrate towards the delicate lining epithelium (Laine *et al.*, 2008).

Other regions of the stomach mucosa contain different proportions of these cells or additional specialized cell types. So, for example, the glands of the cardiac region of the stomach (nearest the oesophagus) contain mostly chief cells, whereas the pyloric region (nearest the duodenum) consists mostly of mucus-secreting cells admixed with specialized gastrin-secreting neuroendocrine cells which are important in regulating the sensation of fullness and activating the secretory processes associated with digestion (Fig 7.14).

The submucosa and muscularis propria of the stomach are modified to assist its function. The submucosa in the body of the stomach is thrown into thick folds, termed rugae, that allow the organ to expand and accommodate the entry of food. To ensure the mixing of food with enzymes and acid, the stomach contents must be churned together; this is facilitated by the presence of a third, oblique layer of smooth muscle on the inner aspect of the muscularis propria.

Although the stomach is predominantly a secretory organ, it does have the ability to absorb as well. Water uptake occurs in dehydrated individuals and a small proportion of amino acids are also absorbed. The stomach is also a significant site for the absorption of alcohol and some pharmaceutical drugs (e.g. aspirin).

7.5 Small intestine

Representing the longest section of the alimentary tract, the small intestine's primary role is nutrient absorption. From the pyloric valve of the stomach (see Box 7.8), the duodenum is the shortest division of the small intestine at about 25 cm in length. This is followed by the jejunum (2.5 m) and then by the ileum (3.5 m) that together takes the total length of the small intestine to about 7 m.

BOX 7.8 Functional junctions

Junctions between the parts of the alimentary canal are characterized by a sharp transition from a histology characteristic of one area to the histology of the next area. Examples include the stratified squamous epithelium of the oesophagus changing to the glandular epithelium of the cardiac region of the stomach. There may also be other specializations such as the valve structure at the ileocaecal junction regulating flow of content from ileum to caecum.

While the **duodenum** features all the mucosal characteristics of the small intestine, its main purpose is to ensure that the chyme released from the stomach is ready for absorption by the epithelium of the jejunum and ileum. The submucosa of the duodenum is filled with branched tubulo-alveolar seromucous **Brunner's glands**. These produce neutral and alkaline mucins (Figure 7.15) together with hydrogen carbonate ions at a pH of 8–9 in a mucus secretion that empties into the lumen; this has the effect of neutralizing the acid chyme. This neutralization process is assisted by the presence of hydrogen carbonate ions in the pancreatic juice delivered via the **common bile duct** that is also the conduit for bile from the gall bladder. These secretions, which also include the proteolytic enzymes trypsin and chymotrypsin together with amylase and lipase from the pancreas, empty into the duodenum via the **ampulla of Vater**. This delivery is made in response to the presence of chyme in the duodenum that stimulates the release of cholecystokinin and secretin from epithelial neuroendocrine cells (see more later in this section).

The small intestine is adapted to provide maximal epithelial surface area for absorption. This is achieved by the presence of submucosal folds (the **plicae circulares**) that proceed in a sometimes corkscrew fashion along its length. The epithelium is divided into crypts and villi, with the latter providing the second amplification of the absorptive surface, while the microvilli present on the absorptive epithelial cells further increases the absorptive area (Figure 7.16). Thus, the total surface area is approximately 400 m^2.

The crypts, often termed **crypts of Lieberkühn**, and **villi** (Figure 7.17) are lined with simple columnar epithelium. The villi are predominantly populated with absorptive (enterocyte)

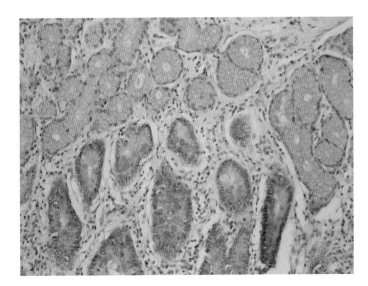

FIGURE 7.15
Brunner's glands (top) containing neutral mucin (yellow) and crypts of Lieberkühn (bottom) containing goblet cells with predominantly acid mucin (blue) (Short Maxwell stain; low-power magnification).

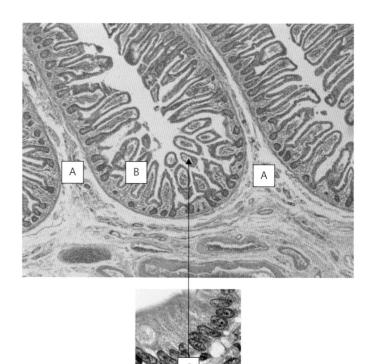

FIGURE 7.16

Increasing the epithelial absorptive area in the small intestine. The low-power view shows the plicae circulares (A), villi (B) and, at high power, microvilli (C) on the absorptive epithelial cells (H&E stain).

and goblet cells. In the jejunum the absorptive cells predominate, while in the distal portion of the ileum the goblet cells represent the majority, reflecting the primary requirement of lubrication of the luminal contents. Maturing absorptive and goblet cells are also present in the crypts, together with stem cells and Paneth cells at the base of the crypts (Shaker and Rubin, 2010). Scattered throughout the crypts and villi are neuroendocrine cells. Owing to the presence of incompletely digested food and a hostile bacterial environment, the epithelium needs constant renewal with villus tip cells needing replacement every few days.

At the entrance to the jejunum, the chyme has largely been reduced to its constitutive elements and can now be absorbed (Table 7.2). The majority of the nutrients are actively absorbed via glycocalyx-coated microvilli on the **absorptive cells**. Thus, there is a concentration of enzymes

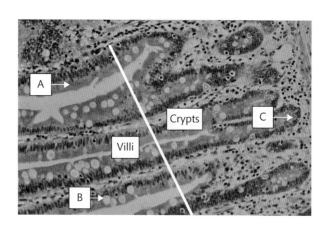

FIGURE 7.17

Survey of epithelial cells of the small intestine. Absorptive cells (A), goblet cells (B) and Paneth cells (C) (H&E stain; medium-power stain). The white line indicates transition between villi and crypts.

TABLE 7.2 Processes for absorption of nutrients in the small intestine.

Nutrient	Pancreatic enzyme	Absorptive cell enzyme	Product
Small peptides		Intracellular peptidases	Amino acids
Large peptides	Proteases	Membrane peptidases	Amino acids
Starch	Amylase	Membrane maltase and iso-maltase	Glucose
Lactose		Membrane lactase	Galactose and glucose
Sucrose		Membrane sucrase	Fructose and glucose
Lipids	Lipases		Triglycerides

that are situated just below this layer (Figure 7.18). Amino acids and monosaccharides enter the capillaries and eventually are transported via the hepatic portal vein to the liver. Triglycerides, as chylomicrons, are transferred to lacteals in the lamina propria via channels that run between the absorptive cells and enter the circulation via these extensions of the lymphatic microvasculature that empty via the thoracic duct into the blood stream. Given the active role of the absorptive cells, their cytoplasm contains abundant mitochondria and endoplasmic reticulum. The nucleus is situated in the basal region of these cells.

Goblet cells are so-called because of the accumulation of a mixture of acid and neutral mucopolysaccharides, termed mucin, above their basally positioned nuclei. On release, mucin dissolves into the water present in the luminal fluid and forms a protective mucous coating for the epithelial cells, as well as facilitating the free movement of the luminal contents by peristalsis. Owing to the production of mucin, a prominent Golgi apparatus is situated between the nucleus and the mucin goblet, while rough endoplasmic reticulum is also found around the nuclei of these cells (Figure 7.19).

Forming a crescent at the base of the crypts, the **Paneth cells** contain prominent acidophilic secretory granules between their apical surfaces and the basally-located nuclei (Figure 7.20). Lysozyme and glycoproteins are secreted, and the cells that are also rich in

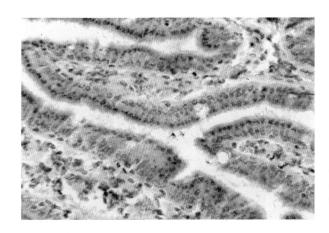

FIGURE 7.18
Acid phosphatase (red precipitate) demonstrated in the apical regions of jejunal absorptive cells (Frozen section, simultaneous naphthol capture method; high-power magnification).

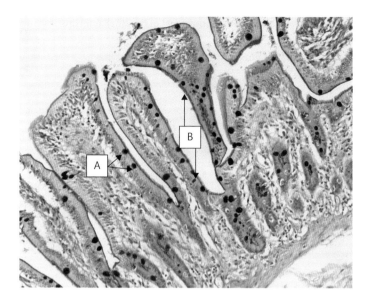

FIGURE 7.19
Goblet cells (A) in crypts and villi, together with glycocalyx (B) (Alcian Blue/periodic acid Schiff stain showing combined acid and neutral mucin [purple] and neutral mucin alone [magenta]; low-power magnification).

arginine, proteins and zinc. Paneth cell secretions also include defensins that contribute to the regulation of intestinal bacterial flora via innate immunity mechanisms (Elphick and Mahida, 2005).

The **neuroendocrine cells** secrete peptide hormones into the blood circulation via the capillary network in the lamina propria. These hormones provide local paracrine control of the digestive processes. Each neuroendocrine cell produces one hormone type and these vary according to location in the small intestine (Table 7.3). The application of special stains identifies the cells (Figure 7.21), while the use of immunohistochemistry permits the identification of specific hormone cell types and therefore assessment of their distribution. Due to the orientation of hormone secretion, the granules are located towards the base of the neuroendocrine cell, with the nuclei in a superior position.

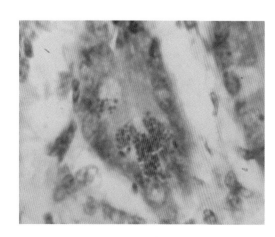

FIGURE 7.20
Paneth cells granules (red) (Phloxine tartrazine method; high-power magnification).

TABLE 7.3 A selection of peptide hormones of the small intestine. (Adapted from Holdcroft, 2000)

Hormone	Function	Comment
Cholecystokinin	Gall bladder and pancreatic secretion	Also produced by nerve plexi
Gastric inhibitory peptide	Inhibits gastric acid secretion	Produced in duodenum and jejunum
Gastrin	Stomach acid and glucagon secretion	Minor duodenal production (cf stomach)
Neurotensin	Delays gastric emptying	Wide distribution and function outside the alimentary tract
Peptide tyrosine	Promotes satiety	
Secretin	Stimulates secretion from pancreas (major) and gall bladder (minor)	
Vasoactive intestinal peptide	Stimulates secretion from pancreas and gall bladder	

Lymphoid aggregates known as **Peyer's patches** are frequently present in the jejunum and ileum (Figure 7.22). They are embryologically fixed structures that span the submucosal and mucosal boundaries and form an important component of GALT (see Box 7.1). When the aggregates penetrate the mucosa they may reach to just below the epithelial surface that is reduced to a layer of covering columnar cells only, the **follicle-associated epithelium (FAE)**. In this covering, some of the epithelial cells will be **microfold (M) cells**. These cells are specialized for sampling the luminal 'antigens' and transporting them for potential immune response via the underlying lymphoid aggregates.

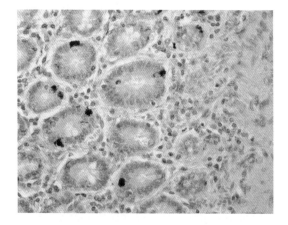

FIGURE 7.21
Neuroendocrine cells. Note the orientation of the secretory granules (black) towards the basal regions of the cells (Grimelius silver stain; medium-power magnification).

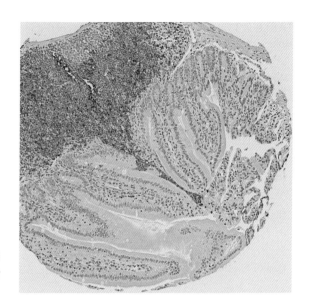

FIGURE 7.22
B lymphocytes within a Payer's patch (top left) in murine small intestine. (Tissue microarray core, anti-CD79b IHC; low-power magnification).

7.6 **Large intestine**

The large intestine (large bowel) receives the mostly digested products from the caecum and is primarily responsible for concentrating the faeces prior to defaecation.

7.6.1 **Caecum**

The caecum is a blind-ended pouch connecting the ileum with the colon and serves to expand and receive digested food before delivering it to the colon. Important for bacterially mediated cellulose digestion in many mammals, in humans this ability has been lost and therefore they have a correspondingly smaller caecum. Histologically, the caecum is very similar to the colon described below.

7.6.2 **Appendix**

The appendix is a blind-ending protrusion from the caecum. It is 6–10 cm long and gradually involutes after adolescence and eventually may be reduced to scar tissue only. The mucosal structure is similar to that in the colon with glands containing abundant goblet cells. The glands are more loosely ordered than in the colon and the lamina propria contains many lymphocytes. There are many lymphoid aggregates within the submucosa, often with visible germinal centres. These aggregates, like the Peyer's patches of the small intestine, also penetrate the mucosa, reducing the epithelial layer to FAE (Figure 7.23). The appendix is now regarded as an important contributor to the development of innate immunity, especially during early life, and is sometimes referred to as the tonsil of the lower alimentary tract (see Box 7.1).

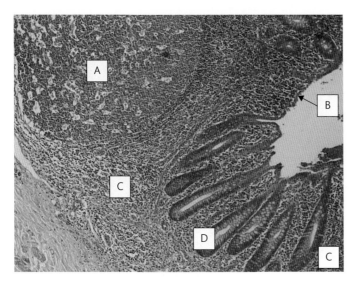

FIGURE 7.23

Mucosa and submucosa of appendix. Note prominent germinal centre (A), follicle-associated epithelium (B), lymphocyte infiltrate in submucosa and lamina propria (C) and mucosal glands lined with goblet cells (D) (H&E stain; low-power magnification).

7.6.3 Colon

The colon is up to 1.5 m long and organized into characteristic bulges (teniae). The colon is divided according to its anatomical orientation into ascending, transverse, descending and sigmoid regions.

The main function of the colon is to maintain water balance and store indigestible remnants of food before expulsion. The liquid digested contents of the ileum move along the colon where water is absorbed, concentrating and solidifying the faeces. In this way water homeostasis is maintained. The colon is also important for the extraction of electrolytes and uptake of fat-soluble vitamins such as vitamins K, B_{12} and thiamine. These various functions are reflected in the structure of the colon. There is a specialized and predominantly secretory simple epithelium producing copious lubricating and protective mucus, a comparatively thick muscularis mucosae to expel the mucin, and a powerful muscularis propria for the movement of the faecal mass. Macroscopically, the surface of the mucosa forms a smooth, flat surface punctured by numerous microscopic glandular openings.

Microscopically, the epithelium is organized into close-packed, regularly tall, simple glands that straddle the full thickness of the mucosa (Figure 7.24). With mucus production the prime function of these glands, it is of no surprise that goblet cells are the most abundant cell type. These produce a mucus composed of Alcian Blue-positive 'acid' mucin (see Box 7.9). The luminal aspect of the epithelium is composed of absorptive columnar colonocytes that effect water transport and ion exchange into the lamina propria and hence into the local circulation. While it is traditionally accepted that the secretory and absorptive functions are clearly demarcated between the two cell types, there is increasing evidence that goblet cells are also important in absorption (Geibel, 2005).

Stem cell niches of the colon can be found at the base of the glands supported by a small number of Paneth cells. The neuroendocrine cell component contributes some five per cent of population. These cells are important for the production of serotonin, a chemical signal important in driving peristalsis and activating secretion throughout the alimentary tract.

SELF-CHECK 7.4

Explain the functional basis of the physiological risk of prolonged diarrhoea.

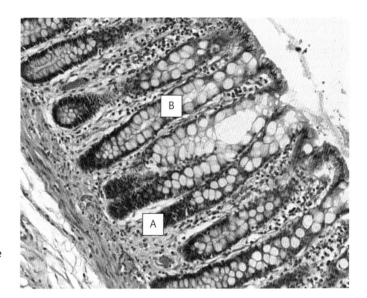

FIGURE 7.24

Mucosa of large intestine. Note the closely packed straight epithelial glands (A) and the preponderance of goblet cells (B) (H&E stain; medium-power magnification).

SELF-CHECK 7.5

One noted feature of the intestinal mucosa is its high rate of cellular turnover. Why is this important for the proper functioning of the intestine?

The colon is richly populated by a diverse and specialized microflora. The role of these microbes is to aid the breakdown of remaining undigested carbohydrate and fibre by fermentation to release additional energy used by the bacteria and the cells lining the colon. This can represent up to ten per cent of an adult's daily energy requirement. This microflora also represents a protective ecology that out-competes potentially pathogenic bacterial infections. The thick mucus secreted into the mucosa acts as a barrier against microbial entry into the epithelium and also guards against damage from bacterial toxins. Further protection of the mucosa is achieved by isolated patches of GALT distributed as B-cell-rich lymphoid follicles throughout the length of the colon. These have a role not only in immune surveillance but also in supporting epithelial regeneration (Sipos and Muzes, 2011). You will read more about the normal alimentary tract microflora in the following chapter.

BOX 7.9 Mucus in the intestine

Mucus is a gel formed from hydrated mucins secreted by epithelial cells. The mucins comprise multiple polysaccharide chains attached to a protein core (derived from one of 12 *MUC* genes) and it is these side chains that hold the water shell and determine the viscosity of the mucus. Classically, histologists have split mucins into two types, acid mucin (charged mucins that stain bright blue with Alcian Blue dye at pH 2.5) and neutral mucin (uncharged mucins that stain red with the PAS technique but do not react with Alcian Blue). The mucus of the stomach and Brunner's glands is mostly composed of neutral mucins, while mucus derived from the oesophageal glands or found in the goblet cells of the small and large intestine is a mixture of acid and neutral mucins. This property can be exploited in histopathology as altered patterns of mucus production can be indicative of underlying pathology; for example, intestinal metaplasia of the stomach is characterized by the abnormal presence of Alcian Blue-positive goblet cells.

FIGURE 7.25
Anal Squamocolumnar junction (H&E stain; high-power magnification).

7.6.4 Rectum and anal canal

The rectum and anal canal present a complex transition of mucosal and valvular organization, reflecting the need to collect and void faeces in a controlled manner. Essentially, the rectum is mostly (up to 17 cm) lined by glandular epithelium similar to that of the colon but with more widely spaced glands predominantly composed of goblet cells. The lower part of the rectum (3–4 cm) transitions into the upper part of the anal canal, which is lined by a non-keratinized, stratified epithelium, and the lower half of the anal canal lined by keratinized stratified epithelium (Figure 7.25). This becomes epidermal as it merges with the perineum. The opening of the anus for defaecation is regulated by the involuntary intrinsic muscle, arising from a thickened muscularis propria and the striated (voluntary) extrinsic muscle.

7.7 The intestine and cancer

There is accumulating evidence that the types of food passing through the intestine can have an impact on the relative risk of developing cancers of the alimentary tract. For example, high consumption of alcohol is linked to an increased risk of developing cancers of the oral cavity and oesophagus, while the consumption of diets containing a lot of processed meat is linked to an increased risk of developing colon cancer. On the other hand, a diet rich in fruit and vegetables appears to reduce the risk of developing oesophageal, stomach and colon cancer (Cancer Research UK, 2013).

 Chapter summary

Over the course of this chapter we have introduced the major anatomical structures of the alimentary tract, from mouth to anus.

- A common pattern of organization of the supporting structures is maintained throughout the alimentary tract but shows histological specialization according to function.
- The pattern of organization is related in a linear fashion to the processing of ingested solids and liquids into physiologically important entities. In order to appreciate and aid recall

of the complexity of the organization of the intestine, it is important to integrate structure and function of the specific structures and the cellular relationships found throughout.

- The luminal surface of the intestine is effectively the 'outside on the inside' and, much like the skin, the epithelium is exposed to a hostile environment.

- There are two important protective processes in play at the luminal surface: cellular replication and immune surveillance. Epithelial cells of the mucosa are constantly being replaced by amplifying cohorts of dividing cells (derived from a small population of stem cells) that ultimately differentiate into the relevant specialized cell types. The interface of the intestine with the 'outside world' exposes the mucosa to antigenic challenge. This is met with continuous lymphoid activity mediated by individual cells or strategically located lymphoid aggregates.

- While the epithelium provides a protective, secretory or absorptive barrier, the supportive structures beneath consist of connective tissue and muscular arrangement. This ensures that food is masticated, churned and propelled from the mouth to the anus.

- Embedded within the connective tissue are blood vessels (and lacteals in the small intestine) that collect the products of digestion and supply the resources required for the high-energy demands of the digestive process.

- The majority of the muscle is of the smooth type and therefore under autonomic control, while the extremities of the alimentary tract are under the conscious control of voluntary muscle.

- Nervous control is supplemented by neuroendocrine cells embedded in the epithelium that respond to luminal stimuli and secrete peptide hormones and amine-based compounds that typically elicit paracrine responses that aid the digestive process.

- The simple interface that is maintained by the delicate tracery of epithelium draped over the underlying connective support tissues is able to perform many functions and yet maintain its own integrity.

- Many of the pathologies of the alimentary tract involve loss of function or integrity of the epithelium, and aspects of these are introduced in the *Histopathology* volume in this *Fundamentals of Biomedical Science* textbook series.

Further reading

- Camilleri M, Brown ML, Malagelada JR. Relationship between impaired gastric emptying and abnormal gastrointestinal motility. *Gastroenterology* 1986; **91**: 94–9.

- Corr SC, Gahan CC, Hill C. M-cells: origin, morphology and role in mucosal immunity and microbial pathogenesis. *FEMS Immunol Med Microbiol* 2008; **52**: 2–12.

- Cancer Research UK. *Diet, alcohol and cancer in the UK* 2013.

- Elphick DA, Mahida YR. Paneth cells: their role in innate immunity and inflammatory disease. *Gut* 2005; **54**: 1802–9.

- Fitzgerald RC. Barrett's oesophagus and oesophageal adenocarcinoma: how does acid interfere with cell proliferation and differentiation? *Gut*, 2005; **54**: i21–i26.

- Geboes K, Geboes KP, Maleux G. Vascular anatomy of the gastrointestinal tract. *Best Pract Res Clin Gastroenterol* 2001; **15**: 1–14.

- Geibel JP. Secretion and absorption by colonic crypts. *Annu Rev Physiol* 2005; **67**: 471–90.

- Goyal RK, Chaudhury A. Physiology of normal esophageal motility. *J Clin Gastroenterol* 2008; **42**: 610–9.

- Groschl M. The physiological role of hormones in saliva. *Bioessays* 2009; **31**: 843–52.

- Holdcroft A. Hormones and the gut. *Br J Anaesth* 2000; **85**: 58–68.

- Humphrey SP, Williamson RT. A review of saliva: normal composition, flow and function. *J Prosthet Dentistry* 2001; **85**: 162–9.

- Kawai T, Fushiki T. Importance of lipolysis in oral cavity for orosensory detection of fat. *Am J Physiol- Regul Integ Compar Physiol* 2003; **285**: R447–R454.

- Kraehenbuhl JP, Neutra MR. Epithelial M cells: differentiation and function. *Annu Rev Cell Dev Biol* 2000; **16**: 301–32.

- Laine L, Takeuchi K, Tarnawski A. Gastric mucosal defense and cytoprotection: bench to bedside. *Gastroenterology* 2008; **135**: 41–60.

- Metcalf AM, Phillips SF, Zinsmeister AR, MacCarty RL, Beart RW, Wolff BG. Simplified assessment of segmental colonic transit. *Gastroenterology* 1987; **92**: 40–7.

- Orchard G, Nation B eds. *Histopathology*. Oxford: Oxford University Press, 2012.

- Rescorla RA. Pavlovian conditioning. It's not what you think it is. *Am Psychologist* 1988; **43**: 151–60.

- Roper SD. Taste buds as peripheral chemosensory processors. *Semin Cell Dev Biol* 2013; **24**: 71–9.

- Shaker A, Rubin DC. Intestinal stem cells and epithelial–mesenchymal interactions in the crypt and stem cell niche. *Transl Res J Lab Clin Med* 2010; **156**: 180–7.

- Sipos F, Muzes G. Isolated lymphoid follicles in colon: switch points between inflammation and colorectal cancer? *World J Gastroenterol* 2011; **17**: 1666–73.

- Weaver IT, Steiner H. The bowel habit of young children. *Arch Dis Child* 1984; **59**: 649–52.

- Wedel T, Roblick U, Gleis J *et al.* Organization of the enteric nervous system in the human colon demonstrated by wholemount immunohistochemistry with special reference to the submucous plexus. *Ann Anat - Anatomischer Anzeiger* 1999; **181**: 327–37.

Discussion questions

7.1   Discuss the role of mucus in the protection of the alimentary tract.

7.2   Why might diet have such an effect on the relative risk of developing cancers of the alimentary tract?

Answers to self-check questions and discussion questions, hints and tips are provided on the book's Online Resource Centre.

 Visit www.oxfordtextbooks.co.uk/orc/orchard_csf/.

8

Cells and microbial flora of the gastrointestinal tract

Kathy Nye

Learning objectives

After reading this chapter you should be able to:

- Describe the development and complex interactions of the cells and microbial flora of the gastrointestinal tract.
- Understand the importance of the normal microbial flora in health and disease.
- Understand the normal variation in the microbial flora.
- Appreciate what happens if either the cells or microbial flora are unable to function normally.
- Have insight into how the microbial flora can be manipulated for health.

The gastrointestinal (GI) tract has a surface area of between 250 and 400 square metres and processes around 60 tons of food during an average lifetime. It is the largest and most complex environment in the human body and is subject to continuous challenges.

Around 10 billion human cells, which are replaced every three to four days, interact with an even greater number of microorganisms—a ratio of around one cell to ten microorganisms. This interaction is key to the proper development and functioning of the GI tract and the immune system, as well as to the assimilation and production of essential nutrients, although scientists are only just beginning to understand the full extent of this vital, symbiotic relationship.

In this chapter, you will discover more about these complex interactions, in health and disease, and how an understanding of these mechanisms may lead to improvements in health and the management of some types of illness.

8.1 General principles and the development of the normal microflora

The GI tract begins at the mouth and runs for around ten metres via the oesophagus, stomach, and small and large intestines to the anus (Figure 8.1).

In the moments before birth, the entire GI tract is sterile; however, during the birthing process and thereafter, for the rest of the individual's life, microbes will be swallowed and become part of the tract's rich microflora. All the major groups of microbes will be involved, including protozoa, fungi and viruses, but bacteria form by far the largest group of organisms.

The type of birth and length of the birth process affect the initial colonization. In a vaginal delivery, faecal and genital tract organisms from the mother may be detected in the newborn's

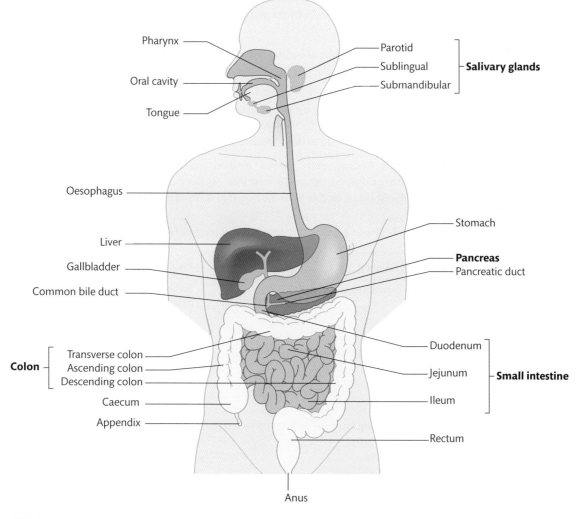

FIGURE 8.1

Diagram of the human gastrointestinal tract.
© Oxford University Press 2010.

stomach within ten minutes of birth. However, after a Caesarian section, the first organisms are acquired from the equipment used and from the attending staff.

In infancy and early childhood, the majority of acquired organisms come from the mother and of these many may persist into adulthood. In this way, specific microbes are passes from one generation to the next.

Obviously, less sanitary conditions increase the exposure to, and acquisition of, microorganisms. In the developed world, the development of the normal GI tract microflora is relatively delayed and involves fewer species compared with developing countries.

Not all the microbes to which the baby is exposed will remain in the GI tract as colonizers, although the ones to which the child is first exposed are more likely to become established than would be the case in adults who already have a stable microflora. In the infant, specific organisms become established at different stages of development—a principle termed 'microbial succession'.

The greatest changes in GI tract microflora occur in the first year of life, as dramatic changes in feeding take place. However, by the second year, the normal flora generally resembles that of the adult.

While there is huge individual variability, four main phases of acquisition are recognized, as shown in Table 1.

The first colonizing organisms in an infant delivered vaginally are *Escherichia coli* and *streptococci*, which originate from the maternal GI tract. It is thought that they help to create an environment of low oxygen tension that favours colonization by anaerobic species such as *Bacteroides*, *Clostridium* and *Bifidobacterium*.

During breast feeding, other species diminish so that *Bifidobacterium* predominate. *Staphylococcus*, *Corynebacterium*, *Lactobacillus*, *Micrococcus* and *Propionibacterium* are also acquired during breast feeding from the maternal skin and milk ducts.

TABLE 8.1 Phases of microflora acquisition during early life.

Phase	Timing	Predominant microorganisms
1	First two weeks of life	*Escherichia coli* *Streptococcus* species *Lactobacillus* species *Bifidobacterium* species
2	During exclusive breast feeding	*Bifidobacterium* species *Escherichia coli* *Lactobacillus* species *Bacteroides* species
3	During formula-feeding and weaning	*Bacteroides* species *Peptococcus* species *Streptococcus* species *Bifidobacterium* species *Clostridium* species
4	Weaned child receiving mixed diet (Adult-type flora)	*Bacteroides* species *Bifidobacterium* species *Peptococcus* species *Streptococcus* species

Clostridium difficile

This is an anaerobic, Gram-positive bacillus which forms part of the normal GI tract flora in around two-thirds of children under two years old and in at least three per cent of adults. Carriage may be much higher in the elderly or hospitalized. Some strains can produce toxins which give rise to diarrhoea, and this often seems to be triggered by the disruption of normal flora caused by the use of broad-spectrum antibiotics.

In formula-fed infants, Enterobacteriaceae other than *Escherichia coli* are more common, and the proportion of anaerobic bacteria is greater compared with breast-fed babies. *Clostridium difficile* is also more commonly present, although it does not normally cause illness.

Once breast-fed infants start to wean, the GI tract microflora begin to resemble that of formula-fed infants, with *Bifidobacterium* decreasing in numbers to be replaced by other anaerobes.

By the time the child is fully weaned, the various microbial genera and species have increased greatly, although anaerobic species, including *Bifidobacterium*, continue to predominate.

While the greatest changes in normal flora occur during the first year, change continues throughout life and is dependent on factors (Figure 8.2) that include:

- diet—most food and drink contains various microorganisms.
- environment—organisms ingested directly (particularly in children who put objects in the mouth), from contact with other humans, and indirectly via hands to mouth.
- drugs—antibiotics have a major effect on the microflora, but many other classes of drug have unexpected antimicrobial effects.

The indigenous microbes colonize the GI tract stably over time so that a steady state is reached whereby microbial multiplication matches losses due to washout/elimination via peristalsis along the GI tract.

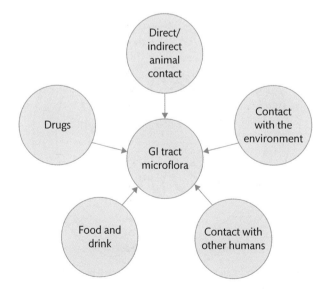

FIGURE 8.2

Factors affecting the normal GI tract microflora.

Most exogenous microbes are acquired only transiently, or do not even survive the passage through the intestine; however, if the stability of the normal flora is disturbed (e.g. when some of the resident microorganisms are killed by antibiotic treatment) then they may find a vacant niche and are much more likely to colonize.

SELF-CHECK 8.1

What is meant by 'microbial succession' and why might this differ in breast-fed and formula-fed infants?

8.2 Regions of the GI tract and their normal microbial flora

Conditions vary considerably along the length of the GI tract, but essentially there are four main habitats (Figure 8.3) at any one point, and these are the:

- GI tract lumen
- mucus covering the epithelium
- deep layer of mucus in the crypts between the cells
- surface of the epithelial cells.

8.2.1 Mouth and oesophagus

The mouth and oesophagus are lined by non-keratinized, stratified squamous epithelium (i.e. epithelial cells are continually formed at the deepest basal layer of the epithelium and move nearer to

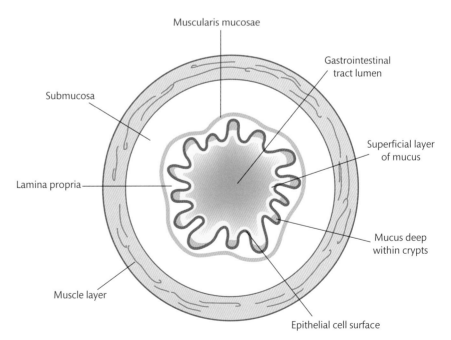

FIGURE 8.3
Indicated on the right of the diagram are the four main GI tract microflora habitats.

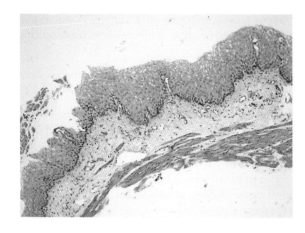

FIGURE 8.4
Normal oesophagus showing non-keratinized, stratified squamous epithelium in which cells grow in layers, becoming progressively flatter towards the lumen
(courtesy Dr M-A Brundler).

the surface as each new layer is formed, becoming flattened and more differentiated as they do so [Figure 8.4]). In contrast to skin, the cells remain moist and do not form a tough layer of keratin. As cells reach the uppermost layer, they are rubbed off by the passage of intestinal contents.

From birth to the eruption of teeth, *Streptococcus salivarius* forms around 98 per cent of the oral microflora, with *Lactobacillus* also being an important component. However, once the primary teeth appear then a wider range of genera and species are able to colonize. These organisms seem able to adhere to specific carbohydrate and protein residues on epithelial cells. *Streptococcus sanguis* and *Streptococcus mutans* are unable to colonize desquamating surfaces (i.e. the epithelial surface constantly being rubbed off and renewed from the layers below). Teeth form a suitable surface to which they can adhere, and these species persist for as long as the teeth remain. Other, mainly anaerobic, species also find a suitable niche in the gingival crevices where teeth emerge from the gums.

The cells of the mouth are kept moist by saliva produced by the salivary glands. This contains digestive enzymes (e.g. amylase) which break down starch into simple sugars, providing nutrients not only for the host but also for the oral bacteria. In turn, these bacteria stimulate the local production of secretory antibodies—mainly IgA—that help to prevent the colonization of pathogenic species. They also produce substances such as bacteriocines and peroxide that inhibit non-indigenous species.

Unfortunately, the normal flora can also have a damaging effect if teeth are not cared for properly and can cause dental caries and periodontal disease.

The microflora of the oesophagus is transient and highly variable, consisting entirely of swallowed organisms originating from food or saliva, although *Lactobacillus* and *Streptococcus* species may adhere to the epithelium.

8.2.2 Stomach

At the junction of the oesophagus and stomach, an abrupt change in the type of epithelium occurs as the non-keratinized, stratified squamous epithelium gives way to glandular epithelium. Glandular epithelium, specialized according to area, continues along the entire length of the GI tract as far as the anus where a transition to keratinized, stratified squamous epithelium occurs.

The gastric epithelium (or gastric mucosa), like the majority of the gastrointestinal tract, is covered by a single layer of columnar cells, many of which are specialized to produce specific substances. Below this layer is the lamina propria, which contains capillary blood vessels, lymphatics and connective tissue, and is surrounded by a layer of muscle tissue, the muscularis mucosae.

The epithelium is thrown up into complex folds at both the macroscopic and microscopic level, such that the surface area for contact with food and liquid is maximized.

The gastric mucosa has a honeycomb appearance when viewed closely and this is due to the presence of numerous glands that pit the surface. These glands are lined by specialized cells. In the part of the stomach closest to the oesophagogastric junction, known as the cardia, the glands are lined by mucus-secreting cells. This mucus forms a layer that protects the epithelium from the hydrochloric acid produced by parietal cells, which are found mainly in glands in the body of the stomach. Here, oxyntic glands predominate and are lined by 'chief' cells that produce pepsinogen, a precursor of the enzyme pepsin that is involved in the breakdown of proteins.

Various other cell types are found in the gastric mucosa, including A cells that produce glucagon required to break down glycogen stored in the liver, G cells that produce the hormone gastrin responsible for the stimulation of acid production by parietal cells, enterochromaffin cells that produce serotonin to stimulate smooth muscle contraction, and enterochromaffin-like (ECL) cells that produce histamine, which is also involved in promoting acid secretion. The pH varies between pH 1 and pH 4, which is optimal for protease enzyme function.

The stomach is therefore a very hostile environment for most bacteria, and the majority of organisms swallowed will not survive, leaving acid-tolerant strains such as *Lactobacillus* and *Streptococcus* as the predominant organisms. *Helicobacter pylori*, which is implicated in the causation of stomach ulcers also forms part of the normal microflora in many people.

8.2.3 Small intestine

After food is mixed thoroughly with acid and digestive enzymes in the stomach, it is ejected into the first part of the small intestine, the duodenum, via the pyloric sphincter—a ring of muscular tissue that contracts to close off the stomach.

The lining of the small intestine shows subtle variation along its length, but essentially it is the site of the most active digestive processes. The predominant cell type is the tall, columnar enterocyte, or absorptive cell, through which basic nutrients released from food are absorbed. Mucosal folding is accentuated and, at the microscopic level, the columnar epithelium forms small, fingerlike processes called villi that surround a column of blood vessels and a lymphatic lacteal allowing rapid absorption of nutrients from the GI tract lumen. In addition, the luminal surface of the epithelial cells is finely folded into microvilli, thereby maximizing the absorptive surface area, and the enterocytes secrete glycoprotein enzymes at the surface, including peptidases and disaccharidases, enabling local breakdown of proteins and carbohydrates.

The second most abundant cell type is the mucus-secreting goblet cell. Once again, the mucus helps to protect the epithelial cells from chemical damage, not only from acidic stomach contents but also from the highly alkaline bile and enzyme-rich pancreatic juice that empty into the duodenum.

In the duodenum, submucosal Brunner's glands are a prominent feature. These contain specialized cells that produce a mucus-rich alkaline secretion containing bicarbonate in order to protect the mucosa from the acidic stomach contents and to raise the intestinal pH to enable the proper functioning of other digestive enzymes such as trypsin.

At the bases of the villi, simple tubular glands are found, called the crypts of Lieberkuhn, in which are found two important cell types. Paneth cells secrete the antibacterial enzyme lysozyme as well as the glycoproteins and zinc that are essential for the function of certain enzymes. They are also able to phagocytose some bacteria and protozoa. Enteroendocrine cells secrete substances that inhibit gastric acid production and stimulate emptying of the gall bladder and secretion of pancreatic enzymes and biliary bicarbonate.

Other cells are intimately concerned with immune function. Microfold cells are found in the epithelium overlying large aggregates of lymphoid tissue. These cells do not possess microvilli but are involved in the endocytosis and presentation of antigens. Lymphocytes, believed to have a role in antigen processing, are often present in the epithelial intercellular spaces, and lymphocytes, plasma cells and macrophages are abundant in the deeper lamina propria, contributing to the gut-associated lymphoid tissue (GALT).

Key points

Gut-associated lymphoid tissue (GALT) forms the largest collection of lymphoid tissue in the body and comprises lymph nodes, Peyer's patches (lymphoid aggregates in the intestinal wall) and diffusely distributed lymphocytes. As the surface area of the entire mucosa is around 300 square metres, this presents the major site of antigen exposure and is crucial to the optimal development of all aspects of immune function.

The small intestine is also a hostile environment for microorganisms that, having survived the gastric acid, are now faced with pH 6–7.4, toxic bile salts and highly active pancreatic enzymes as well as phagocytes and an enhanced immune response. It is not surprising, therefore, that relatively low numbers of bacteria are found (around 100 000 to a million per mL contents). The species, again, include *Lactobacillus* and *Streptococcus*, although, at this point *Enterococcus faecalis* predominates. Further along the small intestine, a wider variety of species is found, including coliforms such as *Escherichia coli* and some anaerobes (e.g. *Bacteroides* species).

8.2.4 Large intestine

The mucosa of the large intestine is much smoother than that of the small intestine and possesses no villi. Instead, it has large numbers of straight, tubular glands lined by columnar epithelium, the cells of which are principally the mucus-secreting goblet cell type or absorptive cells (Figure 8.5). Goblet cells increase in number towards the rectum and anus, the mucus produced helping the passage of intestinal contents, which are of a more solid consistency than in the small intestine, as this is the site where reabsorption of fluid and electrolytes takes place. The absorptive cells have much shorter microvilli and secrete no digestive enzymes. There are no Paneth cells and very few enteroendocrine cells. Gut-associated lymphoid tissue is particularly abundant in the lamina propria, due to the larger range of microorganisms present in this part of the intestine, as well as the presence of toxic by-products of the metabolic processes found in the faecal material.

The environment of the large intestine is much less hostile for microorganisms, with a more stable pH, fewer toxic digestive enzymes and chemicals, and a much slower transit of contents. Bacteria are therefore found in much greater numbers, around 10^{11}/g of contents, and with a greater diversity of genera and species.

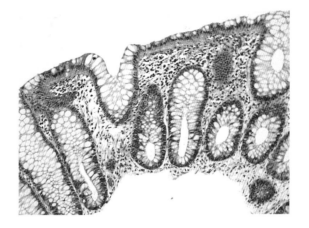

FIGURE 8.5
Normal colonic mucosa showing a flattened surface and many mucus-secreting goblet cells
(courtesy Dr M-A Brundler).

Around 400 species are represented, of which 99 per cent are anaerobes due to the decreased oxygen tension in this region. These are mainly *Bacteroides*, *Lactobacillus* and *Bifidobacterium* species, which outnumber enterobacterial species such as *Escherichia coli* by between 1000 and 10 000 to 1.

Escherichia coli and *Enterococcus faecalis* form part of the microflora of the large intestine in all humans and can be used as indicators of faecal contamination in the environment. Other organisms, such as *Clostridium* species, *Staphylococcus aureus* and *Pseudomonas aeruginosa*, vary widely in incidence, while others are found only transiently, and many bacterial pathogens fall into this last group.

SELF-CHECK 8.2

What are the main differences between the microbial populations of the stomach and large intestine?

8.3 Functions of the normal flora

Experimentally reared, germ-free (gnotobiotic) animals have been shown to have reduced immunity to infection and a poorly-developed GI tract which is not very efficient at processing food and nutrients. Although the GI tract microflora is extremely complex and incompletely understood, it is clear that its health is directly related to the overall health of the host.

The normal microflora is known to be essential for the development of immunity and protection against infection by pathogenic microorganisms. It also plays an important role in nutrition and the efficient working of the intestine (Figure 8.6)

8.3.1 Protection

The intestine is one of the most important routes of entry into the body for microbial pathogens, but most need to be able to adhere to the epithelium before they can initiate the disease process. The normal flora is already adherent to the mucosa and prevents colonization by pathogens by competing for attachment sites. When the normal flora is disrupted, a much lower infective dose of pathogens (e.g. *Salmonella*) is required.

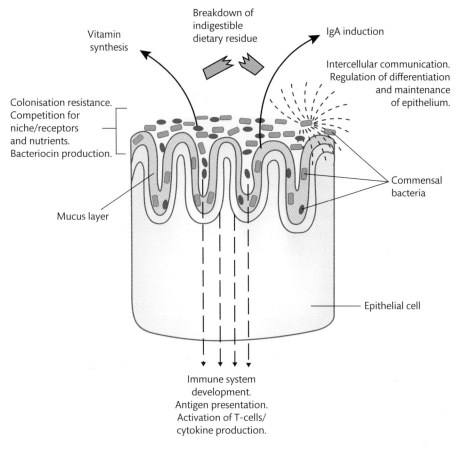

FIGURE 8.6
The normal intestinal microflora plays an essential role in the development of immunity, protection against infection by pathogenic microorganisms, and in nutrition and the efficient working of the intestine.

Cells of the normal flora may also produce antimicrobial substances such as bacteriocines, peroxides and nitric oxide to kill or damage non-native species. In addition, nitric oxide is involved in promoting mucus secretion, the regulation of intestinal motility and blood circulation, and the stimulation of immunity.

Other protective mechanisms include competition with pathogens for essential nutrients, the stimulation of peristalsis and the creation of restrictive physiological environments for non-native organisms.

8.3.2 Immunity

The microbial flora helps to define the host immune system, and exposures in the first year of life are crucial as they determine early immune responses. Apart from stimulation of the immune response to bacterial antigens and the subsequent production of antibodies, in particular secretory IgA at the mucosal surface, the normal flora also promotes the development of GALT. Abnormal exposures, particularly during the first year of life (e.g. microflora disruption due to antibiotics, delivery by Caesarian section, formula feeding and prematurity) are associated with an increased incidence of allergy and asthma.

Certain bacteria appear to be able to induce the production of CD4 helper cells and cytokines. In mice, it has been shown that particular strains of *Acinetobacter, Lactococcus* and *Lactobacillus* can direct T-cell maturation towards specific types of cytokine secretion (e.g. interleukin 12, interferon-γ), that diminish allergic responses and improve immune defences.

8.3.3 Nutrition

The normal GI tract microflora helps to break down complex dietary nutrients into easily absorbable components (e.g. microbial proteases and ureases aid in the breakdown of proteins to amino acids). Others help to metabolise otherwise indigestible dietary polysaccharides. However, the microbial flora will adapt to the diet of the host. In high-fat or carbohydrate-rich diets, the relative numbers of *Bacteroides* and *Bifidobacterium* species alter to favour those organisms that can best use the available carbon sources.

Certain organisms are also able to synthesize vitamins, notably B group vitamins and vitamin K. For example, lactic acid bacteria synthesize riboflavin (vitamin B_2) and *Propionibacterium* species synthesize thiamine (vitamin B_1), pantothenic acid and folic acid. These are produced in small amounts and would only compensate partially for a deficiency in the diet.

In certain situations, however, such as so-called blind-loop syndrome, where part of the intestine is bypassed in a surgical procedure, food residue cannot flow properly and bacterial build-up occurs to such an extent that vitamins, usually vitamin B_{12}, are used up by the bacteria, leading to deficiency and malabsorption in the host.

SELF-CHECK 8.3

What are the main mechanisms by which the normal microflora protects the host from infectious disease?

8.3.4 Intercellular relationships

The intimate relationship between the cells of the GI tract and its microflora is one of mutual benefit. However, scientists are just beginning to understand that this is a much more active and communicative relationship than previously imagined.

It is now known that signalling systems exist between microbial and mammalian cells. While eukaryotic mammalian cells communicate by means of hormones, prokaryotic microbial cells communicate via hormone-like molecules known as autoinducers. When the concentration of these autoinducers reaches a certain level, the bacterial population responds by altering its gene expression. This mechanism is known as quorum sensing.

Mucosal epithelial cells act as a mechanical barrier that separates the host's internal systems from the external environment, and they also have a specialized function in the transport of ions, fluid absorption and secretion. These cells generate and transmit signals between the microbes in the GI tract lumen and the host's immune cells in the underlying mucosa. These signals may activate mucosal inflammatory and immune responses via chemical signals producing cytokines, prostaglandins and nitric oxide. Bacterial pathogens such as enterohaemorrhagic *Escherichia coli* effectively hijack these communication systems to allow themselves to gain a foothold.

Epithelial cells are also able to sense the 'normal' microflora, and there is evidence that this ability allows them to provide signals to the host, regulating the growth, development and function of cells in the adjacent and underlying mucosa.

Key points

Quorum sensing is a mechanism by which bacteria communicate with each other. Each bacterium produces small amounts of signalling chemical which will increase in concentration as the bacterial population increases. The bacterial community, within and between species, is able to sense this increased concentration and responds by altering the gene expression and therefore the behaviour of the entire population, triggering, for example, increased virulence and spore formation. There is now good evidence to show that mammalian cells can also sense and respond to these bacterial signals.

8.4 Normal flora in disease states

In the same way that a normal flora is essential to good health in the host, disruption to the normal flora can have extremely damaging effects. The two most obvious effects are caused by the administration of drugs and by a restricted diet.

8.4.1 Drug effects

While antibiotics have an innate antimicrobial activity and will always exert a deleterious effect on the normal flora, it is often not appreciated that around 50 per cent of all medicines have some gastrointestinal side effects. Tricyclic antidepressants, antihistamines, allopurinol and cytotoxics are just a few of the non-antibiotic drugs known to have antimicrobial properties. Radiation therapy has similar effects.

The broader the antimicrobial spectrum of the agent and the more of the active drug present in the intestine, the more catastrophic the effects on the normal microflora. After cessation of antibiotic treatment in a previously healthy adult, it will usually take around four weeks for the microflora to return to normal, although some species may not have returned to pretreatment levels by six months.

This damage to the normal flora allows pathogens to enter much more easily, and more resistant species such as *Clostridium difficile* (Figure 8.7) are more easily acquired and can result in toxin-mediated diarrhoea. Children are particularly susceptible, with 11 per cent of treated infants under the age of two being affected by antibiotic-associated diarrhoea.

When pathogens such as *Salmonella* species or *Clostridium difficile* are acquired, the absence of normal flora also increases the propensity for these organisms to be excreted in very large numbers (so-called super-shedders) so that patients pose a greater risk of infection to others.

More resistant strains (e.g. Enterobacteriaceae, pseudomonads and Micrococcaceae) that normally form a small part of the GI tract flora will also overgrow and it is these genera that are particularly associated with opportunistic infections in patients with haematological malignancy.

As exposure to antimicrobials is selective for more resistant strains, repeated or prolonged exposure may result in the remaining bacteria acquiring a high degree of resistance that may

Cross reference

See *Medical Microbiology* in this *Fundamentals of Biomedical Science* series for more information on the investigation of gastrointestinal disease

FIGURE 8.7
Resected colon showing
pseudomembranous colitis
caused by *Clostridium difficile*
infection.

be transmitted between closely-associated bacteria in the intestine. Plasmids may code for several resistance mechanisms in individual bacteria, and these can spread to other susceptible patients in a healthcare setting who then have few treatment options.

Disruption to the microflora (dysbiosis) down-regulates the expression of genes involved in antigen presentation and therefore impairs the general immune response. It may also lead to an imbalance in the populations of T cells that can result in abnormal inflammatory responses directed against the normal flora, and it has been suggested that this may be linked to the development of inflammatory bowel diseases such as ulcerative colitis.

CLINICAL CORRELATION

Ulcerative colitis

This is an inflammatory disease affecting the colon and rectum. The mucosa becomes inflamed and ulcerated, leading to bloody diarrhoea and abdominal pain. In severe cases, the colon may need to be removed surgically. It mainly affects young adults and, although a cause is unknown, it has a genetic predisposition. More information may be found at www.nhs.uk/conditions/ulcerative-colitis.

8.4.2 Effects of diet

Apart from the effects of diet in the first year of life, as discussed earlier, dietary variations cause significant diversity in the normal flora, which may have unexpected effects on the wellbeing of the host.

Gastrointestinal tract microorganisms derive their nutrients from the contents, so it is unsurprising that dramatic changes in microflora occur in situations of starvation. This occurs not only in famine but also in critically ill patients, especially those receiving intravenous (parenteral) feeding, and in those on low-residue, chemically-defined diets (e.g. astronauts), where absorption of nutrients is designed to take place in the first part of the small intestine in order to reduce the quantity of faecal matter. Dramatic changes in the microflora occur within 24 hours of cessation of oral feeding.

In starvation, mucosal IgA production decreases quickly, reducing the protective barrier to bacterial adherence. Simultaneously, the bacteria respond to the lack of nutrients by increasing their ability to adhere. In these circumstances, Gram-negative bacteria seem to be better able to adhere than the less-pathogenic flora, and an increase in GI tract-derived, opportunistic infections is seen. Studies have shown that enteral feeding in post-surgical patients with abdominal trauma resulted in 76 per cent fewer episodes of sepsis compared with those on parenteral feeding, and similar outcomes have been shown in related patient groups.

Astronauts returning to earth after a period of space flight have been shown to have lost the *Lactobacillus* component of their intestinal flora, to be replaced, at least in part, by Enterobacteriaceae, pseudomonads and Micrococcaceae. Stress is also believed to have effects similar to enteral starvation, but is less well understood.

Diets rich in particular elements may profoundly affect the microflora, such that there is wide variation between populations. For example, the traditional Japanese diet, which is low in protein but rich in carbohydrate, favours the presence of the more antibiotic-resistant *Enterococcus faecium* rather than the *Enterococcus faecalis* found in those on higher protein diets. A similar shift in the enterococcal subpopulation is seen in western patients on low-protein diets.

A high fat diet, which is relatively difficult to digest, results in a decrease in *Bacteroides* and *Bifidobacterium* species and promotes other bacteria such as *Clostridium* species, which are able to metabolize simpler carbon sources such as glucose and sucrose. The improved metabolism of these sugars and the increased energy harvested from the available food is then associated with an increase in body fat. It is, therefore, true to say that obese people may continue to gain weight while eating considerably less than a lean person.

SELF-CHECK 8.4

How long do the effects of prescribed antibiotics on the GI tract microflora last and what adverse effects may ensue?

8.5 Therapeutic manipulation of the normal flora

While exposures in the first year of life play a crucial role in the development and maintenance of the immune system, and determine predisposition to disease development later in life, abnormalities of the adult microflora have also been shown to be associated with disease states such as allergy, inflammatory bowel disease and psoriasis. A possible link with bowel cancer has also been postulated.

Most of these diseases are associated with a decrease in bacterial diversity, with loss of key groups such as *Lactobacillus* and *Bifidobacterium*, and an increase in organisms such as the Enterobacteriaceae, *Clostridium* species and yeasts. It is thought that this imbalance stimulates T cell subgroups to produce high levels of cytokines that initiate a chronic, self-perpetuating inflammatory response.

As discussed earlier, parenteral nutrition, stress in critical-care situations and the use of broad-spectrum antibiotics have unintended but dramatic effects on the normal flora, risking antibiotic-associated diarrhoea, *Clostridium difficile* infection and a diminished immune response.

As a better understanding of the normal GI tract flora and the mechanisms of its relationship with the host has developed, the possibility of new therapeutic options has appeared. These are particularly important to consider given the increasing problems of antimicrobial resistance.

The reconstitution or redirection of the microflora to improve health by means of microbial manipulation has been attempted for over a century, but only recently has it been studied in greater depth. Three main approaches are used and these involve feeding the patient probiotics, prebiotics or synbiotics.

8.5.1 Probiotics

Probiotics are microorganisms which can survive to reach the small and large intestines in sufficient numbers that enable them to influence the existing flora and potentially exert beneficial effects. Various organisms have been used; most commonly bacteria of the *Lactobacillus* and *Bifidobacterium* genera, but *Enterococcus* species and yeasts such as *Saccharomyces boulardii* have also been studied.

These organisms, frequently found in fermented dairy products like yoghurt, form part of the normal healthy GI tract flora and have been associated with protection against disease and modulation of the immune system, as discussed in more detail above. Studies have established a potential role for probiotics in the prevention and reduction of duration of rotavirus – and antibiotic-associated diarrhoea, the prevention and alleviation of irritable bowel syndrome, the prevention and alleviation of allergies in children, and in the induction of remission in inflammatory bowel disease. There is also some evidence for a role in the reduction of duration of carriage of some faecal pathogens such as verocytotoxin-producing *Escherichia coli*, and the prevention of urinary tract infections. Some authors have suggested that probiotic supplements may reduce the risk of bowel cancer, ischaemic heart disease and autoimmune disease, but much work remains to be done in this area.

A more dramatic, but proven, method of restoring the healthy normal flora has been used in a number of centres worldwide for the treatment of intractable *Clostridium difficile* diarrhoea. In this case, a suspension of faecal material from a healthy donor, usually a close relative, is infused into the intestine of the patient. This may be performed via a nasogastric tube or endoscope, and aims to recolonise the intestine with a healthy microflora. Despite its unpleasant connotations, so-called faecal transplantation has been shown to be very effective when treatment with antimicrobials has failed.

8.5.2 Prebiotics

Prebiotics are generally indigestible dietary fibres such as inulin, oligofructose and some galactooligosaccharides, which, as they are fermented in the large intestine, allow environmental changes to occur that favour normalization of the flora towards a larger population of *Bifidobacterium* and related beneficial species. The net effect is similar to that achieved by the use of probiotics, with the addition of improved bowel habit, bowel cancer prevention and reduced absorption of lipids, thereby lowering serum cholesterol.

8.5.3 Synbiotics

Synbiotics are essentially combinations of probiotics and prebiotics, and are hypothesized to act synergistically to maximize their beneficial effects.

There remains much to learn about the precise nature and influence of the GI tract microflora in health and disease, but it is already apparent that the development and maintenance of a healthy flora, from the time of birth onwards, is vital for the health and well-being of the host. As knowledge increases, exciting possibilities for optimizing and manipulating this complex ecosystem to prevent and treat a wide variety of human diseases are emerging.

SELF-CHECK 8.5

Name three beneficial effects of probiotic and prebiotic supplements.

Chapter summary

In this chapter we have:

- Described the cells of the gastrointestinal tract and their function.

- Described the development of the normal microflora of the gastrointestinal tract.

- Described the differences in microflora in the different regions of the intestine.

- Investigated the relationship between the microbial and intestinal epithelial cells.

- Discussed the relationship between the microflora and human health/disease.

- Examined the health effects of disruption to the normal microflora.

- Explored some of the ways in which the microflora of the gastrointestinal tract may be manipulated to promote health.

Further reading

- Bengmark S. Ecological control of the gastrointestinal tract. The role of probiotic flora. *Gut* 1998; **42** (1): 2–7.

- Clarke MB, Sperandio V. Events at the host-microbial interface of the gastrointestinal tract III. Cell-to-cell signalling among microbial flora, host and pathogens: there is a whole lot of talking going on. *Am J Physiol Gastrointest Liver Physiol* 2005; **288** (6): G1105–9.

- Ford M ed. *Medical Microbiology* 2nd edn. Oxford: Oxford University Press, 2014.

- Fujimura KE, Slusher NA, Cabana MD, Lynch SV. Role of the gut microbiota in defining human health. *Expert Rev Anti Infect Ther* 2010; **8** (4): 435–54.

- Kagnoff MF, Eckmann L. Epithelial cells as sensors for microbial infection. *J Clin Invest* 1997; **100** (1): 6–10.

- Mackie RI, Sghir A, Gaskins HR. Developmental microbial ecology of the neonatal gastrointestinal tract. *Am J Clin Nutr* 1999; **69** (5): 1035S–1045S.

- *Mandell, Douglas and Bennett's Principles and practice of infectious diseases* 7th edn. Philadelphia: Churchill Livingstone Elsevier, 2010.

- Orchard GE, Nation BR eds. *Histopathology*. Oxford: Oxford University Press, 2012.

 Discussion questions

8.1   What advice would you give to someone who wants to eat healthily and/or lose weight?

8.2   Do you think there is a place for routine dietary advice in modern medicine?

8.3   How might you try to improve the eating habits of the general population?

Answers to self-check questions and discussion questions, hints and tips are provided on the book's Online Resource Centre.

 Visit www.oxfordtextbooks.co.uk/orc/orchard_csf/.

Cells of the cardiovascular and lymphatic systems

Andrew Blann

Learning objectives

After studying this chapter you should confidently be able to:

■ List the different types of cell that make up the cardiovascular and lymphatic systems.

■ Explain how these cells come together to form the vessels and organs of these systems.

■ Outline relationships between the structure and function of arteries, veins, capillaries and lymphatics.

■ Describe the structure and function of the heart and lymph nodes.

■ Comment on the diseases that affect these tissues.

This chapter outlines the cells and organs of the cardiovascular and lymphatic systems, and how together they form the circulation. Almost all of the cells that make up the vessels and organs are endothelial cells, smooth muscle cells, heart muscle cells (cardiomyocytes), macrophages and fibroblasts. In various combinations they make up vessels (arteries, veins, capillaries and lymphatics) and the organs of the heart, thymus, spleen and lymph nodes. The chapter will conclude with a discussion of the different types of disease that affect the cells, vessels and organs of the cardiovascular and lymphatic systems.

9.1 **Overview of the cardiovascular and lymphatic systems**

A minimum requirement for all animal life, especially in the short term, is oxygen and a source of energy such as glucose. In Chapter 6 we read how the cells of the lungs are modified to absorb oxygen and facilitate the release of carbon dioxide, and in Chapter 4 we noted that red blood cells are highly specialized to carry and deliver oxygen to the tissues. In the mammalian cell the key energy substrate is glucose, one of the many different sources of energy absorbed from the diet by the alimentary canal. Oxygen, glucose and thousands of other molecules in the blood are moved around the body by the vascular system, a network of 60 000 miles of tubes comprising **arteries, arterioles, capillaries, venules** and **veins**. These blood vessels also remove the waste products of metabolism that are carried in the blood, such as carbon dioxide and urea. These are excreted by the lungs and the kidneys, respectively.

Cross reference

The alimentary canal is discussed in Chapter 7.

A second transport system is the **lymphatic system**, which has two functions. First, it facilitates the return of interstitial tissue fluid (lymph) to the heart, in which case it is acting as a drainage system. The second role of the lymphatics is to enable the movement of **lymphocytes** between different **lymph nodes** and other reticulo-endothelial tissues. In this respect, the lymphatics are also important in immune responses and inflammation. The lymphatics are not a circulation—the content of the lymphatic system is not actively pumped in the way that blood is driven by the heart. Instead, it is a closed system of capillary-like vessels that merge with the venous circulation near the heart.

Cross reference

Lymphocytes, macrophages, inflammation and immunity are described in Chapter 4.

The blood vessels, lymphatic vessels and organs are composed of different proportions of **endothelial cells, smooth muscle cells** and **pericytes**, as well as fibres of connective tissue such as **collagen**. The heart is a highly modified muscle, and the specialized muscle cells (myocytes) of this organ are termed **cardiomyocytes**. There are also a small number of other cells, such as **macrophages, adipocytes** and **fibroblasts**.

We will begin our exploration of the cardiovascular and lymphatic systems by first looking at the different types of cells present. We will then move to a discussion of how these cells come together to form the different transport vessels, and relate structure to function. Finally, we will look at which cells make up the lymph nodes and the heart, and how together they ensure this most crucial latter organ provides the body with blood.

Figure 9.1 illustrates the circulation and the major vessels and organs.

SELF-CHECK 9.1

Name the major cells that make up the vessels and organs of the cardiovascular system.

9.2 Cells of the cardiovascular and lymphatic systems

The majority of the cells that make up the vessels and organs of the cardiovascular system are endothelial cells, smooth muscle cells, cardiomyocytes, macrophages, fibroblasts and adipocytes. Strictly speaking, lymphocytes are not part of the cardiovascular system; they merely use the lymphatics as a transport system. However, lymphocytes are the major cell type found in lymph nodes. The nervous system is also involved in the cardiovascular system, but as a partial regulator of the rate at which the heart beats, and of blood pressure.

9.2.1 Endothelial cells

Endothelial cells are flattened, orthogonal cells that line all blood and lymphatic vessels, and also the inside of the heart, forming a continuous layer, the endothelium. In the adult it consists

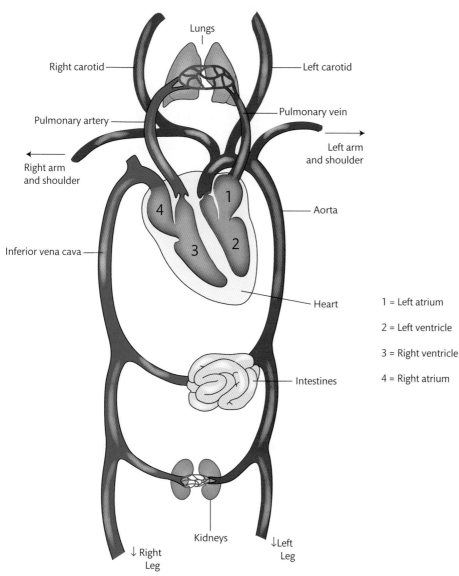

The head

Lungs

Right carotid

Left carotid

Pulmonary artery

Pulmonary vein

Left arm
and shoulder

Right arm
and shoulder

Aorta

Inferior vena cava

Heart

1 = Left atrium

2 = Left ventricle

3 = Right ventricle

Intestines

4 = Right atrium

Kidneys

↓ Right
Leg

↓Left
Leg

Cross references

The lung is discussed in
Chapter 6.

Cells of the intestinal tract are
described in Chapter 7.

FIGURE 9.1

**The heart and circulation
(organs not to scale).**

of $1-6 \times 10^{13}$ cells, weighs approximately 1 kg and covers a surface area of 4000–7000 m^2. Each cell is anchored to an underlying basic elastic lamina while individual cells are attached to their neighbours by specialized junctions in the gaps between the cells. These gaps regulate the passage of various cells and substances moving between the blood and the tissues.

Endothelial cells are highly complex, and have many diverse roles. Although a simple layer, this mechanical lining has a number of crucial properties. It can respond to changes in blood flow (such as shear stress) and receives instructions from various molecular messengers in the blood, such as **cytokines**. In the control of blood pressure, the endothelium releases key molecules that act on smooth muscle cells. Other endothelial molecules are active in the process of coagulation (**haemostasis**), both in clot formation (**thrombosis**) and clot removal (**fibrinolysis**), and in the regulation of the passage of certain white blood cells (leucocytes) from the blood and into the tissues. The endothelium also regulates the movement of fluid into the tissues,

Cross reference

Full details of haemostasis, thrombosis and fibrinolysis can be found in the *Haematology* volume in this Fundamentals of Biomedical Science series.

acting as a semipermeable barrier. In many cases, endothelial cells receive instructions from low molecular weight regulators acting on specific receptors in the cell membrane.

At the subcellular level, the endothelial cell can be defined by the presence of specific **Weibel-Palade bodies** in the cytoplasm. These organelles can be observed only by electron microscopy.

Blood pressure

The role of the endothelium in blood pressure regulation involves the synthesis and release of factors that act on the smooth muscle cells of the middle part of arteries (the media), resulting in constriction or dilation. The balance between the molecules that mediate constriction and dilation of the vessel has a profound effect on blood pressure. **Vasodilators** include nitric oxide (NO; Box 9.1) and certain prostenoids (e.g. prostaglandin I_2 [prostacyclin] and prostaglandin E_2), while **vasoconstrictors** include endothelin, angiotensin, certain catecholamines, vasopressin, thromboxane A_2 and prostaglandin $F_{2\alpha}$. Indeed, high blood pressure (hypertension) can be the result of the failure of the endothelium to regulate vascular tone correctly, either by over-activity of vasoconstrictors (too much endothelin) and/or under-activity of vasodilators (too little NO).

Cross reference

Chapter 5 discusses the nervous system.

However, the endothelium is not the only factor influencing blood pressure. The sympathetic nervous system acts on the smooth muscle cells via nerves that feed from the outside of the blood vessel. Indeed, an entire class of drugs that interferes with this innervation (beta-blockers) are used to treat hypertension.

Haemostasis

A healthy endothelium resists clot formation because it has an essentially anticoagulant nature. An example of this is the expression of molecules (e.g. heparin) at the cell membrane and the release of others (e.g. tissue factor pathway inhibitor). However, in certain circumstances, such as in acute blood loss where thrombosis is desirable, endothelial cells can become procoagulant and release molecules (e.g. **von Willebrand factor** [vWF]) to promote clot formation and so minimize blood loss.

BOX 9.1 Nitric oxide

Vasodilation is mediated through two distinct pathways, dependent or independent of the powerful vasodilator, formerly called endothelium-derived relaxing factor and later discovered by Furchgott and Zawadzki to be nitric oxide (NO). This molecule can be synthesized in response to shear stress, and is produced by the activity of the enzyme nitric oxide synthase (eNOS) on its substrates L-arginine and oxygen. As NO has a very short half-life, it must be generated continuously by eNOS. A small pleiotropic free radical, NO was first discovered in cardiovascular and neural systems as a vasodilator and a neurotransmitter, respectively.

Once generated, NO diffuses to the smooth muscles, where it activates guanylate cyclase, which in turn produces cGMP. Increased cytoplasmic levels of this small cyclic nucleotide set off a chain of reactions that result in changes to the structure of the actin and myosin complex, and thus smooth muscle relaxation. By this pathway, the endothelium adapts to changes in cardiac output to maintain blood pressure and organ perfusion. Apart from changes in blood flow, several molecules (e.g. bradykinin, adenosine, vascular endothelial growth factor and serotonin) can all induce the endothelium to upregulate eNOS and so increase levels of NO. Failure of the ability of the endothelium to generate NO may be a contributory factor in the development of hypertension and atherosclerosis.

von Willebrand factor is the major product of the endothelium that is released into the bloodstream. It is stored preformed in Weibel-Palade bodies (WPB), and so large amounts can be released rapidly in response to an emergency such as haemorrhage. However, it is also produced directly by direct protein synthesis that by-passes WPBs. Lack of vWF results in von Willebrand disease, which in its most severe form closely resembles the **haemorrhagic** disorder **haemophilia**, a disease caused by lack of coagulation factor VIII. This is because one of the functions of vWF is to act as an essential co-factor for factor VIII; therefore, absence of vWF leads to lack of functional factor VIII, and so the pseudo-haemophilia that is von Willebrand disease. Conversely, therefore, a large concentration of vWF may promote thrombosis by promoting the function of factor VIII in the coagulation pathway.

A key constituent of the membrane of the WPB is the adhesion molecule P-selectin. Thus, exocytosis of the WPB results not only in release of vWF but also in the appearance of P-selectin at the cell surface. This molecule promotes the adhesion of certain white blood cells and so possibly their passage into the vessel wall. von Willebrand factor is also secreted into the vessel wall where it may help the adhesion of the endothelial cells to the basement membrane of the intima. This is because vWF also has binding sites for connective tissue components (e.g. collagen and elastin) found in the basement membrane.

In addition to the release of vWF, endothelial cells also contribute to haemostasis through the expression of several components of the coagulation pathway, such as tissue factor. This molecule is the receptor for coagulation factor VII and so is procoagulant. It is inhibited by tissue factor pathway inhibitor, which is synthesized mainly by endothelial cells under basal conditions and is bound to the endothelial cell surface. Tissue factor expression leads to the activation of factor X, which then combines with factor Va to convert prothrombin to thrombin.

Thrombin is a multifunctional protein with some important haemostatic anticoagulant effects as well as procoagulant activity. It binds to thrombomodulin expressed on the endothelial cell surface, which is the major physiological buffer for the procoagulant effects of thrombin in normal vessels. Thrombomodulin binds to the same site on thrombin as do fibrinogen, platelets or factor V, in effect blocking procoagulation. The thrombin-thrombomodulin complex activates protein C and initiation of the activated protein C pathway, which is augmented by the endothelial protein C receptor (EPCR). Activated protein C must dissociate from EPCR before it can bind to protein S and function as an effective anticoagulant through the inactivation of factor Va. Thrombin is also involved in the process of inflammation and up-regulates endothelial cell P-selectin expression via WPB degranulation, which also releases vWF. Endothelial cells also produce ectonucleotidases that dephosphorylate ADP via AMP to adenosine. As ADP activates platelets, the effects of these enzymes inhibit platelet aggregation and, in turn, thrombosis.

Leucocyte migration

An important response to inflammation or infection in tissues is the movement of white blood cells such as monocytes and neutrophils from the blood into the tissues to combat the presumed or actual pathogen. The endothelium is strategically located at the blood-tissue interface, and is essential to the immune and inflammatory response specifically by producing and reacting to various low molecular weight protein mediators. These factors include bacterial endotoxin, colony-stimulating factors, growth factors and cytokines, examples of the latter being interleukins, tumour necrosis factor and interferon. These stimulators produce various changes in the endothelial cell, one of which is the up-regulation of adhesion molecules.

Endothelial cells promote the transfer of certain leucocytes (and possibly platelets) by increasing the expression of adhesion molecules on their cell membrane. Leucocytes cooperate in this process by expressing their own complementary adhesion molecules and other receptors involved in cell-cell adhesion. These adhesion molecules include E-selectin, L-selectin, intercellular adhesion molecule (ICAM) and vascular cell adhesion molecule (VCAM). All these adhesion molecules are present constitutively at low levels at the endothelial surface, but, as we have seen, P-selectin appears at the surface only after exocytosis of the WPBs.

Barrier function

Endothelial cells form a barrier to regulate interstitial fluid movement between the blood and tissues. This may be directly through the cell itself, involving endocytosis and pinocytosis, but also by the regulation of the gaps between adjacent cells. Normally, cells have co-called 'tight' junctions that minimize the exposure of the blood to the subendothelium. If cells are called upon to retract from each other, this can leave an intercellular space (i.e. gap junction) that provides a route of access for blood constituents to the vessel wall. It may also allow the movement of leucocytes into the blood vessel wall and beyond, independently of the endothelium.

Activation of endothelial cells

A resting endothelial cell can become physiologically activated (i.e. undergoes a change in function) by various factors. These include the rate of blood flow and blood viscosity, both of which act by changes in shear rates at the endothelial surface. The consequences of endothelial cells activation include increased release of coagulation factors, WBP exocytosis (and therefore release of vWF and the appearance of P-selectin at the cell membrane), the increased expression of adhesion molecules, and possible changes in barrier function.

Vasoactive mediators such as bradykinin and thrombin act to relax or constrict blood vessels to regulate vascular tone. Immunological activation can be facilitated by the presence of certain inflammatory mediators (e.g. interleukins) which may be secreted by, for example, macrophages or certain lymphocytes, while haemostatic activation may be mediated by molecules such as histamine and thrombin. Histamine also increases vessel permeability to bloodborne products by rapidly and reversibly 'detaching' the endothelial cells from each other.

Cross reference

Chapter 4 discusses the roles of macrophages and lymphocytes.

The endothelium also has the capacity to self-activate. Some of these factors act on distant endothelial cells (i.e. have endocrine activity), while others act on nearby endothelial cells (i.e. have paracrine activity). One such molecule, vascular endothelial growth factor (VEGF), can be released by endothelial cells and has endocrine, paracrine or autocrine (i.e. acting on itself) activity in promoting endothelial cell growth and the development of new blood vessels, a process known as **angiogenesis**. Indeed, endothelial cells about to participate in angiogenesis upregulate receptors for VEGF and other growth factors such as angiopoietin.

An important aspect of endothelial cell activation is that it is reversible when stimulation ceases. However, if the endothelium is activated for a long time period it may not recover its normal function and so becomes damaged.

Damage to endothelial cells

The endothelium is the target for the disease process in atherosclerosis, and is attacked by the four risk factors for this disease (i.e. smoking, hypertension, dyslipidaemia and diabetes) and by pathogenic microbes. The consequences of this are endothelial dysfunction, which manifests as failure to regulate vascular tone correctly and failure to regulate haemostasis. A damaged endothelium is less likely to be able to respond to those stimuli that control blood pressure, one of which is failure to up-regulate eNOS and so generate and release NO.

Damage to the endothelium also results in the increased exocytosis of WPBs and so increased vWF that promotes thrombosis. The disease process may also cleave thrombomodulin from the cell surface, so that this molecule is no longer able to sequester thrombin, which therefore remains active. A damaged endothelium also expressed less EPCR. All these changes promote thrombosis.

It seems likely that a damaged endothelium will be less likely to be able to regulate fluid passage correctly into the vessel wall and tissues, possibly leading to oedema. A further consequence of damage to the endothelium is that cells will lose their ability to adhere to their basement membrane and thus may be driven off the vessel wall by the blood flow. Indeed, circulating endothelial cells can be detected in the blood of patients with atherosclerosis and other conditions associated with endothelial damage, including inflammatory conditions such as vasculitis. Loss of endothelial cells exposes the subendothelium to flowing blood, which will permit the passage of fluids into the tissues, leading to oedema, and may also promote the adhesion of platelets and thus the establishment of a thrombus.

There may also be organ-specific changes. Damage to endothelial cells of the cerebral circulation may lead to cerebral oedema, and possibly contribute to stroke (cardiovascular accident [CVA]), while damage to the endothelium of the glomerulus is likely to lead to renal failure and proteinuria. Loss of function of endothelial cells of the retina and its blood supply may cause visual disturbances and, ultimately, blindness.

Figures 9.2 and 9.3 summarize the functions of, and molecules released by, endothelial cells.

SELF-CHECK 9.2

Describe the major functions of endothelial cells.

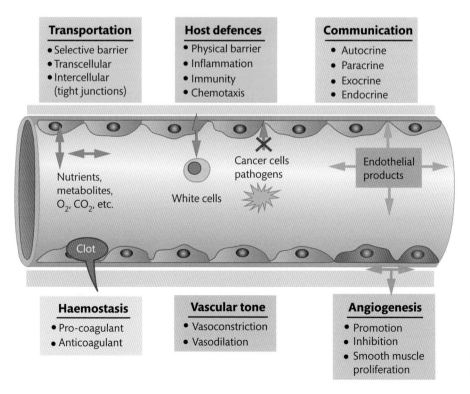

Transportation
- Selective barrier
- Transcellular
- Intercellular (tight junctions)

Host defences
- Physical barrier
- Inflammation
- Immunity
- Chemotaxis

Communication
- Autocrine
- Paracrine
- Exocrine
- Endocrine

Nutrients, metabolites, O_2, CO_2, etc.

White cells

Cancer cells pathogens

Endothelial products

Clot

Haemostasis
- Pro-coagulant
- Anticoagulant

Vascular tone
- Vasoconstriction
- Vasodilation

Angiogenesis
- Promotion
- Inhibition
- Smooth muscle proliferation

FIGURE 9.2
Functions of the endothelium.

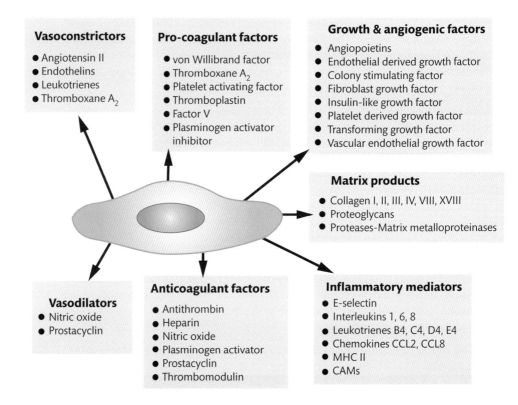

Vasoconstrictors
- Angiotensin II
- Endothelins
- Leukotrienes
- Thromboxane A$_2$

Pro-coagulant factors
- von Willibrand factor
- Thromboxane A$_2$
- Platelet activating factor
- Thromboplastin
- Factor V
- Plasminogen activator inhibitor

Growth & angiogenic factors
- Angiopoietins
- Endothelial derived growth factor
- Colony stimulating factor
- Fibroblast growth factor
- Insulin-like growth factor
- Platelet derived growth factor
- Transforming growth factor
- Vascular endothelial growth factor

Matrix products
- Collagen I, II, III, IV, VIII, XVIII
- Proteoglycans
- Proteases-Matrix metalloproteinases

Vasodilators
- Nitric oxide
- Prostacyclin

Anticoagulant factors
- Antithrombin
- Heparin
- Nitric oxide
- Plasminogen activator
- Prostacyclin
- Thrombomodulin

Inflammatory mediators
- E-selectin
- Interleukins 1, 6, 8
- Leukotrienes B4, C4, D4, E4
- Chemokines CCL2, CCL8
- MHC II
- CAMs

FIGURE 9.3
Products of the
endothelium.

9.2.2 Smooth muscle cells

Cross reference
Chapter 10 considers skeletal muscle.

There are three different types of muscle cell (myocytes): smooth muscle cells, skeletal muscle, and cardiomyocytes, the key features of which are shown in Table 9.1. Cardiomyocytes are the subject of the next section, while skeletal muscle, being a key component of connective tissues, is discussed in Chapter 10.

These cells are termed 'smooth' because they lack the striations present in skeletal muscle and in cardiomyocytes. Other major features are that their contractions and relaxations are

TABLE 9.1 Features of muscles cells

Smooth muscle cells	Cardiomyocytes	Skeletal muscle
Non-striated	Striated	Striated
Require a stimulus to contract	Contract spontaneously	Require a stimulus to contract
Present widely throughout the body and in epicardial arteries and veins	Specific to the heart	Widely distributed, connecting bones
Relatively mitochondria-poor	Relatively mitochondria-rich	Mitochondria supply intermediate
Non-branched	Branched	Non-branched

involuntary (i.e. they must be stimulated to do so), and that they are unbranched. Individual cells are spindle-shaped and, compared to skeletal and cardiac muscle, are slow to contact.

Smooth muscle cells are found in the walls of many organs, such as in the gastrointestinal tract (oesophagus, stomach, small and large intestines [where they drive peristalsis]) and genitourinary system (ureters, bladder, urethra and uterus). They are also found in the skin, being responsible for erecting hair. The media of arteries and (if present) veins is composed mostly of (vascular) smooth muscle cells, and these cells provide physical strength to the vessel, and contract and relax to regulate blood pressure and the flow of blood.

The cytoplasm of the individual smooth muscle cell is dominated by two proteins: actin and myosin. These proteins are arranged in sheets and chains, interspersed with dense bodies, and have the capacity to contract, according to the 'sliding filament' theory. By fixing sections to the cell membrane (the sarcolemma), the entire cell will contract. Other proteins are calmodulin, tropomyosin, calponin and caldesmon, of which the first is involved in regulation of contraction. Instructions for the actin/myosin complex (and thus the cell) to contract come from the autonomic nervous system, from catecholamines such as adrenaline (as smooth muscle cells have alpha-adrenergic receptors) and from the endothelium (endothelin). Upon receipt of these signals, intracellular second messengers such as calmodulin regulate levels of calcium. These in turn influence the action of kinase and phosphorylase enzymes that ultimately provide the adenosine triphosphate (ATP) needed to enable the actin and myosin fibrils to contract.

Relaxation of smooth muscle cells can be mediated by soluble factors such as nitric acid and prostacyclin, which by different routes lead to a reduction in intracellular calcium. This leads, via a convoluted pathway, to the uncoupling of the actin-myosin filaments, which slide apart and thus allow the cell to contract.

Apart from their contractile properties, smooth muscle cells (like fibroblasts) have the capacity to secrete numerous connective tissue components, including elastin, collagen and proteoglycans. Fibres of these structural proteins also provide physical strength and elasticity to the media.

Cross reference
Additional details of smooth muscle cells can be found in Chapter 10.

9.2.3 Cardiomyocytes

In common with smooth muscle cells, cardiomyocytes are involuntary and, in common with voluntary (skeletal) muscle, have striations. However, cardiomyocytes have two unique features: they are branched and can expand and contract spontaneously, even when isolated and grown in tissue culture. Unlike skeletal and smooth muscle cells, cardiomyocytes are permanently contracting and relaxing, and so demand considerable resources of oxygen and a source of energy, mostly fats and glucose. They are also characterized by the presence of many more mitochondria than are found in other muscle cells, and there is only one nucleus per cell.

BOX 9.2 Language and vocabulary

To the novice, many words being used seem incomprehensible, but in fact they derive from a comprehensive system based on Latin and/or Greek. For example, the prefix *myo* tells us the subject is muscle, while *sarco* refers to flesh, and *cyte* means a cell. The suffix *itis* means inflammation, *lemma* is a sheath, while *ase* tells us the molecule is an enzyme. Hence, *myositis* is inflammation of muscle.

The plasma membrane of the cardiomyocyte is called a sarcolemma, and the cytoplasm is referred to as the sarcoplasm. The major contractile material within the sarcoplasm comprises bundles of myofibrils, themselves composed of myofilaments, which are arranged in blocks called sarcomeres and which are separated from each other by Z bands. Within each sarcomere, areas where the myofibrils are dense are called A bands, whereas an I band marks an area where the myofibril density is low. This combination of alternating A and I bands gives the cardiomyocytes its striated appearance.

Myofibrils are composed of a complex of strands of the proteins actin, myosin, tropomyosin and troponin, of which there are three types. The combination of these proteins produces a helical structure that allows them to slide over each other. As the filaments are anchored to the Z lines, this has the effect of drawing the Z lines together, which produces the contraction of the cell.

SELF-CHECK 9.3

List major features of the two types of muscle cell of the cardiovascular system.

9.2.4 Pericytes

Despite pericytes being first described well over 100 years ago, the defining features of this cell remain unclear. One of the principal reasons for this is that pericytes are present in small numbers in different tissues and are uniformly difficult to isolate and so characterize. However, the proteoglycan NG2 may be a marker of immature pericytes. They are thought to be derived from a **mesenchymal** cell precursor that has the capacity to differentiate into vascular smooth muscle cells, depending on the direction of growth factors. However, pericytes may be involved in different aspects of vascular physiology; for example, in the cerebral circulation, pericytes may be important in the integrity of the blood-brain barrier, and actively communicate with other cells of the neurovascular unit (e.g. endothelial cells, astrocytes and neurons).

A role in angiogenesis

The presence of cell membrane receptors for molecules such as platelet-derived growth factor (PDGF), transforming growth factor (TGF)-beta, VEGF and angiopoietin (another growth factor) on the pericytes surface implies a role or roles in angiogenesis. This may be linked to the presence of Notch signalling pathways in pericytes, which suggests that these cells may be involved in remodelling an immature primary vascular system into a more mature vascular bed. Further evidence comes from *in vitro* and animal models, and demonstrates the requirement by pericytes for endothelial-derived signalling via PDGF, TGF-beta, VEGF and NO.

A role in blood pressure regulation

The presence of smooth muscle alpha-actin, non-muscle myosin, tropomyosin and desmin in some pericytes suggests involvement in blood pressure control, and so a possible role in the regulation of vascular tone. Indeed, pericytes can be induced to contract *in vitro* by the addition of vasoactive mediators (e.g. histamine, angiotensin, bradykinin and serotonin) that also vasoconstrict vascular smooth muscle cells *in vivo*.

Transformation to other phenotypes

There is also evidence that other pericytes can transform into fibroblasts or macrophages, if required, whilst others suggest that pericytes have stem cell activity and can transform into muscle, bone and skin cells, giving them possible roles in tissue regeneration after injury. One possible explanation for this multitude of functions is that pericytes should not be considered as a single cell type, but (as with macrophages) as cells that can have a number of diverse phenotypes. Under certain conditions, pericytes can transform into vascular smooth muscle cells and fibroblasts. Some commentators suggest that pericytes are a sourcee of adult pluripotent stem cell, although this needs to be confirmed.

The functions of pericytes are summarized in Table 9.2.

TABLE 9.2 Potential roles of pericytes.

- Endothelial cell proliferation and differentiation (angiogenesis)
- Stabilization and permeability of capillaries
- Regulation of contractibility and tone
- Maintenance of the basement membrane

9.2.5 Leucocytes

Various white blood cells are found in vascular tissues, and for decades it was believed their presence simply reflected some non-specific immunological or inflammatory function. However, there is increasing evidence that certain leucocytes (principally macrophages and lymphocytes) are in fact an important part of blood vessel physiology.

Macrophages

These cells arise in the bone marrow (as do all blood cells) as monocytes and move through the blood to various tissues, where they are described as macrophages. These cells have various functions.

Macrophages are involved in phagocytosis of microbial pathogens (both bacteria and viruses). Indeed, the word macrophage derives from the Greek for large eater. In this respect, they are also involved in the removal of the dead and dying cells in all organs and tissues, not only of the cardiovascular system, where they are found in the wall of large arteries and in the heart.

A second aspect is that different types of macrophage have other important roles in immunity; for example, in supporting the production of antibodies by lymphocytes, in which cases they are referred to as antigen-presenting cells, and in inflammation, through the release of pro-inflammatory cytokines such as interleukin-1, interleukin-6 and tumour necrosis factor.

A third function of macrophages is the secretion of proteolytic enzymes (e.g. elastase and matrix metalloproteinases); a fourth is the production of supportive connective tissue such as collagen and actin. Together, these last two functions are described as remodelling. However, not all macrophages perform all these functions in all locations. Some of these cells can be differentiated by the presence of molecules at the cell surface. For example, CD14 is the receptor for bacterial lipopolysaccharide, and CD16 is the receptor for the Fc domain of immunoglobulin. Another, CD11a, complexes with CD18 to form the adhesion molecule lymphocyte function-associated antigen-1 (LFA-1). CD11a has wider roles as it interacts with CD54, also known as intercellular adhesion molecule-1.

Monocytes/macrophages are also participants in haemostasis by generating tissue factor and coagulation factor V.

Subtypes of macrophages are found in specific locations:

- Kupffer cells are present in the liver, where they may take part in immunosurveillance, but may also be involved in the recycling of effete red blood cells.

- Langerhans cells are found in various tissues, frequently the skin, and may act as antigen-presenting cells.

- Osteoclasts are important in bone turnover (perhaps related to their ability to release bone-dissolving enzymes) but seem to have minimal or no immunological or inflammatory activity.

These specialized cells often lose the typical monocyte/macrophage morphology, but some bear particular CD molecules (e.g. CD19, CD23 and CD83 on dendritic cells, CD1a, CD14 and CD83 on Langerhans cells, and CD53, which is shared between leucocytes, dendritic cells and osteoclasts). Use of CD markers demands care because of poor sensitivity and specificity; CD19 is also found on B lymphocytes and CD16 is also found on neutrophils.

Lymphocytes

Cross reference

Additional details of leucocytes are provided in Chapter 4.

Lymphocytes dominate lymph nodes, making up over 90 per cent of all cells present, the remainder being macrophages and fibroblasts. The two major functions of lymphocytes are in the production of antibodies (B lymphocytes and T helper lymphocytes) and in the destruction of cells infected with viruses (T cytotoxic lymphocytes). However, lymphocytes must be assisted in these functions by modified macrophages acting as antigen-presenting cells.

9.2.6 Fibroblasts

Fibroblasts belong to the connective tissue family of cells, arising from mesenchymal tissue. Historically, they have been viewed as being uniform, with similar functions regardless of their particular anatomical site, be it the skin, bone marrow, lung or the heart. However, it is now becoming clear that fibroblasts are phenotypically diverse and have a number of different properties depending on the particular tissue and organ in which they are found. However, fibroblasts are often defined on morphological criteria; that is, they are flat, spindle-shaped cells with multiple processes growing out of the main body of the cell, making contact with neighbouring cells.

With regard to the cardiovascular system, fibroblasts (like some macrophages) secrete various connective tissue molecules such as collagen, fibronectin and elastin, all of which provide support. Cardiac fibroblasts can be differentiated from other cells of the heart (i.e. myocytes, endothelial cells, vascular smooth muscle cells) by the heart-specific expression of a collagen receptor named DDR2. Other potential fibroblast markers include fibroblast-specific protein-1, fibroblast activation protein, and cadherin-11. In addition to cytoplasmic and cell surface markers, genes highly expressed by cardiac fibroblasts include those for vimentin, beta-1-integrin, fibronectin and connexins.

9.2.7 Adipocytes

Adipocytes, often present near cells and organs that are metabolically very active, are able to synthesize and store various lipids. It seems likely that the fats and lipids stored within adipocytes are a source of energy. Indeed, there is increasing evidence that the heart draws a significant amount of energy from the fat stored in adipocytes, so that modest deposit of fat around the heart may be physiological and not, as once thought, pathological.

SELF-CHECK 9.4

What other cells are found in the cardiovascular system?

Having looked at different types of cells, we now examine how these individual cells come together to form the functioning vessels and organs of the cardiovascular and lymphatic systems.

9.3 Vessels of the cardiovascular and lymphatic systems

Figure 9.1 shows a simple layout of the heart and circulation. The two types of vessel of the cardiovascular system are those that carry blood and those that carry lymph. Blood vessels carry blood around the body in two independent systems. In the pulmonary circulation, oxygenated blood leaves the lungs and passes via the **pulmonary vein** to the left side of the heart. The heart then drives this oxygenated blood, in a continuous cycle of beats, around the body and head via the principal artery, the **aorta**. Some blood passes into the digestive organs where nutrients are collected, while other blood is transported to the brain. As blood passes through the kidneys the proportions of different ions, and pH, are regulated and waste material is excreted. This organ also regulates blood volume, which largely involves the amount of water in the blood.

Deoxygenated blood returns from the lower part of the body and the upper part of the body and the head to the heart via veins—the **inferior vena cava** and **superior vena cava**, respectively. The circulation is completed by blood moving from the right side of the heart to the lungs via the **pulmonary artery**. The vessels of the lymphatics do not form a circulation but effectively drain the tissues of interstitial fluid. In addition, the lymphatics also act as a transport system for lymphocytes, and also link lymph nodes.

9.3.1 Arterial circulation

The aorta is the major artery, taking oxygenated blood away from the heart to the head and body. The first crucial branch gives rise to the coronary arteries that feed the myocardium. Other arteries branch off the top part of the aorta, becoming the left and right carotid arteries (running up the neck to feed the brain) and the left and right subclavian arteries, feeding the shoulders and arms. The aorta moves down through the thoracic cavity and into the abdomen, where the left and right renal arteries feed the kidneys and the mesenteric arteries feed the intestines.

In the lower abdomen the aorta divides to give the left and right iliac arteries. As each iliac artery feeds its particular leg, it becomes, in turn, the femoral, popliteal and tibial artery, and ultimately the small arteries of the foot. In the pulmonary circulation, deoxygenated blood leaves the heart via the right ventricle and moves to the lungs via the left and right pulmonary arteries. As the major arteries move further away from the heart to deliver blood to the organs and tissues of the body, they divide many times and become progressively smaller. These small arteries are termed arterioles.

9.3.2 Arteries

The major driving force of the heart beat is delivered by the left ventricle, and to enable this it has greatest concentration of cardiomyocytes of all the chambers. Blood leaving the left chamber through the aortic valve and into the aorta will be at a high velocity and pressure. This high pressure must be maintained if the blood is to be distributed to the body. Accordingly, the aorta and its branches have a thick muscular and elastic wall to enable this high pulse pressure to be maintained as the arteries become increasingly narrow and transform into arterioles. Blood leaving the heart from the right ventricle is at lower pressure, but the relevant vessel, the pulmonary artery, still needs a thick wall. The largest arteries—those closest to the heart, such as the aorta—are likely to have a diameter of up to 20 mm. Arteries further away from the heart, such as the femoral arteries in the legs, may have diameters of 5-10 mm.

We can view the structure of a typical artery as having three distinct layers. The innermost, the **intima**, interfaces with the blood. The outer layer is the **adventitia**, and between intima and adventitia lies the **media** (Figure 9.4).

Intima

The intima consists of a single layer of endothelial cells, beneath which is a basement membrane (the internal elastic membrane, also called a basal lamina) composed of connective tissue components such as collagen, actin and elastin. The internal elastic lamina provides not only support but also a barrier to minimize the unregulated passage of fluids into the artery wall and (in capillaries) the tissues. As we have seen, the interface between the blood and the artery wall, the intima, has key roles in haemostasis, leucocyte migration and in regulating blood pressure.

Media

The thickest layer of the artery, the media, consists of smooth muscle cells interlaced with fibres composed of collagen and elastin. The arrangement of these cells and fibres (some of

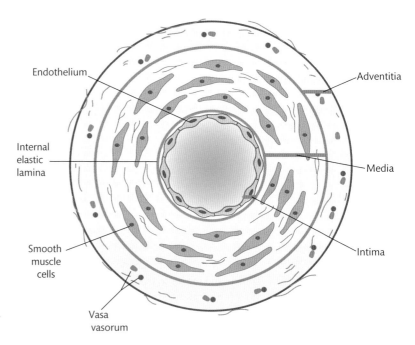

FIGURE 9.4
An artery.

which form sheets, or lamellae) is in concentric circles. These can lay perpendicular to the flow of blood, although some smooth muscle cells may lay parallel with the direction of blood flow. The degree to which these smooth muscle cells constrict and dilate, in response to vasoconstrictors and vasodilators, respectively, and also to sympathetic innervation, has a direct effect on blood pressure, and this is a process known as 'tone'.

The smooth muscle cells and elastic fibres together confer strength, elasticity and contractability, as arteries (and arterioles) need to be able to respond to the differences in blood pressure during the cardiac cycle and upon factors such as exercise. During systole, when blood is forced from the heart by the contraction of the ventricles, the extra pulse of blood must be accommodated by the arteries. This is enabled by the elastic properties of arteries; a feature known as **compliance**.

Vascular tone is controlled not only by the small molecular vasoconstrictors and vasodilators released by the endothelium, but also by the effects of sympathetic innervation by the autonomic nervous system. The interaction between neurons and the blood vessels is mediated in most cases by synapses bearing beta-type receptors (alpha receptors are common in other tissues such as the eye and the ureter), the activation of which results in vasoconstriction and thus the maintenance of vascular tone. Beta-blockers are an important class of drug used after a myocardial infarction (a heart attack) to reduce blood pressure. These inhibit the messages passed from beta-receptor to the smooth muscle cell, so that the latter fails to constrict as firmly, which results in lower blood pressure and thus less stress on the heart.

The external surface of the media of many arteries is bound by an external elastic membrane.

Cross reference
Additional details of the smooth muscle cells of other tissues and organs are provided in Chapter 10.

Adventitia

The outer layer of the artery is the adventitia. It is composed of a loose arrangement of connective tissue, such as elastic and collagen fibres (possibly secreted by fibroblasts), which may bind to nearby tissues and so provide physical support. In places, this layer of connective tissue may be sufficiently substantial to form an external membrane. Adipocytes may also be present, and it is tempting to speculate that these serve as reservoirs of fat destined to be a source of energy for the smooth muscle cells of the media. Indeed, non-esterified fatty acids are the second most important energy substrate of the heart, but it is unclear if the same is true for arteries. The definition of the outer boundary of the adventitia is arbitrary.

In small arteries, the smooth muscle cells of the media are likely to be able to obtain oxygen and nutrients, and dispose of waste material, passively by diffusion. In the larger arteries, however, the substantial mass of smooth muscle cells will not be able to obtain all its requirements or remove waste by simple diffusion alone. The problem is solved, curiously, by large arteries having their own blood supply. This consists of arterioles and venules found in the adventitia, termed the **vasa vasorum**. These form branches that burrow into the media, and so deliver blood to the smooth muscle cells. The nervous supply to the smooth muscle cells of the media, which also partly regulates contraction and relaxation, and thus vascular tone, is also found in the adventitia.

9.3.3 Arterioles

As arteries move further away from the heart, they become progressively smaller and typically have a diameter of 0.1–0.5 mm. This is reflected in their more refined structure, which has far fewer smooth muscle cells in the media—perhaps only two or three layers. However, these smooth muscle cells still contribute to the regulation of blood pressure. In the smallest arterioles there may only be a single layer of smooth muscle cells, and these are less likely to form the typical concentric layers seen in the larger arteries, but may instead be arranged in a spiral pattern. There is also much less adventitia, although the endothelium remains present. Endothelial cells of arterioles retain the ability to influence vascular tone (perhaps under the control of sympathetic innervation) and to participate in haemostasis. The vasoconstriction of arterioles may be regulated, in part, by pericytes.

As the arterioles subdivide they progressively lose the media of smooth muscle cells, and ultimately give rise to capillaries.

SELF-CHECK 9.5

Describe the structure of an artery.

9.3.4 Capillaries

Capillaries are the smallest of blood vessels, with a diameter of perhaps 4–15 μm, and consist only of a tube of endothelial cells on a basement membrane, with the occasional pericyte. Lacking smooth muscle cells, capillaries do not contribute to vascular tone but are still able to express and/or release molecules important in haemostasis and in regulating leucocyte migration into tissues. With only a single cell thickness (i.e. the endothelial cell) nutrients are able to pass easily from blood into tissues. Similarly, waste products such as carbon dioxide can pass in the reverse direction, from tissue to blood.

The capillary endothelium regulates the passage of certain components of plasma from blood into tissue; the components form what is called tissue fluid, or interstitial fluid. Capillaries may not always be present as single vessels, as are larger vessels, but often form large networks of vessels, and as such are described as a capillary bed. Sometimes described as the microcirculation, the major capillary beds are found in the liver, skin, lung and brain. The capillary bed has site-specific functions; for example, the cells of the capillary bed in the lung must allow the passage of oxygen and carbon dioxide.

We currently recognize three different patterns for the structure of a capillary:

- Type 1 is the most frequent type where the endothelium is continuous and forms a complete layer of cells over the basement membrane. In this setting, the capillary is essentially an arteriole without a media. Adjacent endothelial cells often form 'tight junctions' between themselves to resist, or perhaps control, the flow of material into the tissue.

- In type 2, there are occasional gaps, or pores, between individual endothelial cells (fenestrations) which facilitate the passage of materials into the tissue. In these situations, the basement membrane is exposed to flowing blood. Fenestrations, which are perhaps 70–100 nm in diameter, may therefore be a well-regulated method for controlling the movement of tissue fluid.

- A third structural variant, type 3, aids further the free passage of material, and is characterized by the absence of a basement membrane. Blood therefore interfaces directly with the underlying cells and there is no barrier to the transfer of nutrients or the interaction with the cells of the blood (Figure 9.5).

All three classes of capillaries may be supported by pericytes, but the precise role of these latter cells is unclear. Nevertheless, the intimal relationship between endothelial cells and pericytes implies a role such as regulation of endothelial function, such as in angiogenesis. There is no evidence that pericytes are involved in any haemostatic activity of capillary endothelial cells.

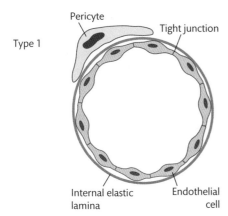

Type 1

Pericyte

Tight junction

Internal elastic lamina

Endothelial cell

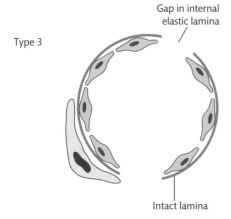

Type 2

Gap between endothelial cells

Internal elastic lamina

Endothelial cell

Type 3

Gap in internal elastic lamina

Intact lamina

FIGURE 9.5

The three types of capillary.

SELF-CHECK 9.6

How do capillaries differ from arterioles?

9.3.5 Venous circulation

The venous circulation carries blood from the tissues to the heart (Figure 9.1). In the pulmonary circulation, oxygenated blood leaves the lungs via the pulmonary vein, and is delivered to the left atrium. In the peripheral circulation, deoxygenated blood from below the heart (the lower chest, abdomen and legs) is delivered to the right atrium by the interior vena cava. Blood returning from above the level of the heart (the upper chest, arms and head) is delivered by the superior vena cava. Having fed oxygen- and nutrient-rich blood to the heart via the coronary arteries, deoxygenated and nutrient-depleted blood returns from the myocardium via coronary veins, culminating in the coronary sinus that empties into the right atrium. In parallel to the arterial circulation, these large veins are fed by many small veins, themselves formed by the confluence of many even smaller veins, the venules.

Venules

Venules are formed when several capillaries merge. Those venules closest to the capillaries are composed of an intima of endothelial cells lying on an elastic internal membrane, with an external adventitia. The functional status of these endothelial cells in relation to haemostasis and leucocyte migration is unknown. As several venules merge and the lumen increases in diameter, the larger size of the vessel needs to be supported by a media of smooth muscle cells.

Having passed through a capillary bed, the force of blood pressure has been spent and the flow of blood is weak. As additional venules join together, blood flow can be a problem as, in the absence of blood pressure, blood may actually flow backwards (retrograde flow) from the venules to the capillaries. At the point at which this becomes haemodynamically important, retrograde flow is prevented by valves, and the large venules therefore become veins. Pericytes have been described in post-capillary venules but their function remains to be clarified.

Veins

Veins essentially are large venules with valves and a thicker wall. As the vein increases in diameter, the media becomes thicker, with more smooth muscle cells. However, the media lacks the elastic fibres that contribute to the strength of the arterial walls, as venous blood pressure is low. With greater blood flow, the presence of valves becomes crucial in preventing retrograde blood flow, especially in the veins below the level of the heart as gravity must be countered. Conversely, veins above the level of the heart (e.g. the jugular veins of the neck) have few, if any, valves as blood can return to the heart by gravity. The valves are composed of fibrous tissue with modified smooth muscle cells covered by endothelial cells. Venous endothelium appears to have little influence on smooth muscle cells, but does have a role in haemostasis (Figure 9.6).

As with other large vessels, veins are held in place by an adventitia fixed to nearby tissues. However, because there are fewer smooth muscle cells in the media, and those that are present are less metabolically active, the vasa vasorum, if present, is less well developed than in arteries. With increasing age, and possibly because of mechanical factors, the valves become weak and incompetent. This, accompanied by a reduction in the strength in the walls, leads to tortuous and dilated vessels (varicose veins), especially in the legs, and sluggish or even static blood flow. Indeed, many instances of deep vein thrombosis are anchored near valves. Key features of arteries, veins and capillaries are summarised in Table 9.3.

SELF-CHECK 9.7

Describe the architecture of veins.

TABLE 9.3 Key features of arteries, capillaries and veins.

	Arteries	Capillaries	Veins
Blood pressure	High	Low	Low
Structure of the wall	Thick	Thin	Thin
Direction of blood flow	From the heart	Intermediate	To the heart
Oxygenation	Oxygenated	Intermediate	Deoxygenated
Valves	Absent	Absent	Present

9.3.6 Lymphatic system

Blood consists of cells and plasma (effectively, water in which is dissolved ions and molecules of various size and function). In health, the cells (red blood cells, platelets and most white blood cells) remain in the blood. However, a portion of the plasma moves across the capillary wall and into tissue, often driven by the physics of high blood pressure. This filtered fluid, rich in nutrients and oxygen but lacking large proteins, is called interstitial fluid. Owing to the relatively high pressure at the arteriolar ends of the capillaries, this tissue fluid is constantly being formed. Some fluid returns to the circulation via a special network of capillary-like vessels called the lymphatics, and once inside the vessels this fluid is called lymph. Lymphatic vessels consist of endothelial cells on a basement membrane (as are capillaries), and the vessels may be anchored to nearby tissue by filaments of connective tissue, such as collagen, which are likely to be secreted by fibroblasts (Figure 9.7).

The lymphatics drain lymph from the tissues, and, in a manner analogous to capillaries, venules and veins, converge to form larger vessels. In the lower part of the body, the lymphatics merge to form a large vein-like vessel, the thoracic duct, which eventually empties into a vein near the heart. Lymphatics draining the right side of the head, neck and the right side of the chest merge into the right lymphatic duct, and this also empties into a vein near the heart.

In common with veins, lymphatics also have valves to prevent retrograde flow. However, they differ from blood vessels in that they pass along a chain of small bodies known as lymph

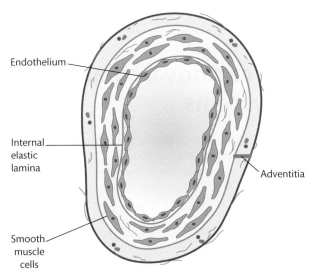

Endothelium

Internal elastic lamina

Adventitia

Smooth muscle cells

FIGURE 9.6

Vein.

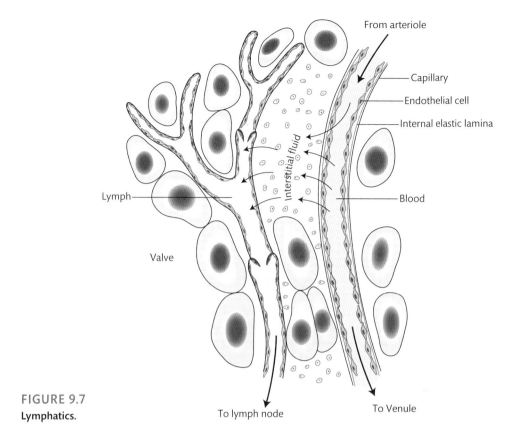

FIGURE 9.7
Lymphatics.

nodes, which are often fed by several lymphatics that drain a particular anatomical region (see section 9.4). Lymph nodes are filled with lymphocytes and some macrophages and fibroblasts. Indeed, the lymphatics form not only a conduit for lymph but also a pathway for lymphocytes moving from node to node. Consequently, lymphatics and lymph nodes play a crucial role in inflammation and immunity. In the gastrointestinal tract, lymphatics carry lymph rich in fats.

Cross reference

Further details of lymphocytes and lymph nodes, and their role in inflammation and immunity, are provided in Chapter 4.

SELF-CHECK 9.8

What are the key features of the lymphatics?

9.4 Organs of the cardiovascular and lymphatic systems

The most important single (and largest) organ in the cardiovascular system is the heart. However, there are many considerably smaller organs, the lymph nodes, scattered all over the body, some of which are in crucial anatomical positions. The thymus gland, liver and spleen comprise the remainder of the cardiovascular organs.

9.4.1 Heart

The heart is undoubtedly the most important organ in the body and, unlike almost all other organs, failure to perform its function can lead rapidly to death. A hollow, muscular body lying in the thoracic cavity and protected by the sternum, it weighs around 280–340 g in the adult male and 230–280 g in the adult female. The purpose of this organ is to provide the lungs, head, and body with over 3500 L of blood daily by the action of over 100 000 separate beats.

Cells and structure of the heart

Almost all of the heart is composed of millions of highly specialized muscle cells, the cardio-myocytes, which together make up the large mass of muscle we call the myocardium. Scattered within the myocardium are macrophages and fibroblasts. The external surface of the myocardium is bound by a layer of fibrous connective tissue, the epicardium. This layer interfaces with the pericardium, a fibrous, non-cellular sac that also helps to keep the heart in place within the thoracic cavity (Figure 9.8). A small amount of pericardial fluid minimizes friction between these fibrous layers as the heart beats. The internal surface of the myocardium, which interfaces with and so lines the chambers, is composed of a single layer of endothelial cells, and is called the endocardium. These endothelial cells are continuous with the inner linings of the blood vessels that leave the heart, and also cover the valves and the tendons that hold the valves in place.

The heart is, of course, a highly dynamic functioning organ in itself, and so needs to be supplied with blood. The high metabolic demands of a constant supply of oxygen and an energy substrate (e.g. glucose), and the removal of waste products (e.g. carbon dioxide), cannot be met by simple diffusion alone, so must be delivered by a separate coronary circulation. This blood is supplied by arteries, arising from the aorta, that run across the external surface (the epicardium) of the heart, and then divide and burrow into the myocardium. The walls of these coronary arteries include 'conventional' smooth muscle cells. The circulation is completed by cardiac veins on the epicardium, which carry the deoxygenated, nutrient-depleted, waste product-rich blood back to the right atrium. A healthy heart also has modest fat deposits, formed from clusters of adipocytes, on the outside of the heart, through which the epicardial arteries and veins run. This adipose tissue may be a reservoir of fat that can serve as a source of energy.

Although cardiomyocytes contract spontaneously, individually or in bundles, when grouped together by the million in the myocardium, their contractions must be tightly regulated and coordinated in order for the heart to beat correctly. The stimulus for the heartbeat comes from two sets of highly specialized tissue located on the epicardium—the **sinoatrial node** and **atrioventricular node**. From the cell biology viewpoint, these cells are difficult to classify accurately, but it seems likely that they developed from nervous tissues. At the top of the heart are the left atrium and the right atrium (the atria), below which lay the left ventricle and right ventricle. The two ventricles are separated by a septum.

SELF-CHECK 9.9

Describe the different types of cell that make up the heart.

Function of the heart

It is convenient to explain the function of the heart by looking at one complete cycle of a heart-beat, often called the cardiac cycle. The regulation of the beat is complex as the four chambers must act in a well-defined sequence. The stimulus for the first part of the cycle is activation of the sinoatrial node, which is spontaneous and associated with an electrical impulse that results in the controlled contraction of muscles of the right and left atria. These chambers will have been provided with blood from the superior and inferior vena cavae and from the pulmonary veins, respectively. Atrial contraction causes blood to pass through valves (the tricuspid on the right and the bicuspid, or mitral, on the left) into the right and left ventricles.

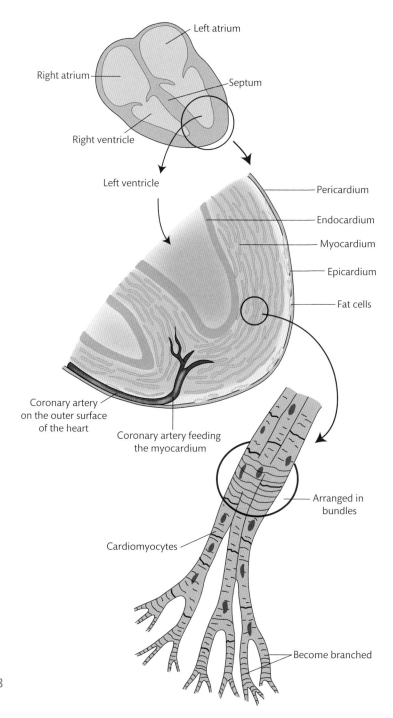

FIGURE 9.8
The heart.

The sinoatrial node acts not only on the myocardium of the atria but also on a second special-ized group of cells located near the junction of the right atrium and right ventricle. This second cluster is the atrioventricular node and, once activated, it in turn passes an impulse to the **bundle of His**. The impulse is then transmitted into the top of the septum, and the fibres then divide into the right bundle branch and the left bundle branch. Thus, the impulse message is transmitted down the septum that separates the left and right ventricles, to the muscles of the right and left ventricles.

The terminal sections of the bundle branches—those that contact the cardiomyocytes—are called **Purkinje fibres**. Not to be confused with the Purkinje cells of the cerebellum, these are specially differentiated cardiac muscle fibres. The atrioventricular node, the bundle of His, its left and right limbs and the ventricular Purkinje fibres are surrounded by a connective tissue sheath that insulates them from the neighbouring myocardium.

Upon receipt of the impulse, the cardiomyocytes of the right and left ventricles contract, forcing blood out of the right side of the heart through the pulmonary valve into the pulmonary artery, and from the left side of the heart through the aortic valve into the aorta. Once the ventricles have relaxed, the cardiac cycle is complete, and another cycle begins. Systole is the term given to that phase of the cardiac cycle when the chambers contract, while diastole is the term used when the chambers relax. These two phases give us expressions for measuring blood pressure in arteries of the arm: systolic blood pressure and diastolic blood pressure.

We use the term sinus rhythm to describe the normal sequence by which the four chambers work together, the key being the microsecond delays between the impulse messages that cause the regulated contractions. Failure of this regulation can easily lead to life-threatening cardiac disease. One such problem happens when the normal rhythm of the heart (sinus rhythm) is replaced by an abnormal pattern of beats, called an arrhythmia, the most common of which is **atrial fibrillation** (AF) (Box 9.3)

Cross reference

Further details about how warfarin provides protection from stroke and other clots, such as in the lungs and the legs, can be found in the *Haematology* volume in this Fundamentals of Biomedical Science textbook series.

SELF-CHECK 9.10

Describe how the cardiac cycle is regulated.

Control of heart rate

The sinoatrial node is the major pacemaker, which spontaneously delivers impulses at a rate of 60-100 per minute to control the rate at which the heart beats. However, this rate is also influenced by the action of nerves. A branch of the Xth cranial nerve (the vagus nerve), part of the parasympathetic nervous system, innervates the heart and acts on the sinoatrial and atrioventricular nodes to decrease the rate and strength of contractions. A sympathetic nerve supply, the cardiac nerve, which arises from the cardiac plexus, acts to increase the rate and strength of contractions. Further regulators are groups of cells embedded in the wall of the aorta and carotid artery. When faced with increasing blood pressure, these **baroreceptors** act to reduce the rate and force of the beating of the heart.

Cross reference

Further details of the nervous system can be found in Chapter 5.

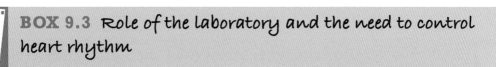

BOX 9.3 Role of the laboratory and the need to control heart rhythm

A major problem with atrial fibrillation (AF), especially in the elderly, is that the extra turbulence caused by the arrhythmia promotes clot formation. Should these clots run up the carotid arteries in the neck, they may become lodged in the brain and could cause a stroke that may be fatal. Fortunately, drugs such as amiodarone can help to bring the heart back to sinus rhythm, as can a controlled electric shock. In many patients, however, these treatments are not very successful, and AF can return, along with the risk of stroke. Therefore, patients with AF need to be protected from stroke, and one of the most effective ways of doing this is with the drug warfarin, which reduces clot formation. However, warfarin is a complex drug, and patients need to have a blood test about once a month to ensure the dose is effective and safe. Biomedical scientists in haematology are foremost in providing this care.

CLINICAL CORRELATION

A note on cardiac innervation

Surgeons transplanting a heart take care to ensure that the relevant arteries and veins are connected correctly, but are unable to re-attach nerves. Therefore, it is a testament to the continued functioning of a transplanted heart that it is fully capable of delivering blood to the body in the absence of these regulating nerves.

Cross reference

Cells of the nervous system are described in Chapter 5.

The combination of heart rate and the volume of blood expelled from the left ventricle (perhaps 70 mL) gives the cardiac output. So, if a heart beats 75 times a minute then the cardiac output is 5.25 L/minute.

Heart rate can also be influenced independently of the nervous system by other factors such as gender (the heart beats faster in women compared to men), temperature (higher when the temperature rises), age (slower with increasing age), thyroid hormones, adrenalin (as we experience during the 'fight or flight' response, when a surge of adrenaline causes heart rate to increase), and the concentration of certain ions in the blood. If heart rate is too high (tachycardia), the chambers may not have enough time to fill with sufficient blood, the result being a poor blood supply to the circulation. Conversely, when heart rate is too slow (bradycardia), although the chambers may fill with blood, the same insufficiency of supply can arise due to the small number of beats per minute. Both conditions need to be addressed (Box 9.4).

9.4.2 Lymph nodes

These small, bean-sized bodies (0.1–0.5 cm diameter), bound by a connective tissue capsule, are repositories of immunologically active cells, almost all of which are lymphocytes, with a small number of specialized macrophages. The structure of a typical lymph node (Figure 9.9) is maintained principally by fibroblasts that secrete connective tissue such as reticulin.

Cross reference

Further details about the importance of potassium and calcium (and many other ions) can be found in the *Biochemistry* volume in this Fundamentals of Biomedical Science series.

Tissue fluid draining various tissues is fed into the outer aspect of the lymph node (the cortex) by afferent lymphatics. From an immunological perspective, this fluid may carry viral and bacterial pathogens from infections in those tissues that the lymphatics drain. The internal structure is arranged into discrete areas around the cortex, at the centre of each is a lymphoid follicle. This structure is the primary area of immunological activity, the germinal centre being the area where foreign material is processed by macrophages and which stimulates the production of antibodies by lymphocytes. This follicular structure is also found in other tissues of the reticulo-endothelial system where antibodies are generated (e.g. spleen).

BOX 9.4 Role of the laboratory and the need to control heart rate

Tachycardia and bradycardia can result from several different processes, such as problems with the way that sinoatrial and atrioventricular nodes coordinate the contraction of the chambers. Some tachycardias can be treated with drugs such as lignocaine, digoxin and adenosine, and bradycardia with atropine. Both can be treated by an artificial pacemaker, which may be inserted surgically into the chest wall. High or low levels of potassium and calcium in the blood can also cause problems with heart rate and cardiac cycle, and this can prove fatal. Biomedical scientists in biochemistry, where these ions are measured, need to be fully aware of this and act urgently if abnormalities are found.

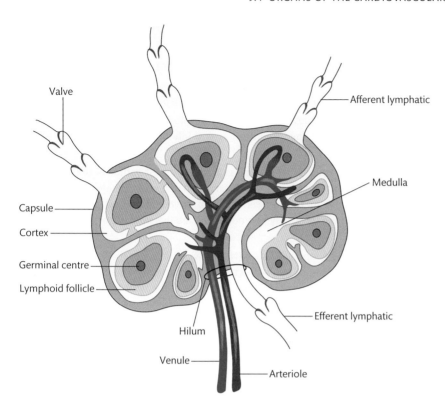

FIGURE 9.9
Lymph node structure.

A single efferent lymphatic leaves the lymph node via the hilum, and the region between this area and the cortex is called the medulla, which is dominated by draining tubules from the follicles. The hilum is also the region through which blood vessels enter and leave the lymph node: an arteriole bringing oxygen and an energy source, and a venule carrying away antibodies and possibly some lymphocytes. The efferent lymphatic carries lymph and lymphocytes to other lymph nodes, which are arranged in clusters and chains.

The location of lymph nodes in the body is not random; they are sited at key anatomical points through which the body may become infected (Table 9.4), and accordingly have focused

TABLE 9.4 Location and function of selected lymph nodes.

Location	Area defended and function
Tonsils	Throat: protection from pathogens in food and drink
Pulmonary, bronchial and tracheal	Airways: inhaled pathogens (including fungal spores)
Abdominal	Intestines: pathogens in food and drink
Axillary (armpit)	Arms: pathogens moving up the arms, perhaps from infections in the hands
Inguinal (groin)	Legs: pathogens moving up the legs, perhaps from infections in the feet
Para-uterine and cervical	Female reproductive tract: sexually transmitted diseases (NB: these lymph nodes are also enlarged during pregnancy)

functions. These tissues, described collectively as mucosal associated lymphoid tissues (MALT) include:

- Bronchus-associated lymphoid tissues (BALT), consisting of lymph nodes attached to the bronchus, although some commentators include lymph nodes in the lungs and lining the trachea. These tissues are likely to protect against airborne pathogens.

- Gut-associated lymphoid tissues (GALT), such as mesenteric and gastric lymph nodes, and the appendix. These seem primed to protect against intestinal pathogens that have been ingested alongside food.

- Peyer's patches are aggregates of lymphoid tissues in the lowest portion of the small intestine. It could be argued that they are not classical lymph nodes, as they are unconnected to the lymphatic circulation, but they are certainly rich in lymphocytes, which are organized into follicles that resemble those in lymph nodes and are therefore important in immunological defence.

- Nasopharynx-associated lymphoid tissues (NALT), which includes the tonsils and lymph nodes of the nose and pharynx, also protect against airborne pathogens. It has also been suggested that the cervical and clavicular lymph nodes of the front and sides of the neck should be part of the NALT family.

There are also clusters of lymph nodes in the armpit (axillary region) and groin (inguinal region) which protect against infections moving up the lymphatic vessels of the arms and legs, respectively.

9.4.3 Other cardiovascular tissues

The heart and lymph nodes, and the vessels that connect them, comprise almost all of the cardiovascular system. The remaining players are the thymus, spleen and liver.

Thymus

This organ is situated in the front of the chest, between the sternum and the heart, and is linked to nearby arteries and veins. Notably, there are no lymphatics entering the thymus, only lymph vessels that leave. The thymus grows consistently from birth, reaching its maximum size at puberty, and then slowly involutes so that in the elderley it may be rudimentary. There are two lobes, enclosed in a capsule, and each lobe is composed of a number of irregular follicles, each of which is composed of a medulla and a cortex. In this respect, there are several similarities between the thymus and a lymph node.

Cross reference

Full details of lymphocytes are found in Chapter 4.

The thymus is required for the complete development of T lymphocytes that are essential for defence against viruses and (indirectly) in the generation of antibodies. Immature cells (pre-T cells) arrive at the thymus from the bone marrow, and become thymocytes. As these cells mature and acquire immunological function, they express the T-cell receptor for antigen, and the markers CD3 and either CD4 or CD8. It seems likely that this process of education requires the participation of dendritic cells (a type of macrophage). The maturation process begins in the cortex and is completed in the medulla (Figure 9.10). The medulla contains characteristic Hassall's corpuscles, concentric layers of epithelial cells, that may produce certain growth factors.

Mature T cells leave the thymus: those bearing CD4 help B lymphocytes to produce antibodies, generally in the lymph nodes, while T cells bearing CD8 attack and kill cells infected with viruses, wherever they may be found.

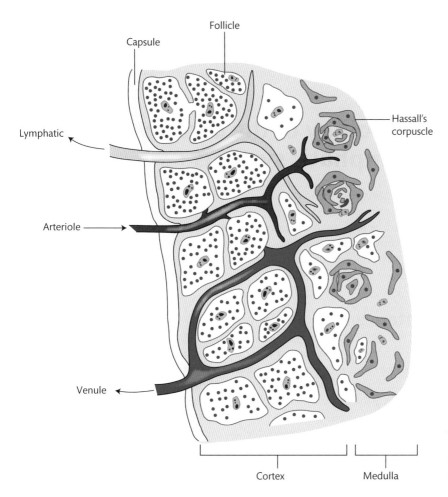

FIGURE 9.10
Structure of the thymus.

Spleen

This unusual organ, found on the left of the abdomen immediately below the diaphragm, is supplied by a splenic artery and vein but, like the thymus, only efferent lymphatics. The spleen has several different functions. It can be viewed as an emergency reservoir of red blood cells, should they be needed urgently to replace those lost by disease or accident, but it is also the site of the destruction of many old, effete and damaged red cells. Indeed, removal of the spleen is a treatment for certain diseases of those red cells that are marked out for (otherwise) premature destruction (e.g. sickle cell disease and thalassaemia).

The spleen also has a role in immunology. Within it are follicles that resemble those areas of lymph nodes where B lymphocytes generate antibodies, and other areas, called periarteriolar lymphoid sheaths, that are rich in T lymphocytes. Notably, those people who have lost their spleen to accident, surgery or disease are prone to certain bacterial infections, and therefore often require antibiotics. This is presumably because the spleen is home to specialized B lymphocytes specific for this organ that produce antibodies to these particular pathogenic bacteria.

Liver

This large organ, found on the right upper side of the abdomen, is dealt with in detail in Chapter 11, but deserves a brief mention here as it has a role in immunology. Like the spleen,

Cross references

Cells of the immune system are described in Chapter 4.

The liver is described in Chapter 11.

it can be the site of antibody production, and also generates many acute-phase reactants that are important in inflammation. The liver is supplied by the hepatic artery (a branch of the aorta) and also by the hepatic portal vein, which brings blood principally from the spleen, pancreas, and intestine. Blood leaves the liver by the hepatic vein, which joins the inferior vena cava.

9.5 Pathology of the cardiovascular and lymphatic systems

Cardiovascular disease is the leading cause of mortality and morbidity in the UK, Europe, North America, and in the developed and developing world. The common feature is the inability of the heart to deliver sufficient blood to other organs and tissues, and in many cases this is due to atherosclerosis. A second major disease process is inflammation, and this can also cause dysfunction. Although inflammatory disease has a completely different pathophysiology to atherosclerosis, there are instances where the two processes overlap.

9.5.1 Arteries

The process of atherosclerosis can in theory occur in any artery, although in practice those of the shoulders and arms are spared. As discussed, the endothelium is the primary target organ, damage to which leads to development of an atherosclerotic plaque. This lesion is characterized by invasion of the arterial wall by lymphocytes and macrophages, the latter ingesting oxidized low-density lipoprotein cholesterol to become foam cells. There can also be disruption of the smooth muscle cells in the media. The late consequences are narrowing of the lumen, leading to reduced blood flow and thus less oxygen, resulting in ischaemia of the tissues fed by the particular vessels.

If the atherosclerotic plaque ruptures, its contents, atheroma and thrombus, will be released into the circulation, be swept downstream and inevitably become lodged in, and so occlude, an arteriole. Therefore, the tissue distal to this blockage will be denied oxygen and nutrients, and undergo a process called infarction.

Weakness of the media of arteries may lead to a distension, which in turn leads to turbulence and the development of thrombus. Such a distension is called an **aneurysm**, and is most frequently found in the aorta as it passes through the abdomen. Should this aneurysm grow, the artery wall will become stretched, thinner, and will eventually burst (as would a balloon). This will have catastrophic consequences with massive and rapid blood loss, and carries a very high risk of mortality.

Inflammation of the blood vessels is called vasculitis, and of the arteries is arteritis. In common with other inflammatory disease, the aetiology of these diseases is of attack and destruction by leucocytes of those tissues perceived to be foreign. This may happen if the blood vessel harbours a pathogen (e.g. a hepatitis virus), but may also happen in the absences of a clear pathogen, in which case the vasculitis is termed autoimmune. Vasculitis can be a serious consequence of chronic inflammatory diseases such as rheumatoid arthritis and systemic lupus erythematosus.

9.5.2 Heart

There are many facets to heart disease but the most common can be classified into one of three groups: disease of valves, of the muscle cells, and of the blood vessels of the heart. The

BOX 9.5 *The laboratory and the diagnosis of myocardial infarction*

One of the many enzymes involved in extracting energy from glucose is creatine kinase (CK), which is found in many muscles. Damage to muscles leads to increased levels of CK in plasma, as is found after a myocardial infarction. Fortunately, we can distinguish myocardial damage from that of other muscles as a subtype of CK, CK-MB, is found only in heart muscle. Thus, raised plasma CK-MB is indicative of a heart attack, and there is a relationship between increased CK and CK-MB and the mass of heart muscle that has been damaged, so that this biochemistry blood test effectively tells us of the severity of the heart attack. Although not an enzyme, a small portion of heart protein, troponin, is often found in the blood after a heart attack.

four valves of the heart, two on each side, regulate blood movement in to and out of the four chambers. If the valves fail to open and close correctly, or leak, blood flow will be impeded, leading to poor heart function. Disease of the cardiomyocytes (**cardiomyopathy**) can be caused by factors such as an insufficient oxygen supply (**ischaemia**), drugs, excess alcohol consumption, inflammation and valve disease. Cardiomyopathy leads to weakness of the cardiac muscle, the consequences of which include arrhythmia and failure of the chambers to pump adequately and so deliver sufficient blood to the lungs and the body—a condition known as **heart failure**. The blood test for brain natriuretic peptide (BNP) has proved useful in the diagnosis and management of heart failure.

Ischaemia of the heart muscle may be due to disease of the coronary arteries on the external surface of the heart (i.e. coronary atherosclerosis). As alluded to above, rupture of plaque within a coronary artery will deny cardiomyocytes oxygen and glucose, and so they will fail to beat correctly and may eventually die. This process is called **myocardial infarction** (heart attack); if severe, this will lead to cardiac arrest and death. Damage to the myocardium by an infarction can contribute to heart failure.

9.5.3 Brain and legs

Similarly, narrowing of the carotid arteries of the neck leads to insufficient oxygen delivery to the brain, and thus to cerebral ischaemia. If atheroma from carotid atherosclerosis becomes dislodged and passes into the cerebral circulation, it may block an arteriole, damage the brain tissue and cause a stroke. If so, it is called a **cerebral infarction**. However, a clot arising in the heart passing up 'clear' carotid arteries may also cause a stroke. The parallel problem in arteries of the leg (e.g. femoral atherosclerosis) leads to pain in the muscles of the thigh and calf. This can be noted in a patient who is limping, and is described as intermittent claudication. If severe, this may result in lack of oxygen and nutrients to the foot, and can lead to ulceration and gangrene of the toes, which may need to be amputated.

9.5.4 Veins

Atherosclerosis rarely attacks veins, possibly because the blood pressure within is low, and as a result of differences in the structure of the vessel wall. However, veins do suffer disease, one

BOX 9.6 *The laboratory and the treatment of atherosclerosis*

Most of the thrombus formed by atherosclerosis is composed of platelets, and therefore this cell is the target of treatment. The principal antiplatelet drugs are aspirin and clopidogrel, and their use is mandatory in those who have had a heart attack or stroke, or who have heart failure. The laboratory does not have a role in the monitoring of these drugs, but biomedical scientists in biochemistry are involved in assessing two major risk factors for atherosclerosis (the role of the laboratory in cardiovascular disease is summarized in Table 9.5). High levels of certain lipids promote the formation of atheroma, while high blood glucose leads to diabetes, which attacks endothelial cells. Damage to endothelial cells leads to high blood pressure (hypertension) and thrombosis.

of which is the propensity for clots to form around valves. When this happens to veins deep within the leg, such as the saphenous vein, the clot is called a **deep vein thrombosis (DVT)**. From an anchor point near a valve, DVTs can extend up the vein and produce various signs and symptoms, including pain in the leg.

Although rarely fatal, DVTs can fragment—a process called embolization—and allow small clots (emboli) to move through the venous bloodstream, and around the right side of the heart. When these emboli reach the lung they can block a branch of the pulmonary artery. This is called a **pulmonary embolus (PE)**, and is analogous to blockage of an arteriole by a portion of atheroma or a thrombus. Thus, a PE prevents blood perfusing a portion of the lung, and this results in symptoms such as chest pain. The clinical distinction between PE and myocardial infarction, both of which produce chest pain, can be difficult. The collective name for DVT and PE is **venous thromboembolism (VTE)**. Clots rarely appear in other veins, such as those of the shoulders and arms.

TABLE 9.5   The laboratory and cardiovascular disease.

Arterial disease (atherosclerosis)
• Blood glucose, HbA1c
• Total cholesterol, high-density lipoprotein cholesterol

Venous disease (VTE, consisting of DVT and PE)
• D-dimers
• Monitoring the effect of warfarin on the coagulation system using the international normalized ratio (INR)
• Monitoring the effect of unfractionated heparin on the coagulation system using the activated partial thromboplastin time (APTT)
• Monitoring the effect of low molecular weight heparin on the coagulation system using the anti-factor Xa test

Heart disease
• Creatine kinase, creatine kinase MB, troponins (in myocardial infarction)
• Brain natriuretic peptide (in heart failure)
• Measuring levels of therapeutic drugs (e.g. digoxin)

BOX 9.7 *The laboratory and the treatment of venous thromboembolism*

In contrast to arterial clots, which are dominated by platelets, most clots in veins are the result of the over activity of clotting proteins such as fibrinogen. The most common drug used in the prevention and treatment of venous clots is warfarin, an agent we have already met as it is used to reduce the risk of stroke in atrial fibrillation (the role of the laboratory in cardiovascular disease is summarized in Table 9.5).

Weaknesses in veins, particularly those of the leg, can lead to a tortuous appearance under the skin. These unsightly vessels, often appearing at the surface of the back of the thigh and calf, are called varicose veins, and can lead to a DVT.

9.5.5 Lymphatics and lymph nodes

As these vessels do not carry blood, they are not the target of atherosclerosis, and neither do they suffer VTEs. However, lymphatics can become inflamed or blocked, and this is associated with the build up of fluid in the tissues, leading to oedema.

When involved in ongoing immune or inflammatory response, lymph nodes become engorged with fluid and cells—a feature called **lymphadenopathy**. If near the surface, these swollen lymph nodes can be felt, and a good example of this is in the case of tonsillitis, where lymph nodes in the front of the neck become swollen, often due to a bacterial infection of the throat. Another model of lymphadenopathy is infectious mononucleosis, where the lymph nodes of the neck become enlarged as part of the response to infection by microbes such as the Epstein-Barr virus. In all cases, once the immune response has subsided, and the pathogen eliminated, the lymphadenopathy resolves and the node returns to its normal size.

However, lymphadenopathy may also happen when nodes are infiltrated with cancer cells, such as in leukaemia or tumour metastasis, or in the case of primary cancer of the lymph node (lymphoma).

Cross reference

Malignancies of white blood cells (leukaemia, lymphoma and myeloma) are described fully in the *Haematology* volume in this Fundamentals of Biomedical Science series.

SELF-CHECK 9.11

What are the major diseases of the circulation?

9.5.6 Diseases of the thymus, spleen and liver

The primary malignancy of the thymus is a form of lymphoma, which, if present, results in impaired cell-mediated immunity. The same outcome is present in DiGeorge syndrome, the leading congenital thymic disease, where the organ is markedly reduced in size, or absent. However, these conditions are exceedingly rare.

Far more common is enlargement of the spleen (splenomegaly) and liver (hepatomegaly). These are often found in late-stage leukaemia and lymphoma.

Chapter summary

- Together, the cardiovascular and lymphatic systems form the circulation, its function being the delivery of blood rich in oxygen and glucose to the tissues, the transfer of waste products of metabolism to excretory organs, and the drainage of fluid from the tissues and its return to the venous circulation as lymph.

- For the cardiovascular system to perform its functions, blood is driven around the body by the heart, which is a highly complex muscle.

- Regulated contractions of four chambers in the heart propel blood around the pulmonary and peripheral circulations.

- Almost all of the heart is composed of muscle cells (cardiomyocytes), with a smaller number of endothelial and other cells.

- The arterial circulation (i.e. arteries and smaller arterioles) carries blood away from the heart. Blood leaving the heart is at high pressure, and vessels need a thick wall, dominated by smooth muscle cells that relax and contract to maintain this pressure through the circulation.

- Endothelial cells form the inner lining of arteries and arterioles, while the outer layer, the adventitia, is a loose amalgam of cells and connective tissue.

- Linking the arterial and venous circulations are capillaries. These consist of a single layer of endothelial cells on a basement membrane, and perhaps some pericytes.

- The venous circulation (i.e. veins and venules) carries blood to the heart. These vessels have thinner walls than do arterial vessels, and potential backflow of venous blood is prevented by valves.

- The lymphatic system is a closed network of capillary-like vessels that carry lymph and lymphocytes through a chain of lymph nodes, emptying into the circulation near the heart. The major diseases of the cardiovascular and lymphatic systems are atherosclerosis (involving arteries), VTE (involving veins) and tumour-associated lymphadenopathy (affecting lymph nodes).

- The laboratory has little role in the treatment of atherosclerosis, but is involved in assessing and monitoring the treatment of two of its major risk factors, hyperlipidaemia and diabetes.

- Venous thromboembolism is prevented by, and treated with, the drug warfarin, which requires close laboratory monitoring. The role of the laboratory in cardiovascular disease is summarized in Table 9.5

Further reading

- Ahmed N. *Clinical biochemistry*. Oxford: Oxford University Press, 2011.

- Baum J, Duffy HS. Fibroblasts and myofibroblasts: what are we talking about? *J Cardiovasc Pharmacol* 2011; **57** (4): 376–9.

- Braunwald E. *Heart disease: A textbook of cardiovascular medicine* 4th edn. Philadelphia: WB Saunders, 1992.

- Fuster V, Alexander RW, O'Rourke RA. *Hurst's the heart* 10th edn. New York: McGraw Hill, 2001.

- Moore G, Knight G, Blann A. *Haematology*. Oxford: Oxford University Press, 2010.

- Pate M, Damarla V, Chi DS, Negi S, Krishnaswamy G. Endothelial cell biology: role in the inflammatory response. *Adv Clin Chem* 2010; **52**: 109–30.

- Ribatti D, Nico B, Crivellato E. The role of pericytes in angiogenesis. *Int J Dev Biol* 2011; **55** (3): 261–8.

- Tortora G, Anagnostakos. *Principles of anatomy and physiology* 5th edn. New York: Harper and Row, 1987.

- Valentijn KM, Sadler JE, Valentijn JA, Voorberg J, Eikenboom J. Functional architecture of Weibel-Palade bodies. *Blood* 2011; **117** (19): 5033–43.

- Weibel ER, Palade GE. New cytoplasmic components in arterial endothelia. *J Cell Biol* 1964; **23**: 101–12.

 Discussion questions

9.1 Explain how damage to the endothelium promotes hypertension.

9.2 Describe how different cells come together to form the blood and lymphatic vessels.

9.3 Outline the extent to which damage or disease of different cells leads to heart disease.

Answers to self-check questions and discussion questions, hints and tips are provided on the book's Online Resource Centre.

 Visit www.oxfordtextbooks.co.uk/orc/orchard_csf/.

10

Musculoskeletal system

Suha Deen

Learning objectives

After studying this chapter you should confidently be able to:

- Outline skeletal muscle development after birth.
- Explain the profile of muscle fibres that contributes to their macroscopic appearances.
- Describe the microscopic and ultrastructural features of muscle fibres.
- List types of myofibres and explain how they relate to muscle performance.
- Describe the physiological factors that determine the size of skeletal muscle.
- Describe the internal bone organization.
- Comment on bone structure and how it aids in coping with external stress.
- Describe different types of ossification.

The musculoskeletal system provides the body with mechanical support. It is composed of skeletal muscles, tendons, bones, joints and ligaments, with all the different components working together as a unit. The main functional attribute of bone is its specialized extracellular matrix, which is hardened by the deposition of calcium, enabling it to function as a rigid lever, articulating with other bones through joints that are kept in relationship by ligaments. Bone also contains blood-forming marrow. Skeletal muscles act as contractile levers and are connected to bones by tendons. Muscles always act globally; electromyographic studies show that the first muscle to contract when lifting the arm is a muscle in the foot.

This chapter deals with the skeletal system, including skeletal muscles, bones and cartilage. The engineering behind body shape and the composition of skeletal muscle, bone and cartilage, and how they are deployed to serve this adaptable nature, is discussed in this chapter.

10.1 **Skeletal muscle**

The skeletal muscles are responsible for voluntary movements. Essential for physical movement of the joints, muscle also participates in assisting vital activities such as respiration and swallowing. The musculoskeletal system is the largest organ in the body, comprising approximately 650 muscles. Muscle contraction and stretching result in an orderly alignment of the constituent cells. Skeletal muscles work in a highly coordinated manner. They respond and adapt to our lifestyle by modifying their content and shape. Severe injury can result in muscular atrophy in a short time, but with training, muscle size can be regained in a relatively short period

---

Key points

From the arrangement shown in Figure 10.1, a labyrinth of connective tissue sheath can be visualized consisting of infinite folds and endings that bifurcate (fork) to encase different partitions, starting from the outermost surface to surround the innermost microscopic compartment, forming a scaffold to carry smaller branches of nerves and capillaries down to single muscle fibres.

In order to appreciate the morphological features of muscle fibres, it is essential to explore the embryogenesis of muscle fibres.

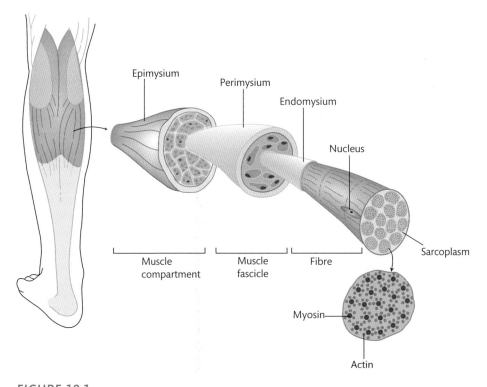

FIGURE 10.1
Diagram to show the relationship of the **epimysium**, **perimysium** and **endomysium** of skeletal muscle fibres.

10.1.1 Embryogenesis

In brief, during embryogenesis, skeletal muscle cells are seen in the form of primitive mesenchymal cells (**myoblasts**). Fusion of many individual myoblasts ultimately forms a syncytium of elongated cells that is the mature multinucleated skeletal muscle fibre (discussed later). These cells are extremely large and their growth depends on receiving motor stimulation.

Key points

A **sarcomere** is the contractile unit of a myofibril (sarcomeres are repeating units, delimited by Z lines along the length of the myofibril); the sarcolemma is the membrane covering a striated muscle fibre; and the sarcoplasm is the cytoplasm of a striated muscle fibre.

10.1.2 Stages of development

In utero *development of skeletal muscles*

Myoblasts are the most immature muscle cells and are identified morphologically using haematoxylin and eosin (H&E) staining as strap-shaped cells with a single nucleus, prominent nucleoli and abundant pink-staining cytoplasm. Myoblasts are capable of going through mitosis to divide. They have no myofilaments in their cytoplasm. Ultrastructurally, myoblasts are attached to each other by filopodia and are joined by gap junctions.

Myoblasts recognize each other through a plasma membrane glycoprotein that facilitates fusion of myoblasts to form **myotubes**, which is the next stage in the embryological development of muscle.

Myotubes have multiple nuclei and have started to show features of specialization in the form of cytoplasmic filaments. Initially, the filaments are formed in the peripheral part of the **sarcoplasm**. Filaments stain positive for vimentin and desmin. At this stage, groups of myotubes and myoblasts are enclosed by a common basement membrane and separated by interstitial cells. Myotubes continue to mature with an increasing number of nuclei that appear central in cross sections. Evidence of contractile activity is demonstrated by the development of more prominent **myofilaments**. Having acquired these features, the myotubes give rise to **muscle fibres** as a next step in development.

At the **muscle fibre (myofibre)** stage, myoblasts and myotubes stop fusing and start to develop diffusely distributed **acetylcholine receptor** protein on the cell surface. Later, this protein becomes localized into hot spots as the foundation for **motor end-plates**.

Morphologically, muscle fibres are more specialized. Their nuclei are peripheral and their filaments are organized into sarcomeres, which are the smallest functional units of the muscle fibre.

A myofibre consists of a continuous stretch of approximately 10 000 sarcomeres.

After the twenty-first week of gestation, the number of myotubes continues to decline, whereas the muscle fibres undergo histochemical differentiation with variable ATPase and oxidative enzyme activity, setting the scene for future type 1 and type 2 fibres.

Muscle development after birth

After birth, muscle fibres continue to increase in size, mainly in the following forms:

- Increase length of sarcomeres. The reserve in sarcomeres allows muscle fibres to elongate during spurs of rapid growth. This accounts for up to 25 per cent increase in muscle fibre length.
- Addition of new sarcomeres seen predominantly at the myotendonous junction at the end of the muscle fibres.
- There is evidence to suggest a gradual increase in the number of muscle fibres after birth. Dividing stem cells may be the origin of these new muscle fibres.

10.1.3 Morphology of muscle fibres

Under the light microscope

Muscle fibres are the building blocks of skeletal muscles. Each skeletal muscle fibre (cell) is extremely long (up to 10 cm in length) and hence is often described as a fibre rather than a cell. However, even though muscle fibres are long cells, they do not extend the full length of the muscle, which sometimes may reach 40–50 cm in length. Muscle cells are usually arranged in overlapping bundles, with the force of contraction being transmitted through the arrangement of the support tissue.

The muscle fibre is a multinucleated long and narrow cylinder. In a cross section, normal adult muscle fibres are polygonal with multifaceted profiles that fit together like pieces of a jigsaw puzzle (Figure 10.2). Normally, the size of each muscle fibre is determined by factors such as gender, training and/or age. In a random cross section, four to six nuclei are usually identified, located under the sarcolemma in each muscle fibre. This equates to approximately 30 nuclei for each millimetre of fibre length. The nuclei measure 5–12 μm in length and 1–3 μm in width, with a fine chromatin pattern and small, usually invisible, nucleoli; they are flat and slender, running parallel to the longitudinal axis of the muscle fibre. Hence, in a cross section, the nuclei are seen as small, dark purple bodies under the sarcolemma (Figure 10.2a). The cytoplasm (sarcoplasm) is abundant and contains contractile protein in the form of myofilaments as well as numerous intervening mitochondria and abundant glycogen. Muscle stains light pink with haematoxylin and eosin (H&E), while the mitochondria appear as tiny red granules with Gomori's rapid trichrome (Figure 10.2c). Muscle fibres can be divided into red and white fibres and individual fibres can be typed using enzyme histochemical techniques (Figure 10.2d).

Under the electron microscope

The sarcoplasm contains a highly complex organized apparatus of contractile **myofilaments** of 1–2 μm diameter. The myofilaments are classified mainly into thin and thick filaments, which are predominantly **actin** and **myosin**, respectively. Each myofilament is arranged in contractile functional units (sarcomeres) separated by **Z lines**. Each sarcomere is composed of actin and myosin myofilaments (Figure 10.3). Once contraction is initiated, actin myofilaments are stretched and slide over myosin myofilaments, whereas the latter remain unchanged. The alternating pattern of arrangement of the thin and thick myofilaments provides the skeletal muscle with a striated appearance that is seen best in longitudinal section.

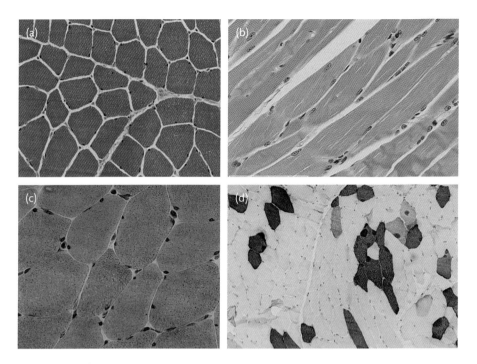

FIGURE 10.2

The muscle fibre is a multinucleated long and narrow cylinder. In cross section (a) normal adult muscle fibres are polygonal with a multifaceted profile fitting together in a jigsaw puzzle with the adjacent muscle fibres (H&E stain). Characteristic cytoplasmic cross striations are seen in longitudinal section (b) as alternating dark and light bands (H&E stain). The cytoplasm (sarcoplasm) is abundant and contains contractile proteins in the form of myofilaments as well as intervening numerous mitochondria and abundant glycogen. It stains with Gomori's rapid trichrome (c), a method used on frozen sections, in which mitochondria appear as tiny red granules. Muscle fibres can be broadly divided into type I (pale) and type II (dark) fibres (d) on ATPase histochemical staining.

Ultrastructurally, a longitudinal section of muscle demonstrates the different zones of the sarcomere in variable shades, reflecting the different overlapping layer arrangement of myofilaments (Figure 10.4). The central part of the sarcomere shows a darkly stained band (A band) representing thick myosin myofilaments. The A band is darker where actin myofilaments overlap myosin myofilaments at each side, leaving a central column (H zone) of lighter staining, representing myosin myofilaments without overlapping actin myofilaments. The A band is flanked each side by a light band (I band) representing actin myofilaments. The I band extends from the A band of one sarcomere to the A band of the next sarcomere. The I bands of adjacent sarcomeres are separated by a Z line. The distance between two Z lines represents the sarcomere's length. The Z line is an accessory protein (actinin), the function of which is to stabilize the thin actin myofilaments in place and separate two adjacent sarcomeres. Myosin myofilaments are stabilized by another accessory protein (myomesin), which is seen as a central dark line within the H zone. A cross section under the electron microscope demonstrates a regular hexagonal pattern repeat of a central thick filament surrounded by six thin filaments.

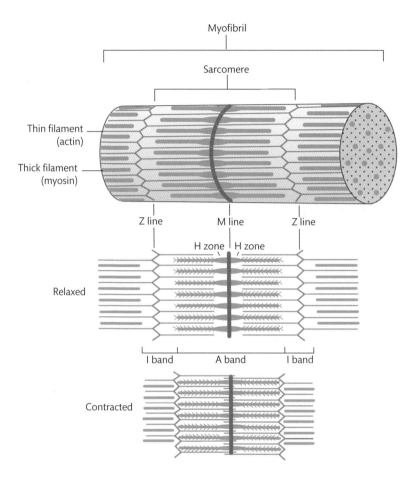

Myofibril

Sarcomere

Thin filament (actin)

Thick filament (myosin)

Z line   M line   Z line

H zone   H zone

Relaxed

I band   A band   I band

Contracted

FIGURE 10.3

Muscle sarcoplasm contains a highly complex organized apparatus of contractile myofilaments of 1 -2 μm diameter. The myofilaments are classified mainly into thin and thick filaments, which are predominantly actin and myosin, respectively. Each myofilament is arranged in contractile functional units (sarcomeres) separated by Z lines. Each sarcomere is composed of actin and myosin myofilaments. Once contraction is initiated, actin myofilaments are stretched and slide over myosin myofilaments, whereas the latter remain unchanged.

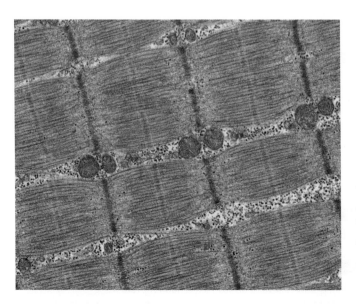

FIGURE 10.4

The alternating pattern of arrangement of the thin and thick myofilaments provides skeletal muscle with a striated appearance that is seen best in longitudinal section, as in this electron micrograph of relaxed muscle.

Reproduced with kind permission from Dr Liz Curtis, Queen Elizabeth Hospital Birmingham.

Macroscopic appearances of muscle

Red muscle is characterized by the following:

- greater mitochondrial and lipid content
- greater capillary density
- relies on aerobic respiration
- designed for postural function or sustained activity
- greater **myoglobin** protein contents than white muscle, which gives the muscle its red colour.

10.1.4 Muscle function

Skeletal muscles act in pairs whereby the contraction of one muscle is balanced by a relaxation of its paired muscle or group of muscles. This antagonistic action assists joint movement. Skeletal muscles that support the skull, vertebrae and rib cage, where power rather than fine, coordinated movement is required, are called axial or proximal skeletal muscles, whereas the muscles of the limbs are called peripheral muscles. The fibres of proximal muscles are larger than those of peripheral muscles. Individual muscle fibres are arranged in groups to form anatomically distinct muscles, enabling them to execute physical movement. Other features that enable muscles to deliver their function include the tendons, blood supply and innervation.

10.1.5 Tendons

Muscles are anchored from each pole to the underlying bones by highly organized, thick connective tissue called tendons which are encased by a lubricated connective tissue sheath (synovial sheath) to reduce friction and facilitate sliding over bone. Attachment of the tendon to the bone is called the origin of the muscle, while the other end of the tendon is the insertion of the muscle. Tendons are cylindrical structures of poorly vascularized, tightly packed collagen fibres with intervening scanty fibrocytes. This, therefore, indicates low oxygen demand in order to strengthen and aid the supportive function of the tendons.

SELF-CHECK 10.1

What do you think red muscle is best used for?

White muscle is characterized by the following:

- contains fewer mitochondria than red muscle
- relies on anaerobic respiration as it contains abundant glycogen stores
- better suited to sudden and intermittent contraction, as it depends on anaerobic respiration.

SELF-CHECK 10.2

What do you think white muscle is best used for?

SELF-CHECK 10.3

Why do you think healing is slow following an Achilles' tendon rupture?

10.1.6 Blood supply

Muscles are provided with a rich blood supply as they have high metabolic demands. Generally, skeletal muscles are served by several arteries. This arrangement makes them less likely to suffer from ischaemia, which may result from occlusion of a single vessel. Large arteries penetrate the epimysium and divide into a number of primary intramuscular arteries that branch into the epimysium and perimysium. These secondary arteries terminate in a rich anastomosing capillary network that runs into the endomysium (Figure 10.1). It is estimated that a single muscle fibre is surrounded by an average of 1.7 capillaries.

10.1.7 Innervation

Skeletal muscle is the link between the nervous system and bones. The muscular and skeletal systems work in a coordinated manner as a lever system with joints acting as hinges to carry out instructions from the nervous system. Skeletal muscles implement the instruction from the brain to accomplish an activity. The performance of muscle fibres and fibre types is controlled by sensory and motor innervations. At this stage, it is appropriate to introduce the characteristics of type I and type II myofibres.

Different muscles are characterized by different physiological and metabolic properties. Different constituent myofibrils of the muscle cells determine the physiological and metabolic properties of individual skeletal muscles and reflect their histochemical properties. On this basis, myofibres are divided mainly into type I and type II fibres, but not all muscles have the same proportions of these two types. In general, muscles with a role in maintaining posture (e.g. calf muscles) have a high proportion of type I fibres, while muscles used for short bursts of power contain abundant type II fibres. Depending on the anatomical location and function of the muscle, the ratio of type I to type II varies. For example, in the vastus lateralis muscle in the thigh, the most commonly biopsied muscle, more than 50 per cent of the fibres are type II, while in the biceps of the upper arm the balance typically favours type I (Table 10.1).

Type I myofibres are characterized by the following:

- slow twitch action
- slow contraction time following electrical stimulation
- generate less force than type II myofibres
- first to be recruited in response to the application of gradually increasing loads
- used for sustained, low-level activity—to assist this function, they are equipped with numerous large mitochondria and abundant intracellular lipid for oxidative metabolism.

Type II myofibres are characterized by the following:

- fast twitch action
- rapid contraction time following stimulation
- late to be recruited in response to the application of gradually increasing loads
- used for brief duration, intense activity and for carrying heavy loads
- specialized for anaerobic metabolism
- contain smaller, less-numerous mitochondria, less lipid and have larger glycogen stores.

TABLE 10.1 Types of muscle fibre.

	Type I	Type II
Twitch	Slow	Fast
Metabolism	Aerobic	Anaerobic
Myosin heavy chain	Type I	Type II

SELF-CHECK 10.4

Which activity would be more likely to create an oxygen debt: swimming laps or lifting weights?

SELF-CHECK 10.5

Which type of muscle fibre would you expect to predominate in the large leg muscles of someone who excels at endurance activities, such as cycling or long-distance running?

10.1.8 Muscle fibre alignment

Stretching of muscle fibres decreases the overlap of actin and myosin myofilaments, allowing muscle fibres to elongate. As the applied tension increases, the surrounding connective tissue fibres align along the same line of force as the tension. The length of stretched muscle depends on the number of stretched muscle fibres: the more fibres stretched and aligned, the longer the muscle. Muscle fibre alignment is the key to muscle rehabilitation following injury.

10.1.9 Muscle spindle

The muscle spindles or stretch receptors (Figure 10.5) are scattered in the fleshy bellies of the skeletal muscle, encapsulated in a layer of connective tissue located parallel to the muscle fibres. Each muscle spindle is composed of adapted skeletal muscle cells, sensory and motor neuron endings. Muscle spindles detect any change in physical movement or position of the body, and they control normal tone and muscle length coordination during movement.

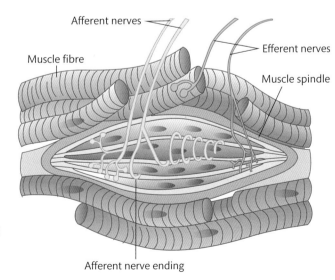

FIGURE 10.5
Diagram showing a muscle spindle.
Modified from original image: Mosby's Medical Dictionary, 8th edn.
© 2009, Elsevier.

When muscle fibres lengthen, so do muscle spindles. The receptors of the muscle spindle are triggered as a result of the change in muscle length and the rate of change in muscle length. It sends signals to the spinal cord which attempts to resist the change in muscle length by causing the stretched muscle to contract—a feedback system known as the spinal stretch reflex arc. This system helps to maintain muscle tone and protect the body from injury. The muscle contraction response is proportional to the suddenness of the change in muscle length.

Key points

Holding a stretch for a prolonged period of time trains the muscle spindle to become accustomed to the stretched position and reduces its signalling, gradually allowing further lengthening of the muscles. This training is useful for muscle rehabilitation and healing of scarred tissue. However, there is a risk of lack of response when injured.

10.1.10 Motor nerves and motor end plates

When nerves supplying individual muscles enter the surface of the muscle, they are accompanied by one or more major arteries. The penetrating nerves contain motor and sensory axons. As the distal motor axon approaches the muscle fibre, it is transformed into the terminal axon. This is the proximal end of the **neuromuscular junction**.

10.1.11 Contraction

Changes at the neuromuscular junction are the first steps in skeletal muscle contraction. As explained earlier, each muscle fibre is controlled by a neuron at a single neuromuscular junction. Neuronal axons within the perimysium branch into finer twigs ending at the synaptic terminal of the muscle fibre where the cytoplasm is especially rich in mitochondria and acetylcholine-rich vesicles. Acetylcholine, a **neurotransmitter**, once released by exocytosis into the synaptic cleft, alters the permeability of the sarcolemma, causing depolarization of the muscle fibre and increases the level of calcium ions in the sarcoplasm. This triggers an action potential which, if significant, causes muscle fibre contraction (Figure 10.6)

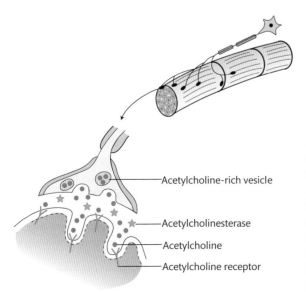

Acetylcholine-rich vesicle

Acetylcholinesterase

Acetylcholine

Acetylcholine receptor

FIGURE 10.6

Acetylcholine, a neurotransmitter, once released by exocytosis into the synaptic cleft, alters the permeability of the sarcolemma, causing depolarization of the muscle fibre and increasing the level of calcium ions in the sarcoplasm.

The part of the sarcolemma that contains acetylcholine membrane receptors is called the motor end plate, which has deep compact folds to increase the surface area and hence increase the number of acetylcholine receptors. The level of acetylcholine is kept in checked by **acetylcholinesterase** in the synaptic cleft.

10.1.12 Relaxation

The released acetylcholine in the synaptic cleft is rapidly broken down by acetylcholinesterase to its inactive form, putting an end to muscle contraction (relaxation). However, continuous production of acetylcholine into the synaptic cleft produces a series of action potentials and thus repeated contraction cycles.

Key points

The physiological factors that determine muscle fibre size are:

- *Gender:* muscle fibres are larger in males.
- *Function:* muscle fibres around the shoulder girdle or pelvis provide power support and hence are of large diameter, whereas fibres from smaller peripheral muscles required for coordination or those around the orbit are fine and of small diameter.
- *Training:* exercise causes muscle hypertrophy in men and women.
- *Age:* generally, muscle fibres are larger in young, active adults than in young children or the elderly.

Key points

Rigor mortis is a post-mortem physical state of intense muscle contraction. After death, blood circulation ceases and muscle is deprived of oxygen and hence ATP. This state leads to paralysis of the calcium pump mechanism and accumulation of calcium within the sarcoplasm, leading to sustained contraction. The individual becomes 'stiff' because all skeletal muscles are involved in this process. Rigor mortis lasts until lysosomal enzymes are released during autolysis, resulting in the breakdown of myofilaments 15–25 hours later. This timing, however, depends on several environmental and physical factors.

CLINICAL CORRELATION

Myasthenia gravis

This disease is characterized clinically by progressive muscle weakness. Muscle weakness is due to the lack of sarcolemmal depolarization at the neuromuscular junction, and in the case of myasthenia gravis is precipitated by developing autoantibodies to the acetylcholine receptors on the sarcolemma. These autoantibodies prevent the action of acetylcholine on the end plate. Curtailing the degradation of acetylcholine by inhibiting acetylcholinesterase is one way of treating this disease. Enhancing the level of acetylcholine provides a greater chance of binding to residual receptors and triggering muscle contraction.

CLINICAL CORRELATION

Gender

Type II fibres are usually larger than type I fibres in men, whereas type I fibres are of equal or greater diameter than type II fibres in women. Why is there a difference between males and females with regard to muscle fibre size?

- Men are generally bigger built than females.
- Men have a larger muscle mass for body size.
- Men are generally more active and are often engaged in more strenuous physical exertion.
- Testosterone increases muscle mass.

CLINICAL CORRELATION

Increased physiological or pathological demand

Normal adult skeletal muscle cells are not capable of dividing and therefore any increase in physiological (e.g. exercise) or pathological (e.g. compensation for a weakness) demand is supported by increase in fibre size (hypertrophy). However, following physical damage or surgical excision, there is evidence of muscle regeneration, although this cannot replace large defects completely. Following muscle damage, it is believed that a reserve of stem cells is available that can divide and differentiate into muscle fibres.

CLINICAL CORRELATION

Physical exercise

The effect of exercise on skeletal muscle has achieved a high profile in sports medicine. It is believed that any kind of exercise and training causes an increase in muscle fibre diameter. Anaerobic activities promote hypertrophy of type II fibres; this is seen commonly in sprinters or weightlifters. In contrast, type I fibres tend to be larger in long-distance runners, where aerobic respiration is essential.

So, does the composition of muscle fibre type change following a long period of exercise and training? Is there is a causal relationship between the types of sport and this change in muscle fibre type? Does the proportion of type I to type II fibres change according to the type of exercise, or is it more to do with having certain distribution of the two types of muscle fibres that determines why a particular sport is chosen? People with genetically determined skeletal muscle composition containing more type II fibres are more likely to be good at sprinting, while those genetically determined to have a larger number of type I fibres are better at long distance running. However, different pathological processes alter the ratio of the myofibre types and their distributions in the muscle, and this may affect the size of one type in relation to the other, or affect both types equally.

CLINICAL CORRELATION

Ageing

A functional and structural decline in skeletal muscle mass starts in the sixth decade and accelerates after the age of 70 years, resulting in 30–50% decline in muscle strength. The integrity of muscle fibres is mainly related to a full range of activities and the maintenance of its blood supply. Therefore, the reduction in the size of muscle fibres with ageing is mainly due to:

- change in flexibility and reduced full range mobility with a corresponding reduction in muscle volume and contractile strength.
- Insidious damage to the motor units in elderly people, specifically to the anterior horn cells in the spinal cord, causes atrophy of muscle fibres.

10.2 **Bone and cartilage**

Bone (osseous) tissue is a highly specialized connective tissue with a unique architecture. The presence of mineralized extracellular matrix that is organized in this ingenious form of outer cortical framework with internal spongy arrangement gives the bone the rigidity and strength of cast iron but with weight as light as that of wood. Bone markings in the form of either depressions or grooves are seen along bone surfaces, elevations where tendons and ligaments attach, or as tunnels where neurovascular bundles enter the bone. Within bone there is a complex, highly organized communications system between the cellular component to orchestrate the composition of the surrounding mineralized matrix and in response to physical stress.

10.2.1 **Primary functions of bone**

- *Mechanical support*: bones provide a structural framework for the body (e.g. ribs, vertebrae).
- *Locomotion*: bones act as attachment points for tendons and work together with muscles as simple mechanical lever systems to facilitate body movement (e.g. long bones).
- *Protection of internal organs*: the skull, for example, contains and protects the brain.
- *Metabolic reservoir for mineral salts, especially calcium and phosphorus*: bones play a vital role in mineral homeostasis.
- *Blood cell production*: a function of the bone marrow.

Key points

All bones have similar composition: an outer layer of periosteum, a cortex made of compact cortical bone, and a medullary cavity composed of a mixture of cortical and trabecular bone, fatty tissue, haematopoietic marrow, blood vessels and nerves.

10.2.2 **Bone structure**

Figure 10.7 shows the anatomical composition of a long bone. The **epiphysis** extends from the articular surface to the growth plate. It is wider at the metaphysis and present on either end of a long bone in articulation with other bones. The **metaphysis** is where the diaphysis and epiphysis meet. It extends from the growth plate to where the narrow part of the diaphysis starts. The **diaphysis** extends from one metaphysis to the opposing metaphysis. It is composed of a wall of compact bone and a central space—the medullary cavity or marrow. In contrast, the anatomical parts of a flat bone (e.g. the parietal bone of the skull) resembles a sandwich of spongy bone in between two layers of compact bone

---

Key points

The bone matrix surrounds long pillars that conduct the vascular supply (osteons). Osteocytes, stress sensors, sit within lacunae. Well-connected canaliculi form pathways for osteocyte processes so that blood vessels can exchange nutrients and wastes. The dense matrix contains deposits of calcium salts. The periosteum covers the outer surface of bone but it is not present in the regions covered by articular cartilage. The periosteum consists of an outer dense fibrous layer containing bone-forming cells. The endosteum is the tissue that lines the inner surfaces of bone.

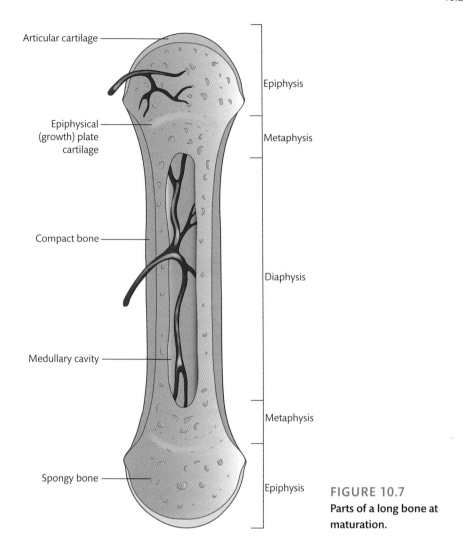

FIGURE 10.7
Parts of a long bone at maturation.

10.2.3 Bone classification

Bones are classified by shape as follows:

- *Sutural bones*: small, irregular bones found between the flat bones of the skull.
- *Irregular bones*: these have complex shapes; for example, spinal vertebrae and pelvic bones.
- *Short bones*: small, thick bones, such as in the ankle and wrist.
- *Flat bones*: thin with parallel surfaces, found in the skull, sternum, ribs and scapulae.
- *Long bones*: long and thin bones found in arms, legs, hands, feet, fingers and toes.
- *Sesamoid bones*: small and flat, they develop inside tendons near joints of the knees, hands and feet.

Bones are classified according to their internal organization. Depending on the pattern of collagen arrangement that forms the osteoid, bone can be classified into two types:

- *Woven bone*: this is the first bone formed at any site and is characterized by haphazard organisation of collagen fibres, hence it has a mechanically weak structure.

- *Lamellar bone*: this replaces the woven bone and is present in all the bones of a healthy adult—it shows a regular parallel alignment in the form of sheets (lamellae), hence its mechanical strength.

Woven bone can be found in fetal bones and in adult bones. In fetal bone, the rapidly produced osteoid is characterized by an irregular, loosely entwined pattern of collagen deposition, and this is the precursor of woven bone formation. Collagen fibres are gradually remodelled, forming a more resilient lamellar bone that replaces the pre-existing woven bone. In adult bone the excess development and rapid production of osteoid creates woven bone when there is need for new bone formation. For example, healing fracture starts with woven bone formation that is remodelled and replaced by lamellar bone. In contrast, woven bone formed in Paget's disease of the bone persists, leading to weak bone structure and bone deformity.

CLINICAL CORRELATION

Paget's disease of bone

Paget's disease of the bone is commonly found in England, United States, New Zealand, Canada, South Africa and France. The disease is rare in Asia and Scandinavia. Paget's disease affects women and men equally and increases in prevalence with age, especially in those over 50. Current evidence supports an autosomal mode of inheritance. Skeletal deformity, mostly seen in the skull and lower extremities, is the common feature of Paget's disease. With cranial enlargement, it is not uncommon for the patient to suffer from hearing loss.

The primary pathology of Paget's disease is a localized increased risk of osteoclastic bone resorption with concomitant active plump osteoblasts that are found in large numbers on bone surfaces. The osteoclasts are larger than normal and contain up to 100 nuclei. Also, there is an increase in the number of osteoclasts present in Howship's lacunae both in cortical (compact) and trabecular (spongy) bone. As a consequence of progressive osteoclastic activity, there is a reduction in bone volume. At the same time, masses of woven bone are deposited as a result of osteoblastic activity. This is followed by fusion of individual osteons to form confluent seams of bone, and the replacement of the medullary cavity by new fibrous tissue formation to substitute for the resorbed bone.

Ultrastructural studies of bone in Paget's disease reveal the presence of nuclear, and occasionally cytoplasmic, microfilaments that have the same structural features as nucleocapsids of viruses of the *Paramyxoviridae* family in osteoclasts, but these are not seen in osteoblasts or osteocytes. This family of RNA viruses is responsible for measles and respiratory syncytial virus (RSV)-related lung infections.

Interestingly, despite increased bone resorption, there is no rise in serum or urinary calcium levels and also fairly normal serum parathyroid hormone levels. This has been explained by the associated increase in bone formation. This explanation is supported by kinetic analysis of plasma disappearance rates and skeletal uptake of radiocalcium.

As a reflection of increased disease activity, there is an increase in the level of total serum alkaline phosphatase, which is released from the cytoplasm of active osteoblasts. Enzyme assays have little cross-reactivity with non-skeletal alkaline phosphatase. Recently, more specific biomarkers have been identified, such as measurement of serum procollagen type 1-N-terminal peptide level. Also, recent research demonstrated that sclerostin, a protein that inhibits bone formation produced by osteocytes, is elevated in patients with Paget's disease. However, neither of these tests has been used in routine practice.

10.2.4 Coping with external physical stress

Bones are hard and rigid enough to cope with external physical stress, and there are four main factors that provide the bone with its hard, rigid structure: architecture, structure, biochemical structure and cell composition.

Architecture

The way the basic architecture of bone is designed (Figure 10.8) can be compared to a truss in architecture where one or more triangular units are constructed with straight members the ends of which are connected at joints (nodes). External forces act only on nodes. In a similar way, the bone has an outer frame (cortical bone) forming a rigid outer shell to resist deformation and an inner complex system of truss composed of a trabecular meshwork. This arrangement provides maximum resistance, with the minimum possible bone mass, to the normal physical stress put on the bone during everyday activity. The spaces between the trabecular meshwork are occupied by bone marrow.

Structure

Bone is composed of an outer cortical or compact zone and an inner trabecular or spongy zone.

Cortical (compact) bone The mature skeleton is composed of lamellar bone, having replaced the pre-existing woven bone. The thickness of the lamellar bone depends on its location and the mechanical stress place upon it, being thickest in bones that undergo torsional and weight-bearing stress, and thinnest near articular surfaces and in vertebral bodies. **Osteons** or **Haversian systems** (Figure 10.9), the basic units of lamellar bone, assist the bone in withstanding mechanical stress. They are long cylindrical canals ranging from 25 to 125 μm in diameter. They are wider near the medullary cavity, and are created by osteoclastic activity on the circumferential lamellae. The number of osteons in a particular bone varies depending on age, exercise and body weight. Numerous cellular cytoplasmic processes extend into the bone to establish contact via nexus (gap) junctions with cytoplasmic processes of other osteoblasts and osteocytes.

Cortical bone is composed of three different lamellar patterns (Figures 10.9 and 10.10):

- *Circumferential lamellae*: these extend parallel to the long axis of the bone and make the outer and inner layers of the cortex.
- *Concentric lamellae*: as the mechanical stress on the bone increases, some of the circumferential lamellae are replaced by concentric lamellae of the Haversian systems.
- *Interstitial lamellae*: these irregular units of lamellar bone fill the gaps between the Haversian systems and help to anneal them together.

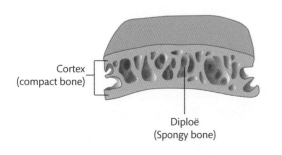

Cortex (compact bone)

Diploë (Spongy bone)

FIGURE 10.8

The basic architecture of bone can be compared to a truss in architecture where triangular units are constructed with straight members whose ends are connected at joints (nodes).

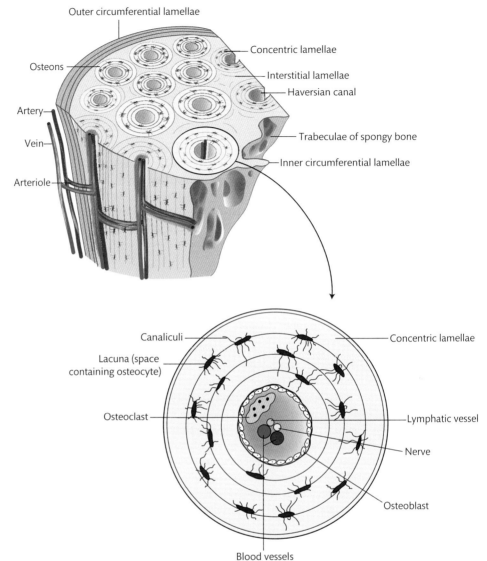

FIGURE 10.9
Osteons or Haversian systems are the basic units of lamellar bone.

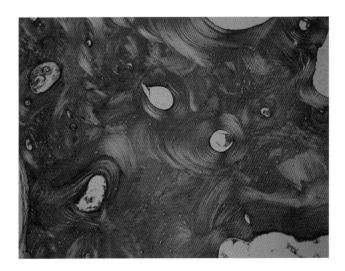

FIGURE 10.10
Cortical bone is composed of three different lamellar patterns: circumferential, concentric and interstitial (H&E stain).

Trabecular (spongy) bone Trabecular bone is fenestrated and composed of interconnected plates and struts of lamellar bone located within the medullary cavity. The lamellae are orientated parallel to the long axis of the bone. To provide added support, trabecular bone is deposited according to the lines of mechanical stress to distribute large weight-bearing forces in different directions. Therefore, it is ideal to have trabecular bone abundant in the weight-bearing end of bones (e.g. in the epiphyses).

Key points

The osteon is the basic unit of bone and comprises bone lamellae, including embedded osteocytes, arranged in a concentric pattern with a central canal containing blood vessels. Perpendicular to the central canal, there are perforating canals that carry blood vessels into bone and the marrow space.

Biochemical structure

Calcification gives the bone its hardness and thus its resistance to twisting. With the help of collagen fibres, bone resists tensile stress. Failure of adequate calcification occurs in diseases such as rickets, which is characterized by bone deformity as a result of relative softness. Large amounts of deposited osteoid are seen in cases of rickets compared to what would be expected in normal bone.

Osteoid (bone-like) tissue is the first step in bone formation prior to calcification. It forms one-third of the bone matrix and is composed of non-mineral organic matrix of 70% collagen (90% type I) and glycosaminoglycans, glycoproteins, osteonectin (anchors bone mineral to collagen), osteocalcin (calcium-binding protein) and proteoglycans (less than in cartilage). Crystallisation of salts occurs in the spaces between collagen fibres and together with deposition of minerals in the extracellular matrix this forms calcified bone (mineralized osteoid). Inorganic mineral salts are deposited within the matrix, with two-thirds of the bone matrix being calcium phosphate, which reacts with calcium hydroxide to form crystals of hydroxyapatite. Only 25% of bone is water.

Cellular components

Cells make up only two per cent of bone mass and four main types are present:

- *Osteoprogenitor cells*: maintain bone growth
- *Osteoblasts*: produce the osteoid seen on the surface of bone
- *Osteocytes*: nourish osteoid and are seen within small spaces (lacunae) in the bone substance
- *Osteoclasts*: remodelling cells.

Osteoprogenitor cells Osteoprogenitor cells are derived from mesenchymal stem cells and have the capacity to produce osteoblasts. They are seen as flattened cells closely adherent to bone surfaces within the periosteum, the Haversian system and in the endosteum (the inner layer of the periosteum). Under the light microscope, osteoprogenitor cells appear as spindle cells with no discriminating morphological features.

Osteoblasts Osteoblasts (the word 'blast' means to form something) are immature cells that, when surrounded by mineralized bone, mature into osteocytes. In osteogenesis, osteoblasts have an essential role in:

- production and arrangement of the organic matrix components in the form of osteoid: once calcified, osteoid changes to bone.

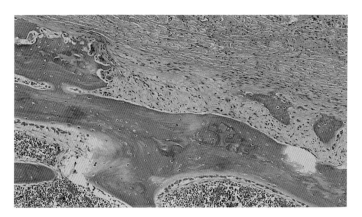

FIGURE 10.11
Osteoblasts are found in direct contact with bone (H&E stain).

- initiation and regulation of osteoid mineralization.
- controlling the activity of other osteoblasts, osteocytes and osteoclasts through autocrine and paracrine pathways.

Under the light microscope (Figure 10.11), osteoblasts are found in direct contact with bone. The morphological features of osteoblasts reflect their physiological activity. Inactive osteoblasts have a spindle shape whereas those that are metabolically actively are large (10–80 μm) polygonal cells with large nuclei situated away from the bone surface. The nuclei contain clumped chromatin and conspicuous nucleoli. The cytoplasm is abundant amphophilic or basophilic, reflecting the presence of a considerable amount of rough endoplasmic reticulum, which indicates an active metabolic state. Also, sometimes a prominent perinuclear halo is seen in osteoblasts, representing a well-developed Golgi apparatus. Numerous cellular cytoplasmic processes extend into the bone to establish contact, via nexus (gap) junctions, with cytoplasmic processes of other osteoblasts and osteocytes. Under electron microscopy, the cytoplasm of an active osteoblast is rich in rough endoplasmic reticulum, mitochondria, lysosomes and large, prominent Golgi apparatus.

Osteocytes Osteoblasts surrounded by osteoid become osteocytes and these have a half-life that may be up to 25 years. They are found in spaces (lacunae) between layers of matrix called lamellae. Osteocytes are connected by ramifying cytoplasmic extensions, through canaliculi in lamellae via gap junctions, with processes of other osteocytes and osteoblasts in order to maintain intercellular communication. These canaliculi are used to transport nutrients and waste. The cytoloplasmic processes usually run parallel to the osteoid collagen fibre. The arrangement of osteocytes varies depending on the type of bone. In woven bone, the arrangement of osteocytes appears disorganized as they lie parallel to the random arrangement of collagen fibres in different directions. In contrast, osteocytes appear organized in lamellar bone where collagen fibres are laid in the same direction. Osteocytes contain scanty cytoplasm with small nuclei and are located within lacunae. The nuclei show dense chromatin pattern and invisible nucleoli. Due to their small size, the nuclei are not seen in every plane of section. Therefore, in tissue sections it is not unusual to identify empty-appearing lacunae. Osteocytes are considered as the 'stress sensors' because they maintain the protein and mineral content of bone matrix and also help repair damaged bone. They are mechanosensory cells that detect mechanical stress and send messages through their cytoplasmic processes to stimulate osteoblasts and start the process of remodelling. The fundamental and highly complex communication system of osteocytes and their cytoplasmic extensions serves a vital role in mineral homeostasis. They provide an efficient response to changes in calcium ion concentration and mediate the exchange of mineral ions between bone matrix and extracellular fluid.

Osteoclasts Osteoclasts are large cells (40–100 μm) and appear as multinucleated giant cells containing between four and 20 nuclei (in certain circumstances there may be 100 nuclei) with abundant amphophilic cytoplasm. Where active resorption is taking place, osteoclasts are seen within the resulting crater, attached to the bone surface at the site of erosion. The crater is called the resorption bay or Howship's lacuna. Osteoclasts, so-called bone breakers, are derived from stem cells that produce macrophages. They are mobile cells that have a half-life of several weeks. The cytoplasm adjacent to the resorbing bone surface is rich in acid phosphatase, carbonic anhydrase and membrane-bound lysosomes. Secreted acids and protein-digesting enzymes facilitate erosion of the mineralized bone matrix, and the stored minerals are released into the circulation.

10.2.5 Osteogenesis

Osteogenesis (bone formation) begins during fetal life and continues after birth throughout adulthood; it is the process of replacing cartilage by bone tissue. It occurs in two main forms, intramembranous ossification and endochondral ossification, giving rise to membranous bone and cartilage bone, respectively.

Intramembranous ossification

The great majority of flat bones (e.g. bones of the skull) are formed by intramembranous ossification. In addition, it is responsible for the formation of the cortices of bone from periosteal osteoblasts between pre-existing membranes. The stages of intramembranous ossification are as follows:

- *The first appearance of intramembranous bone in prenatal life is flat, well-vascularized, membrane-like mesenchymal tissue.* The process includes stimulation of the primitive mesenchymal cells within the fibrous membrane, which is present at the site of the future bone, to become osteoprogenitor cells that contain abundant rough endoplasmic reticulum, indicating a metabolically active form. They eventually transform into osteoblasts.

- *Bone matrix is secreted within the mesenchymal membrane.* Osteoblasts start to secrete osteoid, and mineralization begins within a few days. Trapped osteoblasts mature and change to osteocytes.

- *Woven bone and periosteum are formed.* The produced osteoid is deposited within the mesenchyme between fetal blood vessels. Calcium starts to be deposited within the laid osteoid, transforming cartilage into bone. Once mineralized, it forms a random network of woven bone trabeculae. Condensation of vascularized mesenchymal tissue on the outer surface of woven bone forms the periosteum.

- *Bone collar of compact bone forms, and red marrow appears.* The trabeculae just beneath the periosteum thicken to form a woven bone collar that is later replaced by lamellar compact bone. Spongy bone consists of distinct trabeculae and its vascular tissue becomes the red marrow. Later, spaces within the spongy bone are filled with bone matrix, forming compact bone. The osteoblasts continue to deposit calcium within the bone matrix until the cells are totally surrounded by mineralized bone. At this stage, osteoblasts change into osteocytes encased in lacunae

FIGURE 10.12

Cartilage at the epiphyseal plate continues to grow, whereas cartilage next to the diaphysis starts to degenerate (H&E stain).

Endochondral ossification

During the initial stages of fetal life, newly formed cartilage containing no nerves or blood vessels vaguely resembles the shape of the future bone. The process of replacement of this cartilaginous model by bone is called endochondral ossification. Using the fetal cartilage skeleton as a model to form the new bone means that the bone forms while the cartilage it is replacing is broken down. Initially, the bone begins to form at ossification centres. With continuous cell growth and deposition of calcium in bone matrix, the ossified area expands to fuse together until the whole bone is formed. In effect, the bone replaces the cartilage (Figure 10.12). Most of the skeletal bones are formed in this manner and are called endochondral bones. During the third month of gestation, the perichondrium that surrounds the cartilage model starts to show increased vascularity with increased osteoblastic cells, changing into a periosteum. The osteoblasts form a collar of compact bone around the diaphysis. At the same time osteoblasts start to migrate to the centre of the diaphysis where the pre-existing cartilaginous tissue starts to disintegrate. As a result, spongy bone replaces the cartilage, forming the primary ossification centres. Ossification surrounding these centres continues to expand towards the ends of the bones. The medullary cavity is created with the help of osteoclasts by eroding the newly formed spongy bone.

The epiphysial cartilage continues to grow in order to allow for the increasing length of the developing bone. Secondary ossification centres are formed in the cartilaginous epiphyses. The process of ossification at the epiphyses is similar to that seen in the diaphyses. However, the spongy bone is retained and no medullary cavity is formed. When secondary ossification is complete, the hyaline cartilage is totally replaced by bone except in two areas, the articular cartilage and the cartilage between the epiphysis and diaphysis (epiphyseal plate or growth region).

10.2.6 Bone growth

Bone growth in length occurs at the epiphyseal plate by a process similar to endochondral ossification (Figure 10.13). This process of bone growth continues until adulthood when the growth of the cartilage declines and eventually stops. At this stage, the bone can no longer

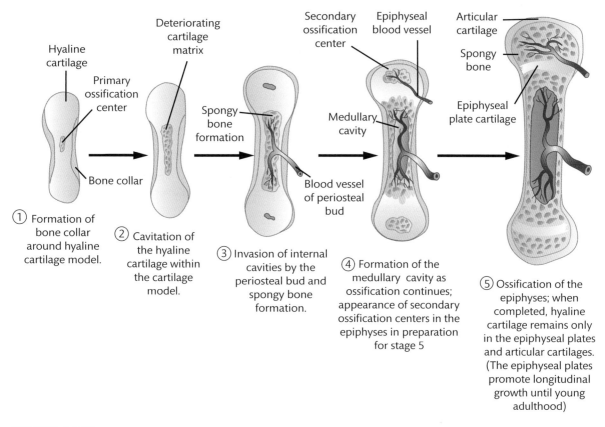

FIGURE 10.13

Bone growth in length occurs at the epiphyseal plate by a process similar to endochondral ossification, and continues until adulthood when the growth of the cartilage declines and eventually stops.

Modified from original image: Copyright © 2004 Pearson Education, Inc., Publishing as Benjamin Cummings.

grow because the epiphyseal plate is completely ossified, leaving only a thin epiphyseal line. However, bones can continue to increase in thickness and diameter throughout life in response to physical activity or under the influence of secreted growth hormone. The process whereby bones increase their thickness is called appositional growth. Here, two synchronized processes occur at the same time to increase the diameter of the bone while maintaining a light weight— osteoblasts create compact bone around the external bone surface while osteoclasts erode the bone in the endosteum on the internal bone surface around the medullary cavity.

SELF-CHECK 10.6

Explain how bones detect if they are undergoing mechanical or pathological stress.

Bone is a dynamic tissue, being formed and destroyed continually under the control of hormonal and physical factors. This constant activity permits the process of remodelling. Bone turnover is normally low in adults, but in babies and children it is high to allow for the growth and active remodelling required to cope with new demands (e.g. beginning to walk). In adult bone, turnover can increase from its normal basal level to meet increased demand (e.g. to repair a fracture). In addition, increased bone turnover can be the result of the need for bone remodelling.

Bone remodelling

Bone is a dynamically active metabolic organ that undergoes continuous remodelling throughout life to meet mechanical stress and growth demands. This process serves to facilitate several important functions, as follows:

- adjusts bone architecture to meet the physical demands of growth or a response to local mechanical stresses.
- provides a refined process to maintain micro-damages of the bone due to wear and tear.
- a crucial role in maintaining plasma calcium homeostasis.

Bone remodelling is a highly complex and finely balanced process involving bone resorption by osteoclasts and a counterbalanced process of laying down osteoid by osteoblasts to replace the resorbed bone.

In a normal, healthy adult, continuous fine adjustment of the balance between osteoblastic and osteoclastic activity is maintained in order to achieve constant bone mass. However, in cases of increased demand, such as bone fracture repair or undergoing a physical exercise training programme, osteoblastic and osteoclastic activity reserve is available. Bone remodelling is the process of maintaining a balance between calcium deposition in the osteoid and bone resorption to release minerals into the circulation through the coordination of the activities of osteoblasts and osteoclasts. The remodelling units, composed of these two cells, are monitored by the osteocytes (stress sensing cells). Remodelling occurs on the outer or the inner side (lining the marrow cavity) of compact bone. It is an active metabolic process that, in addition to calcium, requires adequate amounts of phosphate, vitamins A, C and D, and protein.

Remodelling is a response to mechanical forces where increased physical activity is positively correlated with the rate of remodelling. For example, weight training induces mechanical stress on bones, which is followed by increased osteoblastic activity to lay down new bone matrix and hence increased bone thickness; whereas, during the immobilization stage of a healing fracture osteoblasts migrate to repair the defect by laying down new bone matrix, while lost bone mass is a result of limited physical activity. However, this process is reversed once physical activity is regained, stimulating osteoblasts to lay down bone matrix.

SELF-CHECK 10.7

Explain what happens to the bones of astronauts if they stay in the Space Station for a year.

Remodelling is a continuous process regardless of physical activity. Calcium content cycles between bones and blood due to the effect of the constituent cells and specific hormones, which work by transporting free calcium ions between the bones, kidneys, intestines and blood. Maintaining a constant regular level of calcium is essential for survival. Therefore, the highly synchronized activity of osteoclasts to release calcium into the circulation is counterbalanced by osteoblast activity to keep bone resorption in check.

The two mechanisms that control bone density are intimately associated. For example, increased physical stress during exercise results in increased bodily cellular demand for calcium, a process to which the body responds by stimulating the secretion of parathyroid hormone (PTH). This hormone is secreted by the parathyroid glands, which monitor changes in free calcium level in blood and control it by a negative feedback mechanism once the level is normalized. One of the mechanisms that PTH uses to increase the level of calcium in the blood is the activation of osteoclasts, which start to erode bone matrix and release calcium into the circulation. At the same time, due to the mechanical stress on bone, messages are sent to indicate the need for increased calcium deposition in bone. As a result, calcitonin, a hormone released by C-cells in the thyroid that decreases calcium levels in the blood and is opposite in effect to PTH, is secreted to stimulate osteoblasts to migrate and lay down bone matrix and also initiate deposition of calcium in bone. From this example, it is evident that the net result in areas undergoing physical activity is either neutral or positive in favour of bone deposition. This contrasts with other areas that are under less mechanical stress where the balance is in favour of bone resorption. Therefore, it is important to appreciate that remodelling occurs at different rates in different bones.

SELF-CHECK 10.8

What is the role of calcium as it relates to the skeletal system?

Abnormal bone resorption

Osteoclasts undergo abnormal stimulation in some bone diseases. While osteoclasts appear to be paralyzed, having lost their mobility, they retain their enzyme resorption function. While remaining static, they digest the adjacent bone, drilling deeply into the bone at one locus. This is called tunnelling resorption. Parathyroid hormone effect on bone is not usually detectable histologically unless there is prolonged and excessive parathyroid hormone secretion.

Failure of homeostasis

Imbalance between osteoclast and osteoblast activity can occur, and is seen in two main clinical conditions, osteomalacia and osteoporosis. Failure of calcium deposition results in osteomalacia (soft bones), whereas failure to lay down osteoid, due to reduced osteoblast activity, results in osteoporosis (very brittle bones), as adequate osteoclastic activity ensures continued bone resorption.

CLINICAL CORRELATION

Osteoporosis

Primary osteoporosis (bone mass loss due to ageing or loss of gonadal function, and not associated with any other illness) is a metabolic bone disease characterized by low bone mass and micro-architectural deterioration of bone tissue, resulting in bone fragility and increased fracture risk. Importantly, there is normal mineral-to-collagen ratio. In osteoporosis, there is an imbalance between osteoclasts and osteoblasts and a failure to keep up with the constant microtrauma to trabecular bone.

Osteoclasts require weeks to resorb bone, whereas osteoblasts need months to produce new bone. Therefore, in periods of rapid remodelling (e.g. after the menopause) the net result of remodelling is in favour of bone loss, leaving temporarily unfilled gaps. This is complicated further by the fact that new woven bone is mechanically fragile and therefore at increased risk of fracture.

The surface area of the trabecular bone far exceeds that of cortical bone, therefore trabecular bone (e.g. vertebral bone) is more severely affected if there is imbalance in bone remodelling. During the accelerated period of bone loss immediately after the menopause, and in comparison to cortical bones, trebecular bone loss shows a three-fold increase.

CLINICAL CORRELATION

Oestrogen deficiency and loss of gonadal function

Oestrogen deficiency leads to excess bone resorption and inadequate bone formation. Accelerated bone loss can be seen in women and men, and oestrogen receptors are identified on osteoblasts, osteocytes and osteoclasts.

10.3 **Cartilage**

Cartilage is a resilient and pliable solid connective tissue, which owes its characteristics to a ground substance that is rich in proteoglycans linked to long, rigid molecules of hyaluronic acid. The matrix also contains fine collagen fibres (collagen type II) that are not visible in standard histological preparations. Collagen fibres are linked to other molecules by electrostatic interactions and cross-linking glycoproteins. The resilient nature of cartilage is largely attributable to the high water content of the matrix (60–80 per cent); however, it is poorly vascularized and obtains the necessary nutrients by diffusion—it should be no surprise that, if cartilage is damaged, the repair process is slow. Under the light microscope, chondrocytes are seen to sit in spaces (lacunae) within the cartilage matrix (Figure 10.14). In fresh tissue, chondrocytes completely fill the lacunae, but after fixation the cells frequently shrink, giving the appearance of some empty spaces.

10.3.1 **Growth of cartilage**

The perichondrium is composed of an inner and an outer layer. Cellular activity in these two layers contributes to the growth of the cartilage. Appositional growth occurs as a result of the secretion of collagen type I by the spindle fibroblasts of the outer fibrous layer of the perichondrium, whereas collagen type II and other components of the ground substance are secreted by the chondroblasts attached to the inner layer of the perichondrium. Once these chondroblasts are completely surrounded by matrix, they change into chondrocytes.

Cartilage stained with H&E demonstrates differences in eosin uptake, being most intense around the cartilage cells, and this reflects differences in the distribution of proteoglycans.

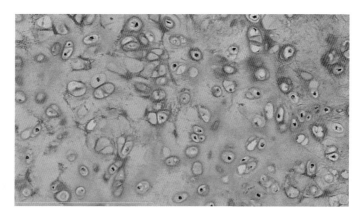

FIGURES 10.14

Chondrocytes sit in spaces (lacunae) within the cartilage matrix (H&E stain).

In addition, metachromasia (i.e. production of a colour different to that of the original dye) may be seen with certain dyes, due to the presence of negatively charged groups in the glycosaminoglycan, forcing the dye molecules to form aggregates of dimers or polymers. As a consequence of these change in physical properties, staining with toluidine blue, for example, changes from blue to pink.

10.3.2 Types of cartilage

Hyaline cartilage is the most abundant type of cartilage in the fetal skeleton. In adults, hyaline cartilage is found in the nose, parts of the respiratory tract, at the ends of ribs, and on the articular surfaces of bones.

Elastic cartilage has a matrix that is rich in elastic fibres; otherwise, its structure is similar to that of hyaline cartilage. It is found in the external ear, the walls of the external auditory canal, the Eustachian tubes, the epiglottis and the larynx.

Fibrocartilage is a mixture of a small amount of cartilage and abundant dense connective tissue. It is found where there is a need for resistance to compression and shear forces; therefore, it is found in intervertebral discs, the symphysis pubis, articular discs of the sternoclavicular and temperomandibular joints, and menisci of the knee joint.

 Chapter summary

In this chapter we have:

- Described the development of skeletal muscle through embryogenesis, childhood and adult development.

- Examined how light and electron microscopy are used to relate structure to function.

- Considered the roles and functions of related structures such as the vasculature, tendons, connective tissue and nervous system.

- Outlined the functions of the different muscle types and how these impact on various forms of exercise.

- Examined the physiology of muscle contraction and relaxation.

- Described the development of bone and cartilage and explained the significance of structural characteristics.

- Described the biochemical structure of bone and its cellular components.

- Explained the intricacies of bone growth and remodelling, and its role in homeostasis.

- Examined the growth of cartilage and how various types perform different roles in the body.

Further reading

- Alikhan MM. Paget disease (http://emedicine.medscape.com/article/334607-overview).

- de la Roza. G. Histology of bone (http://emedicine.medscape.com/article/1254517-overview).

- Esiri MM, Perl D. *Oppenheimer's diagnostic neuropathology: A practical manual* 3rd edn. Boca Raton: CRC Press, 2006.

- Lateva ZC, McGill KC, Johanson ME. Electrophysiological evidence of adult human skeletal muscle fibres with multiple endplates and polyneuronal innervations. *J Physiol* 2002; **544** (2): 549–65.

- Mills SE. *Histology for pathologists* 4th edn. Philadelphia: Lippincott Williams & Wilkins, 2012.

- Orchard GE, Nation BR eds. *Histopathology*. Oxford: Oxford University Press, 2012.

- Ross MH, Pawlina W. *Histology: a text and atlas*. Philadelphia: Lippincott Williams & Wilkins, 2011.

- Seidman RJ. Skeletal muscle–structure and histology (http://emedicine.medscape.com/article/1923188-overview).

- Suvarna KS, Layton C, Bancroft JD eds. *Bancroft's theory and practice of histological techniques* 7th edn. Edinburgh: Churchill Livingstone, 2013.

- Tran T, Andersen R, Sherman SP, Pyle AD. Insights into skeletal muscle development and applications in regenerative medicine. *Int Rev Cell Mol Biol* 2013; **300**: 51–83.

 # Discussion questions

10.1 Why would a sprinter experience muscle fatigue before a marathon runner?

10.2 How do bone cells coordinate their function to adjust the balance between bone formation and bone resorption?

10.3 Explain why weight training contributes to bone density whereas swimming does not?

Answers to self-check questions and discussion questions, hints and tips are provided on the book's Online Resource Centre.

 Visit www.oxfordtextbooks.co.uk/orc/orchard_csf/.

11

Liver

Anne Rayner and Alberto Quaglia

Learning objectives

After reading this chapter you should be able to:

- Describe the anatomy and location of the liver and gall bladder, including blood supply.
- List the major functions of the liver.
- Describe the cells and microscopic structures found within the liver.
- Describe the main types of pathology affecting the liver and biliary tract.

The liver is a large, solid organ situated in the upper right quadrant of the abdomen, below and partly behind the lower rib cage, separated from the thoracic cavity by the diaphragm. Anatomically, it comprises four lobes (left, right, caudate and quadrate). It is also subdivided into eight segments following the distribution of the internal vascular supply and drainage. The surface of the liver is covered by a layer of collagenous tissue, the Glisson's capsule.

At the microscopical level, the liver parenchyma consists of a main epithelial cell population (hepatocytes) arranged in cords (hepatic plates) in close contact with a network of capillary-like vascular structures (sinusoids) containing a mixture of oxygenated arterial blood and venous blood carrying nutrients from the gastrointestinal tract. Hepatocytes are also connected to the biliary system through a network of canaliculi structures that allow them to excrete various products via the biliary tree into the small intestine.

The hepatocyte plates are supported by a fine meshwork of reticulin fibres that can be demonstrated using ammoniacal silver stains. Variations in the reticulin pattern are a very useful indicator of liver disorders and tumours. Other cell components include biliary epithelial cells lining the biliary tree ducts, endothelial cells lining the sinusoids, stellate cells involved in vitamin storage and fibrogenesis, and lymphocytes and macrophages providing an immunological barrier against circulating antigens. During fetal development, the liver is the site of haemopoiesis.

The hepatocytes have a complex homeostatic function involving protein, fat and glucose metabolism, vitamin and element storage, detoxification and bile production. The liver has a remarkable regenerative potential, as it is able to recover to full size following severe parenchymal loss (e.g. toxic injury) or surgical resection.

Key points

The liver is the largest solid organ of the body, and occupies a substantial area of the upper abdominal cavity. In adults it weighs 1.0–1.5 kg, and is approximately 2 per cent of the whole body mass.

Key points

The liver has both an arterial and venous blood supply.

Major functions of the hepatocytes are as follows:

Metabolism of fats

- Oxidation of triglycerides for energy production
- Plasma lipoprotein synthesis
- Cholesterol and phospholipid synthesis

Metabolism of carbohydrates

- Carbohydrates are converted into fatty acids and triglycerides
- Regulates the concentration of glucose in the blood–glycogenolysis–gluconeogenesis

Metabolism of proteins

- Synthesis of amino acids, plasma proteins, e.g. albumin and clotting factors
- Detoxification of waste products of metabolism–deamination of amino acids and urea production

Detoxification

- Drugs
- Alcohol

Secretion of bile

- Contains products of metabolism (bilirubin / biliverdin)
- Acts as a detergent to emulsify fats.

BOX 11.1 Regeneration of the liver

The ability of the liver to regenerate after an injury has been known for so long that it is even mentioned in Greek mythology. Prometheus was punished by Zeus for stealing fire and giving it to mortals. The punishment consisted of binding Prometheus to a rock. An eagle would peck out and eat Prometheus' liver every day, and the liver would grow back every night.

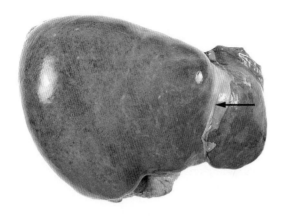

FIGURE 11.1
Anterior surface of the liver showing the
falciform ligament (arrowed).

11.1 Segmental anatomy, blood supply and biliary drainage

The **falciform ligament** on the anterior liver surface divides the liver into the right and left lobe (Figure 11.1). The **quadrate** and **caudate lobes** are visible on the posterior inferior surface of the liver.

The liver is subdivided into eight segments (segment IV is further subdivided into IVa and IVb), with each segment having their own vascular and biliary supply and drainage. This is the most important subdivision as it allows the surgeon to remove one or more segments of the liver, minimizing injury to the portion of liver left behind (Figure 11.2).

Key points

The segments of the liver have their own vascular and biliary supply and drainage, and surgeons remove parts of the liver following this subdivision.

The liver has both an arterial and venous blood supply. Products of digestion are transported to the liver with venous blood coming from the gastrointestinal tract via the **portal vein**. The liver receives oxygenated blood from the **hepatic artery**. The hepatic artery is a branch of the coeliac trunk, which originates from the abdominal aorta. The blood is drained via the hepatic veins into the inferior vena cava, and then to the right cardiac atrium (Figure 11.3).

SELF-CHECK 11.1

Describe the blood supply to the liver.

SELF-CHECK 11.2

How many segments make up the liver?

The liver produces **bile**, which is drained through the bile ducts that form the biliary tree, to the **duodenum**, where it emulsifies fat in preparation for digestion. The microscopic bile ducts form an intrahepatic biliary tree that emerges in the form of the common hepatic duct at the

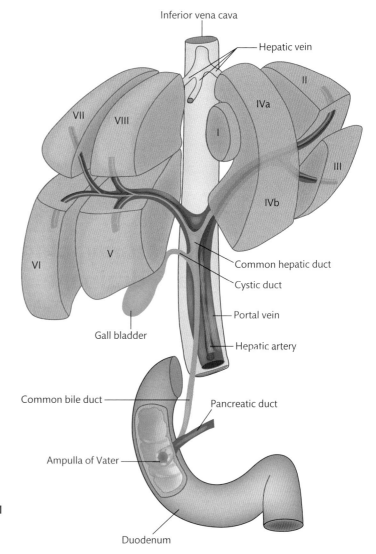

FIGURE 11.2

Segmental anatomy of the liver, vascular perfusion and biliary tree (note: the detailed segmental perfusion and drainage is not illustrated in this picture).

porta hepatis, located on the inferior surface of the organ where all the biliary, vascular structures and nerves feed into, and exit, the liver (Figure 11.4).

The biliary tree is connected to the **gall bladder**, a saccular structure attached to the inferior surface of the right liver lobe. It stores bile and squeezes it back into the biliary tree by contracting in response to hormones secreted by the duodenum in response to the fat content of a meal.

The gall bladder connects to the common hepatic duct via the cystic duct to form the common bile duct. The common bile duct joins the pancreatic duct to form the **Ampulla of Vater** before draining into the duodenum (Figure 11.5).

The portal vein, hepatic artery and common hepatic duct are closely aligned to each other, and their branches and tributaries maintain this anatomical relationship all the way to the periphery of the organ, and down to its microscopical level.

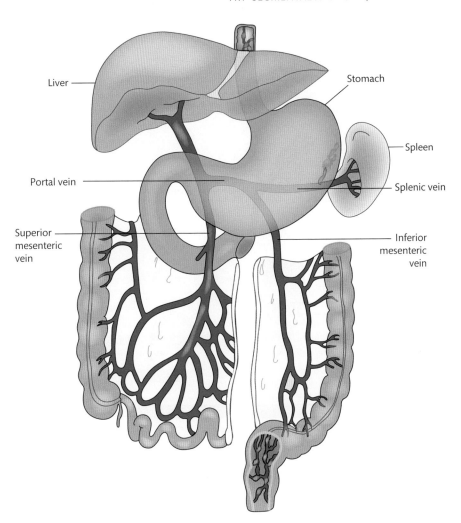

FIGURE 11.3
Venous blood supply and blood drainage.

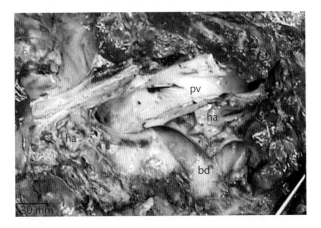

FIGURE 11.4
Biliary and vascular structures at the porta hepatis (pv = portal vein; bd = common hepatic duct; ha = hepatic artery).

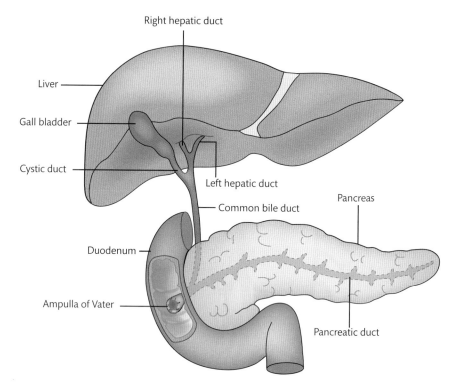

FIGURE 11.5
The biliary tree.

What is the practical application of knowledge of the anatomy of the liver segments?

Why might a liver biopsy not be sufficient to identify disorders affecting the larger portal structures?

Name six major functions of the liver.

> ## Key points
>
> The liver is supplied with venous blood from the intestine and spleen via the portal vein, and arterial blood via the hepatic artery.

11.1.1 The relationship of the liver to other organs

Owing to its size, position and function, the liver has a close anatomical relationship with many organs. The diaphragm separates it from the heart and lungs. It is largely protected by the lower part of the right rib cage. Its posterior and inferior surfaces are in contact with the right kidney and adrenal gland, the stomach and abdominal part of the oesophagus, duodenum, and the hepatic flexure of the colon.

11.2 **Microscopic structure**

The reciprocal proximity of a bile duct, portal vein and hepatic artery at the porta hepatis is maintained throughout the anatomy of the liver, from entry to the liver at the porta hepatis, right down to the microscopical level.

Portal tracts contain microscopic branches of the portal vein, hepatic artery and the smallest peripheral biliary tributaries (Figure 11.6). The portal tracts are surrounded by hepatocyte plates and sinusoids roughly orientated in a parallel way around the centrilobular venule, where the blood flowing from the portal structures is drained into the hepatic veins and then to the inferior vena cava. Portal tracts, and the structures they contain, are larger the closer they are to the porta hepatis.

Key points

The triad of portal vein, hepatic artery and bile duct is seen from the porta hepatis all the way down to the microscopic level in the portal tracts.

CLINICAL CORRELATION

Portal tracts

This structure is relevant because many disorders tend to affect the larger rather than smaller portal structures, and can be identified on imaging (e.g. computed tomography [CT] scan or ultrasound), but may not be represented on liver biopsy specimens, which sample the periphery of the liver.

11.2.1 **Liver lobule and acinus**

At the microscopic level, the parenchyma can be subdivided according to different anatomical/functional models. For example, the term **lobule** refers to an approximately hexagonal area of liver parenchyma with a hepatic venule in the centre and surrounding portal tracts to

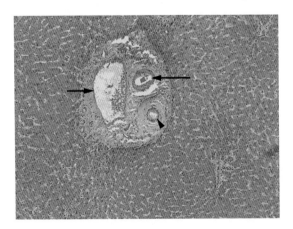

FIGURE 11.6
Portal tract showing branches of the artery (long arrow), vein (short arrow) and bile duct (arrowhead).

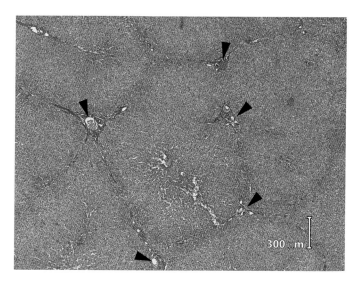

FIGURE 11.7
Liver lobule showing a roughly hexagonal shape (arrowheads show portal tracts).

form its external boundaries. (Figure 11.7). The term **acinus** refers to an approximately triangular area, the base of which is formed by the portal vein radicles branching into the parenchyma, with the nearby centrilobular venule at the apex.

The lobule is further divided into centrilobular, midzonal and periportal regions.

The acinus is further divided into zones 1, 2 and 3. Zone 1 is the most proximal to the portal vein branches, while zone 3 is closest to the centrilobular venule (Figure 11.8).

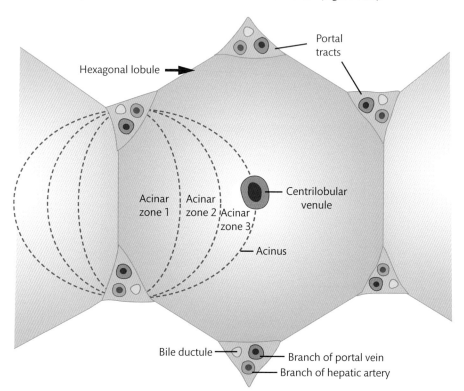

FIGURE 11.8
Acinus and lobule.

What is the difference between a liver lobule and a liver acinus?

What is the significance of the liver acinus in liver pathology?

Lobule and acinus

These two models are used because they correspond to the distribution of some patterns of injury or reflect some differences in parenchymal function. For example, the centrilobular region receives less oxygenated blood than the periportal region, which is thought to influence local hepatocyte function. The hepatocytes forming the boundary of the portal tracts constitute the 'limiting plate', the portal-parenchymal interface, which is often the site of injury, particularly in inflammatory conditions. The point of contact between the canalicular system and the inter-lobular bile duct is thought to be the **Canal of Hering** , which is also considered to be the site where progenitor cells reside and become activated in response to liver injury and contribute to regeneration (Figure 11.9).

11.2.2 The hepatocyte plates

Hepatocytes are organized in plates and are in contact with each other to form **canaliculi**, where bile is secreted, and face the sinusoidal space, where blood flows. They have a close

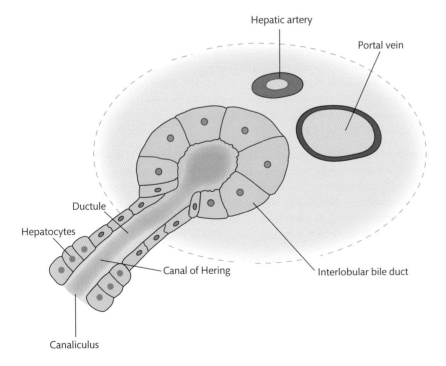

FIGURE 11.9
Canal of Hering.

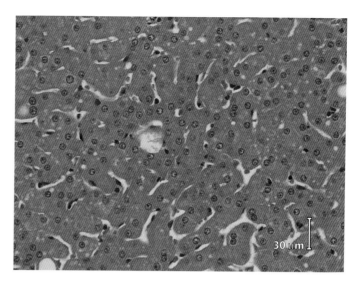

30 μm

FIGURE 11.10
Hepatic plates.

anatomical and functional relationship with other cells, including **sinusoidal endothelial cells, stellate cells, Kupffer cells** and **lymphocytes** (Figure 11.10).

11.3 **Ultrastructure of the hepatocyte**

Cross reference

See the Electron Microscopy chapter in the *Histopathology* volume in this Fundamentals of Biomedical Science series.

The membrane of the hepatocyte can be subdivided into a basolateral aspect facing the sinusoid, and an apical aspect facing the canaliculus. This distinction is not just topographical but has functional applications, as the basolateral aspect is implicated in exchanges with the blood, and the apical aspect is implicated in bile production. **Tight junctions** between cells make hepatocytes adhere to each other and separate the apical from the basolateral aspect, sealing the canalicular space containing bile from the intercellular space, the space of Disse, and the sinusoidal lumen (Figure 11.11).

The **space of Disse** contains extracellular matrix, which forms the collagenous reticulin framework demonstrated in light microscopy sections by ammoniacal silver staining, and fluid (see section 11.4). Gap junctions and desmosomes are also present and contribute to exchanges between hepatocytes. **Microvilli** are present at the basolateral aspect facing the space of Disse and may protrude through the sinusoidal fenestration into the sinusoidal lumen, and at the apical aspect protrude into the canalicular aspect (Figure 11.12).

Mitochondria are abundant in the hepatocyte as there is a great need for production of adenosine triphosphate (ATP). Mitochondria are easily identified by their size and the presence of cristae, and do not show any particular distribution in relation to other organelles. Generation of ATP via the respiratory chain, and involvement in β-oxidation and urea cycles are among the main functions. The endoplasmic reticulum is involved in various functions. The rough endoplasmic reticulum is the site of abundant protein synthesis and is identified by the presence of polyribosomes on the surface. Synthesized proteins transit from the rough endoplasmic reticulum to the Golgi apparatus, from where they are secreted in the sinusoidal space or canalicular lumen, or inserted in membranes. The smooth endoplasmic reticulum is also

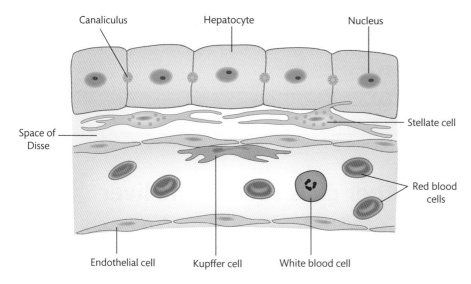

FIGURE 11.11
Ultrastructure of the hepatic plate (drawing) showing hepatocytes arranged around the canaliculus.

involved in various aspects of glucose and lipid metabolism, and in the metabolism of drugs. The endoplasmic reticulum does not have a particular topographic location in the cytoplasm. The Golgi apparatus tends to be located close to the nucleus and the canaliculus. Lysosomes are organelles that consist of a lipid membrane (membrane bound) containing enzymes (acid hydrolases) which tend to be more numerous close to the canaliculus, and are involved in the intracellular digestion of other organelles (autophagy) and by-products (e.g. **lipofuscin**, copper complexes and iron), and digestion of extracellular substances following endocytosis. Peroxisomes are also membrane-bound organelles, smaller in size than mitochondria, which contain enzymes involved in oxidative reactions and the production of hydrogen peroxide.

Lipofuscin
This is a yellow/brown, granular, iron-negative lipid pigment found particularly in muscle, heart, liver and nerve cells; it is the product of cellular wear and tear, accumulating in lysosomes with age. It is therefore sometimes known as the 'ageing' or 'wear and tear' pigment. It can be demonstrated in tissue sections using electron microscopy and shows autofluorescence (orange/yellow) under ultraviolet (UV) light.

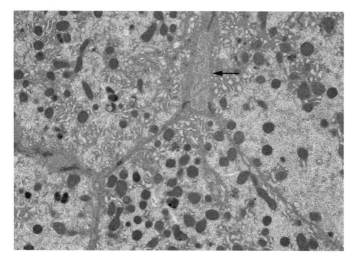

FIGURE 11.12
Ultrastructure of the hepatocyte showing hepatocytes arranged around the canaliculus (arrow).

Hepatocytes are rich in glycogen, the synthesis of which is a major hepatocyte function in maintaining glucose stasis. Glycogen accumulates in the hepatocyte cytoplasm and is seen in the form of minute, dark (i.e. electron dense) particles between organelles. The nucleus is large and usually has a prominent nucleolus.

SELF-CHECK 11.8

Why is the hepatocyte particularly rich in rough endoplasmic reticulum?

11.4 Other cells seen in the liver

The space of Disse is occupied by **stellate cells** that are capable of producing extracellular matrix, contracting and modulating the sinusoidal lumen, as well as storing retinoids. They have a spider-like or star-like shape, with long cytoplasmic processes, and contain droplets of fat due to the accumulation of vitamin A (Figure 11.13). Attached to the sinusoidal lumen are **Kupffer cells**, which essentially are macrophages with the main function of removing microorganisms or other antigens coming through the portal circulation, or removal of cell debris following parenchymal injury (Figure 11.14), and are rich in lysosomes and phagosomes. Also attached to the sinusoidal lumen are lymphocytes that, along with the Kupffer cells, act as an immunological barrier against antigens coming through the portal circulation from the intestine.

The endothelial lining of the sinusoids has some peculiar characteristics. The endothelial cells have holes called fenestrae (windows) that permit direct communication between the sinusoidal lumen and the space of Disse, so that solutes can flow through and come in direct contact with the hepatocytes. Sinusoidal endothelial cells are not bound to each other by cell junctions, they do not rest on a basal membrane, and they show **endocytosis**. Under normal conditions, sinusoidal endothelial cells do not express typical endothelial markers such as CD34 (Figure 11.15).

Endocytosis

This is a process by which cells absorb molecules such as proteins by engulfing them.

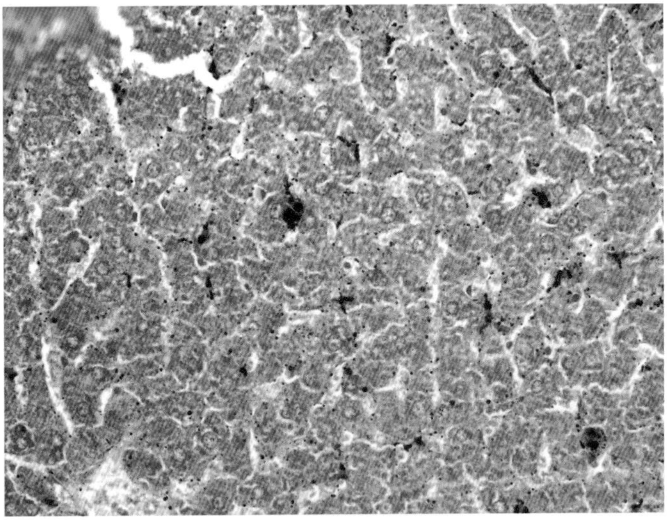

FIGURE 11.13
Mouse liver showing accumulation of vitamin A (Marchi method)
courtesy John Difford, RFH.

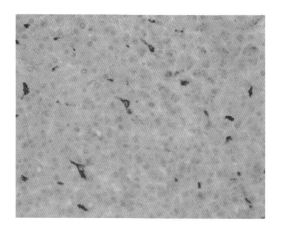

FIGURE 11.14
Kupffer cells (CD68 immunostaining).

FIGURE 11.15
Endothelium (CD34 immunostaining).

What is the function of Kupffer cells in the liver?

TABLE 11.1   Function and appearance of cells of the liver.

Cell type	Function	Appearance
Hepatocyte	Production of bile, protein synthesis and storage, fat metabolism including synthesis of cholesterol, bile salts and phospholipids, glucose metabolism, detoxification	Large polyhedral, eosinophilic cell with round nucleus, peripherally dispersed chromatin and prominent nucleoli
Kupffer cell	Removal of cell debris from spent erythrocytes/organisms by phagocytosis	Large, plump cell with oval nucleus, found among endothelial cells in the sinusoid lining. May show phagocytosed debris on microscopy
Stellate cell (also known as Ito cell)	Production of extracellular matrix and collagen; storage of vitamin A	Not easily distinguished on light microscopy — may contain lipid droplets. Seen among endothelial cells in the sinusoid lining
Endothelial cell	Lines the sinusoids	Flat, with darkly stained nuclei. Thin, fenestrated cytoplasm

11.5 **Bile ducts**

The epithelial lining of the bile ducts can be seen to change with their location, from stratified columnar epithelium in the large hepatic ducts, via simple columnar epithelium in the inter-lobular bile ducts, down to simple cuboidal epithelium in the intralobular ducts seen in the small peripheral portal tracts.

11.6 **Liver development**

The liver starts forming at approximately three weeks of embryonal life. The duodenum gives rise to the **hepatic diverticulum**, which buds into the septum transversum, a structure separating the pericardium from the peritoneum. The hepatic diverticulum is composed of **endodermal cells**, while for the septum transversum it is **mesenchymal cells**. The combination of these two cell types is the origin for the hepatocyte plates and the supporting stroma. The residue of the hepatic diverticulum gives rise to the extrahepatic biliary tree.

11.6.1 Embryonal blood supply

The **coeliac axis** divides to form the left gastric artery, the common hepatic artery and the splenic artery. The left umbilical vein takes oxygenated blood from the placenta and connects with the left branch of the portal vein and directly with the inferior vena cava through the **ductus venosus**. At birth, the ductus venosus closes and becomes the **ligamentum venosum**, while the umbilical vein closes and becomes the **ligamentum teres**. The liver receives blood via the portal vein through its right and left branches and from the hepatic artery.

11.6.2 Development of the biliary tree

The intrahepatic portion of the biliary tree derives from hepatocytes at approximately ten weeks when the hepatocyte precursors (**hepatoblasts**) at the border with the portal tracts form the so-called **ductal plate**, a two-layer structure of hepatoblasts (Figure 11.16). This change is accompanied by changes in cytokeratin expression. Hepatoblasts express cytokeratin 8, 18 and 19. The ductal plate maintains this profile, while the rest of the hepatoblasts lose cytokeratin 19 expression. The ductal plate gradually separates from the hepatocyte plates and moves into the portal tract. A lumen forms within the ductal plate, giving rise to tubular structures that are the precursor of the intrahepatic bile ducts, which start expressing cytokeratin 7 at around 20 weeks' gestation. The part of the ductal plate that does not form tubules then disappears. This development starts close to the liver hilum (with large portal tracts) and later occurs at the periphery (with small portal tracts). Ductal structures maintain the connection between the newly formed intrahepatic bile ducts and the hepatocyte plates, where bile is produced. The point of connection is called the **Canal of Hering**. This structure is also considered to be the site at which cells capable of regeneration reside.

Haemopoiesis
This is the formation of the cellular components of blood.

Another important event in liver development is the beginning of **haemopoiesis** at about six weeks. The liver remains the main haemopoietic organ until approximately five months of gestation, when the bone marrow begins to take over this function.

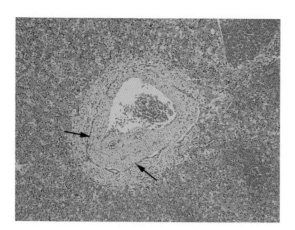

FIGURE 11.16
Ductal plate (arrowed).

When does the liver start to develop in the human embryo?

11.7 **Liver regeneration**

The liver has a very remarkable capacity to regenerate. This is due to the ability of quiescent mature adult hepatocytes to enter the cell cycle and proliferate following damage to the parenchyma in reaction to injury caused by, for example, viral infection or ingestion of a toxin (Figure 11.17). The liver also contains **progenitor cells** that are also capable of entering the

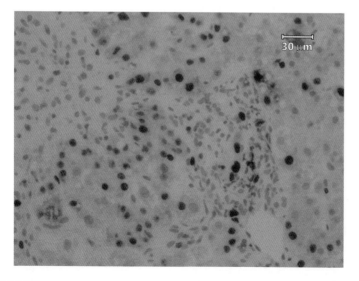

FIGURE 11.17
Hepatocyte proliferation following severe parenchymal injury (Ki-67 staining).

cell cycle, proliferating and transforming into hepatocytes or biliary epithelial cells, as demonstrated in animal models of liver injury, particularly those in which hepatocyte proliferation is blocked by specific agents. It is thought that these progenitor cells reside in the Canal of Hering, at the junction between the canalculi and the biliary tree proper. It is also believed that circulating stem cells of bone marrow origin can contribute to liver regeneration, as they maintain the potential to differentiate into hepatocytes, biliary epithelial cells or other lineages.

CLINICAL CORRELATION

Transplantation

The most commonly used technique is **orthotopic transplantation**, in which the native liver is removed and replaced by the donor organ in the same anatomical location. Other techniques include splitting the donor liver into portions, which each donated to different recipients. **Auxiliary liver transplantation** consists of grafting a liver lobe to a whole or part of an injured liver. The auxiliary graft supports hepatic function while the patient's own liver regenerates. This is used especially where the liver injury is not sustained, such as following **paracetamol** overdose. Subsequently, the grafted liver may be removed as the patient's own regenerated liver resumes hepatic functions.

Liver transplantation

This is the replacement of a diseased liver with a healthy liver from another human being (allograft).

11.8 **Liver pathology**

The liver is a large organ with remarkable spare functional capacity and the ability to recover from injury by regeneration. Signs of liver injury do not simply derive from loss of parenchymal function, but also from the effects on the integrity of the organ structure and anatomical relationships.

Liver pathology revolves around basic morphological patterns of injury and response, clinical manifestation and **aetiology**. The outcome of liver injury depends on the causative factor, the modality of its action, the interplay between the different components of the liver parenchyma (i.e. hepatocytes, endothelial cells, biliary epithelial cells, Kupffer cells, stellate cells and resident lymphocytes and inflammatory influx) and any secondary complications.

The causative agent can affect any component of the liver, and therefore liver disorders can be classified into those affecting primarily the parenchyma, the biliary structures or the vascular structures. Primary injury to one component can affect any other component as a secondary effect. Liver injury can also occur secondary to pathology of other adjacent organs.

Key points

The liver has a large spare functional capacity and the signs or symptoms of liver disease can take a long time to manifest. The liver has a unique regenerative capability and can recover from severe injury. Liver disorders can affect primarily the liver parenchyma, the vascular structures or the biliary structures. The liver can be affected by disorders of other organs.

Liver function tests

A tumour growing in the head of the pancreas can obstruct the nearby common bile duct, resulting in obstruction to bile flow and secondary injury to the hepatic parenchyma. Similarly, in patients with right cardiac failure, the blood may not flow properly out of the liver, and its retention will cause injury to the hepatocytes.

The first line of investigation of liver pathology will involve liver function tests. These determine the levels of certain chemicals and enzymes crucial to the normal functioning of the liver. These tests are employed for a number of reasons. They help to define liver disorders, and monitor their activity and severity, and are also used to determine if there are any side effects on liver function as a result of the use of medication that may cause organ dysfunction. Finally, they can be employed as a screen for potential liver diseases such as hepatitis or the effect of alcoholism.

The tests include:

- *Alanine transaminase* (ALT). This enzyme assists in protein metabolism. In inflammatory liver states (e.g. viral hepatitis), the ALT level usually rises.
- *Aspartate aminotransferase* (AST). If the level of AST increases in the blood it is usually a sign of injury or insult to the liver. This is not specific, however, as elevated AST level can also occur as a result of injury to the heart or skeletal muscle.
- *Alkaline phosphatase* (ALP). This can be elevated in the blood in certain liver disease involving obstruction of the large bile ducts (e.g. intrahepatic cholestasis or infiltration by tumour).
- *Total protein*. Essentially, this is a measure of albumin and other key proteins in the blood, and provides evidence of abnormal protein metabolism.
- *Bilirubin*. Elevated level of bilirubin in the blood is the key reason for the appearance of jaundice (yellow appearance to the skin). Bilirubin is a breakdown product of haemoglobin and is normally processed in the liver. The liver attaches sugar molecules to this breakdown product. This is measured as conjugated bilirubin, raised levels of which occur in a number of liver and bile duct disorders; for example, gallstones blocking the bile duct will cause an increase in conjugated bilirubin level. It is also elevated in long-term alcohol abuse. Raised unconjugated bilirubin is indicative of excessive breakdown of red blood cells containing the haemoglobin molecule and can be an indicator of haemolytic anaemia.
- *Gamma-glutamyltransferase* (GGT). High levels of GGT may be seen in heavy alcohol intake.

Other tests may be performed to confirm a clinical diagnosis, or monitor response to treatment or disease activity. These include tests for antibodies to viruses such as hepatitis B or hepatitis C, blood clotting tests, and tests for autoantibodies. Certain autoantibodies may be associated with particular autoimmune diseases affecting the liver, the most common being:

- primary biliary **cirrhosis** (PBC), which is associated with the presence of antimitochondrial antibodies
- primary sclerosing cholangitis (PSC), which is associated with the presence of antineutrophil cytoplasmic antibodies
- autoimmune hepatitis (AIH), which is associated with the presence of smooth muscle antibodies, nuclear antibodies or liver-kidney microsomal antibodies.

Similarly, high levels of certain proteins in the blood are indicative of specific liver disorders, for example:

- Wilson's disease is associated with a low level of ceruloplasmin
- haemochromatosis is associated with high ferritin levels
- cirrhosis in childhood may be caused by deficiency of α1-antitrypsin.

11.8.1 Morphology of liver disease

The basic morphological patterns of liver injury reflect degenerative processes affecting the cell component primarily involved. Owing to the dominant and complex function of hepatocytes, the most clinically relevant degenerative processes are those affecting these cells. Degeneration is often associated with an inflammatory reaction, **fibrosis** and regeneration. The cause largely influences the predominant process. For example, some drugs can cause impairment of hepatocyte function, and morphological changes to the hepatocytes are the dominant manifestation. Other causes can induce primarily an inflammatory reaction (e.g. to a virus) that indirectly causes injury to hepatocytes. In this case, the inflammatory response is the dominant morphological manifestation.

11.8.2 Microscopic features

Hepatocyte degeneration manifests in changes in their appearance under the microscope. Hepatocytes may become swollen, even 'ballooned', and some of the cytoplasm component may be altered (e.g. clumping of the cytoskeleton which results in **Mallory–Denk bodies**). There may be an accumulation of various products, including fat droplets, iron, copper, lipofuscin and bile. Degenerative changes may result in **necrosis** or apoptosis, depending on the mechanism involved. Necrosis can be limited to individual hepatocytes—this is described as 'spotty necrosis'—or affect large groups of hepatocytes—described as 'confluent necrosis', or may be limited to particular portions of the lobule, such as the centrilobular region following paracetamol-induced injury, or affect multiple lobules. Degeneration and necrosis can occur massively and suddenly, leading to the clinical syndrome of **acute liver failure**, or as an ongoing chronic process of degeneration, inflammation, regeneration and fibrosis. Inflammation is a secondary response to any type of liver injury, but in some cases it is the primary process. Its pattern, and in particular its composition and distribution, can vary and give a clue about the underlying cause. Fibrosis results from the deposition of collagen in an attempt to repair damaged tissue, but this scarring can result in distortion of the liver architecture with impairment of its function and blood flow, ultimately resulting in liver failure and **portal hypertension**. **Cirrhosis** is the extreme form of this scarring, in which bands of fibrous tissue separate the parenchyma into nodular areas (Figure 11.18). This can be demonstrated in the liver tissue using the ammoniacal silver stain for reticulin fibres, which, if left untoned, will show collagen fibres as golden brown and reticulin fibres in black (Figure 11.19).

Cross reference

Further information on staining techniques may be found in the *Histopathology* volume of this Fundamentals of Biomedical Science series.

FIGURE 11.18
Cirrhotic liver (anterior view) showing scarring and nodularity of the parenchyma.

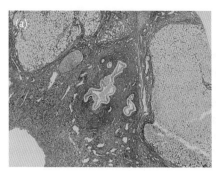
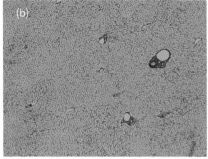

FIGURE 11.19
Reticulin. (a) Cirrhotic liver. (b) Normal liver. Note the increase in collagen (stained brown) in the cirrhotic liver, and the disruption of the reticulin pattern (silver staining).

 METHOD: Staining method

Silver stain for reticulin fibres (untoned)

Reticulin fibres have little natural affinity for silver solutions so they must be treated with acidified potassium permanganate then mordanted with iron alum to sensitise the fibres to silver deposition. Reducing in formalin causes deposition of metallic silver onto the reticulin fibres. Any excess silver in the unprecipitated state is removed by treating with sodium thiosulphate. Any liver section can be used as control material, cut at 4 μm.

Method

1. Dewax sections and take to distilled water
2. Oxidize in acidified potassium permanganate for 5 minutes
3. Wash in distilled water
4. Bleach in 1% oxalic acid until sections are clear
5. Wash thoroughly in several changes of distilled water
6. Sensitize in 2.5% iron alum for 15–20 minutes
7. Wash thoroughly in several changes of distilled water
8. Treat with silver solution for 1–2 minutes
9. Wash in distilled water
10. Develop in 10% formalin in tap water
11. Wash in running tap water
12. Fix in 5% sodium thiosulphate for 3 minutes
13. Wash in running tap water
14. Dehydrate in alcohols, clear in xylene and mount in DPX.

Results

Reticulin fibres   black

Collagen   gold/brown

Solutions

Acidified potassium permanganate

0.5% potassium permanganate: 285 mL

3% sulphuric acid: 15 mL

Mix thoroughly and transfer to a staining bath or jar.

Iron alum

2.5% ammonium ferric sulphate in distilled water

Silver solution

To 5mL of 10% aqueous silver nitrate add concentrated ammonia drop by drop until the formed precipitate redissolves.

Add 5 mL of 3% aqueous sodium hydroxide

- Dissolve the formed precipitate by adding concentrated ammonia dropwise as before, until only a faint opalescence remains.
- Make up the total volume to 50 mL with distilled water.
- Store in a dark plastic bottle.

Formalin in tap water

Concentrated formalin (40%): 10 mL

Tap water: 90 mL

Make solution fresh as required.

Paracetamol overdose

Paracetamol overdose can cause drug-related acute liver failure (ALF), and may be accidental, perhaps as a result of mixing cold and influenza remedies that contain paracetamol, or an intentional act. Damage to the liver occurs because the metabolite of paracetamol, *N*-acetyl-p-benzoquinoneimine (NAPQI) is toxic to the liver, and depletes the antioxidant glutathione, causing hepatocyte damage and necrosis. Histologically, this is seen on haematoxylin and eosin (H&E)-stained liver sections as confluent centrilobular necrosis, spreading outwards from the central vein in the lobule, leading to collapse of the reticulin structure. Treatment consists of limiting the damage to the liver by dosing with acetylcysteine, which neutralizes the effect of NAPQI and protects the antioxidant function. Severe collapse of the liver leads to acute liver failure, which needs to be treated by liver transplantation.

Recent developments in transplantation utilize auxiliary liver surgery, where donor liver is transplanted to support the patient's own liver function. When the patient's own liver has regenerated sufficiently to take over liver function, immunosuppression may be withdrawn and the donor liver shrinks.

11.8.3 Fibrosis and regeneration

Fibrosis is considered to be a dynamic process, based on the balance between deposition and reabsorption of collagen, and the interplay between the various cellular components of the liver. New concepts in liver pathology have put emphasis on the possibility that fibrosis can regress even from a late, advanced stage. Regeneration is a unique characteristic of the liver and is based on the potential for hepatocytes to enter the cell cycle and divide, and of having a back-up regenerative compartment of progenitor cells capable of differentiating into hepatocytes and biliary epithelial cells. These progenitor cells are thought to reside in the Canal of Hering, the junction between the canalicular network and the biliary tree.

Chronic injury, inflammation, fibrosis and regeneration increase the risk of malignant transformation in the liver, as in other tissues. Of note, cirrhosis is the main risk factor for hepatocellular carcinoma, the most common form of primary malignant liver tumour.

CLINICAL CORRELATION

Hepatitis B

Injury to the liver caused by chronic hepatitis B virus infection can result in the development of hepatocellular carcinoma.

SELF-CHECK 11.11

What is the appearance of collagen in an untoned reticulin stain?

11.9 **Classification of liver disease**

Liver disease may be classified according to the area affected and/or the causative agent. Subclassification based on histological appearance can be used, providing more precise information to the clinician. Below is a classification of liver disease based on acquired or inherited disorders divided by the component primarily affected, tumours and developmental disorders.

11.9.1 **Parenchymal disorders**

Diseases that primarily affect the liver parenchyma are termed parenchymal disorders in hepatopathology, and include:

- viral hepatitis (e.g. hepatitis B, hepatitis C)
- autoimmune hepatitis
- drug-induced liver injury (e.g. alcohol, paracetamol, herbal remedies)
- metabolic disorders (e.g. non-alcoholic steatohepatitis [NASH]; Figure 11.20)
- Inherited conditions (e.g. haemochromatosis, Wilson's disease, α1-antitrypsin deficiency).

CLINICAL CORRELATION

Non-alcoholic steatohepatitis

Non-alcoholic steatohepatitis (NASH) is an increasingly common disease with clinical and pathological similarities to alcoholic liver disease, except that it occurs in patients who drink little or no alcohol. It is more common in patients who are obese, have type 2 diabetes or raised cholesterol and triglyceride levels, and is linked with metabolic syndrome. The patient may have no symptoms in the early stages, and the condition is often first diagnosed by the detection of elevated liver enzymes such as alanine aminotransferase (ALT) and aspartate aminotransferase (AST). Ultrasound scan shows fat deposition in the liver. Liver biopsy shows fat accumulation in the hepatocyte cytoplasm, with varying degrees of inflammation, ballooning degeneration of hepatocytes, and pericellular fibrosis. Over time, this may progress to cirrhosis and loss of hepatic function.

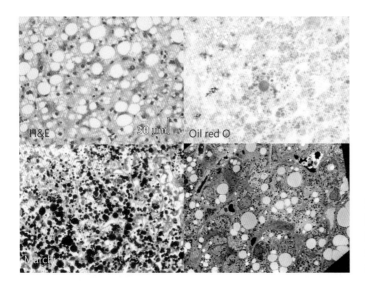

FIGURE 11.20
Fatty liver in non-alcoholic steatohepatitis (H&E, Oil red O and Marchi staining, and electron microscopy).

11.9.2 Biliary disorders

Biliary disorders are subdivided into **intrahepatic** (affecting the bile ducts embedded within the liver; e.g. primary biliary cirrhosis) and **extrahepatic** (outside the liver, adjacent to the gall bladder and pancreas; e.g. primary sclerosing cholangitis, cholelithiasis [gallstones]).

11.9.3 Vascular disorders

Vascular disorders are caused by injury to the hepatic artery, the portal vein, the sinusoids, and the hepatic veins. Injury to the hepatic veins and sinusoids may be caused by ingestion of drugs or herbal remedies, and by certain types of chemotherapy, or may occur in association with haematological disorders that increase the tendency for blood to clot. Damage to the hepatic veins results in accumulation of blood in the parenchyma, with secondary injury to the hepatocytes (venous outflow block).

11.9.4 Tumours

Tumours can be divided into **primary** and **secondary** in nature. Primary tumours can be divided into benign and malignant, and according to their histotype (i.e. resemblance to ambient structures) into hepatocellular, biliary and vascular tumours. Secondary tumours are very common and comprise metastatic deposits spreading to the liver from extrahepatic tumours. Tumours of the gastrointestinal tract are the most common tumours to spread to the liver.

11.9.5 Developmental disorders

Developmental disorders reflect impairment of the normal developmental process, resulting in the creation of aberrant structures. The predominant group is the **ductal plate malformations**, which include congenital hepatic fibrosis, Caroli's disease, polycystic liver disease and autosomal recessive polycystic kidney disease.

11.10 Gall bladder

11.10.1 Structure of the gall bladder

The gall bladder is attached to the posterior inferior surface of the liver, adjacent to the hilum and segment 4 of the right hepatic lobe. It is a roughly pear-shaped saccular structure and in adults measures around 80 mm by 40 mm when fully distended with bile. It is divided into three areas, the neck, body and fundus, and is attached to the liver at the neck, where it connects with the biliary tree via the cystic duct. There is a folding of the mucosa just below the neck (Hartmann's pouch), which is a common site for gall stones to lodge and block the cystic duct. The gall bladder is supplied with arterial blood via the cystic artery, a branch of the hepatic artery.

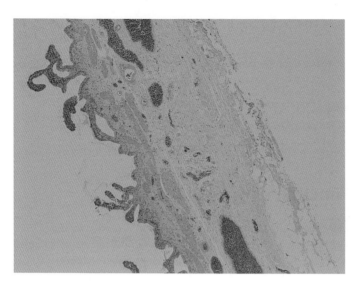

FIGURE 11.21
Gall bladder wall.

The **gall bladder wall** is normally 2–3 mm thick, and its internal surface has a smooth, velvety appearance macroscopically. The gall bladder wall is composed of five layers (Figure 11.21), which, from the inside, are:

Epithelium Tall columnar epithelial cells with basal nuclei, interspersed with 'pencil cells', the function of which remains unclear (Figure 11.22). No crypts are present.

Subepithelial stroma (lamina propria) Loose connective tissue layer which supports the epithelium.

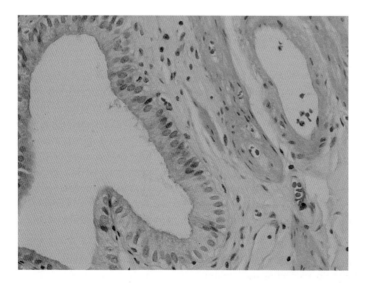

FIGURE 11.22
Tall columnar epithelial cells lining the gall bladder.

Smooth muscle layer Longitudinal and oblique fibres with reticulin, collagen, elastin and fibroblasts.

Perimuscular tissue Fibrous connective tissue supporting branches of arteries, veins and lymphatics.

Serosa External covering originating from the lining of the abdominal cavity.

Unlike the rest of the gastrointestinal tract, there is no muscularis mucosa.

The mucous membrane lining the cystic duct of the gall bladder is folded and forms a 'spiral valve'.

11.10.2 Function of the gall bladder

The gall bladder stores bile secreted by the liver until it is needed for digestion of fats in the duodenum. When lipids pass into the duodenum after a meal, neuroendocrine cells lining the duodenal mucosa secrete the enzyme **cholecystokinin-pancreozymin** (CCK), which causes the gall bladder to contract and squeeze its contents into the duodenum, where the bile emulsifies lipids and begins their digestion.

SELF-CHECK 11.12

What is the function of the gall bladder?

SELF-CHECK 11.13

How many layers make up the gall bladder wall?

> **Key points**
>
> **Gallstones are mainly composed of cholesterol with bile pigments and calcium salts.**

11.11 Pancreas

The pancreas is the largest of the digestive glands, and measures 12–15 cm long and around 2.5 cm thick at its head. It is connected to the duodenum by the pancreatic duct, and is divided into the pancreatic head, neck, body and tail, with the uncinate process (part of the head) extending behind large vascular structures. The pancreatic head is adjacent to the duodenum, lying within the first, second and third duodenal curves. The body tapers away from the left of the head, with the tail lying adjacent to the spleen. These divisions are predominantly made on the basis of anatomical relationships, and there is little functional difference between them, although there are generally more islets of Langerhans in the tail than the head (Figure 11.23). The pancreas develops as an outgrowth from the embryonal primitive foregut, and has endocrine and exocrine components. Further information about the anatomy of the pancreas may be found in Chapter 14 on the endocrine system.

The pancreas is mentioned in this chapter as it has an important relationship with the liver and in the pathology affecting the hepatobiliary system. Tumours of the head of the pancreas often affect the extrahepatic biliary system and, in particular, the common bile duct, causing obstruction and blockage to bile flow resulting in jaundice.

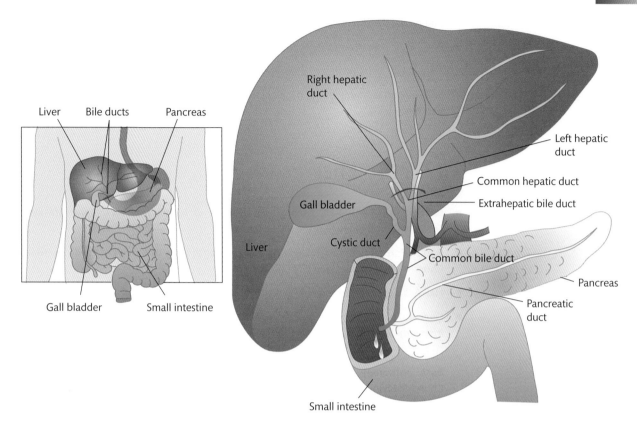

FIGURE 11.23

Diagram showing anatomical relationship between the liver, gall bladder, pancreas and duodenum.
Modified from original image © 2010 Terese Winslow.

SELF-CHECK 11.14

Why is the pancreas important in the pathology of the liver?

 Chapter summary

In this chapter we have described:

- Anatomical structure of the liver: its site in the body, its lobes and segments.

- Microscopical structure: the lobule and acinus, blood supply and drainage.

- Cellular components: hepatocytes, Kupffer cells, stellate cells and endothelial cells.

- Major functions of the hepatocytes, including fat, protein and carbohydrate metabolism, detoxification of drugs and alcohol, and secretion of bile.

- Structure and function of the gall bladder.

- Relationship of the liver to other organs: pancreas, duodenum.

- Embryonal development of the liver and its haemopoietic function.

- The ability of the liver to regenerate after injury, and the regeneration process.

- Liver pathological processes, classification and morphology of liver disease.

- The site and anatomy of the pancreas and its relationship to the liver and hepatobiliary pathology.

 ## Further reading

- Bancroft JD, Gamble M. *Theory and practice of histological techniques* 6th edn. London: Churchill Livingstone / Elsevier, 2008.

- Burt A, Portmann B, Ferrell L eds. *MacSween's pathology of the liver* 6th edn. London: Churchill Livingstone / Elsevier, 2012.

- Hall A, Yates C eds. *Immunology*. Oxford: Oxford University Press, 2010: 162.

- Liver and pancreas. In: Young B, Lowe JS, Stevens A, Heath JW eds. *Wheater's functional histology* 5th edn. Edinburgh: Churchill Livingstone, 2006: 288–301.

- Liver and pancreatobiliary system. In: Young B, O'Dowd G, Stewart W eds. *Wheater's basic pathology, a text, atlas and review of histopathology* 5th edn. Edinburgh: Churchill Livingstone, 2009: 164–76.

- Liver Cancer—NHS Choices.

- Liver function and disease (www.medicinenet.com).

- Liver function tests / health / patient.co.uk (www.patient.co.uk).

- Orchard GE, Nation BR eds. *Histopathology* Oxford: Oxford University Press, 2012: 149, 164-7.

- Standring S ed. *Gray's anatomy: the anatomical basis of clinical practice* 40th edn. London: Churchill Livingstone / Elsevier, 2008.

 ## Discussion questions

11.1 Discuss, with the aid of diagrams, the anatomical relationships between the liver, the gall bladder and the pancreas, and explain how pathological changes in one area can affect the other two.

11.2 Discuss the structural components of the liver

11.3 Discuss how the ultrastructure of the hepatocyte is influenced by liver function

Answers to self-check questions and discussion questions, hints and tips are provided on the book's Online Resource Centre.

 Visit www.oxfordtextbooks.co.uk/orc/orchard_csf/.

12

Kidney and urinary tract

Guy Orchard, David Muskett and Brian Nation

Learning objectives

By the end of this chapter you should be able to:

- Identify the function of the kidney and urinary system.
- Name the cells of the kidney and ureter and relate their structure to their specific function.
- Differentiate kidney into cortex, medulla and interstitium components.
- Identify the cells of the kidney and ureter microscopically.
- Name the key investigations of the kidney and its function.

The kidneys and urinary system provide the basis for homeostatic regulation of the body's water level, controlling electrolytes and fluid balance. The kidneys operate by a series of active and passive transport mechanisms to ensure efficient production of urine. The kidneys have a very rich vascular supply and each day the blood is filtered through the organs approximately 36 times. Nitrogenous waste products in the form of urea are actively removed from the kidneys.

The kidney has numerous hormonal functions; for example, controlling blood pressure through the renin-angiotensin function, the production of red blood cells via the production of erythropoietin, processing vitamin D, and the production of antihypertensive lipids. The kidney removes wastes and extra water from the blood and forms urine, which is discharged to the bladder via the ureters.

The urinary system comprises two kidneys, two ureters, the bladder and a single draining urethra. Anatomically, the kidneys are located on the posterior wall of the abdomen at around waist height, between the transverse processes of the T12–L3 vertebral bodies.

Key points

Humans are born with two kidneys. Average length of each is 10–12 cm, width is 5–7 cm and thickness 2–3 cm. Average weight is around 150 g in men and 135 g in women. The kidneys filter approximately 180 litres of blood a day, which means the kidneys filter the blood around 50 times per day. The functional unit of the kidney is the nephron and there are around a million of these in each kidney. The first successful kidney transplant took place in Boston, in the USA, in 1954, and involved the transfer of a kidney between identical twins—the Herrick brothers.

The kidney and urinary system acts as the filtration system of the body, removing soluble toxins and noxious chemicals from the blood. The key waste product of the body is urea and is removed in urine. The kidneys are very efficient filtration units; each minute the kidney filters approximately 125 mL blood. The kidneys sit at the rear of the abdominal cavity, just below the rib cage. The organs float within the peritoneal cavity and are not tethered to any other organ (Figure 12.1). Congenital anomalies of the kidney are seen where a patient may have no kidney, one kidney on one side of the abdominal cavity, or two kidneys fused to form a single horseshoe kidney. Each kidney receives a blood supply from a renal artery, branching from the descending aorta. The renal artery enters the kidney at the hilum and then splits into a number of branches each of which supplies a sector of the kidney (Figure 12.2). When one of these branches is blocked with a blood clot, renal ischaemia can take place.

SELF-CHECK 12.1

(This is a multiple-choice question.)

Kidneys are supplied with blood by a:

a) single artery and a single vein

b) single artery, a ureter and a vein

c) single artery and two veins.

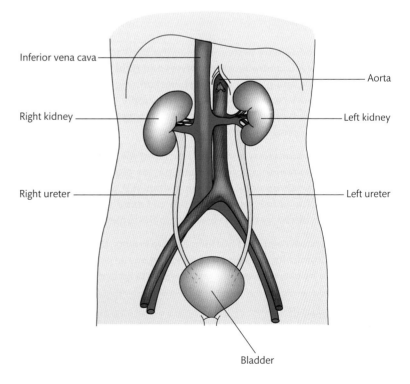

FIGURE 12.1

Anatomical position of the kidneys in the abdomen.

© G. Pocock and C. Richards 2009.

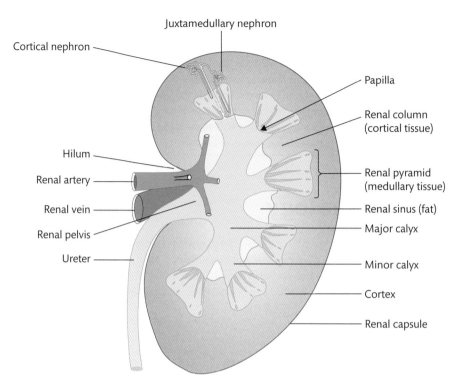

FIGURE 12.2
Diagram of the gross anatomy of the kidney.
© G. Pocock and C. Richards 2009.

CLINICAL CORRELATION

Renal Core Biopsy

As the kidneys float free within the peritoneal cavity, they move when the individual breathes in and out. This can be a problem when a renal core biopsy is being taken. Patients are therefore required to hold their breath when the biopsy is taken under ultrasound guidance.

BOX 12.1 What is urine?

Urine is the fluid waste excreted by the kidneys. Urine is 96 per cent water, while the remaining 4 per cent comprises urea (9.3 g/L), chloride (1.87 g/L), sodium (1.17 g/L), potassium (0.75 g/L), creatinine (0.67 g/L) and other ions and inorganic and organic compounds. The reason urine is yellow is because it contains bile pigments, which are produced by the liver as a result of the breakdown of bile. You read more about bile in Chapter 11. The colour of urine will vary considerably, depending on fluid intake; the more water consumed, the paler the urine will become. In contrast, when engaged in physical exertion resulting in sweating or when simply not drinking moderate volumes of fluid, urine will become darker in colour. The darker the colour, the more indicative it is of dehydration. The pH of urine can vary between 4.6 and 8, with pH 7 (neutral) being normal. The average adult produces around 1–2 litres of urine per day, depending on factors such as hydration states, environmental factors, weight and underlying health conditions. **Polyuria** is the term used to describe excessive urine production (>2.5 L/day), **oliguria** is when urine production falls below 400 mL/day, and the term **anuria** is used in extreme cases where production is <100 mL/day.

Urine as an indicator of disease

The odour of human urine can reflect what has been consumed or indicate a specific disease. A good example is patients who have diabetes mellitus and who may present with a sweet urine odour. This may also be a sign of kidney stones. Similarly, if urine is cloudy (turbid) then this may indicate a bacterial infection or evidence of crystallization of salts (e.g. calcium phosphate). Acidic urine in patients with hyperuricosuria can contribute to the formation of stones composed of uric acid, which may be found in the kidneys, ureters or bladder.

12.1 Embryology

The urinary system represents one of the last organ systems to develop. There are three recognized stages: pronephros, mesonephros and metanephros. All the stages develop from the intermediate mesoderm. The kidneys begin to function at around the 12th week of gestation, initially contributing to the production of amniotic fluid, which is checked as part of the routine first pregnancy scan.

12.2 Kidney function

The important point to make here is that the kidneys play an integral part in body homeostasis. These functions include the regulation of acid–base balance, electrolyte concentration, extracellular fluid volume control and also regulation of blood pressure. The following is a list of the key functions in more detail.

12.2.1 Excretion of waste

The kidneys excrete a wide range of waste products, mostly from the process of metabolism—most significantly including urea, which is produced from protein breakdown, and uric acid, which is produced from the decomposition of nucleic acid and creatinine. It is important to remember that the kidneys also excrete many foreign chemicals, drugs, pesticides and food additives.

12.2.2 Resorption of nutrients

The kidneys reabsorb glucose (100 per cent), amino acids (100 per cent), bicarbonate (90 per cent), sodium (65 per cent), phosphate and water (65 per cent).

12.2.3 Acid–base balance

Acid–base homeostasis is maintained through lung and kidney function. The lungs contribute by regulating carbon dioxide (CO_2) concentration in the body, while kidneys contribute by reabsorbing bicarbonate from urine and by excreting hydrogen ions in urine.

12.2.4 Hormone secretion/regulation

Various hormones are secreted by the kidneys, the most common being **erythropoietin** and **renin**. Erythropoietin stimulates red blood cell formation in the bone marrow, while renin is an enzyme involved in the regulation of aldosterone levels in the body. Aldosterone is a steroid hormone produced by the outer layer of the cortex of the adrenal gland. It is linked to the conservation of sodium, secretion of potassium, increased water retention and increased blood pressure. In addition, **calcitriol**, which is the activated form of vitamin D, stimulates osteoblasts to secrete interleukin-1 (IL-1) and thus osteoclasts to increase bone resorption. The activated form of vitamin D is produced as a result of changes in calcium in the bloodstream as a result of hydroxylation of vitamin D. Prostaglandin hormones (e.g. PGE2 and prostacyclin) are derived from the metabolism of arachidonic acid in various cells of the kidney, and are important in ensuring renal function; however, they have little systemic activity as they are rapidly metabolized once in the pulmonary circulation.

12.2.5 Regulation of blood pressure

Although not a primary regulator of blood pressure, long-term effect is related to kidney activity and involves the maintenance of the extracellular fluid and, in particular, plasma sodium concentration. Renin is also involved here as it acts as a chemical messenger in the **renin-angiotensin system (RAS)**. Principally, changes in renin secreted by the juxtaglomerular cells alter the levels of angiotensin II and aldosterone. Through a complex mechanism, these hormones increase absorption of sodium chloride in the kidney, resulting in expansion of the extracellular fluid (ECF) domains and an elevation in blood pressure. When renin levels are high the levels of angiotensin II and aldosterone increase, resulting in increased sodium chloride resorption, expansion of the ECF domains, and elevated blood pressure. Conversely, when renin levels are low then the levels of angiotensin II and aldosterone decrease, resulting in contraction of the ECF domains and subsequent lowering of blood pressure.

Key points

If the renin-angiotensin-aldosterone (RAS) system is abnormally active, blood pressure will be too high. There are many drugs that suppress or interrupt different steps in this system. These drugs are administered to patients with hypertension, heart failure, kidney failure, or who suffer the harmful effects of diabetes.

12.2.6 Osmolarity regulation

Plasma osmolarity is regulated by the hypothalamus in the brain, which in turn communicates with the posterior pituitary gland. The increase in osmolarity causes the release of antidiuretic hormone (ADH). As a result, water is resorbed by the kidneys and thus urine concentration increases. The ADH binds to cells in the collecting ducts of the kidney that translocate the membranes via **aquaporins**, which allow normally impermeable membranes to become permeable, allowing water to leave the collecting duct cells and be resorbed by the body via the **vasa recta**. The hyperosmotic regulation of the kidney is generally regulated by two principal mechanisms:

- *Recycling of urea*: When the plasma blood volume is low and ADH is released the **aquaporins** become permeable to water and urea. This enables urea to leave the collecting duct and enter the medulla of the kidney, creating a hyperosmotic solution that attracts water. Thereafter, urea can re-enter the nephron and be excreted or recycled, depending on ADH level.

- *Single effect*: This process depends on the fact that the ascending limb of the loop of Henle is not permeable to water but is permeable to sodium chloride (NaCl), creating what is termed a **countercurrent exchange** system. This results in increased ion concentration in the kidney medulla, and this forms an osmotic gradient by which water flows through the aquaporins of the collecting duct opened by ADH.

Key points

Dr Friedrick Henle, an anatomist and physician born in Bavaria in 1809, described and named many anatomical features in the body.

12.2.7 Miscellaneous functions

During periods of prolonged fasting, the kidneys synthesize glucose from amino acids, a process termed **gluconeogenesis**. This process primarily involves the liver but the kidneys and small intestine also play important roles. As previously mentioned, the kidney also contributes to the production of amniotic fluid during pregnancy.

Having considered these functions, how are they achieved? In some cases, the kidneys and urinary system achieve homeostatic regulation independently, such as in processes involving simple filtration, reabsorption and secretion, but also in association with endocrine hormones such as renin, angiotensin II, aldosterone, antidiuretic hormone, and many others.

12.3 Kidney structure

Now that we have discussed the functions of the kidney it will be evident that it has a multitude of homeostatic roles. You will also have been introduced to a number of structural components that make up the kidney. How then does the kidney structure facilitate the performance of these functions? In order to answer this we need to consider the structure of the kidney in a sequence of layers of increasing complexity. Let's start with the kidney's gross anatomy.

12.3.1 Gross anatomy

Cross reference
You can read more about the adrenal gland in Chapter 14.

The kidneys are bean-shaped and sit within a blanket of fat. In relation to other organs, the adrenal (suprarenal) glands sit on top of each kidney, while to the right sits the second part of the duodenum (descending portion) and on the left is the greater curvature of the stomach. The spleen is located anterior to the upper pole of the left kidney (Figure 12.1).

12.3.2 Renal capsule

The outer-most layer of the kidney is the renal capsule. This is a membranous sheath that covers the entire surface of the kidney and is composed predominantly of collagen and elastin fibres. In addition, scanty mesothelial cells secrete ascitic fluid which lubricates the movement of abdominal organs within the abdominal cavity. This layer is important because it provides support for the complex internal structure of the kidney and, more importantly, protection from injury and trauma. Under normal circumstances the capsule is 0.2–0.3 mm thick and is light red/purple in colour, translucent and smooth in appearance. The capsule receives blood from the interlobular arteries, which are small vessels that branch off the main renal arteries. The capsule surrounds the entire kidney and enters a hollow area called the **sinus**.

CLINICAL CORRELATION

Renal cancer

When a kidney containing a malignant tumour is examined macroscopically, inspection of the capsule forms an important part of the procedure. If the tumour is seen to breach the kidney capsule, this represents more advanced disease and thus a worse potential outcome for the patient.

12.3.3 Vasculature and lymphatics

The blood supply to the kidney arises from the renal artery that enters the renal **hilum**. The first branch of the renal artery is called the inferior suprarenal artery, which then subdivides into five branches (posterior, superior, anterior superior, anterior inferior and inferior). These subdivide further into interlobular arteries that branch to form the arcuate arteries that run within the cortex. These extend and radiate into the cortex, becoming the interlobular arteries, and finally afferent arterioles, peritubular capillaries and efferent arterioles.

The renal venous system drains the kidney in a distribution similar to the arterial system, with the renal vein positioned anterior to the artery at the hilum. The left renal vein is longer than the right renal vein because it has to cross the midline to reach the inferior vena cava.

The renal lymphatic system is similar to the venous drainage system. Following departure from the hilum, the left primary lymphatic drainage is into the left lateral aortic lymph node, while on the right it drains into the lateral caval lymph nodes.

12.3.4 Hilum

The renal hilum is the indentation near the centre of the concave portion of the kidney, and is the point at which the renal vasculature, lymphatics and nerves enter and leave the kidney. The hilum is also the point at which the ureter leaves the kidney.

12.3.5 Cortex

The cortex is the outer portion of the kidney and is where the glomeruli (each surrounded by Bowman's capsule) and proximal and distal convoluted tubules can be found. They comprise the functional unit of the kidney, termed the nephron, and you will read more about these during consideration of the microscopic structural features of the kidney.

12.3.6 Medulla

The medulla is the inner portion of the kidney. It has a striated (striped) appearance and is red/brown in colour. Within these areas are the renal pyramids, the bases of which are situated towards the renal cortex, which forms a shell around the medulla. As the cortex dips into the medulla it forms renal columns between the renal pyramids.

12.3.7 Renal calyx

The renal calyx is composed of major and minor parts and carries urine from the renal pyramids of the medulla to the renal pelvis for excretion through the ureters.

12.3.8 Renal pelvis

The renal pelvis is a funnel-shaped cavity that receives urine drained from the nephrons via the collecting ducts and pyramid papillary ducts.

12.4 Microscopic structural features

You can discern a great deal from the gross anatomy visualized with the naked eye. As in a bisected orange, you can recognize the outer skin (renal capsule), segments in the body (renal cortex and medulla), and a central portion (renal pelvis). However, like most things in science, you need to appreciate the structures you cannot see with the naked eye to comprehend how an organ works.

12.4.1 The nephron

The basic unit of the kidney is the **nephron** (Figure 12.3). There are between 800 000 and 1.5 million nephrons in each normal human kidney. Nephrons comprise:

- Renal corpuscle (comprising **glomerulus** and **Bowman's capsule**)
- Proximal convoluted tubule (PCT)
- Descending loop of Henle
- Ascending limb
- Thick ascending limb
- Distal convoluted tubule (DCT)
- Collecting duct.

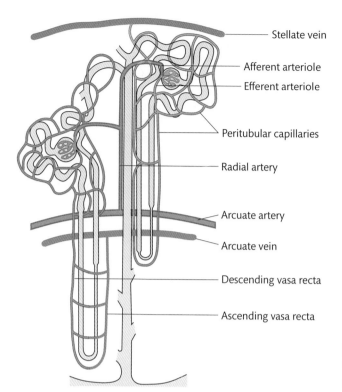

Stellate vein

Afferent arteriole

Efferent arteriole

Peritubular capillaries

Radial artery

Arcuate artery

Arcuate vein

Descending vasa recta

Ascending vasa recta

FIGURE 12.3
The nephron and its blood supply.
© G. Pocock and C. Richards 2009.

The primary function of the nephron is to control the homeostatic regulation of the blood. There are two general types of nephron, **cortical** and **juxtamedullary**. The distinction is based on the length and location of their associated **loop of Henle**. Cortical nephrons have a loop in the renal medulla at the junction with the renal cortex, while in juxtamedullary nephrons the loop of Henle is located near the medulla, but still within the cortex (hence juxtamedullary). Each nephron is around 4.5 cm long.

Key points

The functional unit of the kidney is the nephron. The kidney initially produces weak urine which is concentrated as it passes through the reabsorbing cells of the proximal convoluted tubule, loop of Henle and collecting duct.

Renal corpuscle

The distinctive features of the renal corpuscle are:

- *Bowman's capsule*: A double-walled sac with each wall consisting of a single layer of simple squamous epithelial cells (parietal layer). Inside the Bowman's capsule there are three layers, which, from outside inwards, are:
  - Bowman's space (capsular space): This resides between the parietal and visceral layers, and is where the filtrate enters after passing through the filtration sites.
  - Visceral layer: This layer lies just above the glomerular basement membrane and is composed of cells called podocytes.

- Filtration barrier: This layer is composed of a specialized endothelium, the fenestrated endothelium (which is porous).
- *The glomerulus*: A network of capillaries lying within Bowman's capsule, with the afferent (entering) blood vessel being wider than the efferent (exiting) vessel. It is within the glomerulus that the blood is filtered (Figures 12.4 and 12.5). The glomerular membrane contains pores that are large enough to allow small molecules to pass through, but small enough to prevent the passage of large molecules and red blood cells. Blood arrives in the glomerulus via the afferent arteriole under considerable pressure. The arteriole branches and forms a knot of capillaries, lined by endothelial cells, which sits within Bowman's capsule. Cells that line the outer aspect of the glomerular basement membrane are called **podocytes**, which have foot-like processes that contribute to the kidney's ability to undertake filtration.

The glomerular wall consists of an epithelium which appears to be fenestrated (like a window). It is through these spaces that the glomerular filtrate passes into Bowman's capsule. If these membranes are damaged then various pathological conditions can result.

The glomerular basement membrane consists of three components:

- Glomerular capillary endothelium
- Basal layer
- Podocytes

The glomerular capillary endothelium contains fenestrations that are not supported by a spanning diaphragm. These pores are around 70–100 nm in diameter, which is actually quite large.

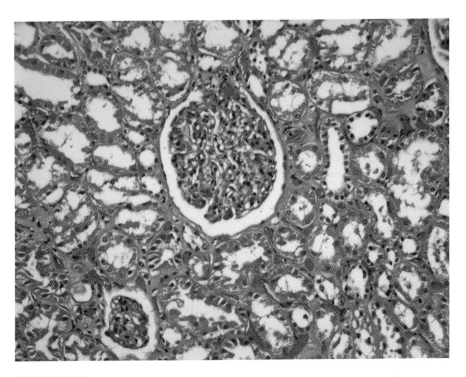

FIGURE 12.4
Glomerular structure (haematoxylin and eosin [H&E]-stained section).

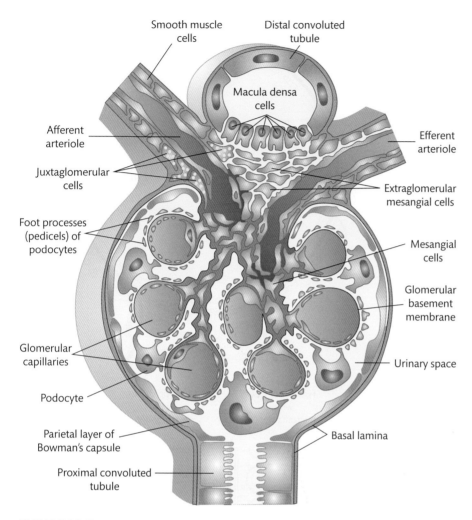

FIGURE 12.5

Diagrammatic representation of the glomerular structure.

Modified from original image: Ross *et al.*, Histology. A Text and Atlas © 2011 Wolters Kluwer Health/Lipincott Williams and Wilkins.

They need to be this size in order to allow the free filtration of fluid containing plasma and associated proteins, but small enough to stop the passage of red blood cells (which are 6–8 μm in diameter (Figures 12.6–12.8).

The basal layer can be divided into three further sublayers:

- The *lamina rara externa* (adjacent to the podocyte processes), composed largely of heparan sulphate.

- The *lamina densa* (the darker zone within the basement membrane complex), composed largely of type 4 collagen and laminin.

- The *lamina rara interna* (adjacent to the endothelial cells), again composed largely of heparan sulphate.

A

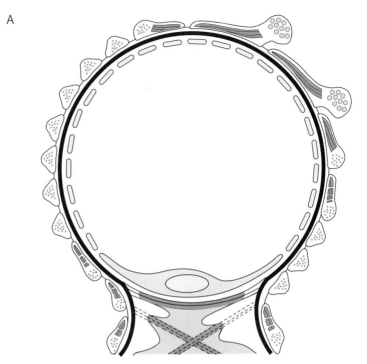

B

Foot processes
with
microfilaments

Major processes
with
microtubules

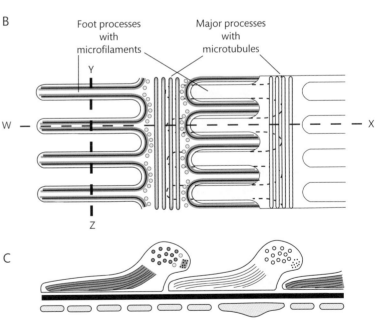

FIGURE 12.6

Diagram of a podocyte.
Modified from Sanden
S. Das Aktinskelett
der Fußfortsätze und
seine Beziehungen
zum Tubulinskelett
der Primärfortsätze–
morphologische und
immunzytochemische
Untersuchungen an den
Podozyten der Rattenniere.
Dissertation 2001
Heidelberg, 1–77. *J Am Soc
Nephrol* 2000; 27:1265–70.

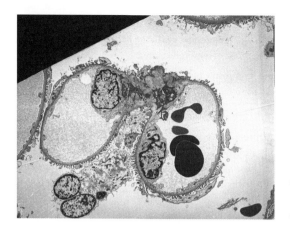

FIGURE 12.7

Glomerular podocyte foot processes (transmission electron micrograph [TEM]).

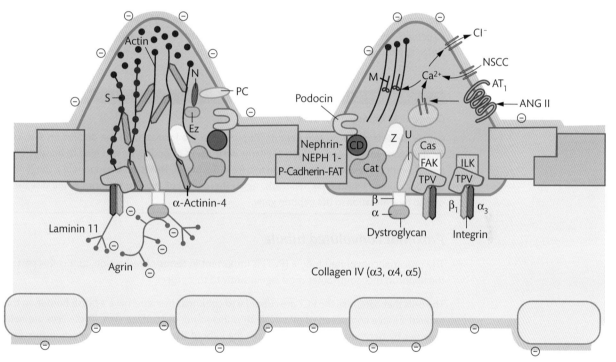

FIGURE 12.8

Interaction of the podocyte and basement membrane.

Modified from Endlich K, Kriz W, Witzgall R. Update in podocyte biology. *Curr Opin Nephrol Hypertens* 2001; 10:331–340. © 2001 Lippincott Williams & Wilkins, Inc.

CLINICAL CORRELATION

Glomerular nephritis

There is a highly complex list of conditions that affect the glomerular basement membrane. Many are classified as glomerular nephritis conditions or disorders. A number of these conditions are autoimmune in nature. Examples include:

■ *Goodpasture's syndrome*: A condition in which the capillaries become inflamed as a result of damage to the basement membrane by antibodies against type 4 collagen.

- *Diabetic glomerulosclerosis*: This is characterized by pronounced thickening of the basement membrane (4–5 times thicker than normal). This interferes with the filtration processes. It can be caused by insulin deficiency or be the result of hyperglycaemia.

- *Nephrotic syndrome*: This comes about as a result of a change in the glomerular filtration mechanism and can result in proteinuria and hyperlipidaemia.

The foot-like processes on podocytes (Figure 12.7), also known as pedicels, wrap around the capillaries, leaving slits between them (filtration slits or slit diaphragms); this increases the cell surface area and enhances ultrafiltration. These slit diaphragms are composed of a number of cell surface proteins, including nephrin, P-cadherin and podocalyxin, which ensure that larger marcromolecules remain within the bloodstream.

Podocytes play a key role in a process known as the **glomerular filtration rate** (GFR). Contraction of the podocytes causes shrinkage or closure of the filtration slits, which decreases the GFR by reducing the surface area available for filtration. At the ultra-structural level, podocytes are seen to contain prominent endoplasmic reticulum and Golgi apparatus, which supports their role in protein synthesis and protein translational alterations.

CLINICAL CORRELATION

Congenital kidney disorders

Genetic defects in the protein nephrin, important in the formation of the diaphragm slits, can result in congenital kidney failure. Similarly, destruction of podocytes can lead to massive proteinuria with the result of large protein loss from the blood. This is what happens in a congenital disorder called Finnish-type nephrosis. Again there is a link here to nephrin, as the disease is caused by a mutation in the nephrin gene.

Proximal convoluted tubule

The proximal convoluted tubule (PCT) is important in terms of resorption, as it is this part of the nephron that contributes most significantly to this process.

The cells that comprise the PCT are cuboidal to columnar in shape, have a brush border on the luminal surface composed of microvilli, thus increasing the surface area. These cells are held together by tight junctions (you will have read about these in Chapter 1), which ensure that no filtrate can pass between the cells. This is important because it prevents nitrogenous waste from returning into the bloodstream. The key purpose of PCT cells is to absorb sodium from the glomerular filtrate. As a rule, whenever sodium is absorbed, water is reabsorbed passively as well. Active transport will also result in the resorption of glucose (Figure 12.9), amino acids, acetoacetate and also vitamins.

So how is this process achieved?

The cells of the PCT need to be able to pick up dissolved solutes. This can only be achieved if the PCT has protein channels in the membrane structure. Many of these materials will 'passively' move across a concentration gradient, sometimes called 'facilitated diffusion'; however, material also needs to be moved against the passive gradient. In this case, the material is moving against the gradient through an 'active' transport channel or 'pump' mechanism.

Essentially, sodium can transport across the apical membrane into the cell through the sodium channel, resulting in it being resorbed from the filtrate in the interstitial fluid and then back into the blood through the peritubular capillaries.

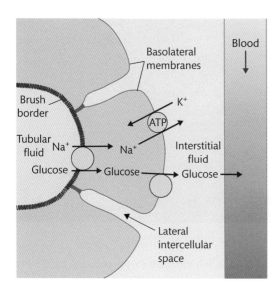

FIGURE 12.9

Illustration of the active resorption of glucose.

© G. Pocock and C. Richards 2009.

Every solute that needs to be resorbed has its own protein channel, and there are therefore specific channels, for example, for sodium, potassium and hydrogen. As these ions are moved against the transport gradient, the filtrate becomes more dilute, or **hypotonic**. This explains why water is resorbed because basic biology dictates that water flows passively through the process of osmosis and the water flows down an osmotic gradient, from a hypotonic to a **hypertonic** solution. By the time the filtrate reaches the end of the PCT it is osmotically balanced (**isotonic**), containing less solute and water. The PCT (Figures 12.10 and 12.11) enables around 60–70 per cent of all required materials to be resorbed.

Loop of Henle

The loop of Henle leads from the PCT to the distal convoluted tubule (DCT), and is named in honour of F. G. J. Henle, who was the first to describe the structure. Its main function is to facilitate a concentration gradient in the medulla of the kidney, which it does by producing a **countercurrent multiplier** system (Figure 12.12). Through electrolyte pumps, the loop of

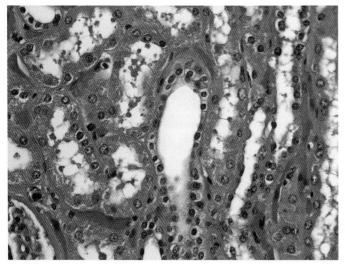

FIGURE 12.10

Proximal convoluted tubule (H&E stain).

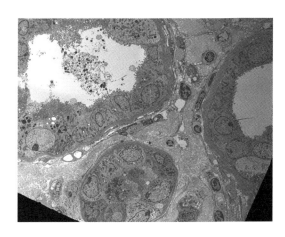

FIGURE 12.11
Proximal convoluted tubule (TEM).

Henle produces high urine concentration deep within the medulla, immediately before the collecting duct. As its name implies, the loop of Henle is U-shaped and produces urine that is far more concentrated than the blood.

The loop of Henle comprises the following:

* *Descending limb of the loop of Henle*: This is a thin limb with limited permeability to ions and urea but high permeability to water.
* *Ascending limb of the loop of Henle*: This portion of the loop is not permeable to water but is permeable to ions.
* *Thicker portion of the ascending limb of the loop of Henle*: This is where ions (e.g. sodium, potassium, chloride) are resorbed by a secondary process of 'active' transport. This is achieved by a sodium/potassium/chloride co-transporter system. The result is that this electrically charged concentration gradient drives the transfer and resorption of ions such as sodium, magnesium and calcium.
* *Thick cortical ascending limb*: This portion of the loop of Henle drains urine into the DCT.

Key points

There are two types of nephron, and, in the majority of the animal kingdom, most animals have a combination of both types. In the juxtamedullary nephron, the glomerulus is situated around the corticomedullary junction, and the loop of Henle travels into the kidney medulla. This means that more water can be resorbed, which would be an advantage in animals living in arid conditions (e.g. a gerbil or desert rat). In contrast, the cortical nephron sits high in the cortex and the loop of Henle barely reaches the medulla, where water is resorbed. Therefore, the urine produced would be copious and highly dilute. In aquatic animals or those living in wet conditions (e.g. beaver), one would expect to find more nephrons of this type.

SELF-CHECK 12.2

What are the two types of nephron and what are their key structural differences?

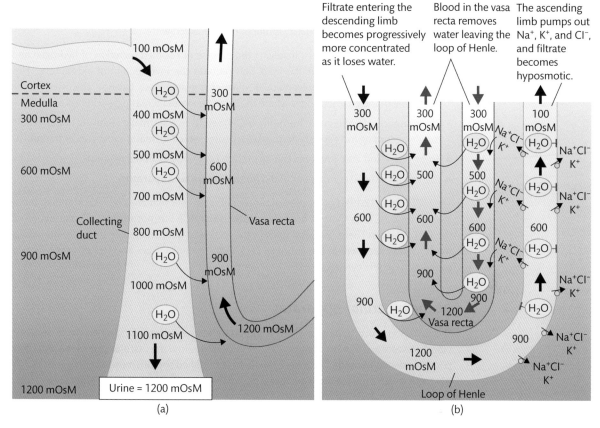

Filtrate entering the descending limb becomes progressively more concentrated as it loses water.

Blood in the vasa recta removes water leaving the loop of Henle.

The ascending limb pumps out Na⁺, K⁺, and Cl⁻, and filtrate becomes hyposmotic.

FIGURE 12.12

Countercurrent diagrams: (a) flow into the vasa recta, and (b) in the loop of Henle.

Modified from original image: Silverthorn, *Human Physiology: An Integrated Approach* © 2004 Pearson Education, Inc., publishing as Benjamin Cummings.

BOX 12.2 Desert dehydration

Dehydration can lead to death fairly quickly in an arid climate, and there are many misconceptions about how to reduce water loss. Here are some tips and clarifications:

- Sucking on a pebble can stimulate production of saliva, which gives the impression of reducing thirst.

- You may think that drinking urine will facilitate water conservation. This is not the case, however, as you are simply reintroducing toxic waste products into the body. As the body becomes more dehydrated so the urine becomes more toxic. Essentially, water that contains approximately 50 per cent salt will increase dehydration.

- Dehydration not only involves the loss of water but also some loss of salt (sodium chloride). Therefore, treating dehydration requires not only the replacement of water but also restoration of the normal salt concentration in body fluid.

- Water deprivation is far more serious than food deprivation. In normal circumstances, 2.5 per cent of total body water is lost per day, mainly (about 1200 mL) in urine. In addition, we also lose water in expired air, by perspiration, and from the gastrointestinal tract. In a desert environment, water loss through perspiration is greatly increased and can result in severe dehydration within only a few hours, which can quickly lead to death.

Distal collecting tubule

The **distal collecting tubule** (DCT) is functionally involved with:

- Reabsorption of sodium ions involving the process of coupled secretion of hydrogen and potassium ions into the tubular fluid. This requires the involvement of aldosterone. This relates to the kidney's involvement in maintaining acid–base balance, discussed at the beginning of the chapter, and is where the urine becomes increasingly acidified.

- The presence of antidiuretic hormone (ADH) results in a normally relatively impermeable tubule becoming increasingly permeable to water, which increases urine concentration.

- Some drugs and also ammonia can be excreted at this point in the nephron.

- It forms an integral component of the **juxtaglomerular apparatus.**

The DCT (Figure12.13) is composed of simple cuboidal cells that are a little shorter than those found in the PCT. As there is a brush border in the PCT but not in the DCT, the lumen of the DCT is larger. There are basal folds in the DCT, and ultrastructurally there are numerous mitochondria seen within the cuboidal cells. Cells of the DCT are also less eosinophilic and generally have less cytoplasm.

Juxtaglomerular apparatus

The juxtaglomerular apparatus (JGA) is important in the kidney's functional role in systemic blood pressure and volume control. The JGA is located within the vascular pole of the renal corpuscle, and is composed of cells best described as modified smooth muscle cells (Figure 12.14). It is these which produce the renin that is part of the renin-angiotensin system/network.

Structurally, the JGA is composed of three components:

- *Juxtaglomerular cells*: These are the modified smooth muscle cells located in the wall of the afferent arteriole. Functionally, they detect changes in blood pressure and secrete the enzyme renin under the appropriate stimulus.

- *Extraglomerular mesangial cells (lacis cells)*: These are flat cells that support the JGA.

- *Macula densa cells*: This area of the JGA is associated with the glomerulus because it loops back into the cortex, where the cells become taller and more densely packed. It is thought

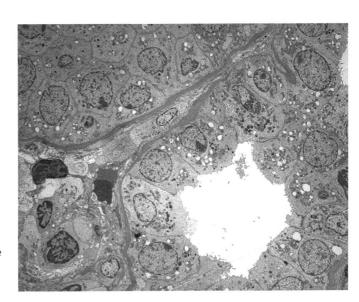

FIGURE 12.13
Distal collecting tubule composed of simple cuboidal cells (TEM).

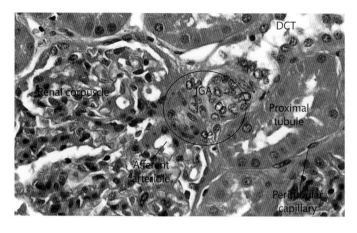

FIGURE 12.14
Juxtaglomerular apparatus (H&E stain).

that these cells can detect changes in sodium ion concentration in the DCT and are able to relay this to the juxtaglomerular cells.

The mechanisms by which the JGA helps to control blood pressure is expanded in the clinical correlation box covering the subject. Furthermore, you can read more about this in Chapter 14 (Endocrine System).

Key points

It is perhaps not so well known that the kidney has a higher level of blood flow than in the brain, liver and heart.

CLINICAL CORRELATION

Control of blood pressure in the juxtaglomerular apparatus

The lacis cells of the juxtaglomerular apparatus sense a decrease in blood pressure and secrete the enzyme renin. Renin circulates in the blood and converts circulating angiotensinogen to angiotensin I and then angiotensin II. Angiotensin II causes the cells of the adrenal cortex to release aldosterone, a hormone that acts on the distal convoluted tubule and the collecting duct of the kidney to retain more sodium and thus water.

SELF-CHECK 12.3

What is the structure of the juxtaglomerular apparatus (JGA) and how does this part of the kidney nephron help to control blood pressure and volume?

12.5 **Renal interstitium**

The renal interstitium is composed of extracellular matrix containing glucosamines and interstitial cells. The fibroblast-like interstitial cells are loosely divided into two types: renal interstitium and medullary renal interstitium. It is the renal interstitium that comprises 5–20 per cent of the total renal mass.

The most significant point to make about the renal interstitium is that the hormone erythropoietin (EPO) is produced by fibroblasts of the interstitial area. It can also be produced by the perisinusoidal cells of the liver. This hormone controls red blood cell synthesis. It is also implicated in brain responses to neuronal injury and generally in the wound healing process.

Key points

Erythropoietin (EPO) is an erythropoiesis-stimulating agent and can be used as a performance-enhancing drug. Testing for EPO in blood is one of the key methods to detect 'drug' abuse in sport, as it is one of the most commonly used substances used to help athletes gain unfair advantage in competitive sport.

We have now described the structure and functions of the kidney, but the urinary system relies on the presence of other key structures to enable waste products to be dispelled from the body.

12.6 **Ureters**

There are two ureters, one leading from each kidney. They are simply tubes composed of smooth muscle fibres that allow urine to be passed from the kidney to the urinary bladder (Figure 12.15). In adults the ureter is 25–30 cm long, and the endothelial lining is composed of transitional cells. This is a specialized form of endothelium that also lines the bladder. The ureter arises from the renal pelvis and in cases where kidney stones are present it is not uncommon to find stones within the ureter, resulting in severe pain (renal colic).

CLINICAL CORRELATION

Hydronephrosis

If lodged within the ureter, a kidney stone can block the flow of urine. This could lead to hydronephrosis, where the kidney becomes swollen due to accumulation of urine. There are three sites where a kidney stone may lodge in the ureter:

- **at the junction with the renal pelvis**
- **as the ureter passes over the iliac vessels**
- **where the ureter enters the bladder.**

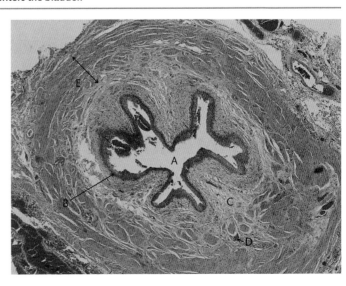

FIGURE 12.15

Transverse section of a ureter (H&E stain, original magnification ×20; A Lumen, B transitional epithelium, C lamina propria, D capillary, E smooth muscle [muscularis]).

Preparation courtesy of Michelle Farrar.

12.7 **Urinary bladder**

The bladder is a smooth muscular structure that receives urine from the left and right ureters (Figures 12.16 and 12.17). It is located within the pelvic girdle and is roughly spherical in shape, although this will depend on the volume of urine it contains. The bladder is able to expand to accommodate and store urine, and when full can contain approximately 750 mL of fluid. Anatomically, the bladder has a triangular internal floor, the **trigone**, which has three openings, one each at its three angles. The ureters attach to the posterior openings while the anterior opening at the apex of the trigone leads to the neck of the bladder and has a funnel-like appearance. It is here that the urethra connects to the bladder. The smooth muscle fibres forming the wall of the bladder are grouped into four bundles and these interlace to form the **detrusor muscle**, which also surrounds the neck of the bladder and forms the internal urethral sphincter. It is this sphincter that prevents urine from escaping the bladder, and is controlled by a parasympathetic nerve within the detrusor muscle. The upper layer of epithelial cells, called **umbrella cells**, provides a watertight and osmotically resistant layer, protecting the other cells of the epithelium from the hostile environment.

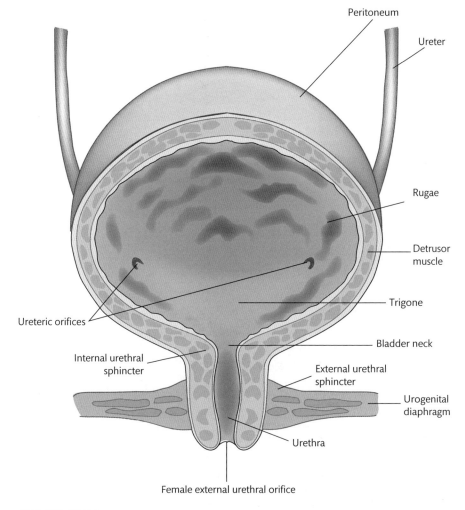

FIGURE 12.16
Diagram of urinary bladder structures.

The bladder wall lining (epithelium) is composed of a special epithelium called **transitional epithelium**. It is composed of multiple layers of cells that can expand and contract (Figure 12.18). They are found only within the bladder, ureters, superior urethra and gland ducts of the prostate. As the epithelium has the ability to expand and contract, this enables it to respond to fluctuations in fluid volume. The epithelium, which contains no glandular structures, must also protect the bladder from the caustic properties that urine at this stage of its journey possesses. It is termed 'transitional' because the luminal layer of cells are large and flattened and have a squamous appearance (umbrella cells), while those in the basal layers are more cuboidal. The umbrella cells are asymmetrical in shape, being thicker on the surface, resulting in a layer that is impermeable to urine. As the bladder fills the cells stretch as they expand. Under the light microscope, the umbrella cells (superficial urothelial cells) are seen to have dense cytoplasm with small, compact nuclei (Figures 12.19).

FIGURE 12.17

Gross specimen of a urinary bladder showing the ureters descending to insert at the trigone on the posterior surface.

SIU/Science Photo Library.

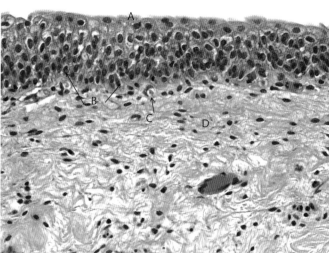

FIGURE 12.18

Transitional epithelium from the bladder: (A) superficial domed cells, cuboidal surface cells (umbrella cells), (B) basal cells, (C) capillary, (D) lamina propria (H&E stain, original magnification ×40).

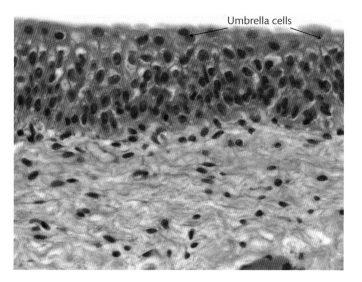

FIGURE 12.19

Umbrella cells (arrowed) form the superficial layer of the urothelium, which is a single layer of cells that are large and elliptical with abundant eosinophilic cytoplasm and often binucleation or prominent nucleoli. One umbrella cell covers several underlying cells and is inconspicuous in the distended bladder (H&E stain, original magnification ×60).

Bladder cancer

Transitional epithelium of the bladder can give rise to bladder cancer (transitional cell carcinoma [TCC]). This is normally recognized following the initial observation of blood in the urine. There is a close association between TCC and cigarette smoking and also workplace chemical solvent exposure.

SELF-CHECK 12.4

(This is a multiple-choice question.)

The bladder epithelium is composed of:

a) Stratified squamous epithelium

b) Transitional epithelium

c) Columnar epithelium

d) Ciliated columnar epithelium.

12.8 **Urethra**

The urethra is a tube that leads from the bladder to the outside of the body. The external orifice is called the **urethral meatus**. In men the urethra is approximately 20 cm long, while in women it is 4 cm long. In men it is divided into three areas, the prostatic urethra, membranous urethra and penile urethra. In men the urethra is important both as part of the urinary and reproductive systems, while in women it is purely part of the urinary system.

12.9 **Urinary system disease**

A brief overview of the urinary system in disease can facilitate an appreciation of the importance of the organs that make up this body system. In the majority of cases, kidney problems can be identified during a general health assessment, following clinical presentation of suspected dehydration, suspected kidney failure, or perhaps before or following treatment with certain drugs that can cause kidney damage or failure.

12.9.1 **Blood tests for kidney function**

The laboratory investigations of kidney abnormalities are broadly classified under the umbrella term 'kidney function tests'. These include:

- *Urea*: High level of blood urea (uraemia) is one of the first indicators that the kidneys are not working normally.
- *Creatinine*: This is a waste product from muscle metabolism. It normally passes out of the body in urine. High levels in the blood indicate kidney failure or disease.
- *Estimation of glomerular filtration rate*: Measurement of creatinine level can assist in eGFR. The normal range for eGFR is 90–120 mL/minute, while levels below 60 mL/minute indicate

kidney failure or disease; however, variation due to gender, age and race must be taken into account.

- *Assessments of dissolved salts*: These salts include sodium, potassium, chloride and bicarbonate, and are broadly referred to as the essential 'electrolytes'. Abnormal blood levels indicate kidney failure or disease. Sodium helps to regulate intracellular and extracellular water level and is important in assisting muscle and nerve function. Potassium helps in the regulation of heart beat, nerve and muscle function. Chloride assists in fluid balance. Bicarbonate has a role in balancing pH.

Following these preliminary investigations, it may be deemed necessary to perform a scan of the kidney or perhaps a biopsy of the tissue.

12.9.2 Scanning the kidney and urinary tract

In relation to investigation of kidney function, the main scanning methods used are nuclear isotope scanning and nuclear magnetic resonance (NMR) imaging. The former involves injection of a radioactive substance into the bloodstream, which travels to the kidneys and is excreted in urine. It is possible to follow what happens to the radioactive tracer substance using a gamma camera. The information gained from this scanning process, and using NMR imaging, is far more detailed than that obtained by conventional X-ray. The main reasons for performing a kidney scan are to check:

- blood flow rates
- transplanted kidneys
- obstructions within the kidney or ureters
- general kidney damage caused by trauma or infection
- cancerous growths.

12.9.3 Kidney biopsy

Kidney biopsy involves the removal of tissue, usually by a special needle, which can then be studied under the microscope following use of various histological staining techniques. Biopsy is normally performed for (i) haematuria (blood in urine), (ii) proteinuria (protein in urine), or (iii) following abnormal kidney function tests.

The histopathological investigations performed encompass routine procedures using special stains (e.g. periodic acid Schiff [PAS], hexamine silver) to demonstrate the basement membrane of the Bowman's capsules and glomerulus (Figures 12.20–12.22). Stains to demonstrate fibrin formation (e.g. Martius scarlet soluble blue [MSB]) are also requested, as are immunocytochemistry methods to demonstrate immune complex formation, including the main immunoglobulin types (i.e. IgM, IgG and IgA), complement C3 and fibrin, and also amyloid components, the deposition of which can cause considerable renal problems.

As there are a host of autoimmune states that can affect the kidney, assessment of ultrastructural features in relation to the glomerular basement membrane and podocyte structure can be extremely useful. You will read more about these techniques in the *Histopathology* volume in this textbook series.

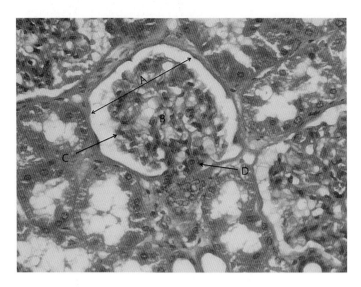

FIGURE 12.20

A glomerular structure showing (A) Bowman's capsule, (B) glomerulus, (C) podocyte, and (D) macula densa (H&E stain, original magnification ×60).

Preparations courtesy of Anne Rayner, photographs courtesy of Dr Alberto Quaglia.

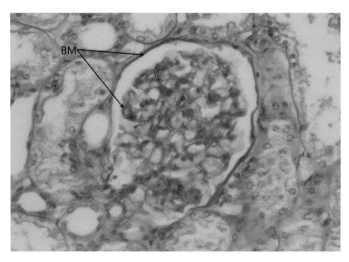

FIGURE 12.21

Bowman's capsule and glomerulus showing the basement membranes (arrowed BM; periodic acid Schiff [PAS], original magnification ×60).

Preparations courtesy of Anne Rayner, photographs courtesy of Dr Alberto Quaglia.

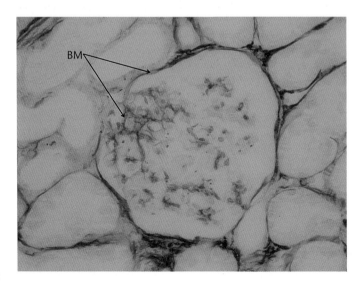

FIGURE 12.22

Bowman's capsule, glomerulus and surrounding tubules showing the basement membranes (Jones methenamine silver, original magnification ×60).

Preparations courtesy of Anne Rayner, photographs courtesy of Dr Alberto Quaglia.

CLINICAL CORRELATION

Key clinical terms in urinary system disease

- *Nephrosis*: non-inflammatory disease of the kidney
- *Nephrolith*: kidney stone
- *Urethritis*: inflammation of the urethra
- *Nocturia*: frequent urination, usually throughout the night
- *Enuresis*: involuntary release of urine, often applied specifically, although not exclusively, to night time bedwetting in children.

Some of these clinical signs may be congenital or acquired.

12.9.4 Kidney disease

Broadly speaking, kidney disease involves damage or some form of attack on the nephron, the functional unit of the kidney. In chronic kidney disease the causative factors may include reaction to prescribed drugs (i.e. iatrogenic change), genetic defects that cause increased susceptibility to kidney problems (e.g. development of autoimmune conditions), or trauma causing a direct injury to the kidney. In terms of acute problems, causes can include (i) cancer, (ii) stones, (iii) cysts, or (iv) infection.

Autoimmune disease of the kidney can have various cell targets, including those within the glomerulus, tubular structures, or even the vasculature. Characteristics of the autoimmune process involve inflammatory changes that develop as a result of the presence of circulating autoantigens that become deposited as immune complexes at specific sites, blocking the normal functional activities of the kidney. Identifying which cell types are involved, which immune complexes are formed, and where they are deposited requires the use of a wide range of laboratory-based investigative procedures. This is a highly specialized area of pathology, but some examples of these diseases include:

- *Antiglomerular basement membrane disease (anti-GBM)*: As the name implies, this disease involves the presence of autoantibodies against the glomerular basement membrane and is clinically defined by a rapidly progressive glomerulonephritis.
- *Lupus*: This relates to the multi-system involvement of the condition systemic lupus erythematosus (SLE). Nearly all patients with SLE will develop renal problems, and renal involvement and subsequent nephritis may progress to end-stage renal disease and eventual death. Several autoantibodies are involved in SLE, including antinuclear antibodies (ANAs) to double-stranded DNA. In investigative terms, antibodies to C1q and nucleosomes are very useful.
- *Vasculitis and glomerulonephritis*: This affects the blood vessels, with immune complexes forming within the vessel wall. It can cause rapid degeneration of kidney function, leading to glomerulonephritis and fibrinoid necrosis. Autoantibodies to neutrophil cytoplasmic antigens (ANCA) are associated with this condition.

12.9.5 Tumours

There are a number of tumours that can arise in the kidney, the most common being renal cell carcinoma (RCC) and urothelial cell carcinoma (transitional cell carcinoma [TCC]) of the renal pelvis. Less common tumours include:

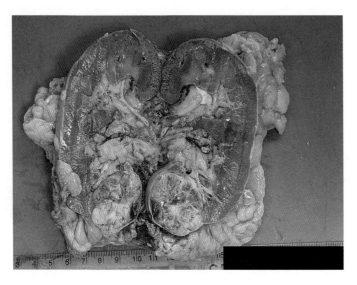

FIGURE 12.23
Renal cell carcinoma in the lower pole of the kidney.

- Squamous cell carcinoma
- Juxtaglomerular cell tumour (reninoma)
- Angiomyolipoma
- Renal oncocytoma
- Bellini duct carcinoma
- Clear-cell sarcoma
- Wilms tumour (nephroblastoma; usually a tumour in young children)
- Mesoblastic nephroma
- Mixed epithelial stromal tumour.

In the bladder, transitional cell carcinoma (TCC) is the most common type of cancer, but it can also affect the kidney and associated organs. It accounts for around five to ten per cent of primary malignant renal tumours and arises from the transitional epithelium. There is a close association with environmental carcinogenic factors as causative agents, including smoking and exposure to petroleum, paint and pigments. Similarly, those individuals who experience long periods of normal urine retention, such as long-haul lorry drivers, have an increased risk of developing TCC.

Worldwide, there are over 200 000 cases of kidney cancer (Figure 12.23) diagnosed each year, and it accounts for approximately two per cent of all cancers. In addition to the increased risk seen in smokers, use of non-steroidal anti-inflammatory drugs (NSAIDs) such as ibuprofen has also been linked to increased risks. General genetic links, obesity and hepatitis C infection have also been linked to an increased risk.

12.9.6 Stones, cysts and infection

Kidney stones are derived from minerals in the diet that are excreted in urine. Stone formation often occurs as a result of low fluid intake combined with high intake of animal protein,

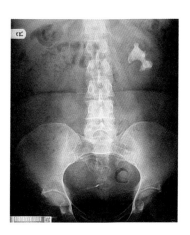

FIGURE 12.24
X-ray of a staghorn calculus in the pelvis of the kidney.

sodium and refined sugars. The most common component of a kidney stone is calcium oxalate; however, sodium may also contribute and fluorination of the drinking water supply has also been implicated. Stones can lodge at various points within the kidney (Figure 12.24), in the ureter and also in the neck of the bladder. In many cases stones pass unnoticed in urine but when they reach 3 mm or larger in size they can cause obstruction. Passing such stones can cause renal colic (extreme pain) and subsequent associated haematuria.

Kidney cysts are extremely common and may be found in 50 per cent of people aged over 50 years. Generally, cysts are fluid-filled sacs but some may contain air or calcium deposits. They are classified as either simple or complex, with the former being more common, and are regarded as benign and mostly asymptomatic. Such cysts remain undetected in most people, only being identified following a conventional ultrasound scan for an unrelated investigation. However, these cysts can increase in size, cause an obstruction, result in infection, hypertension or haematuria, and they can rupture. Cysts in patients who have the genetic abnormality autosomal polycystic kidney disease can develop complications that can lead to renal failure. Complex cysts can progress to cancer and therefore they require close follow-up.

Urinary tract infections are classified as either simple **cystitis** (bladder infection) when it affects the lower urinary tract, or **pyelonephritis** (kidney infection) when it affects the upper urinary tract. The main causative agent is usually the bacterium *Escherichia coli*, although viruses and fungi can also cause infection, albeit rarely. They are more common in women than in men, and recurrent infection is not uncommon. Risk factors include sexual intercourse and a predisposition to such infections within the family. Pyelonephritis normally develops following a bladder infection or a bloodborne infection. Generally, these infections are treated with antibiotics, although infection with antibiotic-resistant microorganisms is increasingly common.

SELF-CHECK 12.5

What are the main types of pathology that can affect the kidney?

12.10 **Kidney transplantation**

Kidney transplantation involves the transfer of a living kidney from a deceased or a living donor and is performed on a patient with end-stage renal disease as an alternative to long-term renal dialysis. It is a complex surgical procedure that usually results in the transplanted kidney being placed close to the bladder in the pelvic area. Transplantation of any donated organ usually requires a close match between the donor and recipient, defined either by a genetic link involving another family member or a non-related person who shows good compatibility in terms of ABO blood group and human leucocyte antigen (HLA) type. The key issue here is an attempt to reduce the risk of kidney rejection by ensuring that compatibility factors are as close a match as possible. However, most kidney transplants only succeed when some degree of immunosuppression is provided in order to reduce the risk that the recipient's immune system will recognize it as foreign and attempt to reject the organ. Thus, for those people receiving a donated kidney, there are associated life-long implications.

 Chapter summary

In this chapter we have:

- Described the structure and functions of the kidney and the associated urinary system.

- Discussed the key features of the cells and structures that comprise the kidney, with specific emphasis on the nephron.

- Considered the importance of the kidney in the body's homeostatic functions, including the following:

  - cleaning the blood

  - maintaining a mineral balance and ensuring essential elements and minerals are resorbed into the body

  - regulating blood composition and pH balance

  - blood pressure control

  - assistance in the process of red blood cell production

  - a role in maintaining healthy bones.

- Considered the differential effects of normal and abnormal events that can occur in the kidney.

- Introduced a range of kidney disease processes.

 Further reading

- Bancroft J, Stevens A. *Theory and practice of histological techniques* 3rd edn. Edinburgh: Churchill Livingstone, 1990.

- Mills SE ed. *Histology for pathologists* 3rd edn. Philadelphia: Lippincott Williams and Wilkins, 2007.

- Moore G, Knight G, Blann A. *Haematology*. Oxford: Oxford University Press, 2010.

- Orchard GE. Immunocytochemical techniques and advances in dermatopathology. *Curr Diagn Pathol* 2006; **12** (4): 292–302.

- Orchard GE, Nation BR eds. *Histopathology*. Oxford: Oxford University Press, 2012.

- Pavenstädt H, Kriz W, Kretzler M. Cell biology of the human podocyte. *Physiol Rev* 2003; **83** (1) 253–307.

- Shambayati B ed. *Cytopathology*. Oxford: Oxford University Press, 2011.

- Young B, Lowe JS, Stevens A, Heath JW. *Wheater's functional histology: a text and colour atlas* 5th edn. Edinburgh: Churchill Livingstone, 2006.

Useful websites

- Distal convoluted tubule (https://en.wikipedia.org/wiki/Distal_convoluted_tubule). Juxtaglomerularapparatus(http://medcell.med.yale.edu/histology/urinary_system_lab/juxtaglomerular_apparatus.php).

- Kidney (https://en.wikipedia.org/wiki/kidney).

- Loop of Henle (www.britannica.com/EBchecked/topic/347799/loop-of-Henle).

- Proximal convoluted tubule. (http://Faculty.stcc.edu/AandP/AP/AP2pages/Units24to26/urinary/PCT.htm).

 Discussion questions

12.1 Discuss the structural features of the kidney nephron and describe how its structure relates to its function.

12.2 Describe how homeostatic mechanisms of the kidney operate.

12.3 Discuss the structure of the glomerulus and how important is it to the integrity of the kidney as a functioning unit in health and disease.

Answers to self-check questions and discussion questions, hints and tips are provided on the book's Online Resource Centre.

 Visit www.oxfordtextbooks.co.uk/orc/orchard_csf/.

13

Reproductive cells and gametogenesis

Andrew Evered and Behdad Shambayati

Learning objectives

By the end of the chapter you should be able to:

- Outline the anatomy and physiology of the male and female reproductive systems.
- Explain the contribution made by the two types of cell division, mitosis and meiosis, to the process of gametogenesis.
- Describe the similarities and differences between germ cells and other cells of the body.
- Explain how the cells of the testes and associated duct system support spermatogenesis.
- Explain how the cells of the ovary and associated ducts and cavities support oogenesis.
- Explain how the structure of spermatozoa and ova facilitates reproduction.
- Describe the structure and function of tissues that comprise the external genitalia.

Two things distinguish this chapter from all others in this book. The first and most obvious point to make is that the male and female reproductive systems are anatomically different and to some extent need to be considered separately. Second, the reproductive system is unique in possessing a population of cells that contain half the amount of genetic material (i.e. 23 chromosomes) normally found in other cells of the body. These germ cells, or gametes as they are sometimes called, are created by a process known as gametogenesis and give humans their reproductive capacity. We will find out about this fascinating and reproductively vital process early in the chapter.

Although the differences between males and females are clear for all to see, one of the aims of this chapter is to draw attention to the many similarities in structure and function shared by cells of the male and female reproductive systems. Perhaps inconveniently for us, cells sharing the same function are given different names in the two sexes. Therefore, to keep any confusion to a minimum and to eliminate unnecessary duplication, we will discuss matters generally, separating the sexes only when necessary.

The chapter is organized as follows. We shall begin with an overview of the anatomy and physiology of the reproductive system, but this part will be kept quite brief because the main aim of the book is to focus on cells rather than organ systems. Next, we examine the process of gametogenesis in general terms without attempting to make any distinction between males and females. This will serve to embed the essential function of the reproductive system firmly in our minds. We will then turn our attention to the specifics of male and female gametogenesis, which will make it clear that, despite the similarities highlighted in the previous section, there are some important and surprising differences between the sexes. This section will also help us get to grips with the terminology that is used to name the cells and cellular processes in the two reproductive systems. Finally, we describe the cells and tissues that make up the reproductive tract. Far from being an extraneous tissue, the reproductive tract exists to support gametogenesis, promote the union of gametes from each sex, and facilitate the subsequent development of a new human being. By the end of the chapter you should have a full understanding of the structure and function of the reproductive system and its component cells.

13.1 The reproductive system— an overview

In simple terms, the purpose of the reproductive system is sexual reproduction and survival of the species. In carrying out such an important function, the reproductive system does not work in isolation but communicates with the **hypothalamus** and **anterior pituitary** via complex hormonal mechanisms. The reproductive system in humans consists of the reproductive organs, or **gonads**, and various ducts, glands and cavities (the **reproductive tract**) that support the gonads in fulfilling their sexual and reproductive functions. In adults, these are located within the pelvic cavity in females and suspended outside the abdominal cavity in males. In both sexes, the gonads perform the dual function of gametogenesis and the secretion of sex hormones.

SELF-CHECK 13.1

What is a hormone?

> **Key points**
>
> Reproductive system = reproduction organs + reproductive tract.
>
> That's all the maths you will see in this chapter!

Figures 13.1 and 13.2 are simple illustrations of the male and female reproductive systems, which you might need for navigational purposes as you read the detail in the rest of this chapter.

The gonads comprise a pair of **testes** (singular **testis**) suspended outside the abdominal cavity in males and a pair of **ovaries** lying within the pelvic cavity in females. It is in the testes that the process of gametogenesis produces male gametes or **spermatozoa** (**sperm**), and the same process in the ovaries produces **ova** (singular **ovum**). Gametogenesis is often referred to as **spermatogenesis** in males and **oogenesis** in females. Gametogenesis is supported by sex hormone; **testosterone** in males and **oestrogen** and **progesterone** in females. Each of the sex hormones is produced by specialized cells within the gonads.

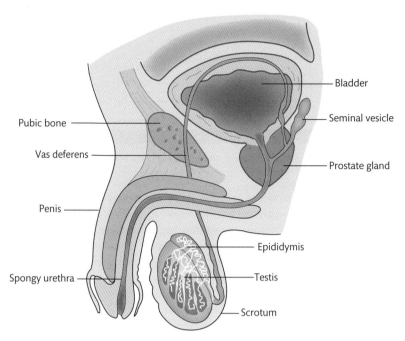

FIGURE 13.1
Lateral view of the male reproductive system.

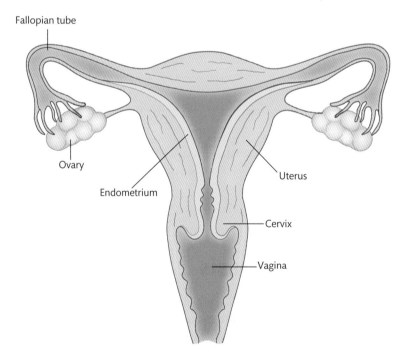

FIGURE 13.2
Front view of the female reproductive system.

Key points

The testes produce spermatozoa by spermatogenesis.

The ovaries produce ova by oogenesis.

SELF-CHECK 13.2

Can you think why, in physiological terms, the testes are outside the abdominal cavity?

Anatomically, the male and female reproductive tracts appear quite different, but in both sexes the tract is responsible for transporting gametes from the gonads to a common site for union. In males, a system of ducts comprises the **epididymis**, **vas deferens**, **ejaculatory duct** and **urethra**, complemented by a suite of **accessory sex glands** comprising the **seminal vesicles**, **prostate gland** and **bulbourethral glands**. Collectively, the male accessory sex glands secrete fluid called **semen** into the duct system during expulsion of sperm (**ejaculation**) that occurs during **sexual intercourse**. Semen is a specialized fluid containing nutrients and other substances essential for the survival and motility of sperm during their delivery to the female. In contrast to the male duct system, the major elements of the female reproductive tract are the **Fallopian tubes**, **uterus**, **cervix** and **vagina**. The Fallopian tubes serve as delivery channels transporting ova from the ovaries to the uterus. The Fallopian tubes are the site where sperm and ovum meet and where **fertilization**, or **conception**, takes place to produce an **embryo**. Subsequently, the embryo implants into the wall of the uterus, which is a thick-walled, hollow vessel responsible for maintaining the embryo, helping it to develop into a **fetus**. Ultimately, after nine months of growth and development (**gestation**), the fully formed fetus is expelled from the uterus by means of muscular contractions of the uterine wall, through the cervix and vagina, into the outside world.

Key points

The reproductive tract exists to unite spermatozoa and ova.

The exterior parts of the reproductive system, the external genitalia, comprise the **penis** and **scrotum** in males and the **clitoris** and **labia** in females. The penis is the organ used to deposit sperm in the female reproductive system during sexual intercourse.

Armed with this very brief summary of reproductive anatomy and physiology, we are now in a position to explore the relevant cells, tissues and processes in more detail.

13.2 Gametogenesis

The process by which gametes are formed is quite different to the way in which cells from other parts of the body (**somatic cells**) grow and divide. Somatic cells are produced by **mitosis**, a type of cell division that conserves the integrity and quantity of genetic material as it divides. By genetic material we mean the DNA molecules that carry the cell's genetic code. This genetic material is not dispersed haphazardly throughout the cell but is carefully arranged in supercoiled, discrete chromosomes within the nucleus of the cell. Normal somatic cells contain 46 chromosomes, 23 of which are derived from the father and 23 from the mother. A cell that contains a full complement of 46 chromosomes is referred to as **diploid**. On the other hand, germ cells are unique in their ability to undergo a specialized form of cell division called **meiosis**. During meiosis, an initial diploid germ cell divides in a way that produces **haploid** daughter cells. In humans this means they contain only 23 chromosomes. In males these haploid progeny are the spermatozoa and in females they are the ova. We will talk more about spermatogenesis and oogenesis in the next two sections of this chapter, where you will discover several important differences between males and females with regard to the number and frequency of gamete production and the time course over which the process takes place. For now we will concentrate on the basic process of gametogenesis.

Key points

Mitosis produces diploid cells. Meiosis produces haploid cells.

Gametogenesis starts with specialized diploid germ cells undergoing a number of mitotic cell divisions in order to produce a population of identical germs cells. These provide the source of future gametes. Figure 13.3 is a simple illustration of the process. The important point to make is that the two daughter cells are identical to the cell from which they arose and that each mitotic division doubles the number of cells that are available for the second phase of gametogenesis.

Chromatids
These are the individual strands of DNA making up a duplicated chromosome.

The second phase of gametogenesis involves two successive meiotic cell divisions to produce haploid gametes. Figure 13.4 illustrates the process.

Here we see that the chromosomes behave differently to the way they behaved at the start of mitosis. Instead of being dispersed randomly throughout the nucleus, chromosomes inherited

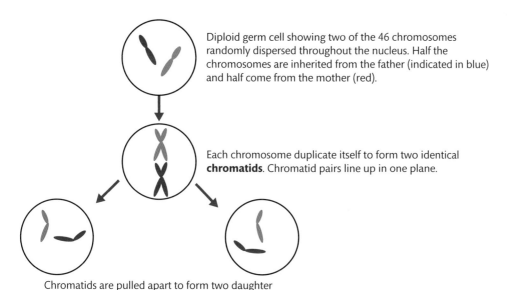

Diploid germ cell showing two of the 46 chromosomes randomly dispersed throughout the nucleus. Half the chromosomes are inherited from the father (indicated in blue) and half come from the mother (red).

Each chromosome duplicate itself to form two identical **chromatids**. Chromatid pairs line up in one plane.

Chromatids are pulled apart to form two daughter germ cells that are identical to the parent germ cell.

FIGURE 13.3
The first phase of gametogenesis requires diploid germ cells to undergo a number of mitotic cell divisions.

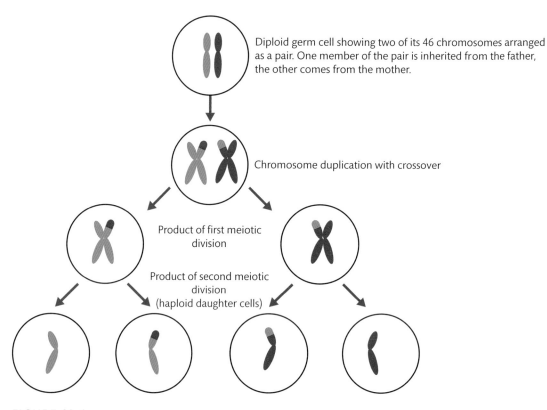

Diploid germ cell showing two of its 46 chromosomes arranged as a pair. One member of the pair is inherited from the father, the other comes from the mother.

Chromosome duplication with crossover

Product of first meiotic division

Product of second meiotic division (haploid daughter cells)

FIGURE 13.4

Two successive meiotic divisions produce haploid gametes.

from the father pair up with their counterpart chromosomes derived from the mother. Each chromosome then duplicates, so for a brief period the germ cell will contain 92 chromosomes, arranged as two sets of 23 pairs. During DNA replication, chromosomal **crossover** occurs, whereby portions of the maternal chromosome are transferred to the paternal chromosome, and vice versa. Crossover represents an opportunity to produce a unique mixture of genetic material in the daughter cells, thus permitting genetic diversity which partly accounts for the huge variation in the physical and behavioural attributes we see in people around the world. Chromosomal replication is soon followed by two successive divisions of the nucleus so that each of the four daughter cells contains only 23 unpaired chromosomes. These daughter cells are the gametes. The unpaired chromosomes in a male gamete ultimately combine and match up with corresponding unpaired chromosomes in a female gamete during fertilization. The end result is a new diploid cell known as a **zygote**, which, by the process of mitosis, becomes the embryo and later the fetus. It should now be clear to you that each cell of the developing embryo contains 23 pairs of chromosomes. One member of each pair is derived from the father and the other from the mother. This mix of genetic material, together with the chromosomal crossover mentioned earlier, is what gives humans (and other living things) their diversity.

SELF-CHECK 13.3

What are the differences (if any) between a gamete, a zygote and an embryo?

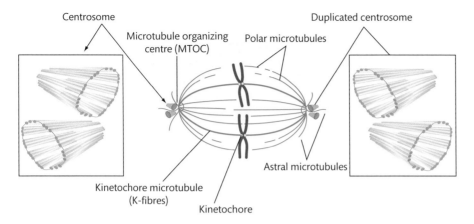

FIGURE 13.5

Two centrioles, each consisting of nine microtubule triplets (boxed). These are arranged at right angles to each other to form the centrosome. The two centrioles are held together by a mass of unknown protein. The microtubule structure of the spindle apparatus, which becomes active during mitosis and meiosis, is shown in the centre. Diagram not drawn to scale.

Now that you have had a chance to grasp the essence of mitosis and meiosis, we need to concern ourselves with the mechanism by which the chromosomes are moved around during this process. Why should we be so interested in this apparently trivial matter? The answer is pretty amazing. All cells contain a microscopic framework of proteinaceous tubules, known as **microtubules**, which are responsible not only for chromosome movement during cell division but also for the motility of fully developed sperm as well as various other processes requiring the movement of cellular components. We shall return to the issue of sperm motility later, but for now let's take a close look at how microtubules act as cellular 'chauffeurs'. The microtubule framework responsible for chromosome segregation during cell division is referred to as the **spindle apparatus**. In this context, microtubules are also sometimes called **spindle fibres**. Throughout the remainder of this section, it is worth noting that the exact mechanism of spindle formation and associated microtubule activity is not entirely clear.

In human cells (and in all **eukaryotic cells** for that matter), microtubules are formed from the **centrosome**, an **organelle** found close to the nuclear membrane. Only one centrosome is present in each cell. The centrosome is composed of two **centrioles**, each of which is composed of nine triplets of microtubules. Look at Figure 13.5, which illustrates the origin and main types of microtubules that make up the spindle apparatus.

Just before the onset of mitosis or meiosis, the centrosome duplicates itself. The duplicated centrosome becomes the **microtubule organizing centre** (MTOC), from which various types of microtubules begin to emanate. **Polar microtubules** grow out from each centrosome towards the centre of the cell and overlap each other. They do not actually have any role to play in chromosome separation, rather their purpose is to push the two centrosomes apart to opposite ends of the cell. The shortest of all spindle fibres are the **astral microtubules** and serve to anchor the centrosomes to the cell membrane. The final spindle fibres to consider are the **kinetochore microtubules**. These extend from the centrosomes and seek out and bind to kinetochores, which are protein structures that form at the **centromeres** of chromosomes at around the same time as centrosome replication. The purpose of the kinetochore microtubules is to move chromosomes to the centre of the cell and align them in a single plane. Look back at Figures 13.3 and 13.4 and you should be able to see why it is so important to arrange the chromosomes in this way before cell division. What happens next is truly stunning, as

Microtubules

These are hollow proteinaceous 'rods' that permit movement of various kinds within eukaryotic cells.

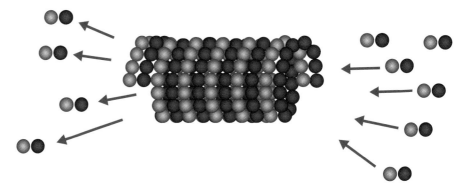

FIGURE 13.6
Molecular structure of microtubules.

anyone who has witnessed time-lapse photography of chromosome segregation during cell division will confirm. Put simply, when all chromosomes are aligned correctly in the centre of the cell, the kinetochore microtubules shorten and pull chromosomes apart, dividing the genetic material of the cell in two.

SELF-CHECK 13.4

Why do you think microtubules are sometimes referred to as the 'cytoskeleton'?

Now, if you were paying attention in the last paragraph, you will be asking an important question: how do microtubules manage to 'push' and 'pull' cellular components around in the way they do? The answer lies in the molecular structure of microtubules, a simple illustration of which is shown in Figure 13.6.

Microtubules consist of spiral arrays of the protein **tubulin** that can lengthen or shorten by the addition (i.e. polymerization) or removal (i.e. depolymerization) of tubulin dimers. Various linker proteins located at the growing ends of microtubules facilitate the attachment of spindle fibres to other cellular components, such as the cell membrane or the chromosome kinetochores, although the exact mechanisms involved remain unclear. In a similar vein, destabilizing proteins at the shortening ends of microtubules are able to trigger the segregation of chromosomes at the appropriate time during cell division.

CLINICAL CORRELATION

Down's syndrome

To illustrate the importance of correct chromosomal segregation during cell division, let's take a brief look at a condition affecting approximately one baby in every 1000 born in the UK. Down's syndrome is characterized by physical disability and mental retardation. The risk of the condition increases with the age at which a woman conceives; above the age of 45 the risk is as high as one birth in 19. Although variants of the disorder exist, the essential condition is caused by failure of chromosome 21 to separate during meiosis, resulting in ova or sperm containing an extra copy of the chromosome (see Figure 13.7). When the defective gamete unites with a normal gamete from the opposite sex, the result is the birth of a child with 'trisomy 21' (i.e. each body cell carries three copies of chromosome 21 instead of the usual two copies). The clinical manifestations of the disorder are related to the extra copies of genes located on this chromosome. A hundred years ago individuals born with Down's syndrome lived for an average of 16 years, but with modern medical care a lifespan of 60 years and beyond is not at all uncommon.

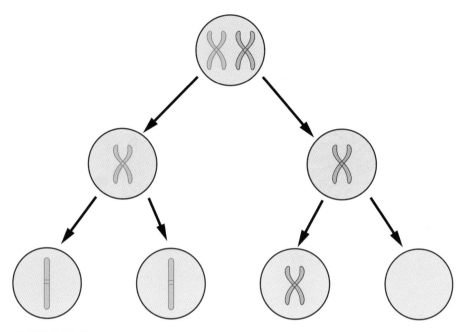

FIGURE 13.7
When chromosome 21 fails to separate during meiosis, a resulting ovum or sperm will contain an extra copy of the chromosome.

After this short digression let us now return to the central theme of the chapter and consider the details of gametogenesis in each of the sexes.

SELF-CHECK 13.5

Can you describe three similarities and three differences between mitosis and meiosis?

13.3 **Testes and spermatogenesis**

The testes are located within a sac of skin (the scrotum) lying outside the abdominal cavity. Each testis is encapsulated by a continuous double membrane forming a fluid-filled cavity called the **tunica vaginalis**. This cavity is lined by **mesothelial cells** and filled with a small volume of serous fluid, acting as a lubricant to permit free movement of the testes within the scrotum (if this description looks familiar then it seems you have already read chapters 6, 7 and 9 describing similar cavities surrounding the lungs, heart and abdominal organs). Spermatogenesis takes place in the seminiferous tubules of the adult testes. Each testis is packed with over 250 metres of seminiferous tubules—that's over half a lap of an athletics track. This provides a massive surface area over which several hundred million sperm are produced daily. The seminiferous tubules are not just haphazardly squashed into each testis but are organized into a series of **lobules**, as illustrated in Figure 13.8.

Lobules are separated from each other by sheets of collagenous **septae**. There are about 250 lobules in each testis, each one converging on a network of channels called the **rete testis**.

Mesothelial cells

These cells are the lining cells of the serous cavities, which, in addition to the tunica vaginalis, include the pleural cavity (lining the lungs), the pericardial cavity (around the heart) and the peritoneal cavity (enveloping the abdominal organs).

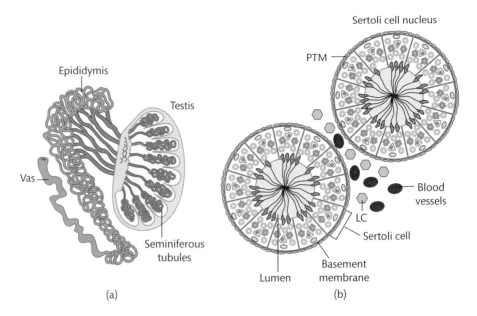

Epididymis

Testis

Vas

Seminiferous
tubules

Lumen

(a)

Sertoli cell nucleus

PTM

Blood
vessels

LC

Sertoli cell

Basement
membrane

(b)

FIGURE 13.8
Cross-section through the testis and seminiferous tubules.
Modified from original image: Cooke and Saunders. Mouse models of male infertility. *Nat Rev Genet* 2002; **3**: 790-801. © Nature Publishing Group.

This complicated system of tubes forms the channels through which spermatozoa begin their journey to the outside world. We will stay with the seminiferous tubules for now and return to the male reproductive duct system in Section 13.5.

Figure 13.8 shows that the seminiferous tubules are very tightly packed inside the testes, and it illustrates a cross-section through one of the tubules. Figure 13.9 shows the intimate relationship between the developing sperm cells and the supporting cells of the seminiferous tubules, the **Sertoli cells**.

Spermatogenesis begins with mitotic proliferation of the undifferentiated diploid germ cells, the **spermatogonia**, in the outermost layer of the seminiferous tubule. At each division, one daughter cell remains in the outer layer to maintain the germ cell line, while the other daughter cell moves towards the lumen of the tubule. During its journey it undergoes two more mitotic divisions to produce four identical **primary spermatocytes**. Each primary spermatocyte then undergoes two consecutive meiotic divisions to produce **secondary spermatocytes** followed by haploid **spermatids**. You will know from reading Section 13.2 that the spermatogonia and their mitotic offspring retain the diploid number of chromosomes and that the spermatids are haploid. You should also be able to work out the number of spermatids produced from each spermatogonium (the answer is 16). Cell division stops at this stage and the spermatids differentiate into the highly specialized mobile spermatozoa, which are held briefly on the luminal surface of the seminiferous tubule before passing into the lumen for storage. The entire development cycle, from spermatogonium to fully mature spermatozoa, takes approximately 64 days to complete.

Key points
Spermatozoa are haploid germ cells produced in the seminiferous tubules of the testis.

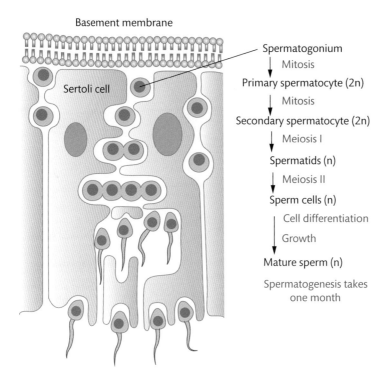

Basement membrane

Sertoli cell

Spermatogonium
↓ Mitosis
Primary spermatocyte (2n)
↓ Mitosis
Secondary spermatocyte (2n)
↓ Meiosis I
Spermatids (n)
↓ Meiosis II
Sperm cells (n)
| Cell differentiation
| Growth
Mature sperm (n)

Spermatogenesis takes
one month

FIGURE 13.9
Developing sperm cells move between Sertoli cells from the basement membrane to the lumen of the seminiferous tubule.

Spermatozoa are incredibly specialized cells and look nothing like the cells from which they originate. Their sole purpose is to deliver a haploid set of chromosomes to the ovum, so it makes sense that a critical ability of sperm cells is their motility. But how do sperm acquire this ability? The answer is that all the organelles within spermatids that are not needed for spermatozoa to carry out their primary function are literally 'stripped away'. The organelles that remain are adapted to leave just three parts: a head, a midpiece and a tail, as shown in Figure 13.10.

The head contains a nucleus consisting of the sperm's genetic material, and an acrosome, which is an enzyme-filled sac at the very tip of the head. The acrosome is derived from remnants of the **Golgi apparatus** and **endoplasmic reticulum** of spermatids and carries out the all-important task of enzymatically 'drilling' through the wall of the ovum to facilitate fertilization. The energy required for the sperm's mobility is generated by numerous mitochondria, which are housed in the midpiece. The tail is quite fascinating. Its internal structure is shown in Figure 13.11, which hints at its motile capabilities.

You will recall that we touched on this topic in discussion of cell division in section 13.2. In fact, you will notice several similarities between the structure and function of the mitotic/meiotic spindle apparatus and that of a sperm tail. Why are they so similar? The answer is that they are derived from the very same organelle, the centriole. To be precise, the sperm's tail is formed from the **basal body**, which itself is derived from a centriole located in the cytoplasm of the spermatid. The basal body is structurally similar to a centriole, consisting of nine microtubule triplets (sometimes called the '9x3' arrangement—refer to Figure 13.5 if you need a reminder)

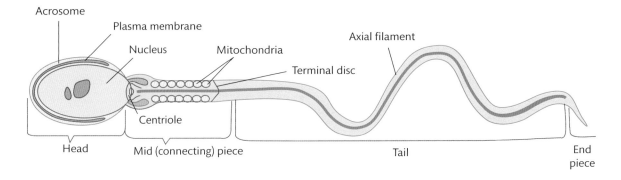

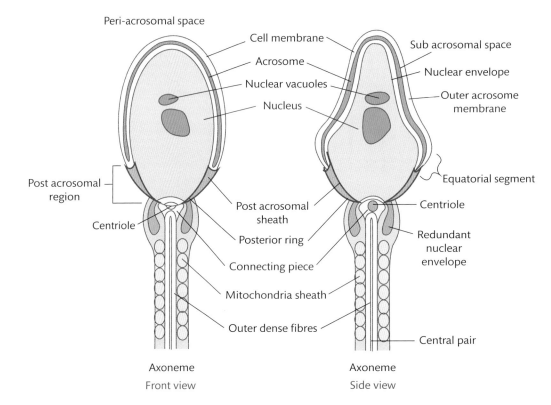

FIGURE 13.10
Structure of spermatozoa.

forming a small cylindrical structure at the base of the sperm tail. However, the microtubules in the tail itself consist of nine fused pairs of microtubules arranged in a ring surrounding a central unfused pair of microtubules, giving the so called '9+2' structure. This ring of microtubules in the tail of the sperm is called the **axoneme**. The arrangement is perhaps best illustrated diagrammatically in Figure 13.12. In simplistic terms, the basal body provides a base from which microtubules grow to form the axoneme. Quite how the 9×3 structure of the basal body transforms into the 9+2 structure of the axoneme is something of a mystery.

We must now ask a similar question to that posed in section 13.2 when we were discussing the motile capabilities of the spindle apparatus. How can we explain sperm motility? The answer

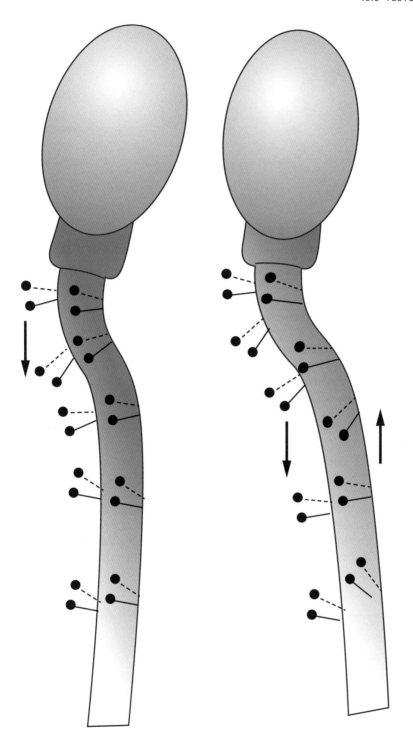

FIGURE 13.11

The sperm tail comprises nine pairs of microtubules (only two of which are shown here) arranged longitudinally. With the help of an accessory 'motor protein' called dynein, microtubule doublets are able to slide past each other rapidly, giving the tail its whip-like motility.

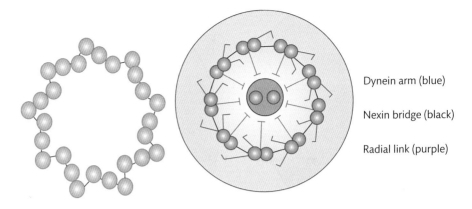

Dynein arm (blue)

Nexin bridge (black)

Radial link (purple)

FIGURE 13.12
9x3 structure of the basal body and centriole (left) and 9+2 structure of the axoneme (right).

lies in a protein called **dynein**, a 'motor protein' attached to the outer microtubule doublets at regular intervals along the length of the axoneme. Another protein called **nexin** connects adjacent microtubule doublets, and yet another protein links the outer doublets to the central pair. The dynein arms have ATPase activity, which means that when ATP is abundant, energy is released to enable the dynein arms of one microtubule doublet to uncouple from an adjacent doublet and rejoin at a point further along the axoneme. In this way, the outer ring of micro-tubules is able to slide past each other, giving the sperm tail its whip-like motility. Of course, all this uncoupling and recoupling of dynein molecules requires careful regulation if the tail is to provide sensible (i.e. unidirectional) propulsion. This is where the nexin bridges and radial links come in. Essentially, these are anchoring proteins that restrict the extent to which micro-tubule doublets are able to slide past one another. To emphasize the importance of these elements, simple laboratory experiments that destroy these bridges and links can result in extreme elongation and malfunction of the axoneme. Incidentally, the relative displacement of paired microtubules is the same mechanism that gives **cilia** their motility in areas such as the respiratory tract and Fallopian tubes. What an incredible example of evolutionary engineering. We could stay with this fascinating topic all day, but alas we must continue with the exploration of spermatogenesis.

Key points

Sperm motility is essential for reproduction.

CLINICAL CORRELATION

Semen analysis

Checking for sperm motility is part of a suite of tests carried out in cases of suspected male infer-tility and for checking the success of male sterilization procedures. Potential sperm donors may also be tested in this way. Other tests on semen might include a sperm count, sperm morphology and measurements of pH and fructose levels.

The sperm's tale is sometimes described as a **'flagellum'**. What are the differences and similarities between **flagella** and cilia?

There two types of cells within the testes that we have not yet discussed but that play a crucial supportive role in spermatogenesis. These are the Sertoli cells and the **Leydig cells**. We touched on Sertoli cells in Figure 13.9. These cells form a ring that extends from the outer basement membrane of the seminiferous tubule to its lumen. Although Sertoli cells are held together firmly by **tight junctions**, the developing sperm cells are quite literally tucked between them. As spermatogenesis proceeds, the tight junctions transiently separate to permit the passage of the sperm cells towards the lumen. As well as providing physical support, Sertoli cells perform several other important functions that maximize the efficiency of spermatogenesis. First, they form an effective **blood–testis barrier**, which essentially filters fluid as it passes from the circulatory system and the tissue spaces of the testis to the lumen of the seminiferous tubules. The filtration process creates a unique fluid containing substances critical for sperm development, but excludes potentially harmful molecules such as antibodies that might otherwise disrupt spermatogenesis. Second, Sertoli cells secrete their own fluid which not only nourishes the developing spermatocytes but also helps to flush fully developed sperm into the lumen of the seminiferous tubule. One of the substances produced by Sertoli cells is **androgen-binding protein**, which, as its name suggests, binds to testosterone and helps to maintain a high local concentration of the hormone. High levels of testosterone are important for sperm development and general maintenance of the reproductive tract. Third, Sertoli cells are sensitive to the hormones testosterone and **follicle-stimulating hormone** (FSH), both of which promote and enhance spermatogenesis. Fourth, Sertoli cells possess phagocytic activity that is important in removing the cytoplasmic fragments shed from developing sperm cells during their maturation. Finally, Sertoli cells secrete their own hormone, called **inhibin**, which serves to regulate the secretion of FSH from the anterior pituitary through a **negative feedback** mechanism.

What is meant by the term 'blood–testis barrier'?

The final piece of this complicated physiological jigsaw is to discuss the source of that all-important male sex hormone, testosterone. This critical sex steroid comes from the Leydig cells, which nestle in the spaces between adjacent seminiferous tubules. By secreting testosterone, they play an indirect but nonetheless crucial role in spermatogenesis. Testosterone is the hormone responsible for giving men their masculine appearance—facial and body hair, enlargement of the larynx, increased muscle mass and so forth. The hormone also facilitates spermatogenesis by stimulating maturation of the spermatids and removal of the last remaining unnecessary organelles and cytoplasm.

Key points

Sertoli cells and Leydig cells provide physical, nutritional and hormonal support for the developing spermatozoa.

It is important to note that spermatogenesis does not result in fully functional spermatozoa; in fact, although they have fully developed tails, they are non-motile at this stage of their development. It is only following their transportation to the epididymis via peristaltic contractions of the seminiferous tubules that they finally gain their ability to swim and therefore fertilize.

We will pause at this stage but shall return to discuss the remaining cells of the male reproductive system after we have considered gametogenesis in the female.

SELF-CHECK 13.8

What are the constituents of semen?

13.4 **Ovaries and oogenesis**

Despite their similarities in terms of the cell division process, you are about to discover a number of important and quite startling differences between gamete production in females (oogenesis) and spermatogenesis. For instance, it might surprise you to learn that the 64-day timescale for the production of mature spermatozoa from spermatogonia is dwarfed by the 11 to 50 years it takes for a mature ovum to develop from its corresponding germ cell! Let's discover how this is so.

Key points

Ova are haploid germ cells produced in the ovaries.

Refer to Figures 13.13 and 13.14 as you read the following text.

During the first few months of fetal life the early ovaries begin to accumulate germ cells (**oogonia**) through mitotic proliferation of the epithelial surface of the ovary. By the fifth

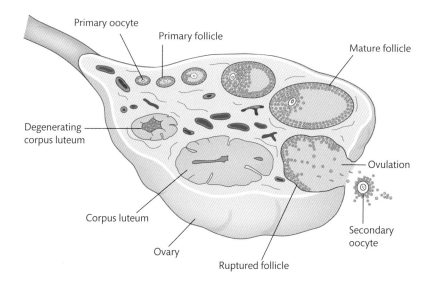

FIGURE 13.13
Oogenesis.
Modified from Purves *et al.*, Life: The Science of Biology © Sinauer Associates and WH Freeman, used with permission.

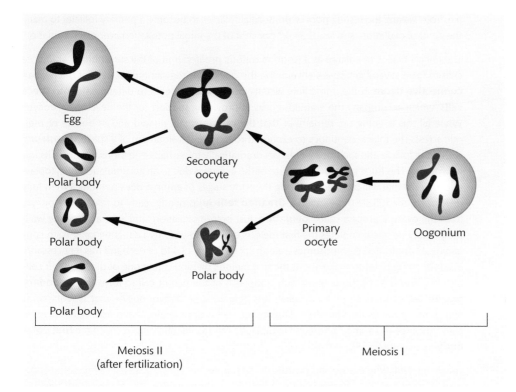

FIGURE 13.14
Illustration of the disproportionate sharing of cytoplasmic contents in daughter cells during oogenesis.

month of **gestation** the fetal ovaries contain up to seven million oogonia. Interestingly, this is the most they will ever contain because almost immediately after they are formed the oogonia begin to degenerate and die. By birth, the ovaries might contain around two million germ cells in the early stages of oogenesis. This number falls drastically to approximately 500 000 by puberty, and by a woman's late 40s only a few remain. The first meiotic division begins during the last part of fetal development, at which stage the oogonia become known as **primary oocytes**. However, this first meiotic division is not completed; pairs of replicated chromosomes are assembled but they do not separate. Instead, primary oocytes that do not degenerate and die remain in a kind of 'suspended animation' for many years, and some remain in this state of meiotic arrest for 50 years or more. At birth, surviving primary oocytes very soon develop a surrounding single layer of protective **granulosa cells** and together they form the **primary follicle**. No new follicles are produced and, of the two million or so primary follicles present at birth, only about 400 will ever develop to maturity.

Each day of a female's life until she exhausts the supply around the time of the menopause, several **primordial follicles** start to grow. Until puberty none of those follicles grow past the preantral stage. At puberty, with the correct endocrine support, follicular growth can be sustained further. If an antral follicle happens to be 2–5 mm in diameter at the time of luteolysis (i.e. end of one **menstrual cycle** and the beginning of the next), and there are usually 10–20 in this cohort, then it will be prompted to enter a rapid phase of growth and might be the one or two follicles that are selected to ovulate approximately 14 days later. It is estimated that it takes a follicle at least six months (perhaps even nine months) from the initiation of growth

Granulosa

These cells surround the oocyte during oogenesis and secrete the sex steroids oestrogen and progesterone during various phases of the menstrual cycle.

(i.e. from leaving the resting pool of primordial follicles to become a primary follicle) to reach the point of ovulation, and less than 0.1 per cent of the initial population ever reach ovulation.

The follicle begins to enlarge as a result of mitotic proliferation of the surrounding granulosa cells to form several cell layers enveloping the oocyte. At the same time, cells of the ovarian **connective tissue** in the immediate vicinity of the follicle begin to differentiate into **theca cells**, which accompany the granulosa cells in providing support for the developing oocyte. While all this is going on, remember that the oocyte is still diploid and in a state of meiotic arrest. The follicle continues to expand through the development of a space, or **antrum**, within the follicle and the oocyte enlarges dramatically to perhaps one thousand times its initial volume. This huge increase in the size of the oocyte is due to an accumulation of cytoplasmic material that will be needed during the early stages of embryo development. When fully developed, the follicle (now called a **Graafian follicle**) is nearly ready to release the oocyte from the ovary, a process called **ovulation**. Just before ovulation, and after a very long wait, the primary oocyte completes the first meiotic division to yield two daughter cells, each containing 23 duplicated chromosomes (refer back to Figure 13.4 to understand this if necessary). The interesting point to make about the first meiotic division is that one of the daughter cells receives almost all of the nutrient-rich cytoplasm of the parent cell to form the **secondary oocyte**. So, contrary to popular belief, it is actually a secondary oocyte that is released at ovulation, not an ovum. The other daughter cell soon degenerates into a non-viable mass of chromosomes and a small amount of cytoplasm. This degenerate mass is called the **first polar body** (see Figure 13.14).

Following ovulation, the second meiotic division only proceeds if a sperm cell penetrates the secondary oocyte. Just like the first meiotic division, the daughter cells of the second meiotic division receive hugely different quantities of cytoplasm from the parent cell. One daughter cell receives almost none at all and becomes the **second polar body**, which soon degenerates. The cell that receives most of the organelle- and nutrient-rich cytoplasm is now the ovum. The 23 chromosomes of the ovum then unite with the 23 chromosomes from the sperm cell to complete fertilization, producing a diploid zygote. The zygote, of course, is destined to become the embryo. Therefore, the ovum only exists during the very short time it takes for the two sets of chromosomes to unite following entry of the sperm into the secondary oocyte. Furthermore, if fertilization does not occur then the second meiotic division of the secondary oocyte is never initialized.

Theca
These cells surround the granulosa cells in two layers (theca interna and theca externa) and provide the hormone precursors required for the production of sex steroids.

Polar bodies
These are non-viable cells produced during meiosis.

SELF-CHECK 13.9

Explain why the formation of polar bodies is a physiological necessity.

At this point, it is worth pausing to reflect on the considerable amount of chromosomal wastage that occurs during oogenesis, and spermatogenesis for that matter. A single phase of mitotic proliferation in the fetal ovaries produces several million oogonia, of which only about 0.01 per cent ever reach maturity. If that sounds like a lot of effort for little return, the male gonads produce an equivalent number of viable sperm every day, the vast majority of which will never even come close to fertilizing an ovum.

CLINICAL CORRELATION

Germ cell tumours

Germ cells, just like any other cell type in the body, can become dysfunctional and form tumours. Germ cell tumours comprise 15-20 per cent of ovarian tumours and approximately 95 per cent of tumours of the testis. They can be benign (i.e. localized to the tissue of origin) or malignant (i.e. disseminated), and some forms have a very strange tissue composition indeed. For instance, benign cystic teratomas of the ovary, known more commonly as dermoid cysts, may contain hair, skin, teeth and even bone. The tumour probably arises from an ovum after the first meiotic division. Such tumours have been a source of fascination for centuries and at one time were blamed on witchcraft and sorcery.

So far in this chapter we have been able to discuss the male and female reproductive systems in fairly general terms, considering the specifics of each gender only when necessary. However, we have reached a point where cell structure and function of the male and female reproductive systems diverge to such an extent that they have to be considered separately. The next two sections deal with the reproductive tracts; that is the ducts, cavities and structures that make up the remainder of male and female reproductive systems.

13.5 Cells of the male reproductive tract

As promised at the end of section 13.3, we now return to consider the male reproductive duct system as the seminiferous tubules emerge from the upper pole of each testis. In this area, the seminiferous tubules converge to form the reti testis, which converges further to form a single duct called the epididymis. The body of the epididymis is approximately 5 cm long and sits alongside the posterior margin of each testis. Despite its relatively short length, the epididymis contains approximately 5 metres of highly convoluted duct. The duct serves several very important functions related to the storage, maturation and movement of spermatozoa. Examine Figure 13.15. Throughout its length, the epidydimis is lined by tall **columnar epithelium**, beneath which is a narrow rim of smooth muscle that gradually thickens towards its distal end. The columnar cells are held together by tight junctions, the purpose of which is to retain an effective blood–sperm barrier.

Key points

The epididymis stores millions of spermatozoa and helps in their final maturation.

The columnar cells are well endowed with surface **microvilli**, providing a massively enlarged surface area over which excess testicular fluid can be reabsorbed, thereby concentrating several million spermatozoa into a relatively small volume of fluid. The muscle layer at the proximal end of the epididymis contracts in a slow, almost constant rhythmic manner to move spermatozoa gently along the duct. The thicker layer of muscle at the distal end of the epididymis contracts much more forcefully to eject sperm at the time of ejaculation, a process that is under sympathetic nervous control (Refer to Chapter 5 if you need a reminder of the sympathetic nervous system).

SELF-CHECK 13.10

Compare and contrast the structure and function of microvilli and cilia.

Emerging from the distal end of the epididymis is the vas deferens, a single non-convoluted tube quite different in structure to the epididymis. The wall of the vas deferens is heavily muscled,

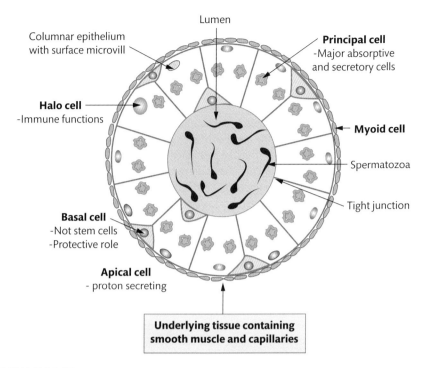

FIGURE 13.15
Transverse section through the epididymis.
Modified and reprinted with permission of the author (Robaire *et al.*, 2003), McGill University and publisher, The Van Doren Co, Charlottesville, VA.

consisting of inner and outer longitudinal layers and a middle circular layer. As you might have guessed, contraction of this muscle greatly supplements the weaker contractions initiated at the distal end of the epididymis at the time of ejaculation. The inner membrane of the vas deferens consists of columnar epithelium very similar to that of the epididymis, but exhibits a curious folded structure that permits expansion of the lumen at the time of ejaculation.

Key points

Powerful muscular contractions in the wall of the vas deferens cause ejaculation.

If you think the vas deferens is simply a muscular tube connecting the epididymis to the urethra then you are mistaken. As you can see in Figure 13.16, three gland-like structures connect with the vas deferens towards its distal end. The first two are the seminal vesicles (one for each vas deferens) which should really be considered as outpouchings, or **diverticulae**, of the vas deferens, rather than true glands. Each seminal vesicle consists of elaborate ducts lined by **secretory epithelial cells** and surrounded by layers of smooth muscle that forcibly expel **seminal fluid** during ejaculation. The epithelial cells secrete a chemically complex fluid that makes up 70–80 per cent of the ejaculate volume. Among the wide range of substances contained in seminal fluid, fructose and other sugars are abundant and serve as a major source of nutrition for spermatozoa.

The tubes emerging from each seminal vesicle are the relatively short ejaculatory ducts, which join at the centre to form the urethra. The point at which the ejaculatory ducts meet the urethra

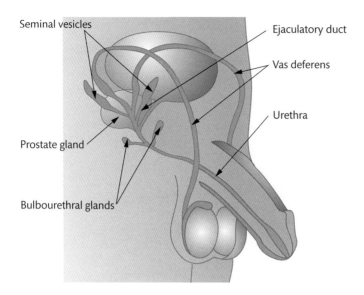

FIGURE 13.16

Interrelation of vas deferens, seminal vesicle, prostate gland, bulbourethral gland and urethra.

John Foster/Science Photo Library.

is where we find the prostate gland, which adds further secretions to the ejaculate during its transition through the urethra. Unlike the secretions from the seminal vesicles, prostatic fluid is rich in citric acid and various enzymes that liquefy the coagulated ejaculate to facilitate sperm motility.

Let's take a close look at the structure and function of this important gland and its component cells. The prostate is an encapsulated gland roughly the size of a walnut, but it can enlarge dramatically with age. Beneath the outer fibrous capsule is a fairly disorganized mass of smooth muscle, with innervations from the sympathetic nervous system. Powerful contractions of this muscle serve to eject prostatic secretions into the lumen of the urethra during ejaculation. The glands within the prostate are best considered as comprising three zones, each separated and supported by a fibrous **stroma**. Although the three zones are not clearly demarcated anatomically, they form concentric rings around the urethra and secrete their contents into its lumen via short ducts, as shown in Figure 13.17. The epithelial lining of the gland consists of a single layer of columnar cells, beneath which is a layer of **basal cells**. The columnar cells are specialized for secretion whereas the basal cells act as **reserve cells**, replenishing dead or damaged secretory cells.

Reserve cells

These are non-specialized epithelial cells that divide and grow to form new tissue when the latter is damaged or ageing. Therefore, reserve cells are important for normal turnover, growth and repair of tissue.

CLINICAL CORRELATION

Benign enlargement of the prostate

By the age of 40, benign prostatic hyperplasia (BPH) affects one man in five; by the age of 60 the proportion rises to seven in ten, and by 80 years of age most men have some degree of benign prostatic enlargement. The precise cause of BPH is unknown but we do know that enlargement of the gland is related to proliferation of stromal cells, and a decrease in the natural rate of death of epithelial cells. Despite the very high prevalence of the condition, only about 50% of men with BPH develop symptoms, the most common of which is reduced urine flow caused by partial obstruction of the urethra. However, in extreme cases the urethra can become completely blocked and require medical or surgical intervention.

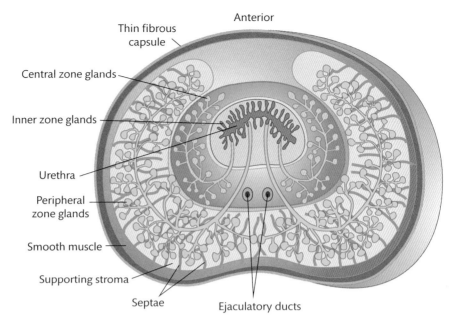

FIGURE 13.17

Human prostate, consisting of three gland zones: inner, central and peripheral.

Modified from Stevens and Lowe: Human Histology © 2004 Elsevier Ltd.

Considering that the secretions from each prostate zone are essentially the same, you might ask why the gland is divided in such a way. The answer lies in the way that pathologists recognize various disease processes. Without going into too much detail, most cancers of the prostate arise in the peripheral zone while many benign tumours originate in the inner and central zones.

The final glandular structures in the male reproductive tract for us to consider are the bulbourethral glands. These pea-sized paired glands sit on either side of the urethra and are connected to it by way of narrow ducts. They secrete a colourless mucoid fluid during sexual arousal, which often precedes the main ejaculate and probably has a lubricating function.

Key points

The seminal vesicles, prostate gland and bulbourethral glands are rich in secretory cells, exuding fluids that help to maintain the viability of sperm.

13.6 Cells of the female reproductive tract

The female reproductive tract consists of the Fallopian tubes, uterus, cervix and vagina, collectively referred to as the **internal genitalia**. The various cells we are about to discuss are, broadly speaking, structurally and functionally adapted to support and maintain female reproductive capacity.

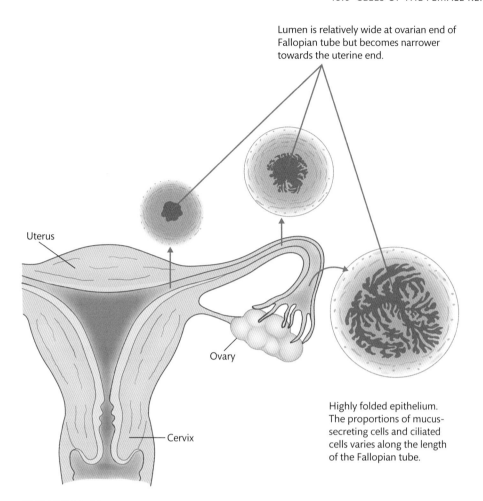

Lumen is relatively wide at ovarian end of Fallopian tube but becomes narrower towards the uterine end.

Uterus

Ovary

Cervix

Highly folded epithelium. The proportions of mucus-secreting cells and ciliated cells varies along the length of the Fallopian tube.

FIGURE 13.18
Transverse sections through a Fallopian tube.

We start where we left off in section 13.4. At ovulation, an ovum (to be precise, a secondary oocyte) is expelled from one of the ovaries into the peritoneal cavity. Rather than floating off into a sea of peritoneal fluid, the ovum is immediately swept into the open end of the Fallopian tube, which lies very close to the ovary. From here it begins its passage along the Fallopian tube to the uterus, a journey of about 12 cm. A close look at the epithelial lining and wall of the Fallopian tube will help to explain how the ovum (which, unlike male gametes, is non-motile) is gently encouraged to move along in the right direction. Figure 13.18 shows transverse sections through a Fallopian tube, but it is worth noting that variations in tube thickness and in the proportion of various cell types along its length, as well as hormonally induced fluctuations, make it difficult to capture the full spectrum of epithelial morphology in this channel.

The open end of each Fallopian tube is known as the **infundibulum** and is thrown into a series of epithelial fronds called **fimbriae**. Some of these may be attached to the adjacent ovary, therefore minimizing the chances of losing the ovum to the peritoneal cavity upon its release. The epithelial cells lining the fimbriae are columnar in shape and a large number have surface cilia. The cilia beat in synchronized waves to create a current that sweeps the ovum into the next part of the Fallopian tube, the **ampulla**. Interestingly, cilia have the same internal structure and mechanism of motility as the axoneme of sperm tails. The 9+2 arrangement of microtubules seems to be popping up quite regularly in this chapter, so look back at section 13.3 if you need a reminder. The ampulla is narrower than the infundibulum and it is here that fertilization usually takes place. As we progress along the length of the Fallopian tube towards the uterus, ciliated epithelium gradually gives way to secretory-type cells with surface microvilli. These cells produce a watery fluid rich in protein and ions, serving to nourish spermatozoa and the ovum while they are in transit. Along its length, the wall of the Fallopian tube contains an inner circular and an outer longitudinal layer of smooth muscle, gradually thickening and acquiring a third layer as it reaches the next section, the narrow **isthmus**. The purpose of the muscle wall is to generate gentle peristaltic contractions, which, along with the synchronized beating of the cilia, assist in the continued movement of the by now fertilized ovum toward the uterus. The final section of the Fallopian tube, the **intramural segment**, is the most muscular part of the channel and opens out into the uterus.

Key points

The Fallopian tube produces muscular contractions and is lined by varying proportions of secretory cells and ciliated columnar cells, all of which help to bring sperm and the ovum together.

CLINICAL CORRELATION

Ectopic pregnancy

Very rarely following fertilization, the embryo does not implant in the uterus but instead grows in the wall of the Fallopian tube. In exceedingly rare cases, the embryo can find its way to the cervix, ovary, or even the peritoneal cavity before implantation occurs. The cause or causes of this unusual event are unknown, but one hypothesis suggests that previous damage to the delicate ciliated lining of the Fallopian tube might impair tubal transportation of the embryo following fertilization. Such damage can be caused by infection or previous gynaecological surgery. Most instances of ectopic pregnancy are non-viable. Approximately half require medical or surgical intervention, while other cases resolve spontaneously by way of a tubal abortion. This means that the embryo is expelled into the peritoneal cavity, which can cause severe pain. Rarely, the accompanying blood loss into the peritoneal cavity can be very heavy, even life-threatening.

The uterus is best described as a hollow, pear-shaped organ approximately 7 cm long that is responsible for nurturing the developing embryo following fertilization. The epithelial lining and muscular wall of this organ are very well adapted for the purpose. During the monthly menstrual cycle, and also during pregnancy, this incredible organ changes in shape, size and histological appearance dramatically.

Unlike the tall columnar cells of the Fallopian tube, the epithelial lining of the uterus, the **endometrium**, consists of shorter **pseudostratified** ciliated cells interspersed with secretory cells. Beneath the endometrium, which is only one cell layer thick, is a highly cellular stroma. A significant proportion of the uterine wall is made up of a thick layer of smooth muscle (the **myometrium**), which, with its capacity for powerful contractions, plays a significant role in expelling the fetus at the end of pregnancy.

SELF-CHECK 13.11

The word 'pseudostratified' appears in descriptions of columnar epithelium in various different body sites. Can you explain what the word means?

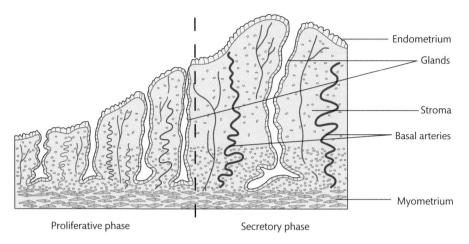

Endometrium

Glands

Stroma

Basal arteries

Myometrium

Proliferative phase

Secretory phase

FIGURE 13.19

Proliferative phase endometrium.

From Gargett *et al*. Hormone and growth factor signaling in endometrial renewal: Role of stem/progenitor cells. *Mol Cel Endocrinol* 2008; **288:** 22-29 © 2008 Elsevier Ireland Ltd.

More than any other epithelium in the female genital system, the endometrium is extremely responsive to the ovarian sex hormones and other hormones secreted during pregnancy. Figure 13.19 shows the uterine wall as it appears during the first and second halves of the menstrual cycle, respectively.

Following the blood loss that occurs between days 1 and 5 of the menstrual cycle (the **menstrual phase**), the endometrium and adjacent stroma begin to regenerate. The stroma gains a rich blood supply, **stromal cells** proliferate and the endometrium becomes increasingly convoluted, forming coiled tubular 'glands' penetrating deep into the stroma. These glands are actually deep invaginations of epithelium and should really be referred to as **crypts**. Full regeneration takes about 10 days and during this time the uterus is said to be in the **proliferative phase**. This all happens under the influence of the hormone oestrogen, which is secreted in increasing quantities from one or more ovarian follicles as the **ovulatory phase** approaches at mid-cycle. (Read section 13.4 again if you need a reminder of the ovarian cycle.) The most prominent change during this time is seen in the stroma, which gains fluid and dramatically increases in volume, thereby preparing the uterus for implantation of a fertilized ovum. In figure 13.19 you can see that the endometrium remains as a single layer of cells throughout the process.

SELF-CHECK 13.12

Describe the difference between a diverticulum, a crypt and a gland.

Following ovulation at around day 14, the ruptured Graafian follicle within the ovary develops into the **corpus luteum**, which continues to secrete oestrogen, albeit at reduced levels. The corpus luteum secretes large quantities of the hormone progesterone, which, in combination with oestrogen, induces further changes in the wall of the uterus. From day 14 to 28 of the menstrual cycle, the uterus is in the **secretory phase**, during which the endometrial glands are rich in secretions, exuding copious quantities of glycogen-rich fluid (hence the name given to this phase of the cycle). At the same time, the stroma becomes **oedematous** and loses collagen, while the stromal cells swell and develop prominent nucleoli, an extensive network of endoplasmic reticulum, lysosomes, glycogen vacuoles and lipid droplets. This stromal reaction is called **decidualization** and the swollen stromal cells are referred to as **decidual cells**. The alterations in cell structure and function within the uterus, which are particularly pronounced during pregnancy but also occur to a lesser extent during each menstrual cycle, are very demanding in terms of their metabolic requirements, and would not occur without good reason. Remembering that the sole purpose of the uterus is to support and maintain pregnancy, these cyclical events are necessary to prepare the uterus for the arrival of a fertilized ovum from the Fallopian tube. In the event that this occurs, secretory-phase endometrium provides the ideal substrate to support the developing embryo until the placenta develops. If fertilization and implantation do not occur, the corpus luteum degenerates and ceases to secrete progesterone. By day 28 of the cycle, the continued absence of progesterone triggers the constriction of the arteries supplying the endometrial stroma, causing tissue **ischaemia**. The loss of a blood supply causes the endometrium to break down, resulting in the menstrual flow mentioned at the beginning of the last paragraph, and a new cycle begins.

Ischaemia

This is a reduction in blood flow to a tissue.

Key points

Stromal cells are found in the connective tissues of all bodily organs.

Key points

The epithelial lining and stroma of the uterus change dramatically during the menstrual cycle, pregnancy and throughout reproductive life.

SELF-CHECK 13.13

How and why must the endometrium respond quickly to fluctuations in ovarian hormones?

The cervix is found at the base of the uterus and protrudes slightly into the vagina. Anatomically it consists of the **internal os** (the inner opening to the uterus), the **external os** (the outer opening to the cervix) and a narrow **endocervical canal** joining the two, as you can see in Figure 13.20.

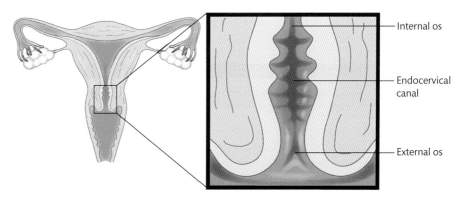

Internal os

Endocervical canal

External os

FIGURE 13.20
Anatomy of the cervix.

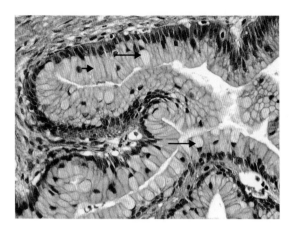

FIGURE 13.21
Mucin-secreting cells (long arrows) interspersed with ciliated columnar cells (short arrow) in the endocervical canal.
PathPedia, LLC © 2006-2014, All Rights Reserved.

Histologically, the endocervix is lined by tall columnar epithelial cells, many of which have a ciliated luminal border. Other cells secrete mucin and may be so distended by secretory vacuoles that they are sometimes called **goblet cells**. Figure 13.21 shows the seemingly random alternation between ciliated columnar cells and secretory cells in a portion of the endocervical canal.

Endocervical epithelium is thrown into tight folds forming crypts, often mistakenly referred to as glands. These deep invaginations are not unlike the highly developed crypts seen in the endometrium during the secretory phase of the menstrual cycle. The combined action of the ciliated cells and secretory cells provide a **mucociliary escalator** that traps microorganisms and other foreign material, preventing their entry to the uterus, which must be kept sterile. You came across this mechanism for expelling unwanted particles from ducts and cavities in Chapter 6, which deals with the respiratory tract. However, unlike the mucociliary escalator in the respiratory tract, which is concerned with transporting inhaled particles away from the lungs, the endocervix must pull off the trick of preventing the passage of microorganisms into the uterus while allowing access to spermatozoa. This seemingly impossible task requires a physiological compromise. For most of the menstrual cycle, cervical mucus forms a thick plug at the external os, which effectively traps and removes foreign particles (including sperm). As the mid-cycle approaches, high oestrogen levels trigger chemical changes in the mucus, making it quite thin and watery. This change in the consistency of cervical mucus is only transient. Very soon after ovulation, cervical mucus once again becomes thick and viscous. Thus, for a brief period, sperm can access the uterus unimpeded to enable fertilization to take place. The downside to this compromise is that during the same period there is an increased risk of ascending infection into the upper genital tract. The perpetuation of our species is not without its risks!

Key points
Goblet cells are found in many epithelia throughout the body, not just the endocervix. Their sole purpose is to secrete mucin, a viscous fluid that, in the endocervical canal, serves to trap foreign particles and prevent their entry into the uterine cavity.

At the external os there is another abrupt histological transition, this time from the delicate columnar epithelium of the **endocervix** to a tougher protective epithelium consisting of

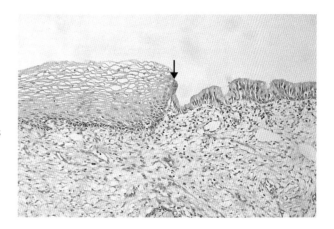

FIGURE 13.22
Cervical squamocolumnar junction at the external os (arrow).
From Mills: Histology for Pathologists, 3rd edition © 2007 Lippincott Williams & Wilkins.

non-keratinizing stratified squamous epithelium on the **ectocervix**. This area is known as the **squamocolumnar junction** and is shown in Figure 13.22. The transition reflects the ability of stratified squamous epithelium to withstand the harsh acidic environment of the vaginal surroundings.

Before we leave the cervix, let us pay brief attention to its subepithelial tissue. Unlike the stroma of the uterus, the cervical stroma contains little smooth muscle and consists mainly of collagen and stromal cells. Although the cervix is less responsive than the uterus to the hormonal variations of the menstrual cycle, nonetheless it undergoes changes in size and shape in a manner that is important both physiologically and pathologically. During each menstrual cycle, under the influence of oestrogen and progesterone, the volume of the cervical stroma fluctuates, causing the squamocolumnar junction to shift its position. As the stroma increases in bulk during the secretory phase, the endocervix everts, exposing delicate endocervical epithelium to the acidity of the vagina. This in turn stimulates **metaplasia**, a physiological process involving the replacement of endocervical epithelium by stratified squamous epithelium. Metaplasia is the natural protective response of endocervical epithelium to a stimulus (low vaginal pH). This zone of metaplasia, called the **transformation zone**, is peculiarly susceptible to malignant change. It is a sad fact that cervical cancer is second only to breast cancer in terms of worldwide female cancer incidence.

Key points

Squamous metaplasia in the cervix is a normal process involving the replacement of columnar epithelium with squamous epithelium at the cervical transformation zone.

SELF-CHECK 13.14

Can you think of examples of metaplasia occurring elsewhere in the body? (Read Chapter 6.)

The vagina is continuous with the cervix and forms a hollow muscular tube approximately 9 cm long. Just like the ectocervix, it is lined by non-keratinizing stratified squamous epithelium, which is predominantly protective in function. However, an important auxiliary function of the vaginal epithelium is to secrete a thin watery fluid that lubricates the vagina. Unlike secretions elsewhere in the female genital tract, this fluid does not originate from secretory cells but is simply a **transudate** from the circulatory system. The secretions increase greatly

CLINICAL CORRELATION

Cervical cytology

Cervical cytology (the study of cervical epithelial cells) is a quick, simple and safe screening test to check for the early signs of cervical cancer. The procedure involves scraping a sample of cells from the epithelial lining of the cervical transformation zone, which is where the majority of cervical cancers begin to develop. This inexpensive test, when properly incorporated into a well-organized screening programme, can result in remarkable reductions in death rates from this disease (Figure 13.23).

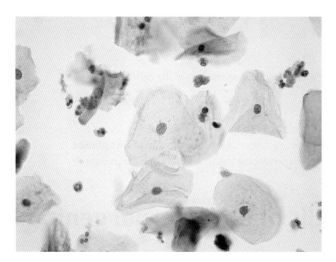

FIGURE 13.23
Intermediate cells from a cervical sample.
© Oxford University Press 2011.

during sexual arousal to facilitate copulation. Immediately beneath the vaginal epithelium is a dense **lamina propria** containing elastic fibres and blood vessels, then a layer of smooth muscle. The combination of elastic fibres in the lamina propria and the muscular vaginal wall permits expansion of the vaginal canal during childbirth. The vagina is no different to the upper genital tract in terms of its responsiveness to hormones. When oestrogen levels rise, the vaginal epithelium becomes fully mature, with small, round basal cells resting on the **basement membrane**, above which are **parabasal cells**, **intermediate cells** and finally mature **superficial cells** at the luminal surface. After the menopause, however, when oestrogen level declines, the epithelium becomes thin and **atrophic**, consisting of just a few layers of immature epithelial cells. Examples of these two epithelial states—full maturity and severe atrophy—are shown in Figure 13.24, but you should remember that these are at the morphological extremes of an epithelial spectrum.

Key points

A **transudate** is a watery fluid that leaks out of blood capillaries and generally has a low cell and protein content.

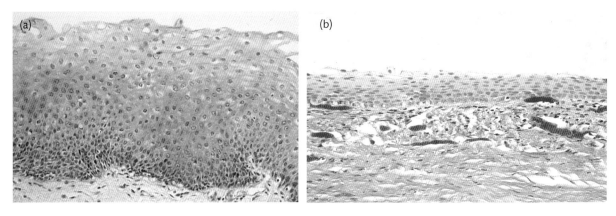

FIGURE 13.24
(a) Mature and (b) atrophic non-keratinizing stratified squamous epithelium of the cervix.
(a) © Ed Friedlander. (b) Ed Uthman/CC-BY.

While the middle and upper regions of the vagina are devoid of secretory glands, the lower vagina contains a few duct openings leading to **Bartholin's glands**. These structures are lined by tall columnar cells producing mucinous lubricating secretions.

13.7 Cells of the external genitalia

13.7.1 Female external genitalia

After a seemingly long journey that began in the ovaries, the female genital tract emerges into the outside world via the external genitalia, consisting of the labia minora, labia majora and clitoris (collectively called the **vulva**). These structures are shown in Figure 13.25.

The labia minora are thin flaps of tissue covered by mildly keratinized stratified squamous epithelium, with a rich supply of blood vessels and **sebaceous glands** opening directly on to the surface. The labia majora are thicker folds of keratinized tissue that lie external to the labia minora. Essentially, the labia majora are skin folds consisting of a pigmented epidermis, hair follicles and subcutaneous fat. Refer to Chapter 15 to learn more about the structure of the skin. Far from being considered extraneous tissue, the labia have an important function in providing a tough external layer of protection. Their complex skin folds lined partly by keratinizing stratified squamous epithelium prevent potentially harmful material from entering and damaging the delicate tissues of the internal genital system.

SELF-CHECK 13.15

Apart from the female external genitalia, where else might you find sebaceous glands? (Read Chapter 15.)

The clitoris is the female equivalent of the penis and rests close to the opening of the urethra. It is packed with sensory nerve endings that have an important role in sexual arousal. Just like the penis, the clitoris contains erectile vascular tissue and is covered with a thin non-keratinized squamous epithelium. An outer **fibrocollagenous** sheath provides a layer of protection when the clitoris is in a non-aroused state.

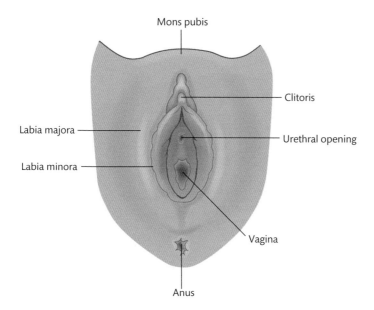

Mons pubis

Clitoris

Labia majora

Urethral opening

Labia minora

Vagina

Anus

FIGURE 13.25
Female external genitalia.

13.7.2 Male external genitalia

The male external genitalia consist of the penis and scrotum. The specialized tissue within the main body of the penis gives it some amazing properties. Figure 13.26 shows a transverse section of the organ containing three roughly circular areas of tissue with a textural appearance quite different to the surrounding tissue.

When viewed in longitudinal section, these areas form three cylinders running along the length of the penis. One of these encircles the urethra and is called the **corpus spongiosum**, while the other two run dorsally and are termed **corpora cavernosa**. All three cylinders consist of **erectile tissue**, so-called because of its incredible capacity to engorge with blood and produce a penile erection during sexual arousal.

Let's look at this in a little bit more detail. On higher magnification (Figure 13.27), it is clear that the erectile tissue consists largely of interconnected vascular spaces.

The spaces are lined by **endothelial cells**, just like ordinary blood vessels. These spaces lie empty when the penis is flaccid, but during an erection they fill with blood to increase the size and rigidity of the penis. All this, of course, is to facilitate copulation and the deposition of semen in the female genital tract. Blood is supplied to the corpora by interconnecting thick-walled arteries called **helicine arteries**, and they are drained via veins lying in the surrounding tissue. At this point you may well ask how an erection is maintained for any length of time—why doesn't blood simply drain away from the corpora as soon as they fill? Put simply, the physical pressure of the engorged corpora causes the thin-walled veins to collapse, thus forestalling venous drainage.

Surrounding and supporting the erectile tissue is a thick layer of **fibrocollagenous tissue** called the **tunica albiginea**, and outside this is a loosely held sheath of skin providing protection for the underlying tissue.

Endothelial cells
These are found on the inner lining of blood vessels and lymphatic vessels. They have complex and unique functions, which you can read about in Chapter 9.

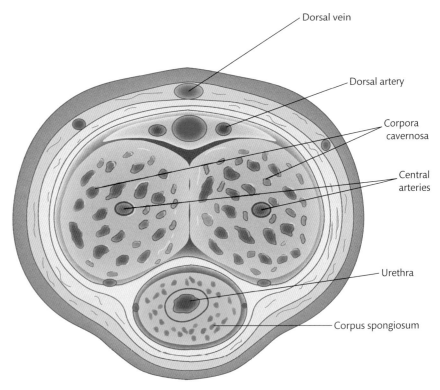

FIGURE 13.26
Transverse section of the penis.
© Elsevier Ltd.

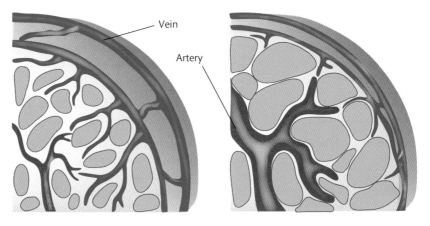

Flaccid penis.
Arteries are narrow. Small volume of blood in vascular spaces.

Erect penis.
Arteries enlarge, allowing increased blood flow into the penis. Vascular spaces fill with blood.

FIGURE 13.27
Erectile tissue of the corpus cavernosa/spongiosum.

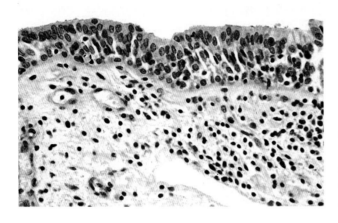

FIGURE 13.28
Transverse section of penile urethra showing stratified columnar epithelium.
© Oxford University Press 2002.

Before we finish this section, let us return briefly to the **urethra**, that all-important channel that serves not only as a conduit for semen but also as a drainage channel for urine. The urethra is a connecting duct that runs from the base of the bladder to the end of the penis, but, as we have already seen, there is an important connection with the ejaculatory ducts en route. The urethral lining is incredibly well adapted to withstand assault from the toxic chemicals found in urine. Where the urethra emerges from the bladder it is lined by classical **transitional epithelium (urothelium)**. You can read more about the structure and function of urothelial cells in Chapter 12, but suffice to say that this type of epithelium forms an almost impenetrable blood–urine and blood–sperm barrier while at the same time accommodating the inevitable stretching forces that occur during urination and ejaculation. Most of the remaining length of the urethra is lined by columnar epithelium, either as a single layer or in stratified formation (Figure 13.28). Towards its distal end and at the external opening (the **urethral meatus**), there is a change to non-keratinizing stratified squamous epithelium, which is continuous with the external epithelium of the **glans penis**.

Key points

The external genitalia are partly covered by keratinized squamous epithelium and hair, helping to protect the delicate internal structures.

 ## Chapter summary

This chapter has covered a lot of ground. It is unlikely that you will absorb all the facts in a single reading, which is only to be expected and is nothing to worry about. After all, the human brain has limited capacity, which no doubt you discovered in Chapter 5. Our strong advice is to take note of the cross references throughout the chapter, and indeed throughout the book, and revisit sections regularly. This will help you to acquire an holistic appreciation of the cells of the human body, which can so easily be lost when chapters are read in isolation.

- The cells of the human reproductive system are diverse in structure and function but share a common grand purpose: to contribute to the creation of new human beings, thereby promoting the survival of our species.

- A human being starts life as a zygote, a single diploid cell created from the union of two haploid gametes. The zygote later becomes the embryo.

- The male and female gonads (testes in the male, ovaries in the female) produce many millions of gametes through the mitotic and meiotic division of germ cells—the process is called gametogenesis.

- At the chromosomal level, gametogenesis in males and females is essentially the same process. At the cellular level, there are several remarkable morphological and physiological differences between male gametes (spermatozoa) and female gametes (ova).

- The male reproductive tract exists to transport spermatozoa to the female genital system. It is supported in this role by secretory cells that produce a nourishing fluid in which sperm can survive during transportation.

- The female reproductive tract acts as a receptacle for spermatozoa, and provides a safe environment for the union of male and female gametes.

- The uterus accommodates the developing embryo, keeping it safe and well nourished during pregnancy. The lining of endometrial cells and the underlying stroma play a crucial role in the process.

- The external genitalia consist of the penis and scrotum in the male, and the vulva in the female. Both are covered by a layer of keratinized skin that protects the more delicate internal genital organs.

 Further reading

- Shambayati B. *Cytopathology*. Oxford: Oxford University Press, 2011.

 Discussion questions

13.1 Compare and contrast gamete production in males and females.

13.2 Consider the physiological reasons for the production of polar bodies during oogenesis but not during spermatogenesis.

13.3 Discuss the contrasting but complementary functions of Sertoli cells and Leydig cells.

13.4 In the female genital system, squamous metaplasia occurs most frequently in the vicinity of the external os. Discuss the physiological reasons.

13.5 Why does meiosis occur only in the gonads?

13.6 Describe the similarities and differences between the '9x3' and the '9+2' arrangement in microtubules.

13.7 Many organelles are shed from spermatids during spermatogenesis. List the organelles that are retained and explain why they are needed.

13.8 Unlike spermatozoa, ova are non-motile. Without intrinsic mobility, explain how a fertilized egg comes to be implanted in the lining of the uterus.

Answers to self-check questions and discussion questions, hints and tips are provided on the book's Online Resource Centre.

 Visit www.oxfordtextbooks.co.uk/orc/orchard_csf/.

14

Cells of the endocrine system

Judy Brincat

Learning objectives

After studying this chapter you should be able to:

- Discuss the functions of the endocrine system.
- Discuss the features of the individual endocrine glands and their cellular components.
- Identify individual endocrine organs by their histological appearance stained by haematoxylin and eosin.
- Discuss the relevance of immunocytochemistry in the identification of cells of the endocrine system.
- Discuss the secretory products of the cells of the endocrine system.
- Discuss the categories of endocrine disease.
- Be aware of the clinical considerations associated with abnormal functions of the cells of the endocrine system.

Cells of the endocrine system are scattered throughout the body, occurring diffusely either within an epithelial surface such as the endocrine cells of the respiratory tract and alimentary canal, or within the stroma of another organ such as the C- or parafollicular cells of the thyroid, or the cells comprising the islets of Langerhans situated among the glands of the exocrine tissue of the pancreas. Endocrine cells also aggregate to form discrete organs or endocrine glands such as the adrenal glands, parathyroids, thyroid, pituitary and pineal gland.

The functions of the endocrine system are essential for maintaining homeostasis and the coordination of body growth and development, and are similar to those of the nervous system. Both systems communicate information to peripheral cells and organs, and because their functions are interrelated they are often referred to as the neuroendocrine system. Both systems can act on the same target cells at the same time, although the endocrine system produces a slower and more prolonged response.

The functional cells of a tissue or organ are known as parenchymal cells (see section 14.2), those of endocrine glands are of epithelial origin. The cells comprising the supportive connective tissue component are known as stromal cells. The stroma of endocrine glands contains a rich vascular supply consisting of a network of fenestrated capillaries. The function of the parenchymal cells of the endocrine system is to synthesize and secrete products called hormones (Greek *hormaein*: to set in motion). A hormone can be described as a biological substance that acts on a specific target to regulate the activities of various cells, tissues and organs.

Communication within the nervous system is via the transmission of neural impulses along nerve cells, and the release of neurotransmitters. Hormones secreted by endocrine glands move to their target tissue via connective tissue spaces and the cardiovascular or lymphovascular system. Over 100 hormones or hormonally active substances are produced by the cells of the endocrine system. They can be steroids, peptides, glycoproteins, biogenic amines or modified amino acids.

SELF-CHECK 14.1

What is a hormone?

14.1 **Hormone action and control mechanisms**

There are three types of hormonal action mechanisms: endocrine, where the hormone is released into the bloodstream and is transported to the target cells; paracrine, where the hormone is released from an endocrine cell into connective tissue spaces and diffuses to hormone-specific receptors on adjacent cells; and autocrine, where the target cell is the cell that has produced the hormone.

Two groups of hormone receptors can be identified. **Cell surface receptors** bind with protein hormone molecules that are too large to enter the cell through the cell membrane, which initiates the production of small intracellular second-messenger molecules activating a cascade of reactions to produce hormone-specific responses in the target cell. **Intracellular receptors** are localized within the target cell, usually inside the nucleus. They are used by steroid hormones which are able to penetrate both plasma and nuclear membranes and bind to DNA. This leads to messenger RNA (mRNA) transcription and the production of new proteins within the cell that results in hormone-specific responses in the stimulated cell.

 METHOD Intracellular receptors

Intracellular receptors for the steroid hormones oestrogen and progesterone may or may not be present in breast cancers. Their presence or absence, intensity of staining and number of positive cells are diagnostic and prognostic indicators. Immunohistochemical techniques for the localization and semiquantitation of hormone receptors are applied to formalin-fixed, paraffin-embedded biopsy and surgical specimens of breast tissue. The presence of receptors for oestrogen and progesterone is confirmed by the presence of a nuclear staining pattern in positive cells (see Chapter 6 of the *Histopathology* volume in this textbook series).

One process by which substances cross the plasma membrane to enter and leave cells is known as vesicular transport. The integrity of the plasma membrane is maintained, and molecules can also be transported between different compartments within the cell. Substances enter the cell by endocytosis, and leave the cell by exocytosis (see Chapter 1). The secretory vesicles of most secretory cells, as well as the synaptic vesicles of neurons, release their secretory products by exocytosis.

The production of hormones by endocrine cells is often controlled by feedback mechanisms from the target cells or organ. The action of the hormone (the stimulus) has an effect on the hormone-secreting cell (the original stimulus). Negative feedback occurs when the response to the stimulus reduces the original stimulus, while positive feedback (less common) occurs when the response increases the original stimulus. The major endocrine glands (ductless glands) are the pituitary, thyroid, parathyroids, adrenals, pancreatic islets of Langerhans, and the pineal gland. The placenta, only present during pregnancy, secretes several hormones, and organs such as the kidney, heart, thymus, ovary, testis and intestine have individual or small groups of cells with endocrine functions (Figure 14.1).

Generally, endocrine glands consist of islands or cords of secretory cells of epithelial origin, closely associated with a very rich supply of fenestrated blood and lymphatic capillaries

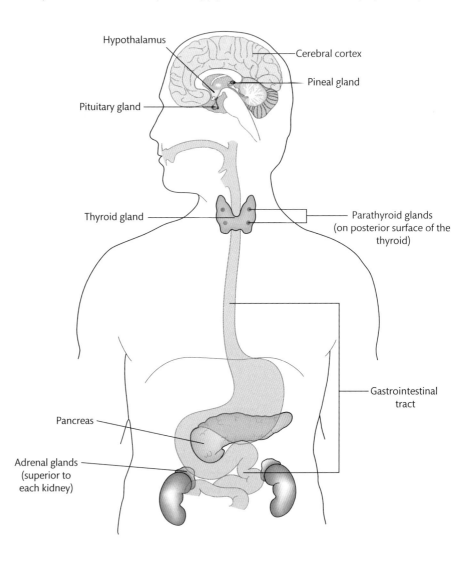

FIGURE 14.1
Organization of the endocrine system.

(see Box 14.2), and a small amount of stroma. Ultrastructurally, endocrine secretory cells are characterized by an abundance of cytoplasmic organelles such as mitochondria, endoplasmic reticulum, Golgi bodies and secretory vesicles, as well as prominent nuclei. This is indicative of their active synthetic and secretory function. Endocrine glands are also referred to as ductless glands because, unlike the exocrine glands that deliver their secretions to the target tissue via ducts, endocrine glands lack ducts, releasing hormones into interstitial connective tissue from where they are able to pass into the blood or lymphatic circulation.

SELF-CHECK 14.2

How do the ultrastructural components of endocrine cells indicate their function?

Key points

Hormones are biological substances that act on specific target cells.

Hormones interact with specific receptors of target cells to alter their biological activity.

Hormones include several classes of compounds.

Hormonal function is regulated by feedback mechanisms.

Hormone secreting cells are present in many organs and have specialized ultrastructural characteristics.

14.2 Pituitary gland

The pituitary gland, also known as the hypophysis, is found at the base of the brain, attached by a short stalk, the infundibulum, to a portion of the brain called the hypothalamus, to which it has extensive vascular and neural connections. The pituitary is the size and shape of a pea (about 1 cm in diameter), weighing 0.5 g in men and up to 1.5 g in multiparous women (i.e. who have given birth twice or more), lying immediately beneath the third ventricle in a bony cavity within the sphenoid bone called the sella turcica (Turkish saddle). This cavity is lined by a thickened extension of the dura mater, and a thin connective tissue capsule encompassing the gland.

The pituitary gland and the hypothalamus are often referred to as the 'master organs' of the endocrine system because the hormones under their influence regulate the physiological activities of many other endocrine glands and tissues. The pituitary gland is essential to life and consists of glandular epithelial tissue and neural secretory tissue. It is divided into two parts, anterior and posterior, which have entirely different embryological origins, functions and control mechanisms (Table 14.1).

CLINICAL CORRELATION

The pituitary

The pituitary has several important anatomical relationships, lying close to the optic nerves and optic chiasm. Pituitary lesions or tumours may encroach on these structures, resulting in visual disturbances or impairment.

The pituitary gland is situated above the sphenoid sinus, which in turn is adjacent to the nasal cavity, affording relatively easy surgical access to the pituitary via the nasal cavity and sphenoid sinus.

TABLE 14.1   Distinctive features of the pituitary gland.

Adenohypophysis	Neurohypophysis
Anterior pituitary	Posterior pituitary
From oral ectoderm	From neural ectoderm
Glandular epithelium	Neural tissue
Pars distalis (anterior lobe)	Pars nervosa (posterior lobe)
Pars tuberalis	Infundibulum
Pars intermedia (intermediate lobe)	Median eminence

The anterior lobe, or adenohypophysis, consists of epithelial tissue, and arises as an upgrowth or 'evagination' from the roof of the primitive oral cavity, which is also known as Rathke's pouch. Simultaneously, a down-growth of neural ectoderm (the infundibulum or infundibular process) from the developing hypothalamus gives rise to the posterior lobe, or neurohypophysis, consisting of neural secretory tissue. The posterior lobe remains connected to the hypothalamus by the infundibulum, also known as the pituitary stalk or infundibular stalk (Figure 14.2).

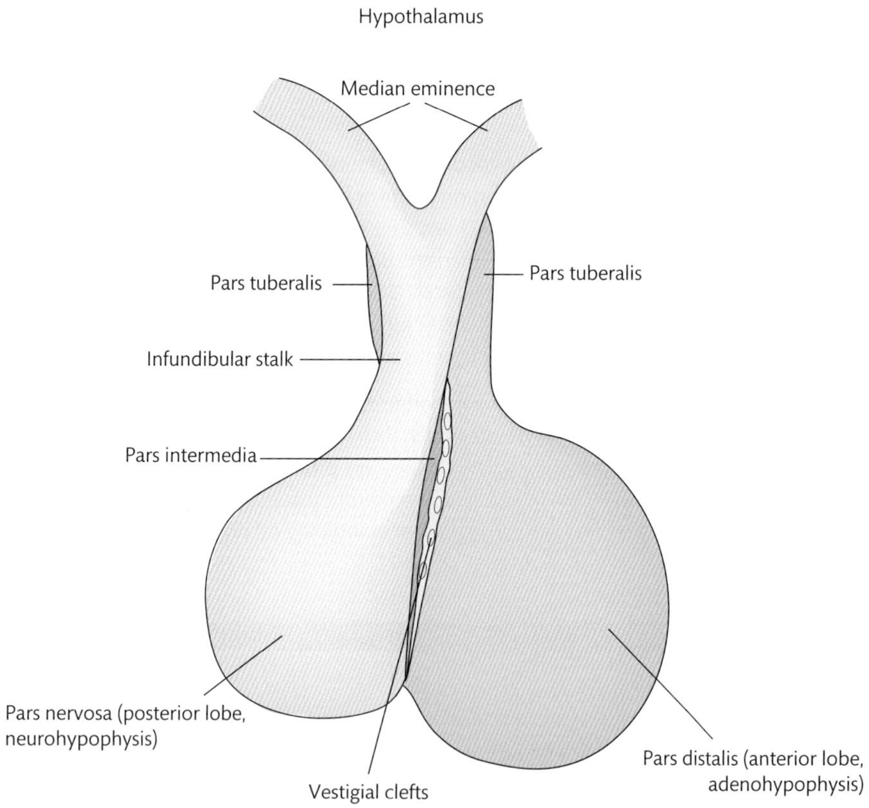

FIGURE 14.2
Diagrammatic representation of the pituitary gland.

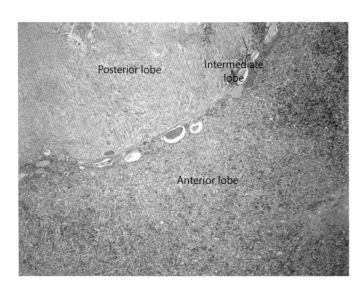

FIGURE 14.3
Pituitary gland (haematoxylin and eosin [H&E] stain, original magnification ×20).

Once mature, the adenohypophysis, comprising glandular epithelium, consists of three derivatives of Rathke's pouch: the **anterior lobe** (pars distalis), which is the largest portion, the **intermediate lobe** (pars intermedia), which is adjacent to the pars distalis, and the **pars tuberalis**, which surrounds the infundibulum like a sheath or collar. The intermediate lobe may contain cyst-like spaces or clefts, which are vestigial remnants of the lumen of Rathke's pouch. The neurohypophysis consists of the **posterior lobe** (pars nervosa), which is the most prominent part and contains neurosecretory axons and their endings, the **median eminence**, which is the upper part attaching to the hypothalamus, and the **infundibulum**, which contains neurosecretory axons (Figure 14.3 and 14.4).

SELF-CHECK 14.3

What are the two components of the pituitary gland?

The pituitary has an unusual and unique blood supply that is critical to its function, derived from two sets of vessels: the **superior hypophyseal arteries**, which arise from the internal

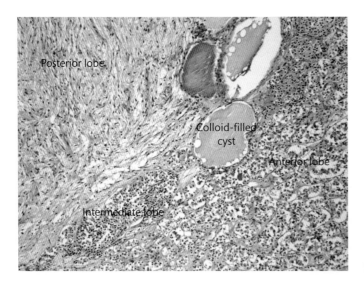

FIGURE 14.4
Pituitary gland (H&E stain, original magnification ×100).

BOX 14.1 Circle of Willis

A circle of arteries at the base of the brain that receives all the blood from the two internal carotid arteries. It is physiologically important because if one of the arteries either in the circle or supplying the circle becomes narrowed or blocked then the blood supply to the brain is not necessarily compromised because blood flow from the other vessels may be sufficient to maintain the supply.

CLINICAL CORRELATION

Intracranial tumours

Some 10–15 per cent of intracranial tumours are benign pituitary tumours (adenomas) of the adenohypophysis, and can be secretory or non-secretory in nature. Acromegaly (in adults) or gigantism (in children) is a condition associated with a pituitary adenoma of somatotrophs, the cells that produce growth hormone.

carotid arteries and the posterior communicating artery of the **circle of Willis** (Box 14.1), supply the pars tuberalis, median eminence and infundibulum; and **inferior hypophyseal arteries**, which arise from the internal carotid arteries and the pars nervosa. It is of note that most of the anterior lobe of the pituitary has no direct arterial supply. The superior hypophyseal arteries give rise to a plexus (network) of fenestrated capillaries, known as the primary capillary plexus, which drain into the **hypophyseal portal veins**, which run along the pars tuberalis before giving rise to a second plexus of sinusoidal capillaries. This second capillary plexus supplies the cells of the anterior lobe with blood, and also delivers neuroendocrine secretions of hypothalamic nerves from their site of release in the infundibulum and median eminence. This is the critical link between the hypothalamus and the pituitary gland and is known as the **hypothalamohypophyseal portal system**, by which the secretions of the adenohypophysis are controlled.

BOX 14.2 Capillaries

The most common type of capillaries are tight or **continuous capillaries**, which are found in all muscle and tissue areas where there is a blood-tissue barrier (e.g. blood–brain barrier in the CNS or blood–air barrier in the lungs). The endothelium is uninterrupted, which reduces permeability and only allows the passage of small molecules between adjacent endothelial cells, thus limiting the movement of material from the capillary lumen to surrounding tissue.

In contrast, **fenestrated capillaries** are highly permeable and are found in tissues such as endocrine glands, the small intestine and the kidneys, where there is a significant amount of molecular exchange between these organs and the blood. The endothelial cells of these capillaries are characterized by the presence of small (60–80 nm in diameter) circular openings, fenestrae (Latin: windows), which perforate the endothelium, creating a channel across the capillary wall. Each fenestra is usually closed by a diaphragm of 6–8 nm in width. Both continuous and fenestrated capillaries are supported by a basement membrane that is patent across the fenestrations of these permeable capillaries.

Sinusoidal capillaries are found in the adenohypophysis and adrenal cortex as well as bone marrow, spleen and liver. They have a wide, irregular lumen (15–20 μm), with large gaps separating the endothelial cells and an incomplete or absent basement membrane, thus permitting the passage of large molecules, fluid and blood cells.

As previously described, the **posterior lobe** of the pituitary is also known as the neurohypophysis, and consists predominantly of the pars nervosa. Histologically, this comprises:

- bundles of **non-myelinated axons** of neurosecretory neurons, the cell bodies of which reside in the hypothalamus
- **pituicytes**, which are irregularly-shaped specialized glial cells with oval nuclei, ensheathing and supporting the axons
- **fenestrated capillaries**
- dilated terminal expansions of the axons, known as **Herring bodies**, in which neurosecretory materials are stored prior to release.

The axons terminate close to the fenestrated capillary network, rather than on another neuron, and contain secretory vesicles in all parts of the cell. The two peptide hormones released by the posterior lobe of the pituitary are oxytocin and antidiuretic hormone (ADH; also known as vasopressin), which are synthesized in the hypothalamus and conveyed to the pars nervosa by the non-myelinated axons. Herring bodies consist of aggregates of membrane-bound neurosecretory vesicles containing either ADH or oxytocin. In a haematoxylin and eosin (H&E)-stained section they appear as amorphous, pale-staining eosinophilic entities, occurring in close proximity to sinusoidal capillaries. This close association facilitates the rapid diffusion of the neurosecretions across the thin vessel wall and into the blood stream (Figure 14.5).

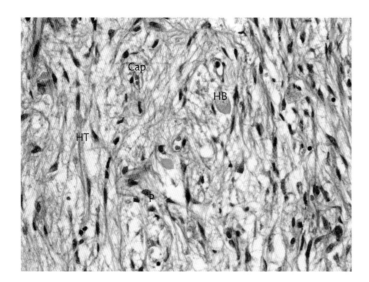

FIGURE 14.5
Posterior pituitary (H&E stain, original magnification ×400; Cap: capillary, HB: Herring body, P: pituicyte, HT: axon bodies of hypothalamohypophyseal tract).

SELF-CHECK 14.4

Which specialized type of blood vessel features in endocrine glands?

Antidiuretic hormone stimulates the contraction of smooth muscle cells in arteriole walls, increasing blood pressure and decreasing urine volume by increasing the amount of water resorbed by the collecting ducts of the kidney. **Oxytocin** stimulates uterine contractions during the later stages of pregnancy and the contraction of myoepithelial cells of the breast to facilitate milk ejection. Oxytocin is sometimes called the 'love hormone'.

BOX 14.3 *Parenchymal cells*

The parenchymal cells release hormones into the vascular network. The cells of the anterior lobe are influenced by releasing or inhibiting factors (hormones) synthesized and released by the hypothalamus and delivered via the hypothalamohypophyseal portal system.

Key points

The posterior lobe of the pituitary is an extension of the central nervous system that stores and releases neurosecretory products from the hypothalamus.

The **intermediate lobe** or pars intermedia is rudimentary in humans, comprising less than two per cent of the adult pituitary, and its function is unclear. In haematoxylin and eosin (H&E)-stained section it is seen to contain scattered clumps of basophilic (blue stained) low columnar-type epithelial cells, often surrounding colloid-filled cysts (remnants of Rathke's pouch). Occasional pale-staining polygonal cells are present. Ultrastructurally, the basophilic cells contain secretory granules similar to those of the corticotrophs of the anterior lobe. Melanocyte-stimulating hormone and α-endorphin or β-endorphin are produced by the cells of the pars intermedia.

The **anterior lobe**, predominantly the pars distalis makes up about 75 per cent of the adeno-hypophysis and consists of dark-staining glandular epithelium comprising a dense population of nucleated parenchymal cells (Box 14.3) forming clumps or cords, closely associated with a network of sinusoidal capillaries.

The cells of the pars distalis vary in size, shape and staining properties. Haematoxylin and eosin staining shows two distinct cell types: larger, intensely staining **chromophils** containing secretory granules, and smaller, paler staining **chromophobes** with less cytoplasm and few or no secretory granules (Figure 14.6). Traditionally, the chromophils of the anterior pituitary were further classified by their affinity for acid or basic dyes, and the staining reaction of their

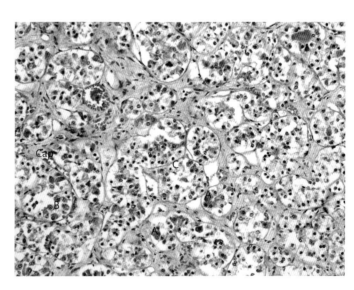

FIGURE 14.6
Pars distalis of the human pituitary (H&E stain, original magnification ×200; Cap: capillary, A: acidophils, B: basophils, C: chromophobes).

TABLE 14.2 Special staining characteristics of cells of the pars distalis.

Cell type		Cell description	MSB	PAS-OG
Chromophils 50%	Acidophils (40%)	Small, with a smooth refractile cytoplasm	Red	Yellow (PAS −ve)
	Basophils (10%)	Larger, more granular cytoplasm	Blue	Magenta (PAS +ve)
Chromophobes 50%		Smaller, less cytoplasm, pale staining	_	Blue-grey

secretory granules (Table 14.2). This classification gives no indication of the function of the different cells of the adenohypophysis. However, the application of antibodies to protein secretions of the cells by immunocytochemistry will identify five cell types and their location based on the localization of hormones within the cells, and correlate structure and function.

METHOD Hormone-specific antibodies

Antibodies are available for growth hormone (GH), thyroid-stimulating hormone (TSH), adrenocorticotrophic hormone (ACTH), luteinizing hormone (LH) and prolactin (PRL). Immunocytochemistry enables us to elucidate the normal distribution of the cells secreting these hormones, correlate structure and function, and can assist in the classification of tumours (Table 14.3).

There are also antibodies available for somatostatin (see the Hypothalamus section), thyroglobulin and calcitonin (see the Thyroid hormones section).

TABLE 14.3 Characteristics of cells of the adenohypophysis.

	Appearance	Hormone	Function	% of cells
Acidophils				
Somatotroph	Medium-sized oval Round central nuclei	Growth hormone (GH)	Stimulates growth of bones, muscles and organs	40–50
Mammotroph or Lactotroph	Large polygonal Oval nuclei	Prolactin (PRL)	Stimulates milk secretion	15–20
Basophils				
Corticotroph	Medium-sized polygonal Round, eccentric nuclei	Adrenocorticotrophic hormone (ACTH)	Stimulates hormone release from the adrenal cortex	15–20
Gonadotroph	Small oval Round, eccentric nuclei	Follicle-stimulating hormone (FSH); Luteinizing hormone (LH)	Stimulates ovarian follicle development in females and spermatogenesis in males.	10
Thyrotroph	Large polygonal Round, eccentric nuclei	Thyroid stimulating hormone (TSH)	Stimulates thyroid hormone synthesis and secretion	5

Pituitary hormones can be classified into two groups depending on whether they act directly on other endocrine glands (trophic) or on non-endocrine tissue (Table 14.4).

TABLE 14.4 Site of action of trophic and non-trophic.

Trophic	Site	Non-trophic	Site
ACTH	Anterior	GH	Anterior
FSH	Anterior	PRL	Anterior
LH	Anterior	ADH	Posterior
TSH	Anterior	Oxytocin	Posterior
		MSH	Intermediate
		α- and β- endorphins	Intermediate

CLINICAL CORRELATION

Pituitary adenomas

Immunocytochemistry is useful in classifying pituitary adenomas on the basis of the hormones produced by the neoplastic cells. The most common is a prolactinoma (30 per cent), a tumour of mammotrophs, resulting in proliferation of cells synthesizing and releasing prolactin. Treatment includes chemotherapy to reduce tumour size and inhibit hormone production, or surgery and irradiation, depending on tumour size (>10 mm in diameter). Untreated, symptoms include erectile dysfunction in males and amenorrhoea and infertility in females.

CLINICAL CORRELATION

Categories of endocrine disease

Endocrine diseases can be broadly classified into four categories: hormone overproduction, hormone underproduction, altered tissue responses to hormones and tumours of endocrine glands.

Hormone overproduction is most commonly caused by an increase in the number of cells producing a specific hormone. This increase can be due to abnormal antibodies mimicking the action of a hormone (e.g. TSH in Graves' disease is over-produced) to a genetic abnormality affecting the regulation of hormone synthesis and release, or to gene mutations resulting in the proliferation of mutant hormone-producing cells.

Hormone underproduction can be caused by the destruction of an endocrine gland by a disease process, or an autoimmune disease where antibodies are produced that target and destroy hormone-producing cells. Under-production can also result from genetic abnormalities that result either in abnormal development of endocrine glands, abnormal hormone synthesis, or abnormal regulation of hormone secretion. Furthermore, medical treatment can result in hormone underproduction (an iatrogenic cause). For example, a thyroidectomy or removal of the parathyroids involves the surgical removal of endocrine tissue.

Altered tissue responses to hormones result from genetic mutations of the hormone receptors.

Tumours of endocrine glands can compress nearby organs or destroy other organs by metastasis, independently of effects on hormone production.

Hormones are used to treat endocrine disease, either administered orally in the case of thyroid and steroid hormones, or by injection for protein hormones such as insulin.

14.2.1 Hypothalamus

The hypothalamus is a portion of the brain situated directly above the pituitary. The hypothalamic paraventricular and supraoptic nuclei contain the cell bodies of neurosecretory neurons which form the neurohypophysis or posterior lobe of the pituitary. The hypothalamus synthesizes ADH and oxytocin, which are transported to the posterior lobe via axoplasmic transport in the hypothalamohypophyseal tract and discharged by exocytosis of neurosecretory granules into the fenestrated capillaries. The hypothalamus also secretes polypeptides that influence the secretion of hormones by the anterior lobe of the pituitary gland (Table 14.5).

The hypothalamic polypeptides accumulate in nerve endings near the median eminence and infundibulum and are released into the capillary plexus of the hypothalamohypophyseal portal system for transport to the pars distalis of the adenohypophysis.

> ### Key points
> The hypothalamus regulates the activity of the pituitary gland and thus the entire endocrine system.

Normally, a hormone itself regulates the secretory activity of the cells in the hypothalamus and pituitary gland that control its secretion. Additionally, most psychological and physiological stimuli reaching the brain also reach the hypothalamus, thereby influencing the activity of the anterior lobe of the pituitary, thus regulating the entire endocrine system.

TABLE 14.5 Hormones regulation of the hypothalamus.

Hormone	Major functions
Corticotrophin-releasing hormone (CRH)	Stimulates adrenocorticotrophic hormone (ACTH) secretion by corticotrophs
Gonadotrophin-releasing hormone (GnRH)	Stimulates luteinizing hormone and follicular stimulating hormone secretion by gonadotrophs
Growth hormone-releasing hormone (GHRH)	Stimulates secretion and gene expression of growth hormone by somatotrophs
Somatostatin	Inhibits growth hormone secretion by somatotrophs; inhibits insulin secretion by β cells of the pancreatic Islets of Langerhans
Thyrotrophin-releasing hormone (TRH)	Stimulates secretion and gene expression of thyroid-stimulating hormone by thyrotrophs; stimulates synthesis and secretion of prolactin.

14.3 Pineal gland

The pineal gland, also referred to as the pineal body or epiphysis cerebri, is a flattened, pine-cone shaped (hence the name) neuroendocrine organ measuring 5–8 mm in length, 3–5 mm in diameter and weighing 100–200 mg, located near the centre of the brain at the posterior wall of the third ventricle. It is connected to the brain by a short stalk containing both sympathetic

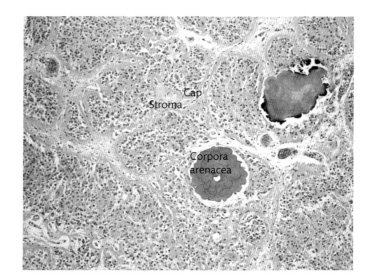

FIGURE 14.7

Human pineal gland. The glandular architecture shows closely packed parenchymal cells arranged in lobules. The stroma supporting the parenchyma contains thin walled capillaries (Cap). Two Corpora arenacea (see Clinical correlation, page 427) are evident (H&E stain, original magnification ×100).

and parasympathetic nerves, many of which communicate with the hypothalamus, and its primary function is to regulate daily body rhythm (circadian: relating to physiological activity occurring every 24 hours).

The pineal gland consists of a thin capsule of pia mater, with connective tissue septae extending inward from the capsule and dividing the gland into poorly defined lobules. The predominant cells of the pineal gland are known as pinealocytes, which have pale pleomorphic nuclei, one or two prominent nucleoli and gold-brown cytoplasmic lipofuchsin pigment. The pinealocytes are arranged in cords or clusters to form a glandular architecture supported by a delicate connective tissue stroma, interspersed with occasional glial cells and a capillary network. Ultrastructurally, pinealocytes show the presence of elongated cytoplasmic processes containing membrane-bound vesicles and parallel bundles of microtubules. The ends of the cytoplasmic processes are closely associated with fenestrated capillaries (Figures 14.7 and 14.8).

Basophilic extracellular concretions, known as corpora arenacea or brain sand, appear in the pineal gland from puberty onwards, increasing in number with age. They consist of concentric layers of calcium and magnesium phosphate in an organic matrix.

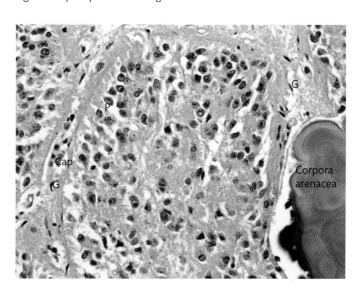

FIGURE 14.8

Human pineal gland (H&E stain, original magnification ×400; P: pinealocytes with cytoplasmic lipofuscin pigment, G: glial cells with elongated nuclei, Cap: capillary).

BOX 14.4 Melatonin

Melatonin is a powerful hormone produced by the pineal gland in response to information about light and dark cycles transmitted from photoreceptors in the retina via a nerve pathway known as the retinohypothalamic tract. During daylight, melatonin production by the pineal gland is inhibited by light stimuli, while melatonin secretion increases during darkness. In humans, the circadian changes influenced by melatonin production play a vital role in regulating daily body rhythms. Melatonin production in mammals also regulates reproductive function by inhibiting the production of steroid hormones by the gonads.

The hormone **melatonin** (Box 14.4, Table 14.6) is produced by the pinealocytes of the pineal gland and released into the capillaries. Smaller, darker interstitial glial cells comprise five per cent of the pineal gland, and these have ultrastructural and staining properties similar to astrocytes and pituicytes. They are best visualized by immunocytochemistry using an antibody to glial fibrillary acidic protein (GFAP).

Key points

The human pineal gland regulates daily body rhythm by relating stimuli from photoreceptors in the eye to endocrine activity.

CLINICAL CORRELATION

Corpora arenacea

Corpora arenacea are present from childhood, increasing with age. They are mineralized and therefore opaque to X-rays. They serve as a useful midline marker for clinicians in radiographic and computed tomography (CT) studies, and hence act as a guide to pathological conditions that may cause displacement of the midline to one side or the other.

CLINICAL CORRELATION

Melatonin as a drug

Most totally blind people (i.e. those with no light perception) have circadian rhythms that are 'free-running', oscillating on a cycle of between 24 and 25 hours. When there is a daily drift from the normal 24-hour cycle, people suffer recurrent insomnia and daytime sleepiness, like jetlag, but continuous. A high daily dose (10 mg/day) of melatonin has been shown to modify free-running circadian rhythms to a more normal 24-hour cycle, thus preventing a debilitating sleep disorder.

TABLE 14.6 Major functions of melatonin.

Source	Hormone	Major functions
Pinealocyte	Melatonin	Regulates: daily body rhythms; circadian cycles; steroid production by the gonads; seasonal sexual activity in animals.

SELF-CHECK 14.5

What is the primary function of the pineal gland?

14.4 **Thyroid gland**

The thyroid gland is an asymmetrical, bi-lobed, butterfly-shaped endocrine gland located in the lower part of the neck. The two lobes, each measuring up to 5 cm in length, 2.5 cm in width and weighing 20–30 g lie on either side of the upper part of the trachea and are connected by a thin band of thyroid tissue known as the isthmus, which passes anteriorly across the trachea. The right lobe is often twice as large as the left. Some 15–40 per cent of the human population have a small pyramid-shaped lobe extending upward from the isthmus, which is a vestige of the embryonic thyroglossal duct from which the thyroid develops.

At week four of embryonic development, thyroid development begins as an endodermal thickening of the floor of the primitive pharynx, from where the main bulk of the gland grows caudally (i.e. like a tail) to form the thyroglossal duct, which moves downwards through the neck tissue, dividing into two lobes once it reaches its final location in front of the trachea.

The mature thyroid is surrounded by a thin connective tissue capsule, from which fine trabeculae or septa extend into the gland, dividing it into irregular lobes and lobules. The septa provide internal support and a pathway for a substantial blood supply as well as lymphatics and nerves.

14.4.1 **Thyroid follicles**

The structural and functional unit of the thyroid gland is the thyroid follicle, which consists of a spherical, cyst-like structure of 50–500 μm in diameter, filled with a homogeneous, gel-like material known as colloid (Figure 14.10). The wall of the follicle is formed by a simple cuboidal or low-columnar epithelium, bounded by a basement membrane. The thickness of the wall, or height of the epithelium, varies with the functional state of the gland. An underactive thyroid has larger follicles, containing large amounts of colloid and lined with cuboidal epithelium. Low columnar epithelium lining smaller follicles is seen in an overactive gland. The apical surfaces of the follicular cells are in contact with the colloid within the follicles, while the basal surface is adjacent to the basal lamina. In H&E-stained preparations, the follicular cells have a slightly basophilic cytoplasm and spherical nuclei containing one or more nucleoli, and the Golgi apparatus positioned above the nucleus. The colloid stains pink.

Ultrastructurally, the follicular cells are shown to contain organelles associated both with secretory and absorptive functions. These include short microvilli on the apical surface which project

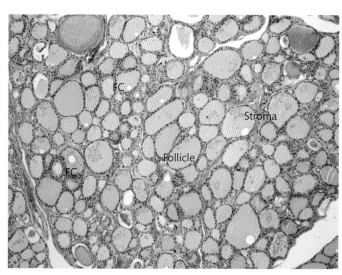

FIGURE 14.9

Human thyroid (H&E stain, original magnification ×100; FC: follicular cells).

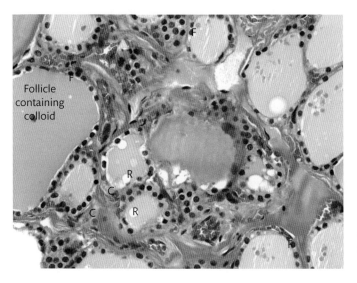

Follicle containing colloid

FIGURE 14.10

Human thyroid gland. Vessels containing red blood cells are evident (H&E stain, original magnification ×400; F: follicular cells, C: parafollicular cells, R: resorption vacuoles).

into the colloid. Junctional complexes, colloidal resorption droplets and lysosomes are also present at the apical pole, while rough endoplasmic reticulum is in abundance at the basal pole.

SELF-CHECK 14.6

What is the structural and functional unit of the thyroid gland?

A second type of cell, the parafollicular cell, is identified within the thyroid follicle, occurring as single cells or in small groups, located either between the basement membrane of the follicles and follicular cells or scattered among the follicular cells. They have no contact with the lumen of the follicle. These cells, with an extensive unstained cytoplasm, are sometimes referred to as C cells and derive from the neural crest (fourth branchial pouch), forming ultimobranchial bodies which migrate towards the developing thyroid during embryonic development, ultimately fusing with the thyroid and dispersing among the follicles within the middle third of the lateral lobes (Figure 14.10).

An extensive network of fenestrated capillaries is present within the delicate connective-tissue septa between the follicles.

> **Key points**
>
> Two types of cells are found within the follicular epithelium: follicular and parafollicular (C cells).

14.4.2 Thyroid hormones

The **follicular cells** of the thyroid follicles synthesize and secrete two hormones: thyroxine (tetraiodothyronine, more simply known as **T4**) and triiodothyronine (**T3**). Prior to secretion, both T3 and T4 are stored within the colloid that fills the follicles as an inactive intermediate secretory product, **thyroglobulin**. This is a unique feature of the thyroid gland in comparison to other endocrine organs, which store their secretory products internally in cytoplasmic vesicles. Thyroglobulin synthesis and storage is stimulated by thyroid-stimulating hormone (**TSH**) released from the anterior pituitary.

BOX 14.5 Thyroid hormone synthesis

Thyroid hormone synthesis involves several steps:

- Thyroglobulin precursor is synthesized by the rough endoplasmic reticulum (RER) of the follicular epithelial cells, packaged into vesicles and secreted by exocytosis into the follicular lumen.

- Iodide from the blood is actively transported into the follicular cell cytoplasm, diffuses towards the apical cell membrane and is transported to the colloid within the follicular lumen where it is oxidized to iodine.

- Iodine atoms are bound to tyrosine residues of thyroglobulin, catalyzed by enzymes in the apical microvilli

projecting into the colloid. T3 and T4 are formed following oxidative coupling reactions and are stored linked to thyroglobulin within the colloid of the follicular lumen.

- In response to stimulation by TSH, follicular cells resorb iodinated thyroglobulin by endocytosis.

- Thyroid hormones follow one of two pathways to traverse the follicular cells from apex to basement membrane.

- T3 and T4 cleave from thyroglobulin, cross the basement membrane and enter the blood in fenestrated capillaries. Tyrosine and iodide are also uncoupled from thyroglobulin and are 'recycled'.

Key points

Iodine from the blood is essential for thyroid hormone synthesis (Box 14.5).

Thyroid hormones act on the cells of the body to:

- increase basal metabolic rate (by increasing the rate of carbohydrate use, synthesis of proteins and their degradation, fat synthesis and degradation)
- increase heart rate
- promote cell growth
- raise body temperature
- enhance cell functions requiring energy.

T3 and T4 also act on thyrotrophs in the anterior pituitary to reduce the secretion of thyroid-stimulating hormone—a negative feedback system. During a normal pregnancy, maternal T3 and T4 cross the placental barrier and play a crucial role in the early stages of brain development, in addition to the fetal thyroid which begins to function and produce its own thyroid hormones during week 14 of gestation.

The **parafollicular cells** of the thyroid follicle synthesize and secrete **calcitonin**, which lowers blood calcium levels by increasing the rate of bone calcification, removing calcium from the circulation and depositing it in the bones, and suppressing the resorption of bone by osteoclasts. Calcitonin acts in opposition to parathyroid hormone (PTH). High serum levels of calcium stimulate calcitonin secretion, low levels suppress it, while the hypothalamus and pituitary have no influence over calcitonin secretion.

CLINICAL CORRELATION

Congenital hypothyroidism

Maternal deficiency of thyroid hormones during fetal development causes irreversible damage to the central nervous system, manifesting as reduced numbers of poorly myelinated neurons, and mental retardation, which is severe if the deficiency is present before the fetal thyroid begins to function.

Iodine deficiency is the most common cause of preventable mental impairment worldwide.

TABLE 14.7 Function of thyroid hormones.

	Hormone	Major function
Follicular cells	Thyroxine (tetraiodo-thyronine, T4) Triiodothyronine (T3)	• regulates tissue basal metabolic rate • raises body temperature • increases absorption of carbohydrates from the intestine • influences body and tissue growth • influences fetal development of the nervous system
Parafollicular cells (C cells)	Calcitonin	Decreases blood calcium levels by: • inhibiting bone resorption • stimulating calcium absorption by the bones

Key points

The cells of the thyroid gland synthesize and secrete three hormones (Table 14.7).

CLINICAL CORRELATION

Calcitonin

Calcitonin deficiency is not associated with a clinical disease; however, several endocrine tumours secrete calcitonin (e.g. medullary carcinoma of the thyroid and pancreatic endocrine tumours) so calcitonin can be used as a tumour marker to monitor the success of surgical resection of the tumour. Calcitonin can also be used to treat disorders of excess bone resorption (osteoporosis and Paget's disease).

SELF-CHECK 14.7

What element is essential for thyroid hormone synthesis?

CLINICAL CORRELATION

Goitre

The most common symptom of thyroid disease is enlargement of the gland, known as a goitre. This can be due to either hyperthyroidism or hypothyroidism.

Hyperthyroidism can lead to many thyroid diseases, the most common being Graves' disease, or toxic goitre. Autoantibodies to TSH receptors on follicular cells result in excessive production of the thyroid hormones T3 and T4, and suppression of TSH production by the adenohypophysis.

Hypothyroidism can result from iodine deficiency or one of several inherited autoimmune diseases such as Hashimoto's thyroiditis. Autoantibodies result in thyroid cell apoptosis and destruction of follicles. This results in low levels of circulating thyroid hormones, thus stimulating the release of excess TSH which increases synthesis of thyroglobulin and causes thyroid hypertrophy.

14.5 **Parathyroid glands**

The parathyroid glands are small ovoid endocrine glands, measuring 3×6 mm (the size of an apple seed) located within the connective tissue of the posterior surface of the thyroid, close to the superior and inferior thyroid arteries, from which they receive their blood supply. There are usually four, arranged in two pairs on either side of the thyroid, and known as the superior and inferior parathyroid glands. Occasionally, individuals possess additional parathyroid glands associated with the thymus.

Embryologically, by six weeks' gestation the inferior parathyroids (and the thymus) develop from the third brachial pouch, and the superior parathyroids from the fourth brachial pouch. Structurally, each parathyroid gland is surrounded by a thin fibrous connective tissue capsule, which separates it from the thyroid, and also gives rise to delicate trabeculae that divide the gland into poorly-defined nodules. The trabeculae also support blood vessels, lymphatics and nerves.

The parathyroid parenchyma consists predominantly of two types of cell, **chief or principal cells** and **oxyphil cells**. Chief cells differentiate during embryonic development, while oxyphil cells differentiate at puberty, and connective tissue becomes more evident in the adult, with the development of adipose cells that increase with age, eventually comprising up to 70 per cent of the glandular tissue. The stromal blood vessels are predominantly sinusoidal capillaries. The more numerous small, polyhedral, slightly eosinophilic chief cells form irregular anastomosing cords. Chief cells are 7–10 µm in diameter, with a central nucleus. The cytoplasm is pale staining, slightly acidophilic with lipofuscin-containing vesicles, lipid droplets and large amounts of glycogen. The staining intensity of the cytoplasm depends on whether or not the cells are actively secreting. Resting chief cells have pale cytoplasm (Figures 14.11 and 14.12).

Key points

Parathyroid parenchyma consists of two cell types, **chief** or **principal** cells and **oxyphil** cells.

Ultrastructurally, chief cells are typical of other polypeptide-secreting endocrine cells. They are linked to adjacent cells by desmosomes, they have prominent nucleoli and Golgi complex, numerous RER and secretory vesicles, scattered ribosomes and a few mitochondria and

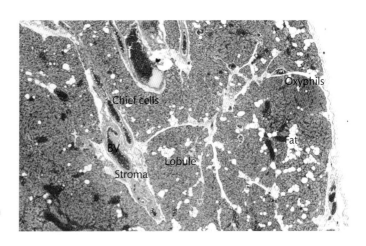

FIGURE 14.11
Parathyroid gland (H&E stain, original magnification ×40; BV: blood vessels).

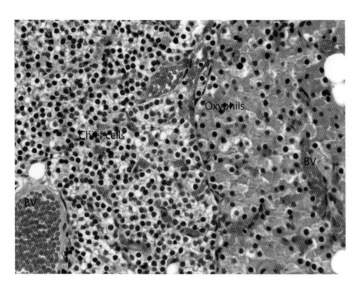

FIGURE 14.12
Parathyroid gland (H&E stain, original magnification ×400; BV: blood vessels).

CLINICAL CORRELATION

PTH and serum calcium levels

Given that the parathyroid glands are literally embedded in the connective tissue of the thyroid gland, some functioning parathyroid tissue must remain after thyroidectomy. PTH, the hormone secreted by the parathyroids, is essential to life because of its role in maintaining serum calcium levels. Without adequate serum calcium levels, muscles (including laryngeal and respiratory muscles) go into tetanic contraction, leading to respiratory failure.

lysosomes. The secretory vesicles are 200–400 nm in diameter, are membrane-bound and are often located near the plasma membrane adjacent to the perivascular space, which surrounds the endothelium of fenestrated capillaries. The secretory vesicles contain **parathyroid hormone (PTH)**, which is synthesized, stored and secreted by the principal cells of the parathyroid.

Key points

Chief cells of the parathyroid synthesize, store and secrete **parathyroid hormone (PTH)**.

Oxyphil cells are irregularly distributed and less numerous than chief cells, occurring singly or in clumps. They are larger and more acidophilic, and ultrastructurally are seen to contain numerous mitochondria, but no secretory vesicles, so are considered to be non-secretory.

Key points

Parathyroid hormone regulates calcium and phosphate levels in the blood, and is essential to life.

TABLE 14.8 The effect of parathyroid hormone.

Source	Hormone	Major functions
Chief (principal) cells	Parathyroid hormone (PTH)	Increases blood calcium levels by: • increasing the rate of osteoclastic activity, thus releasing calcium from bones • increasing renal tubular resorption of calcium ions and inhibiting resorption of phosphate ions from glomerular filtrate • with vitamin D3, increasing the absorption of calcium from the small intestine

---

CLINICAL CORRELATION

Primary hyperparathyroidism

Primary hyperparathyroidism is usually due to the presence of an adenoma, consisting predominantly of principal or chief cells involving one or more parathyroid glands. There is excessive synthesis and secretion of PTH, which causes high levels of calcium in the blood (hypercalcaemia) due to an increase in osteoclastic activity in bone. Kidney stones rich in calcium oxalate and calcium phosphate may form as a result of increased resorption of calcium by the renal tubules. In some clinical situations, the use of ICC for thyroglobulin and PTH can be helpful in categorizing parathyroid lesions encroaching on the thyroid.

Secretion of PTH by principal cells is regulated by the level of calcium in the blood via a simple feedback system. Low serum calcium levels stimulate PTH secretion; high levels inhibit the secretion of PTH. Parathyroid hormone release results in a peak increase in serum calcium level several hours later (Table 14.8), while calcitonin rapidly lowers serum calcium levels, with peak effect occurring in around an hour. Parathyroid hormone is essential for life, calcitonin is not; however, the combined activity of these hormones controls serum calcium level to within very narrow limits.

SELF-CHECK14.8

Which of the two cell types of the parathyroid parenchyma have a secretory function?

14.6 **Adrenal glands**

The adrenal (or suprarenal) glands are roughly triangular, flattened glands measuring approximately $7 \times 3 \times 1$ cm, weighing up to 5 g each, located in the perinephric fat of the upper pole of each kidney. They consist of two distinct zones of endocrine tissue that are functionally different and arise from two distinct embryonic tissues, enclosed in a common connective tissue capsule. The outer, dense fibrous connective tissue capsule, consisting of collagen and fibroblasts, gives rise to trabeculae extending into the parenchyma that conduct blood vessels and nerves. The **cortex** originates from proliferating mesodermal cells of peritoneal epithelium, lies directly beneath the capsule and comprises 90 per cent of the gland. The **medulla** has its embryonic origins in neural crest cells that migrate from the neighbouring sympathetic ganglion to the adrenal cortex and invade it to form the inner medulla.

TABLE 14.9 Overview of the adrenal gland.

Zone	Embryonic origin	Location	Secretory product
Cortex	Mesodermal mesenchyme	Beneath the capsule	Steroids (cholesterol derivatives)
Medulla	Neural crest	Deep within the cortex	Catecholamines (amino acid derivatives)

Key points

The adrenal glands consist of two distinct regions of secretory tissue, with different structure, embryological origins and different types of secretory product (Table 14.9). The cortex is essential to life, the medulla is not.

The adrenal glands have a rich blood supply, initially from the superior, middle and inferior suprarenal arteries, which branch before entering the capsule, to form a plexus of arterioles that penetrate the capsule and give rise to three forms of blood distribution. The system consists of:

- *capillaries* supplying the capsule
- *cortical sinusoidal capillaries* supplying the cortex and draining into fenestrated capillaries in the medulla
- *medullary arterioles* that pass through the cortex within the trabeculae direct to the medulla, transporting arterial blood to medullary sinusoidal fenestrated capillaries and then to collecting veins; hence the medulla receives a dual blood supply.

A large **central vein** conducts venous blood from the cortex and medulla (see Box 14.6) to exit the adrenal at the hilum as the adrenal (or suprarenal) vein. The right adrenal vein drains into the inferior vena cava, and the left adrenal vein drains into the left renal vein.

The adrenal cortex appears yellow macroscopically, and microscopically can be seen to consist of three distinct zones, named according to the arrangement of the secretory cells (Figure 14.13, Table 14.10).

The **zona glomerulosa** is the narrow, often incomplete, outer zone found just below the capsule and constitutes 10–15 per cent of the cortex, containing relatively small columnar or pyramidal cells with densely staining spherical nuclei, arranged in closely packed round to ovoid clusters (Figure 14.14). Each cluster is surrounded by a generous network of fenestrated sinusoidal capillaries. Ultrastructurally, the cells of the zona glomerulosa are seen to contain

BOX 14.6 Moving medullary hormones

In humans the tunica media (see Chapter 9) of the adrenal vein and its tributaries contains prominent bundles of longitudinally orientated smooth muscle cells. The coordinated contraction of these bundles of smooth muscle cells along the adrenal vein and the collecting veins results in a decrease in the volume of the adrenal gland. This in turn accelerates the delivery of medullary hormones into the circulation, much like squeezing a wet sponge.

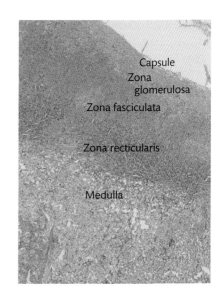

FIGURE 14.13
Zones of the adrenal gland (H&E stain, original magnification ×20).

TABLE 14.10   Cells of the adrenal gland.

Zone	Hormone	Major functions
Adrenal cortex		
Zona glomerulosa	Mineralocorticoids (aldosterone)	Control of blood pressure
Zona fasciculata (spongiocytes)	Glucocorticoids (cortisol or cortisone)	Regulate glucose and fatty acid metabolism
Zona reticularis	Gonadocorticoids (DHEA)	Weak masculinizing effects
Adrenal medulla		
Chromaffin cells	Catecholamines (adrenalin)	Fight-or-flight response

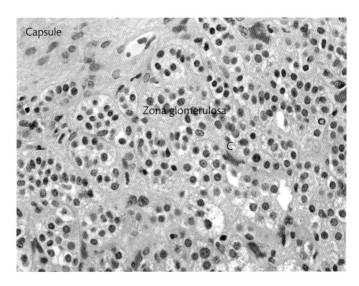

FIGURE 14.14
Cells of the zona glomerulosa (H&E stain, original magnification ×400; C: capillary).

abundant smooth endoplasmic reticulum (SER), many Golgi complexes, large mitochondria, free ribosomes and some RER. The cells of the zona glomerulosa synthesize and secrete **mineralocorticoids** (predominantly **aldosterone**), which regulate sodium and potassium homeostasis and water balance. Aldosterone acts to stimulate sodium and water resorption by the distal tubules of the nephron, the gastric mucosa and the salivary and sweat glands, as well as potassium excretion by the kidney. The predominant effect of aldosterone is to increase blood volume by increasing the resorption of water and ions in the kidney, thereby increasing blood volume and so blood pressure. The feedback mechanism controlling the activity of the zona glomerulosa is known as the renin-angiotensin-aldosterone system (see Chapter 12).

Key points

The secretory cells of the zona glomerulosa synthesize and secrete aldosterone, which is associated with the control of blood pressure.

The **zona fasciculata** is the middle and widest of the three zones, comprising up to 75 per cent of the adrenal cortex. Large, polyhedral secretory cells with abundant, poorly stained cytoplasm and light-staining nuclei form long, straight, radially arranged cords one or two cells thick, separated by sinusoidal fenestrated capillaries. The cytoplasm is acidophilic and appears vacuolated, but in fact it contains numerous lipid droplets that are extracted during the dehydration and clearing parts of tissue processing, resulting in a washed-out appearance (Figure 14.15).

Ultrastructurally, the cells of the zona fasciculata, also known as **spongiocytes**, exhibit typical features of steroid-synthesizing and secreting cells: prominent SER and Golgi apparatus, mitochondria with tubular christae and abundant RER. The lipid droplets in the cytoplasm of spongiocytes are not membrane-bound and contain neutral fats, fatty acids, cholesterol and phospholipids, the precursors to **glucocorticoid** hormones, such as **cortisol**, which are the main secretory products of these cells, although some androgens are also produced. The lipid-soluble hormones produced by spongiocytes are not usually stored, but are synthesized and secreted when required directly into the bloodstream via adjacent fenestrated capillaries. The surface area of the plasma membranes of spongiocytes is often increased by the presence of

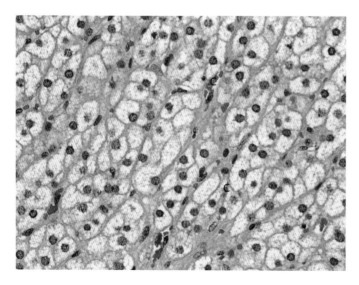

FIGURE 14.15

Cells of the zona fasciculata (H&E stain, original magnification ×400; C: capillary).

short, stubby microvilli to facilitate secretion of hormones into the perivascular space next to the basal lamina of adjacent fenestrated capillaries.

Cortisol acts on many different cells and tissues to increase the metabolic availability of energy sources such as glucose and fatty acids. Circulating glucocorticoids act on neurons in the hypothalamus to stimulate the release of corticotrophin-releasing hormone (CRH) that stimulates the secretion of ACTH by the corticotrophs of the anterior pituitary. Adrenocorticotrophic hormone is essential for cell growth and maintenance, and also stimulates steroid synthesis in the spongiocytes of the zona fasciculata and increases blood flow in the adrenal gland.

Key points

- Glucocorticoids regulate glucose synthesis (gluconeogenesis) and glycogen polymerization (glycogenesis).
- The synthesis and secretion of glucocorticoids by the spongiocytes in the zona fasciculate is regulated by ACTH.

The **Zona reticularis**, the thin innermost zone, comprises 5–10 per cent of the cortex. Both light and dark cells can be identified, with the cytoplasm of the darker cells containing brown lipofuchsin granules. The cells are smaller and more acidophilic, with less cytoplasm and more intensely staining nuclei. They are arranged as an irregular network of short branching cords and clusters, separated by fenestrated capillaries (Figure 14.16).

The ultrastructural appearance is that of steroid-secreting cells, prominent SER, elongated mitochondria with tubular cristae, but insignificant RER, and lipid droplets are few or absent. Consequently, the cytoplasm of these cells is more eosinophilic, and, as there is less cytoplasm, the nuclei are closer together.

The cells of the zona reticularis synthesize and secrete dehydroepiandrosterone (DHEA), a weak **androgen**. At normal serum levels, the effect of DHEA is insignificant. Some glucocorticoids, mostly cortisol, are also produced. The feedback mechanism regulating the secretory activity of the cells of the zona reticularis is the same as for the zona fasciculata.

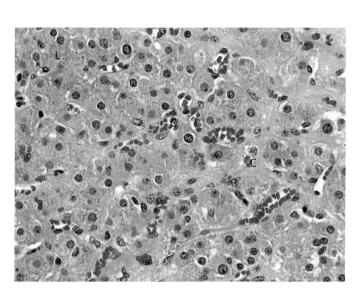

FIGURE 14.16

Cells of the zona reticularis (H&E stain, original magnification ×400; L: lipofuscin pigment in the cytoplasm of slightly darker staining cells, C: capillary).

CLINICAL CORRELATION

Addison's disease

Primary adrenocortical insufficiency, also known as Addison's disease, can occur as a result of:

- incomplete development of the cortex
- destruction of the cortex by autoimmune disease or severe infection (e.g. tuberculosis)
- idiopathic atrophy.

The adrenal cortex is unable to produce sufficient amounts of glucocorticoid and mineralocorticoid hormones, resulting in:

- raised pituitary ACTH levels causing abnormal skin and oral mucosal pigmentation
- muscle weakness and fatigue
- renal fluid and electrolyte imbalance causing lowered blood pressure and circulatory shock.

SELF-CHECK 14.9

What are the zones into which the cells of the adrenal cortex are arranged?

The **medulla** constitutes 10 per cent of the gland and when fresh (unfixed) is red-brown in colour. The medullary parenchyma is composed of large, pale-staining cells with a slightly basophilic granular cytoplasm and large nuclei. The cells of the medulla, also known as **chromaffin cells** (Box 14.7), are arranged in closely-packed ovoid clusters and short interconnecting cords, in close relation to a large network of sinusoidal capillaries (Figure 14.17).

Ultrastructurally, medullary cells contain numerous membrane-bound secretory vesicles, prominent Golgi apparatus and RER. The secretory vesicles store and secrete catecholamines, which are synthesized in the cytosol and secreted only in response to nervous stimulation. Two cell types can be identified by specific histochemical staining. One contains large dense core membrane-bound vesicles, which store and secrete noradrenaline, the other type contains smaller, less-dense vesicles that store and secrete adrenaline. The vesicles also contain soluble proteins called chromogranins, which make up 80 per cent of their contents, and catecholamines (20 per cent). The density of the vesicles appears to be due to the presence of the chromogranins, which are released with the hormones during exocytosis and distributed by lymphatic vessels.

Chromaffin cells of the adrenal medulla have the same embryonic origin as the postganglionic cells of the sympathetic nervous system (see Chapter 5), and are essentially neurons that have been modified to secrete. Numerous myelinated presynaptic sympathetic nerve fibres

BOX 14.7 Chromaffin cells

These are so named because they undergo a specific reaction in response to oxidization by chromic acid salts. When fixed in a solution containing potassium dichromate and formaldehyde (e.g. Regaud's fluid) the cytoplasm of medullary cells is brown. Potassium dichromate oxidizes the formaldehyde to formic acid, and the dichromate reduces to Cr^{3+}.

Cr^{3+} binds with catecholamines and their oxidation products. This is known as the chromaffin reaction, and easily differentiates the parenchymal cells of the medulla from those of the cortex.

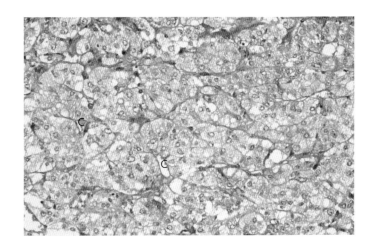

FIGURE 14.17

Chromaffin cells of the adrenal medulla (H&E stain, original magnification ×300; C: capillary).

(ganglion cells) innervate the chromaffin cells and regulate their secretory activity. Acetylcholine released by the presynaptic sympathetic neurons that synapse with the chromaffin cells stimulates the exocytosis of the secretory vesicles and the release of catecholamines directly into adjacent sinusoidal capillaries.

SELF-CHECK 14.10

How do the cells of the adrenal medulla differ from those of the cortex?

Catecholamines are released by the adrenal medulla (Box 14.8) in response to acute physical and psychological stress, and act on receptors throughout the body, as follows:

- heart rate, cardiac output and blood pressure increases
- coronary vessels and vessels supplying skeletal muscle dilate
- bronchioles dilate and respiration increases
- vessels supplying the intestine contract, while digestion and production of digestive enzymes decreases
- vessels supplying the skin contract, and perspiration increases
- glycogen is released into the bloodstream from the liver and skeletal muscle (glycogenolysis) and is converted to glucose
- adipose tissue releases free fatty acids.

Collectively, these physiological actions ensure that there is maximum energy available for maximum physical exertion (the fight-or-flight response).

BOX 14.8 *Conversion of noradrenaline to adrenaline*

As previously described, some of the blood reaching the medulla has passed from the adrenal cortex via continuous sinusoidal capillaries. Glucocorticoids secreted by the cortex stimulate the enzyme that catalyzes the conversion of norepinephrine to produce epinephrine.

Tumours associated with the adrenal medulla

Phaeochromocytoma is a tumour of the adrenal medulla, occurring mostly in adults and arising from chromaffin cells. High levels of circulating catecholamines, mostly norepinephrine, result in potentially life-threatening sustained or intermittent hypertension, cardiac arrhythmia and anxiety. Surgical removal is the treatment of choice.

Neuroblastoma is a malignant tumour of infancy and early childhood and arises from the embryonic neural crest cells that develop into either chromaffin cells of the medulla or postganglionic nerve cells in the peripheral ganglia. **Paraganglioma** develops outside the adrenal medulla within the paraganglia scattered around the body.

> *Key point*
>
> In response to acute physical and psychological stress, the secretion of adrenalin by the chromaffin cells of the adrenal medulla invokes the fight-or-flight response.

14.7 Endocrine pancreas

The pancreas is a major exocrine organ associated with the digestive tract (see Chapter 7). It is an elongated gland measuring 18–20 cm in length and weighing about 100 g, located on the posterior wall of the abdominal cavity. The head of the pancreas sits in the C-shaped curve of the duodenum, the body of the pancreas crosses the midline of the human body, while the tail extends towards the spleen. The pancreas is covered by a thin, loose connective tissue capsule and is joined to the duodenum by connective tissue. The parenchyma of the pancreas is divided into ill-defined lobules by connective tissue septae extending inwards from the capsule. By weight 99 per cent of the pancreas is made up of the exocrine component which consists of secretory acini and associated ducts.

14.7.1 Islets of Langerhans

The endocrine component of the pancreas is found scattered throughout the exocrine glandular tissue, forming isolated, discrete spherical aggregates of cells known as the islets of Langerhans. The islets vary in size (up to 300 µm in diameter), cluster around capillaries, and are predominantly located within the tail of the pancreas. Embryologically, both the exocrine and endocrine pancreas develop as an outgrowth of the foregut, which ultimately becomes the duodenum. The endocrine cells of the islets develop between nine and twelve weeks of gestation, separating from the duct system of tubules that develop into the excretory duct system draining the exocrine acini. The normal human pancreas contains roughly one million individual islets of varying size which in aggregate weigh up to 1.5 g. The islets of Langerhans can contain only a few or many hundreds of closely packed, small, pale-staining polygonal secretory cells arranged in irregular, compact cord-like clusters, surrounded by a delicate connective tissue capsule. The cells of the surrounding exocrine acini are larger and stain more intensely because they secrete large quantities of protein and so contain more rough endoplasmic reticulum (Figure 14.18).

By what name is the functional endocrine unit of the pancreas known?

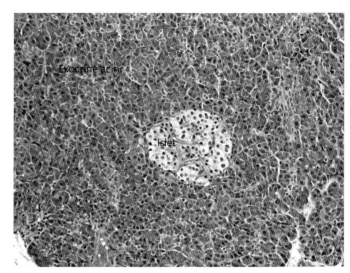

FIGURE 14.18
Pancreas (H&E
stain, original
magnification ×20).

There are several types of secretory cell found within the islets of Langerhans (Table 14.11), and
they are indistinguishable in a routine H&E preparation (Figure 14.19). Prior to the development of immunocytochemical techniques, antibodies specific to the cells' secretory products
and special stains were used to identify three or four cell types, and electron microscopy was

TABLE 14.11   Cells of the islets of Langerhans.

Cell Type	%	Product	Function
Alpha (α; peripheral)	15–20	Glucagon	Stimulates: • release of glucose into the bloodstream • gluconeogenesis and glycogenolysis in the liver • proteolysis Mobilizes fat from adipose cells Inhibits exocrine secretion
Beta (β; central)	60–70	Insulin	Stimulates: • uptake of glucose from the circulation • storage of glucose • glycolysis within cells • exocrine secretion by pancreatic acini
Delta (δ; peripheral)	5–10	Somatostatin	Inhibits both insulin and glucagon Affects gastrointestinal function Inhibits exocrine secretion
D1	<5	Vasoactive intestinal peptide (VIP)	Similar to glucagon Stimulates exocrine secretion
PP (or F)	<5	Pancreatic polypeptide	Stimulates gastric chief cells Inhibits bile secretion Inhibits pancreatic enzymes
EC (enterochro-maffin cells)	<5	Secretin Motilin Substance P	Stimulates pancreatic enzyme secretion Increases gastric and intestinal motility Acts like a neurotransmitter
Epsilon (ε)	<5	Ghrelin	Stimulates appetite

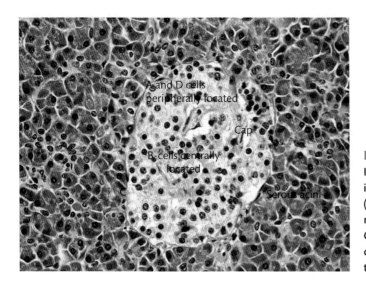

FIGURE 14.19
Islet of Langerhans in the pancreas (H&E stain, original magnification ×400; Cap: capillary, C: delicate connective tissue capsule).

used to measure the size and density of the secretory granules. The hormones secreted by the pancreatic islet cells act systemically, in the gastrointestinal tract (regionally), or in the islet itself (locally), to regulate metabolic functions. Now, using a combination of immunocytochemistry and transmission electron microscopy, three major and three minor islet cell types can be identified, each associated with a specific peptide hormone and each located in a specific site within the islet.

> **Key points**
>
> The hormones synthesized and secreted by the cells of the islets of Langerhans are peptide hormones.

> **Key points**
>
> The β-cells of the islets of Langerhans secrete insulin, the major hormone produced by the endocrine pancreas. α-cells secrete glucagon.

14.7.2 Ultrastructure

The cytoplasm of the centrally-located **insulin-secreting β-cells**, which comprise 70 per cent of the islet-cell population, contains a prominent Golgi complex, numerous membrane-bound secretory vesicles 200–250 nm in diameter, a few scattered ribosomes and small mitochondria, and some RER. The vesicles consist of a loose membrane encompassing a pale matrix that surrounds an electron-dense core of crystallized insulin bound with zinc. The **glucagon-secreting α-cells** are peripherally located within the islet and comprise 15–20 per cent of the cell population. The densely-packed secretory granules within their cytoplasm are 250 nm in diameter and are the site of glucagon storage. The secretory granules present in the cytoplasm of peripherally-located **δ-cells** are 300–350 nm in diameter and contain **somatostatin**. Note that somatostatin secreted by the δ-cells of the islets of Langerhans is identical to that secreted by the hypothalamus, which regulates the release of growth hormone (somatotrophin) from the anterior pituitary.

CLINICAL CORRELATION

Diabetes mellitus

This condition, that results in high blood glucose levels and excretion of glucose in the urine, is due to a complete absence or significantly diminished amounts of insulin. There are two main clinical types, with different causes. Type 1 diabetes (sometimes known as insulin-dependent diabetes) is caused by autoimmune destruction of the β-cells of the islets. This is a progressive condition, starting with a lymphocytic infiltration of the islets, which ultimately fail to produce insulin, and exhibit fibrosis and amyloidosis. Type 2 (non-insulin-dependent) diabetes is caused by insufficient insulin production. The islets appear normal but the insulin receptors on target cells are abnormal.

Key points

Consistent with the islet-cell function of synthesizing and secreting peptide hormones, the predominant cytoplasmic feature is the presence of membrane-bound secretory vesicles that vary in size and electron density.

The pancreatic islet cells are arranged as irregular cords, and adjacent cells are linked by intercellular junctions. The islets of Langerhans are serviced by a generous blood supply, each one being supplied by up to three arterioles at the periphery of the islet, which then branch out into a generous network of fenestrated capillaries. Islet cells usually have one free surface in close proximity to a fenestrated capillary into which the contents of its vesicles are discharged by exocytosis. Up to six venules convey blood to the exocrine acini and interlobular veins. The α- and δ-cells situated at the periphery are infused first before the blood reaches the more centrally located β-cells. Similar portal systems exist in the adrenal and pituitary glands. With the exception of insulin, all the hormones secreted by the various islet cells are also secreted by the endocrine cells found in the gastrointestinal tract.

SELF-CHECK 14.12

Which secretory product of the cells of the islets of Langerhans is unique to the pancreas?

14.8 **Gastrointestinal endocrine system**

Throughout the length of the digestive tract, from the oesophagus to the colon, endocrine cells are found scattered in the mucosa (see Chapter 7). They closely resemble neurosecretory cells of the central nervous system, and many of their secretions are the same. These neuroendocrine cells make up fewer than one per cent of the epithelial cells in the gastrointestinal tract and they are also found in the ducts of the pancreas, the liver (see Chapter 1) and the respiratory system (see Chapter 6), and constitute part of the diffuse neuroendocrine system (DNES). Taken as a whole, these scattered cells, the embryonic origins of which are the foregut, would constitute the largest endocrine organ in the body.

The small pyramidal-shaped enteroendocrine cells are also known as argentaffin, argyrophil, enterochromaffin or APUD (amine precursor uptake and decarboxylation) cells according to

BOX 14.9 *Enteroendocrine cells*

Argentaffin cells are able to reduce silver ions in solution to form metallic silver. **Argyrophil cells** require the addition of an external reducing agent to precipitate metallic silver. Most empirical staining techniques relying on these properties have been superseded in favour of more specific ICC techniques. **Chromaffin cells** are described in the adrenal section of this chapter.

Enteroendocrine APUD cells are derived from the foregut. APUD cells can also develop from the embryonic neural crest and migrate to other sites in the body.

their metabolic properties and staining reactions with silver ions or chromium salts (Box 14.9). Individual cells are difficult to identify in routine H&E-stained preparations, but they can be visualized using ICC and electron microscopy. Different cells can be identified by ICC staining for the peptide and polypeptide hormones and hormone-like regulatory substances that are their secretory products. Ultrastructurally, these cells exhibit similar features: they all rest on the basal lamina and contain small, membrane-bound, electron-dense, basally located secretory vesicles, a single elliptical euchromatic nucleus, a few mitochondria, small Golgi complex and minimal scattered RER. Two distinct types of enteroendocrine cell can be distinguished. Enteroendocrine 'open' cells also feature a thin cytoplasmic extension with microvilli that reaches into the lumen of the tract and allows the cell to function as a chemoreceptor. The cell 'tests' the contents of the lumen and releases secretory product in response to the 'results'. Hence the contents of the tract lumen directly influence the secretion of the open cells. Enteroendocrine 'closed' cells do not extend into the lumen but face the lamina propria, and their secretion is indirectly regulated by the contents of the lumen (Figure 14.20). The secretory products of enteroendocrine cells are released either into the lamina propria or adjacent blood vessels.

Key points
There are two types of enteroendocrine cell, 'open' and 'closed'.

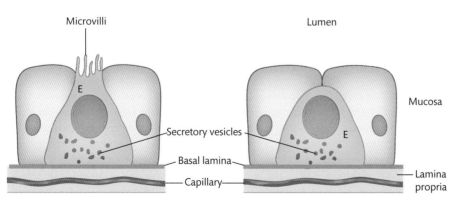

Enteroendocrine (E) 'open' cell   Enteroendocrine (E) 'closed' cell

FIGURE 14.20
Enteroendocrine cells.

CLINICAL CORRELATION

Neuroendocrine tumours

Gastrin stimulates the secretion of gastric acid by the parietal cells of the stomach. Excessive gastrin secretion from a tumour of gastrin-secreting enteroendocrine cells (**gastrinoma**) of the duodenum or pancreatic islets causes continuous stimulation of the gastric parietal cells and results in over-production of gastric acid. The excess acid is unable to be adequately neutralized, and leads to the development of gastric and duodenal ulcers. This condition is known as **Zollinger–Ellison syndrome.**

Neuroendocrine tumours associated with the gastrointestinal tract are referred to as gastroenteropancreatic (GEP) neuroendocrine tumours. The most common site of origin is the appendix, followed by the small bowel. These tumours were originally known as **carcinoid tumours**, secreting a variety of hormonally active substances such as serotonin, which, in the early stages of the disease, are removed from the circulation by the liver. Once the tumour metastasizes to the liver, however, these substances are free to circulate, resulting in a series of symptoms known as **carcinoid syndrome.** Symptoms include:

- flushing of the face, upper arms and chest
- altered blood pressure
- diarrhoea
- asthma-like wheezing due to bronchoconstriction
- heart valve damage.

These tumours can go undetected for many years, and the symptoms are often misdiagnosed.

Over 30 gastrointestinal hormones are secreted by the enteroendocrine cells, including gastrin, ghrelin, motilin, cholecystokinin (CCK), secretin, vasoactive intestinal polypeptide, somatostatin, gastric inhibitory peptide, pancreatic polypeptide, glucagon-like peptide-1, bombesin and histamine.

SELF-CHECK 14.13

What feature of open enteroendocrine cells enables them to test the contents of the lumen?

14.9 Respiratory endocrine system

The enteroendocrine cells of the respiratory system are known as 'small-granule cells' or Kulchitsky cells (K cells). They occur singly, dispersed among other cell types of the respiratory epithelium, and are difficult to distinguish from basal cells in H&E-stained preparations; however, their granules react with silver stains. Ultrastructurally, their appearance is similar to that of the enteroendocrine cells of the alimentary canal. Kulchitsky cells have more abundant cytoplasm than surrounding basal cells and contain small secretory vesicles located towards the basal pole of the cell, RER, Golgi apparatus, mitochondria and numerous cytoplasmic membrane-bound dense-core granules. Occasionally a thin, tapering cytoplasmic process extending into the lumen is present. The nucleus is located close to the basement membrane. One type of small-granule cell synthesizes and secretes a catecholamine, while the secretory products of a second type include the polypeptide hormones serotonin, calcitonin and bombesin.

SELF-CHECK 14.14

Where are Kulchitsky cells found?

14.10 Endocrine cells of the kidney

The kidney is involved in several endocrine functions. The **juxtaglomerular cells (JG cells)** in the tunica media of the afferent arteriole are modified smooth muscle cells (Box 14.10). They have spherical eccentric nuclei, unlike the elongated nucleus of a typical smooth muscle cell, and contain a well-developed Golgi apparatus, secretory vesicles known as juxtaglomerular granules, numerous small mitochondria and RER (Figure 14.21). The membrane-bound vesicles are 10–40 nm in diameter, with an often crystalline, moderately electron-dense core. These vesicles contain renin, an aspartyl protease, or its precursors, and are polarized towards the cell membrane that is adjacent to the lumen of the afferent arteriole. Under certain physiological conditions, such as a decrease in circulating blood volume or low sodium intake, rather than contracting, juxtaglomerular cells secrete renin and release it into the circulation via the lumen of the afferent arteriole thereby activating the renin-angiotensin-aldosterone system (RAAS). Renin catalyzes the hydrolysis of angiotensinogen, a circulating α-globulin synthesized by the liver, into a decapeptide angiotensin I. Angiotensin I is converted to angiotensin II by angiotensin-converting enzyme, which is present on the endothelial cells of lung capillaries. Angiotensin II constricts peripheral blood vessels and stimulates the synthesis and secretion of aldosterone by the cells of the zona glomerulosa in the cortex of the adrenal gland. Aldosterone acts on the collecting ducts to promote the resorption of sodium ions (and therefore water),

BOX 14.10 *The juxtaglomerular apparatus*

The juxtaglomerular apparatus is found near the vascular pole of the glomerulus or renal corpuscle. It derives from the glomerular afferent arteriole and distal convoluted tubule (DCT) of the same nephron. It consists of narrow, closely packed epithelial cells of the DCT, known as the **macula densa**, the modified smooth muscle cells of the afferent arteriole, known as **juxtaglomerular cells**, and pale extraglomerular mesangial cells, known as **lacis cells**.

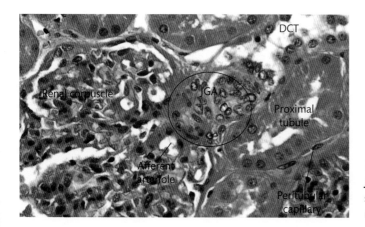

FIGURE 14.21

Juxtaglomerular apparatus (JGA) in relation to surrounding renal structures (H&E stain, original magnification ×400; DCT: distal convoluted tubule).

thus increasing plasma volume and raising blood pressure. In response to aldosterone, the increase in sodium concentration, plasma volume and blood pressure inhibits the release of renin from the juxtaglomerular cells.

Key points

Juxtaglomerular cells are modified smooth muscle cells. They have an endocrine rather than contractile function, which is to synthesize, store and secrete the hormone **renin** into the lumen of the afferent arteriole, thereby activating the renin-angiotensin-aldosterone system to regulate systemic arterial blood pressure.

Cells of the proximal tubules are the site of the hydroxylation of 25-OH vitamin D3 (a steroid precursor produced in the liver) to 1,25-$(OH)_2$ vitamin D3 (**calcitriol**), which is the active hormone. In humans, vitamin D3 is produced in the skin by the action of ultraviolet light on the precursor 7-dehydrocholesterol and also absorbed from the diet by the small intestine. Vitamin D3 is bound to vitamin D-binding protein in the blood and transported to the liver, where it is hydroxylated to form 25-OH vitamin D3, which is released into the circulation. An increase in Ca^{2+} in the plasma initiates the secretion of parathyroid hormone (PTH) by the chief cells of the parathyroid glands, which in turn stimulates the activity of the enzyme 1α-hydroxylase, as does a reduction in circulating phosphates. 1α-hydroxylase is responsible for the conversion of 25-OH vitamin D3 into active 1,25-(OH) vitamin D3 in the proximal tubules of the kidney. Calcitriol stimulates the absorption of Ca^{2+} and phosphate by the intestine, and the release of Ca^{2+} from the bones, and is therefore essential for the normal development of bones and teeth.

The endothelial cells of the peritubular capillaries in the renal cortex synthesize and secrete the glycoprotein hormone erythropoietin (EPO). Erythropoietin acts on surface receptors of erythrocyte progenitor cells in the bone marrow to regulate the formation of red blood cells in response to a decrease in O_2 concentration in circulating blood.

Key points

Under the regulation of parathyroid hormone, which stimulates the activity of 1α-hydroxylase, 25-OH vitamin D3 is converted to hormonally active 1,25-$(OH)_2$ vitamin D3 in the cells of the proximal tubules.

 # Chapter summary

- The endocrine system and the nervous system function together to provide intracellular communication.

- The cells of the endocrine system produce over 100 secretory products, hormones or hormonally active substances, which are generally released into fenestrated capillaries of the circulatory system, to reach the receptors of target cells.

- The glands of the endocrine system include the pituitary, the adrenals, the thyroid, the parathyroids and the pineal gland. Cells of the diffuse neuroendocrine system are located in the pancreas, the mucosa of the gastrointestinal tract, respiratory epithelium and other organs such as the kidney, adipose tissue, the heart, the gonads and the placenta (during pregnancy).

- Most parenchymal cells of endocrine glands are of epithelial origin, the exceptions being the medullary cells of the adrenal gland and the cells of the posterior pituitary.

- The ultrastructural appearance of the cells of the endocrine system reflects their active synthetic function. They are characterized by prominent nuclei and numerous cytoplasmic organelles such as mitochondria, endoplasmic reticulum, Golgi bodies and secretory vesicles.

- The synthesis and secretion of hormones by the cells of the endocrine system is controlled by feedback mechanisms from the target organ.

 # Further reading

- Bancroft JD, Cook HC. *Manual of histological techniques and their diagnostic application*. Edinburgh: Churchill Livingstone, 1994.

- Bancroft JD, Gamble M. *Theory and practice of histological techniques* 6th edn. Edinburgh: Churchill Livingstone, 2007.

- Orchard GE, Nation BR eds. *Histopathology*. Oxford: Oxford University Press, 2012.

- Ovalle WK, Nahirney P. *Netter's essential histology* 2nd edn. Philadelphia: Saunders Elsevier, 2013.

- Ross MH, Pawlina W. *Histology: a text and atlas* 6th edn. Philadelphia: Lippincott Williams & Wilkins, 2011.

- Young B, Lowe JS, Stevens A, Heath JW. *Wheater's functional histology: a text and colour atlas* 5th edn. Edinburgh: Churchill Livingstone, 2006.

 # Discussion questions

14.1 Why are the pituitary gland and the hypothalamus referred to as the 'master organs' of the endocrine system?

14.2 Discuss two examples in which immunocytochemistry is useful in identifying and localizing individual cell types within endocrine glands.

14.3 How do hormones act and how is their production by endocrine cells controlled?

14.4 List the endocrine glands.

14.5 Discuss the four categories of endocrine disease.

Answers to self-check questions and discussion questions, hints and tips are provided on the book's Online Resource Centre.

 Visit www.oxfordtextbooks.co.uk/orc/orchard_csf/.

Acknowledgements

- Professor Richard Williams, Associate Professor Penny McKelvie, and Atha Palios, Anatomical Pathology, St Vincent's Hospital, Melbourne, Australia, for access to tissue samples for photography.

- Dr David Clouston and Neil O'Callaghan, Focus Pathology, Melbourne, Australia, for tissue samples and all photographs.

- Dr Vanessa Fahey, Histopathology, Melbourne Pathology.

15

Skin and breast

Guy Orchard

Learning objectives

After studying this chapter you should confidently be able to:

- Describe the main cell types that compose the skin
- Explain the functional role of the skin as an organ
- Describe the architecture and structural interactions of the skin
- Explain the mechanisms and control functions of the skin
- Understand the importance of the skin in maintaining the integrity of the human body.

The skin represents the largest organ of the human body. It constitutes approximately 15 per cent of total body weight. Like every other organ you have read about in the earlier chapters of this book, the skin has homeostatic and regulatory functions that protect and regulate the body systems. The skin represents the external layer of any animal and its primary role is to separate the body from its external environment. In this context, the protective function means the skin resists frictional trauma, toxic agent damage, temperature fluctuations, excessive water loss, microbial penetration and ultraviolet radiation damage. This is not the only function of the skin, however. It also functions as a regulatory system, most notably reacting to temperature changes, insulating the body by using hair on the body surface to trap heat, and also adipose (subcutaneous fat) offering an insulation layer deep within the skin structure. Heat loss from the body can also be achieved through the process of sweat production to the skin surface, and also the increase of blood flow through the vascular network within the skin matrix. It also regulates the exchange of important electrolytes between its compartments and the rest of the body. The skin also has a sensory perceptive role. It is the largest sensory organ of the human body and has receptors for touch, pain, pressure and temperature. The structures responsible for this role are called corpuscles. You will read more about these later. Finally, the skin also has a metabolic role. The subcutaneous fat provides the body with triglycerides for energy, if required; in addition, the epidermal cells synthesize vitamin D (previtamin D3), which is primarily important in bone structure (Box 15.1).

It may be of some value at this point to use an analogy that will sum up the above functional points pictorially. For this purpose, referring to Figures 15.1a and 15.1b will help you comprehend them. I have used the image of a dry stone wall (Figure 15.1a)

BOX 15.1 *Vitamin D and the skin*

For most people the sun provides the key source of vitamin D. However in the modern world the effects of general chronological aging, skin pigmentation, sunscreen use, time clock variables, seasonal and geographical factors all can have a significant effect on the body's ability to synthesize pre-vitamin D3. It is also the case that dietary food sources of vitamin D are not as well fortified as thought. As a counter indicator it is well established that sunlight, in particular ultra violet light (UVB), is the primary causative agent for skin cancers and in particular malignant melanoma. This would seem to support, in the case of sunlight exposure, the old adage of 'Everything in moderation'.

to depict a strong, secure structure where the stones are locked together. On the top, the stones exist in a slightly loser structure and appear differently to the ones below, they are also angulated and protective of the structure. The stones represent the first compartment of the skin, the epidermis. Moving to the base of the wall, we have the soil, which acts as the contact for the surrounding elements, this represents a transitional and dynamic layer, analogous to the basement membrane. Finally, there is the surrounding grass filled with a wide assortment of components, all of which are integral to supporting the wall structure; this is analogous to the dermis, the second major compartment of the skin. I have used Figure 15.1b to demonstrate what happens in pathological states. Here, you can see that the integrity of the wall structure has been damaged. The stones in the wall are no longer tightly compact but rather loose and in some cases missing. There are breaches in the wall, which affects its rigidity and ability to resist the surrounding environmental elements. This is analogous to what can happen to the skin in any pathological process. It is also worth remembering that the damage to the wall may not relate to the stones (epidermis) themselves, but can also be due to changes in the underpinning soil (basement membrane) or the deeper surrounding grass (dermis).

FIGURE 15.1

(a) A working analogy 'The dry stone wall' representing the status of normal human skin (image selection aided by Anne Lacey). (b) A working analogy 'The broken dry stone wall' representing structural fragility akin to pathological change in human skin (image selection aided by Anne Lacey).

15.1 **Structure of the skin**

The functions of the skin highlighted in the introduction to this chapter are achieved by the fact that it is uniquely designed to support these functions. It is composed of the **epidermis**, which is a constantly changing and regenerating compartment of the skin, replacing and renewing cells and is highly cellular. The outermost layer is laminated and impermeable to toxic or microbial agents. Below the epidermis sits the **basement membrane**, which provides the connection to the second compartment of the skin, the **dermis**. This is a relatively acellular compartment, composed of a largely elastic and collagen-based matrix. Within this matrix reside specialist structures called **adnexae**. Collectively, they are composed of hair and nails, which are primarily protective in function, sebaceous and apocrine glands, whose secretions provide moisture for the skin services, and eccrine glands that also secrete water-based substances which help to regulate body temperature. Finally, below the dermis is the third compartment called the **subcutis** or **subcutaneous fat layer**, which is composed largely of fat-containing cells called adipocytes. Schematically, Figure 15.2 shows how these structures are arranged, and Box 15.2 highlights that skin structure is adapted to its function at a particular anatomical site.

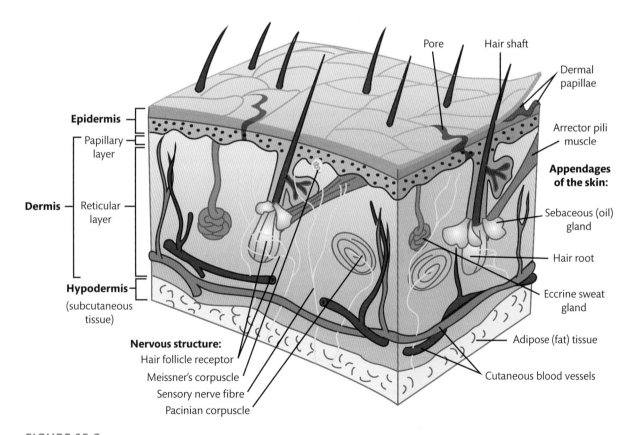

FIGURE 15.2
Schematic of the normal structural components of human skin.
Modified from original image © 2009 Pearson Education, Inc., Publishing as Benjamin Cummings

BOX 15.2 *Structural variation*

Skin structure will vary depending on the anatomical site. Certain parts of the body require a thickened outer epidermal layer; for example, the palms of the hands and soles of the feet. Anatomically, these are the sites where frictional forces have a greater impact on the skin. Conversely, the eyelid skin has a thin epidermis and a very thin outer layer.

These two sites represent the thickest and thinnest layers of skin. Similarly, other sites such as the scalp are rich in adnexae for protective purposes. The important point is that the skin adapts its structure to suit its functional role. Let's look at Figures 15.3a and 15.3b.

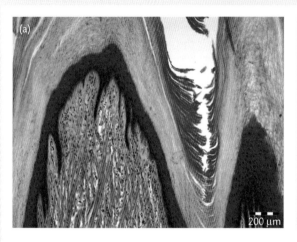

FIGURE 15.3

(a) Normal elephant skin showing a pronounced stratum corneum for protection from extensive abrasion from environmental factors (haematoxylin and eosin [H&E] stain, original magnification ×40). (b) Hippopotamus skin showing a thick epidermal compartment, with a relatively thin stratum corneum but with a densely packed collagen-rich dermis, providing support and protection from the environmental elements. The subcutaneous layer is also very thick. This animal spends most of its time in water, therefore buoyancy is an issue (H&E stain, original magnification ×40).

15.2 **Embryology of the skin**

Epidermal appendages

These include structures such as hair follicles, sweat glands, sebaceous glands and nails.

Some aspects of this section will be have been touched on in Chapter 3. In brief, by just two weeks of gestation the embryo is already covered by a primitive skin. It consists of an outer ectoderm that gives rise to the epidermis and associated **epidermal appendages**. The epidermis sits on a loosely packed mesoderm consisting of the mesenchymal components of the dermis and subcutaneous fat/subcutis. Initially, the ectoderm is composed of a single layer of cuboidal-like cells, but by week six to eight this transforms into two layers, a basal layer and an overlying transient, non-keratinizing layer called the periderm. The periderm has microvilli on its surface and it is believed that these cells have a role in transport or secretion between the embryo and the surrounding amniotic fluid. At around the eighth week, an intermediate layer appears between the periderm and basal layer, which continues to expand and proliferate, forming multiple cells and increases in thickness. By the twenty-third week, keratinization of the outermost cells takes place, the periderm sloughs off and the formation of granular and keratin layers are seen. The intermediate layer then appears as a squamous layer.

Most (90–95 per cent) of the cells within the epidermis are keratinocytes. The remainder include specialist cells such as melanocytes and Langerhans cells, both of which are highly dendritic in nature and move freely within the epidermis. Melanocyte precursor cells migrate to the dermis first from the neural crest and then migrate to the epidermis where they differentiate into melanocytes around 12 weeks of gestation. Langerhans cells are derived from **CD34+** haematopoietic cells of the bone marrow.

A third specialist cell, the neuroendocrine or Merkel cell, is also found in the epidermis. These cells are believed to be specialized keratinocytes. Most of the cells found in the epidermis, such as Langerhans cells and melanocytes, can be seen as early as the twelfth week of gestation; however, Merkel cells are usually seen any time after the sixteenth week of gestation. As the newly formed layers proliferate they extend focally into the mesoderm and these extensions form the pilosebaceous structures (i.e. hair, hair bulbs, hair shafts and associated sebaceous glands), plus the eccrine glands and nails.

The dermis arises from the mesoderm. There are three generally recognized cell types as defined by Breathnach. Type 1 cells, which are stellate and dendritic and are believed to give rise to endothelial cells and smooth muscle cells (periocytes); type 2 cells, which are less dendritic and have round nuclei and are believed to be phagocytic macrophages; and type 3 cells, which have no membrane extensions, are filled with vesicles and are thought to be involved with granulatory/secretory cell types that could be the precursors of melanocytes (melanoblasts) or mast cells. As the dermis matures, there is an increase in the amount of water and glycosaminoglycans that comprise the mucinous 'gluey' matrix of the dermis. This process leads to a reduction in cells and an increase in fibrous content. There are numerous fibroblasts in the dermis by 14 to 21 weeks of development and it is these cells that are responsible for the production of dermal fibres, collagen and elastic fibres. At around 20 weeks there are two dermal compartments that are discernible, the papillary and reticular dermis. They are distinguished largely by the size and density of the collagen fibres present within. Maturation of these compartments continues well after the first or even second year of birth. The subcutaneous fat/subcutis layer forms around week 24 of gestation, with fat cells forming from primitive mesenchymal cells.

The epidermal skin appendages are derived from follicular epithelial stem cells localized in the basal layer of the epidermis. These cells grow both upwards to form the opening of the hair canal and downwards into the dermis to form the dermal papillae of the hair follicle. Subsequently, the hair follicles develop the sebaceous and apocrine glands. The eccrine glands develop from the fetal epidermis in a process independent of the formation of the hair follicles.

> **CD**
> The abbreviation CD stands for 'cluster of differentiation'. This is a universally recognised profile of antibody labelling characteristics for human leucocyte antigens. It enables us to associate a specific CD number with a characteristic labelling profile. You will read more about this in the *Histopathology*, *Cytopathology* and *Haematology* textbooks in this series. Make no mistake, cluster of differentiation is an important concept for students to understand.

SELF-CHECK 15.1

Describe the different types of layer formed during embryological development of human skin, and name the dendritic cells found within the epidermis.

15.3 Structure of the epidermis

The epidermis is composed largely of keratinocytes, with three additional specialist cells: the melanocyte, Langerhans cell and the neuroendocrine cell (Merkel cell). The epidermis is the most densely populated cellular compartment of the skin. The most important point to make clear is that the epidermis is a dynamically evolving and regenerating structure that is continually renewing stratified squamous epithelium that keratinizes and gives rise to the epidermal appendages. This process of regeneration is linked to the biological process known as apoptosis (programmed cell death). The epidermis is between 0.4 mm and 1.5 mm thick depending on the anatomical site. The full thickness of the skin is anything from 1.5 to 4 mm, again depending on the anatomical site. The epidermis extends into the dermis as undulating folds termed rete ridges, and conversely the dermis projects into the epidermis forming finger-like projections called dermal papillae. Essentially there are five layers or zones within the epidermis. Distinctions between these layers are not easy to make using just high-power light microscopy. Changes in keratinocyte maturation can be seen at the ultrastructural level and it is these features that enable the subclassification of the layers of the epidermis.

We mentioned apoptosis earlier and this is the point at which we should expand on this concept. Cells of the epidermis regenerate by a process of division which starts at the basal layer of the epidermis. Approximately one-fifth of the basal cells will be responsible for cell division. Apoptosis is the process by which cells are deleted in normal tissue and also that establishes the normal tissue architecture in the final mature stage of skin development and maintenance. The process of cell division and subsequent apoptosis is essential to the healthy integrity of the epidermal compartment in human skin (Figure 15.4).

The first layer/zone of the epidermis, starting from the bottom and working up is the basal layer or **stratum basale or germinativum**. The cells of this layer are cuboidal and sit on the basement membrane, to which they are attached by numerous hemidesmosomes (Box 15.3). There are numerous cytoplasmic projections that extend through the intercellular spaces and connect to adjacent cells which again are secured by desmosomal junctions. Mitotic figures are frequent in this layer. The stratum basale then migrates into the second layer, the **stratum spinosum**. This layer is also known as the 'prickle cell layer'. The cells are comparatively large compared to those of the stratum basale. The cells are polyhedral with numerous cytoplasmic 'prickles', which again are bound to adjacent cells via desmosomes (Figures 15.5a and 15.5b). This 'prickle cell' appearance is believed to be the result of retraction of the plasma membrane, which occurs during tissue processing for histological evaluation. The cells of this layer also produce a fibrillar protein that is the major component of tonofibrils, which connect with the desmosomal structures. The third layer of cells is the **stratum granulosum**. This layer, one to three cells thick, is characterized by the presence of highly basophilic stained granules within the cytoplasm. These granules are histidine-rich keratohyalin granules that are precursors of the protein filaggrin, which promotes the aggregation of keratin filaments in the cornified layer. The process of keratinization is believed to be a combination of tonofibrils and keratohyalin-forming complexes. The cells in the granulosum layer are also rich in lysosomes. These organelles rupture in the upper cell level and force cells into cell death (apoptosis). The next layer of cells is called the **stratum lucidum**. This layer of cells is really only discernible in specimens of thick skin. It is a homogeneous layer and the cells contain numerous granules (lamellar granules) which are composed of acid hydrolase and neutral sugars conjugated with protein and lipids. Functionally, they are believed to assist in desquamation and also improve

FIGURE 15.4
Epidermis demonstrating a sunburned cell. Keratinocytes divide at the basal layer and then migrate upwards through the epidermal layers, finally becoming anucleate squames before flaking off the skin surface. The arrowed sunburned cell is essentially entering apoptosis earlier than normal due to the detrimental effects of ultraviolet light, essentially killing the cell early in its migratory pathway (H&E stain, original magnification ×100).

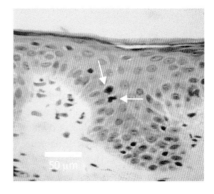

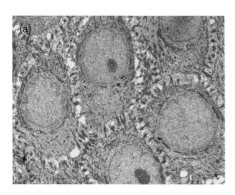
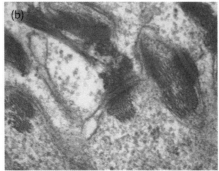

FIGURE 15.5

(a) Cells of the 'prickle' cell layer, otherwise known as the stratum spinosum. Demonstration of retraction of the plasma membranes of the keratinocytes is seen clearly (transmission electron micrograph [TEM], original magnification ×4000. (b) Desmosomal junctions; biological 'rivets' that hold cells together (TEM, original magnification ×7000).

(a) courtesy Richard Mathias).

CLINICAL CORRELATION

Vacuolar changes

Vacuolar changes within the basal cells can lead to subepidermal vesicle formation which gives rise to conditions such as dermatitis and can also be a feature of autoimmune conditions such as graft-versus-host disease and lupus erythematosus (LE).

water preservation properties of the final layer of the epidermis, the **stratum corneum**. The cells of this layer are effectively dead cells that lack nuclei and cytoplasmic organelles. The cells are filled with mature keratin of high molecular weight. The cells are flattened and form a 'basket weave' pattern. As the cells migrate through the cornified layer, the desmosomal junctions become disrupted and loosen, initiating the process of desquamation.

The migration of a basal keratinocyte through the epidermal layers to eventual desquamation can take 22 to 45 days. The variation is due purely to the anatomical site of the skin tissue. The migration process for skin from the palm of the hand and sole of the foot would be nearer 45 days, simply because the skin is thickest at these sites (Figure 15.6b). Conversely, migration in eyelid skin would take nearer 22 days as it is the thinnest skin site in the human body (Figure 15.6a).

FIGURE 15.6

(a) Human eyelid skin, demonstrating a very thin epidermis and a thin 'basket weave' stratum corneum (H&E stain, original magnification ×40). (b) Thumb skin, one of the thickest sites of human skin—that on the sole of the foot is thickest; however, the features are similar. Both palm and sole skin demonstrate an extensive stratum corneum, reflecting the need for protection on sites that experience the effects of friction (H&E stain, original magnification ×100).

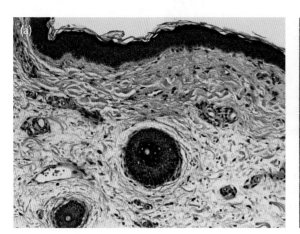
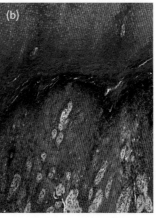

BOX 15.3 Desmosomes

Desmosomes are best regarded as biological rivets. A hemidesmosome is essentially half a desmosome. The structural components of the desmosome are well characterized. First, each cell has a desmosomal plaque which is associated with the internal surface of the plasma membrane. This is composed of six polypeptides: desmoplakin I and II, plakoglobin, desmoyokin, keratocalmin, and band 6 protein. There are keratin filament insertions between cells and into the desmosome structures and these are associated with other interactive proteins such as keratocalmin, which help to regulate the calcium required for assembly and maintenance of the desmosomes. On the external surface or core of the desmosome, there are transmembrane proteins belonging to the cadherin family of proteins (e.g. desmogleins 1 and 3). It is the extracellular domains of these proteins that form part of the core. It is the intracellular domains that then insert into the plaque and link with the intermediate filament keratin as part of the cell cytoskeleton. This then strengthens the process of mooring the cells together. In addition to desmosomes, there are also other intercellular junctions (e.g. gap junctions and adherens junctions). These are quite different in structure and distribution to desmosomes, but offer different mechanisms for cell-to-cell adhesion.

CLINICAL CORRELATION

Pemphigus

Some cutaneous disease states relate to abnormal desmosomal structures. They often result in blistering disorders and lead to exfoliation of skin layers prematurely. An example would be the autoimmune blistering disease pemphigus foliaceus or pemphigus vulgaris. In these two subtypes of pemphigus, patients produce autoantibodies against the extracellular domains of the desmosomal proteins desmoglein 1 and 3. In the case of pemphigus foliaceus it is desmoglein 1, and in pemphigus vulgaris it is demosglein 3. The effect is intraepidermal separation of the keratinocytes within the spinous layer, with the formation of blistering above the basement membrane. The use of **immunofluorescence** (Box 15.4) antibody studies can help to identify and distinguish bullous pemphigus from other autoimmune blistering diseases like bullous pemphigoid (Figures 15.7a and 15.7b). The laboratory-based evaluation of these diseases is discussed in the *Histopathology* volume of this textbook series.

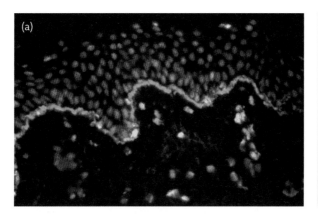

FIGURE 15.7

(a) Direct immunofluorescence using fluorescein isothyocyanate (FITC)-labelled IgG demonstrating a thickened linear BMZ band, resulting in a sub-epidermal blister; this is a characteristic feature of bullous pemphigoid (original magnification ×60). (b) Direct immunofluorescence using FITC-labelled IgG demonstrating intercellular deposits between keratinocytes, effectively forcing cells apart to form an inter-epidermal blister; this is a characteristic feature of pemphigus vulgaris (original magnification ×60).

BOX 15.4 Immunofluorescence

Immunofluorescence (IMF) is a technique used in histopathology for the assessment of patients with blistering disorders of the skin. Evaluation involves tissue (direct IMF) or patient serum for circulating autoantibodies (indirect IMF). In the case of direct IMF, antibodies against IgG, IgA, IgM, complement C3 and fibrinogen (Fb) labelled with fluorescein isothiocyanate (FITC) are applied to sections of affected patient tissue and then evaluated under a fluorescence microscope to detect the presence of immune complexes. Different blistering disorders produce different immune complex patterns, which help to distinguish them diagnostically. You will read more about the use of immunofluorescence in the *Histopathology* volume in this textbook series.

15.4 **Cells of the epidermis**

15.4.1 **Keratinocyte**

You will have already realized the importance of the keratinocyte to the structural integrity of the epidermis; now we will look more closely at its cellular characteristics. First, the keratinocytes contain predominantly the intermediate keratin, which is an α-helical non-glycosylated protein molecule. There are numerous types of keratin—of the order of 30 in the human body. Ten of these are believed to be associated with epithelial cell development and ten are thought to be involved with hair keratins. In terms of molecular weight the range is 40–70 kDa. Understanding why we have so many types of keratin molecule is only just beginning to be clarified. What is clear is that all these various molecular weight keratins have a significant impact on epidermal differentiation and therefore specialization throughout the body, not just in skin tissue but in all epidermal surfaces.

You may wonder how this maturation process of epidermal keratinocytes is regulated. This is a complex process and involves growth factor and proliferation regulators such as epidermal growth factor (EGF) and transforming growth factor-α (TGFα). Similarly, keratinocyte-derived cytokines, interleukin (IL)-1α, IL-6, IL-8 and granulocyte-macrophage colony-stimulating factor (GM-CSF) all have a role in the regulation of the epidermal keratinocyte population. The homeostatic regulation of the epidermis is also linked to several hormones and other factors that control the delicate balance between proliferation, differentiation and apoptosis. Examples of this include vitamin A and its derivative compound, retinoid. This vitamin has an effect on modulating the response of keratinocytes to mitogens. Similarly vitamin D3 has been found to decrease keratinocyte proliferation but increase keratinocyte differentiation in cultured keratinocytes.

15.4.2 **Langerhans cell**

Langerhans cells were first described by Paul Langerhans in 1868. He described a dendritic cell that was non-keratinocytic but which was found interspersed between keratinocytes. It is commonly located in the middle to upper epidermis. The Langerhans cell is approximately the same size as the keratinocyte but has no keratin filaments or desmosomes. The nucleus is described as lobulated. A unique feature of the cell is the presence of cytoplasmic inclusions called Birbeck granules, seen under the electron microscope as 'tennis racket' membrane-bound vesicles (Figure 15.8). The function of these inclusions remains unclear. The dendritic

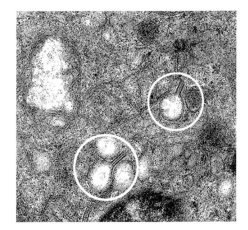

FIGURE 15.8
A Langerhans cell demonstrating the characteristic 'Birbeck granules' that resemble tennis racquets (circled). These structures are classical feature of these cells (HTEM, original magnification ×60 000).
(courtesy Richard Mathias).

nature of these cells cannot be demonstrated with conventional tinctorial stains; however, the use of enzyme histochemical methods such as adenosine triphosphatase (ATPase) or more recently immunohistochemical markers such as anti-S100 protein and anti-CD1a (Figure 15.9) have gained popularity as demonstration techniques. The Langerhans cell is thought to account for 2–4 per cent of the total epidermal cell population. However, there is significant variation in the distribution of these cells throughout the human skin surface. In addition, although the majority are found within the epidermal compartment above the basal layer, they can also be found in the adnexae.

The Langerhans cell is believed to originate in the bone marrow and is the only cell within the epidermis that has receptors for immunoglobulin G (IgG) and complement C3. The dendritic nature of these cells means that they are mobile and will migrate into the dermal compartment and into lymphoid structures. Recent studies on the effects of sunlight on the skin have also implied that the Langerhans cell is photophobic and positively migrates out of the epidermis following sunlight exposure that causes mild sunburn. This is a transient effect and the Langerhans cell will migrate back to the epidermis once the sunburn subsides. This is a significant effect as the Langerhans cell is believed to play a pivotal role in antigen presentation processes in the epidermis and acts as a link at the most peripheral part of the human body's immune system and the outside environment.

FIGURE 15.9
Immunocytochemical labelling of Langerhans cells. The cells are found predominantly above the basal layer and have dendritic processes to assist the process of antigen presentation (anti-CD1a labelling, original magnification ×60).

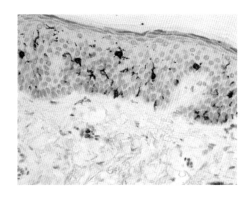

Langerhans histocytosis

The Langerhans cell is significant in its interactions in various inflammatory conditions (e.g. contact dermatitis), but it is also the cell involved in the rare reactive condition Langerhans histocytosis formerly known as histiocytosis X. This comprises several key subset conditions, including Letterer–Siwi disease, Hand-Schuller-Christian disease and eosinophilic granuloma of bone. Although rare, this condition does have significant implications for patients and their subsequent management. Children under the age of two with multisystem disease and organ dysfunction show ≥50 per cent mortality.

15.4.3 Melanocyte

The melanocyte is the cell in the epidermis that is integral to the human body's defence against the effects of ultraviolet (UV) light exposure. It is the melanin pigment that these cells produce that imparts the protective properties of the cells of the epidermis. Embryologically, the melanocyte is derived from the neural crest and it migrates to the epidermis at around 12–16 weeks. Melanin pigment in humans is either brown-black (**eumelanin**) or yellow-red (**phaeomelanin**). Melanin absorbs light at both the visible and UV wavelengths of the spectrum. The synthesis of melanin involves the enzyme tyrosinase and the process takes place in organelles called melanosomes. These organelles are membrane-bound structures located originally in the Golgi apparatus. There are four stages of melanosome maturation, each of which is marked by increasing melanin production. Stage 1 melanosomes have no melanin, stages 2–4 are marked by increasing melanin synthesis along longitudinal lamella-like structures within the melanosomes. By stage 3, melanosomes are marked by prominent melanin deposition (Figure 15.10), while by stage 4 the melanosomes are fully packed with melanin such that their internal structures are obscured. Once synthesized, melanin does not remain within the melanocytes but is transferred to neighbouring keratinocytes via the dendritic cell processes of the melanocyte, from where it is actively taken up by neighbouring keratinocytes. Once in the keratinocytes the melanosomes group together and become enclosed by a membrane, forming the melanosome complex.

It is a fallacy that the number of melanocytes varies in different ethnic groups. The reality is that the number is fairly constant; however, the differentiating factor is that the number of melanosomes, and also their size, will vary from race to race. Thus, in negroid skin the melanosomes are larger and remain singly dispersed within the keratinocytes.

FIGURE 15.10

A stage 3 melanosome showing the characteristic internal lamellar structure (TEM, original magnification ×4000).

(courtesy Tracey de Haro).

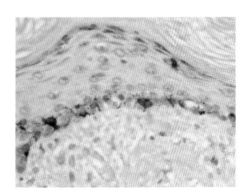

FIGURE 15.11

Immunocytochemical staining of melanocytes in normal skin; the cells are dendritic and located along the basal layer of the epidermis (anti-Melan A labelling, original magnification ×60).

There is some variation in melanocyte distribution depending on anatomical site across the skin surface, and it is worth considering that those races inhabiting hot equatorial climates are often dark-skinned. The melanocytes are located in the basal layer, and histologically cannot be distinguished using routine stains such as haematoxylin and eosin (H&E). Melanin can be impregnated with silver using an **argentaffin** reaction (e.g. Masson Fontana staining); however, the method of choice involves the use of immunohistochemistry to detect S100 protein and a host of melanocyte-associated activation antigens such as HMB-45, Melan A/Mart-1 (Figure 15.11) and tyrosinase.

15.4.4 Neuroendocrine cell (Merkel cell)

Neuroendocrine cells were first described by Merkel in 1875. Round cells with lobulated and irregular nuclei, they are distributed randomly in the basal layer. As with keratinocytes, they do have desmosomal attachments but the keratin filaments are fewer and more loosely packed. They can be grouped together to form clusters, and are often associated with terminal sensory nerve fibres to form slowly adapting mechanoreceptors, which are sensitive to tactile sensations (often called Pinkus corpuscles).

Routine dye staining procedures will not identify Merkel cells; however, using immunohistochemical techniques, the cells are seen to be positive for neuron-specific enolase (NSE), chromogranin, synaptophysin and neurofilament. In addition, they are also positive for the natural killer cell marker CD56. This marker also labels normal nerve tissue and it is probably a shared neural epitope recognized by the same antibody on the Merkel cell that accounts

Argentaffin

These cells take up silver stain. Masson Fontana staining requires no reducing agent (e.g. formalin) to impregnate and demonstrate a tissue component. Conversely, the term argyrophil defines a process in which a reducing agent is required to produce a black deposit. An example of the latter is the silver staining technique of Gordon and Sweet for reticulin fibres.

Cross reference

The demonstration of melanocytes and associated malignant melanoma is discussed in the *Cytopathology* and *Histopathology* volumes in this Fundamentals of Biomedical Science series.

CLINICAL CORRELATION

Malignant melanoma

Melanocytes are the cells that can give rise to malignant melanoma (MM). This is one of the most aggressive cutaneous tumours. If left untreated, MM can kill quite quickly, as it often metastasizes swiftly to other sites in the body. In terms of incidence, malignant melanoma is now one of the fastest growing tumours in the Western world. Its increase in incidence is believed to be associated with the effects of sunlight, particularly UV light (specifically UVB), on the skin. Over recent decades, more leisure time and the opportunity for a wider range of outdoor pursuits have resulted in an increase in sun damage to the skin. Similarly, the increased use of sunbeds has contributed to an increase in the incidence of MM. This has resulted in Cancer Research UK making the statement that 'There is no safe tan unless it comes out of a bottle'.

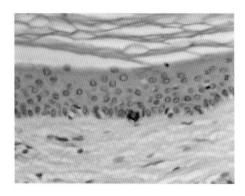

FIGURE 15.12
Immunocytochemical staining of a neuroendocrine cell (Merkel cell) in normal skin; the cells are located along the basal layer of the epidermis and are best regarded as 'specialized keratinocytes'; note that there are no dendritic processes (anti-8/18 cytokeratin labelling, original magnification ×60).

for this positivity. As well as expressing positivity for neural-type markers, Merkel cells are also positive for some cytokeratins. An almost unique finding is the demonstration of a paranuclear dot-like positivity with antibodies to low molecular weight cytokeratins such as cytokeratin 8/18 (Figure 15.12) and cytokeratin 20. It should be remembered that a panel of antibodies is often used when evaluating Merkel cells because the degree of expression with these antibodies will vary from case to case. Under the electron microscope, Merkel cells appear to have scant cytoplasm, invaginated nuclei, cytokeratin filaments and characteristic membrane-bound dense granules (neurosecretory granules; Figure 15.13).

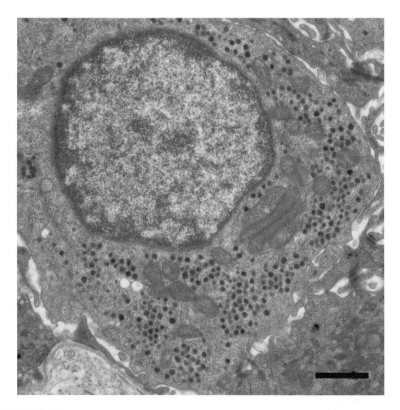

FIGURE 15.13
Neuroendocrine cell of the epidermis demonstrating the characteristic darkly stained neurosecretory granules. The average size of these granules is approximately 100 nm diameter (TEM, scale bar = 1 μm).
(courtesy Patricia Dopping-Hepenstal).

Trabecular carcinoma

Merkel cells can give rise to an aggressive tumour first reported by Toker in 1972 and called a tra-becular carcinoma. The term trabecular relates to the architecture of the tumour cells. They can often appear to have a 'trabecular' pattern. However, this is not a universal finding and the term neuroendocrine carcinoma is now the preferred name for these tumours. Later observations of neurosecretory granules within the tumour cells confirmed the association with Merkel cells. The tumour arises in sun-exposed sites, often in elderly patients. Under the light microscope the tu-mour cells appear as small, round to oval cells with vesicular nuclei. The tumour arises within the dermis, often as sheets and solid nests down as deep as the subcutis. These tumours are, however, relatively rare compared to the incidence of other cutaneous cancers.

15.5 **Basement membrane zone**

Having considered the epidermal compartment in some detail, we now need to understand how this compartment of the skin is attached, or moored, to the dermis. The epidermis rests on a relatively thin but highly complex layer, the basement membrane zone (BMZ). This thin, uniform line of tissue components is not identifiable using conventional H&E staining but requires the use of stains such as periodic acid Schiff (PAS; Figure 15.14). As well as acting as a continuous point of attachment for the epidermis to the dermis, the BMZ also acts to resist the effects of frictional forces on the skin, and plays a central role in the orchestration of the cytoskeleton of basal keratinocytes and acts as a semi-permeable barrier for the transfer of epidermal developmental signals (Figure 15.15). The one point that should be made clear at this point is that the BMZ is a very complex and highly interactive structure. Even to this day,

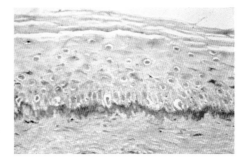

FIGURE 15.14
Basement membrane shown as a thin magenta-coloured line between the epidermis and dermis (periodic acid Schiff [PAS] stain, original magnification ×100).

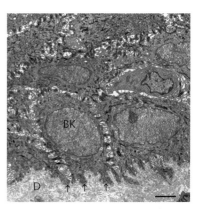

FIGURE 15.15
Epidermal/dermal junction (TEM, scale bar: 2.5 μm; BK: basal keratinocyte, D: dermis; arrows indicate the basement membrane zone composed of the lamina lucida and lamina densa.
(courtesy Patricia Dopping-Hepenstal).

the full supramolecular construction and, more importantly, the associations of these protein complexes to each other, remain to be fully elucidated.

The BMZ is broadly subdivided into three structurally different networks:

- hemidesmosome and anchoring filament complex
- basement membrane proper
- mooring and supporting anchoring fibrils.

The hemidesmosome and anchoring filament complex reside on the plasma membranes of the basal keratinocytes. The plaque portion of the hemidesmosomes consists of several important antigens (e.g. bullous pemphigoid antigen [BP 230] and plectin). The transmembrane component of the hemidesmosome complex also contains important antigens (e.g. BP 180 and α6 β4 integrin).

Just below this complex is the basement membrane proper, which is composed of two distinctive zones. The first appears as an empty electron-lucent area called the **lamina lucida**. This area has anchoring filaments containing an assortment of laminin isoforms, entactin/nidogen and fibronectin. This area of the BMZ is thought to be the weakest part, as separation within the epidermis using salt solutions or heat can be achieved relatively easily through this zone. Below the lamina lucida is the **lamina densa**. Unlike the lamina lucida, the lamina densa is an electron-dense area. This area functions as a barrier/filter that restricts the passage of molecules with a molecular mass >40 kDa. Interestingly, it is penetrated by melanocytes and Langerhans cells as they migrate up and down through the epidermal and dermal compartments. This zone is composed mainly of collagen type IV and some laminins. Below the **lamina densa** we have the third component of the BMZ, the **sublamina densa zone**.

The sublamina densa zone is composed of broad (20–60 nm) and elongated (200–800 nm) anchoring filbrils. These are flexible, fibrillar, come from the lamina densa and are responsible for securing the basal lamina to the dermis. The major component of this area of the BMZ is collagen type VII, although types I, III, V and VI are also present.

Using immunofluorescence, immunoblotting, molecular and electron microscopy techniques to evaluate protein structures present within the BMZ, it is now possible to identify specific protein defects associated with specific blistering diseases of the skin.

SELF-CHECK 15.2

Describe the constituents that make up the structural components (layers) of the epidermis and basement membrane zone.

CLINICAL CORRELATION

Autoimmune blistering

Autoimmune blistering of the skin comprises a wide assortment of diseases, which clinically can have considerable overlap. Distinguishing between these entities and therefore how best to manage the patient's condition is very demanding both scientifically and technically.

Examples of identified antigenic epitope defects include:

- *Bullous pemphigoid*: patients have antibodies against the antigen epitope BP230 associated with the lamina lucida and hemidesmosome complex.
- *Epidermolysis bullosa acuisita (EBA)*: patients have antibodies against an antigen associated with collagen VII, a major component of the anchoring fibrils.

15.6 **Structures of the dermis**

As mentioned earlier, compared to the epidermis, the dermis is relatively sparsely populated with cells. However, it does contain a wide range of structural units that are important to the skin's functional roles. In brief, it is composed of a matrix of fibres, most notably collagen and elastic fibres, and a ground substance mainly composed of glycosaminoglycans (GAGS) (Figure 15.16). The structures within the matrix include the adnexae (epidermal appendages), nerve endings (corpuscles), vessels of lymphatic and blood vessel origin, and finally a selection of specialized cells including mast cells, fibroblasts, histiocytes, dermal dendrocytes and a few lymphocytes. At the very base of the dermis sits the subcutis, which essentially is a layer of subcutaneous fat composed of adipocytes.

15.6.1 **Dermal matrix**

For the purpose of explaining the composition of the fibres and ground substance that make up the dermal matrix, it is necessary to make a distinction between two areas of the dermis:

- *Papillary dermis*: the compartment from directly below the epidermis to the adnexae
- *Reticular dermis*: the remainder of the dermis to the subcutis.

The papillary dermis is composed of poorly organized collagen fibres, mainly type I and III. Mature elastic fibres are not found in this area of the dermis, but fine oxytalan elastic fibres are present. The papillary dermis also has a dense population of fibroblasts and dermal dendrocytes. There are also capillaries that extend from the upper parts of the zone to the epidermis from the dermal papillae (finger-like projections of the papillary dermis that interdigitate with the rete ridges that project from the epidermis into the dermis (Figure 15.17).

The reticular dermis, in comparison, is much thicker and composed of large-diameter collagen fibrils that are interwoven to form large fibre bundles. These are mostly of type I collagen origin. In addition, there are mature elastic fibres, which are often seen to be branching and band-like. These bundles of fibres are often arranged horizontally or parallel to the epidermis. It is the elastic fibres that give the skin its flexibility (Figure 15.18). As we age, degenerative changes affect the elastic fibres; this is termed **solar elastosis** (Box 15.5). The bulk of the structural components of the dermis are found in this area of the dermis, including nerve plexuses and epidermal appendages. In addition, this area also contains a host of specialized cells (e.g. dermal dendrocytes, mast cells and fibroblasts).

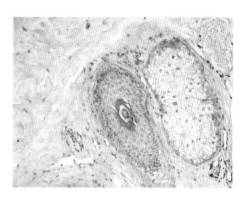

FIGURE 15.16
Glycosaminoglycans (GAGS) within the dermis; invariably there are increased amounts around hair follicles (Alcian Blue [pH 2.5], original magnification ×60).

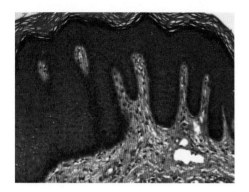

FIGURE 15.17
Rete ridges (pendulous invaginations of the epidermis into the dermis) in normal human skin (H&E stain, original magnification ×40).

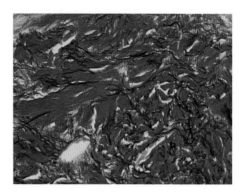

FIGURE 15.18
Normal fine elastic fibres (black) within the dermis (elastic Van Gieson [EVG] stain, original magnification ×60).

BOX 15.5 Solar elastosis

The term solar elastosis is often used to describe the microscopy features that relate to the degenerative changes of elastic fibres. What one sees is a change from organized fibres arranged in bundles running parallel to the epidermis to amorphous globes of tissue, termed 'elastic globes'. Functionally, the process of solar elastosis results in a loss of elasticity in the skin. This can be demonstrated by pinching the skin on the back of the hand of someone with extensive solar elastosis and seeing how slowly the skin returns to its normal resting position. In healthy skin, the tissue springs back almost instantly. Some of these changes are due solely to the effects of 'old age'; however, it can be exacerbated by the effects of sunlight (solar radiation). This can age the skin prematurely and, in terms of elastic degeneration, speed up the process of solar elastosis (Figures 15.19a and 15.19b).

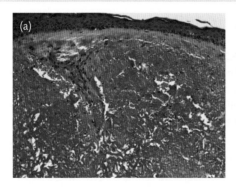
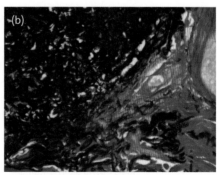

FIGURE 15.19
(a) Extensive solar elastosis in the dermis of sun-exposed skin; note the amorphous appearance of the fibres, no longer resembling fine fibres (EVG stain, original magnification ×20). (b) Extensive solar elastosis; note the black amorphous elastic tissue (EVG stain, original magnification ×60).

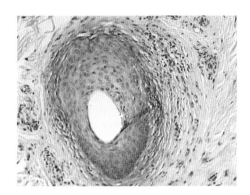

FIGURE 15.20
Hair follicle demonstrating the presence of GAGS (Hale's stain, original magnification ×40).

In addition to the presence of fibres, there is also the dermal matrix ground substance that resides on and between the fibre bundles. This is mainly GAGS or acid mucopolysaccharides composed mainly of hyaluronic acid and sulphated acid mucopolysaccharides (mainly chondroitin sulphate). Cumulatively, GAGS comprise about 0.2 per cent of the dry weight of the dermis. Although low in terms of the percentage of dermal material, they are vital for the general maintenance of the fibres they surround. Hyaluronic acid has been found to regulate the water-binding properties of the dermis, and influence compressibility, proliferation, differentiation and tissue repair processes. Chondroitin sulphate has been found to influence the function of growth factors and cytokines. Generally, the demonstration of GAGS in skin is difficult because they are labile and do not survive as well as other tissue components once processed to paraffin wax and stained for routine histopathological assessment. Stains of choice include cationic dyes such as Alcian Blue or Hale's dialysed iron techniques (Figure 15.20).

15.6.2 Dermal adnexae

The adnexae include the pilar sebaceous unit (i.e. hair follicles, sebaceous glands, arrector pili muscle, eccrine and apocrine glands). We should also include the hairs themselves and the specialized adaptations of these structures, the nails.

15.6.3 Pilosebaceous unit

Pilosebaceous units (Figure 15.21) will vary in distribution over the human body, depending on anatomical, evolutionary and gender differences. For example, the density of such units on the scalp is much higher than on the trunk. Similarly, the distribution of glandular structures such as apocrine glands will also differ in density, with a high density seen in the axilla (armpit). The upper chest will also have a high density of sebaceous glands.

The hair follicle is composed of three sections. At the bottom of the hair shaft is the hair **bulb**. This section develops the hair shaft. It is the growth pressure from the hair bulb that pushes the shaft upwards. The hair shaft (or follicle) has a central core (the medulla), which is surrounded by densely packed keratinocytes (the cortex). There is a thin layer (the cuticle) around the cortex, and a surrounding cylindrical structure (the root sheath). The root sheath also has an inner cuticle and it is the interaction or meshing of the two cuticles that effectively moors the shaft to the sheath. Dendritic melanocytes are present in the upper half of the hair bulb only (Figure 15.22). There are two additional cylindrical layers outside the cuticle of the inner root sheath and these are called Huxley's and Henle's layers, respectively. As the cells are pushed up

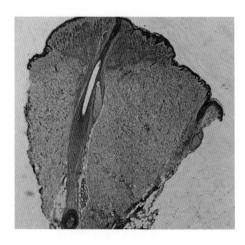

FIGURE 15.21

Pilosebaceous unit demonstrating the hair, hair bulb and shaft plus the associated sebaceous gland and eccrine glands at the base. The hair follicle penetrates an intradermal neavus (mole) before entering the epidermal compartment (H&E stain, original magnification ×4).

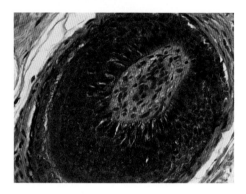

FIGURE 15.22

Hair bulb demonstrating an abundance of dendritic melanocytes and melanin (H&E stain, original magnification ×40).

the hair shaft they become increasingly keratinized. At around the mid-point of the hair shaft, the cells are fully keratinized and this represents the start of the second section of the hair shaft, the **isthmus**. At this point, the arrector pili muscle attaches to the follicle. The arrector pili muscle pulls the hair shaft up when heat needs to be trapped on the skin surface to insulate, and acts to push the hair down on the surface to assist in the dissipation of heat when the body is too hot. At the isthmus, the outer layer of the follicle, the external root sheath, is seen. At the bottom end of the external root sheath there are clear cells, which are glycogen-rich and form what is called the trichilemmoma (Figure 15.23). We then progress to the final section of the hair shaft, the **infundibulum**. Here, the keratinocytes in the hair follicle resemble those

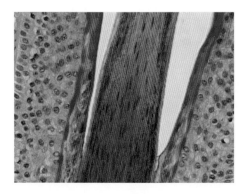

FIGURE 15.23

Bottom of the external root sheath of a hair follicle demonstrating the glycogen-rich tissue termed the tricholemoa (PAS stain, original magnification ×60).

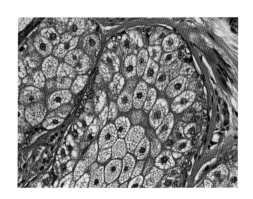

FIGURE 15.24
Sebaceous gland demonstrating lipid-rich sebocytes (H&E stain, original magnification ×40).

seen in the epidermis, and we also see the opening of the sebaceous duct at this point. The hair shaft then continues through to the opening of the hair follicle on the epidermal surface.

15.6.4 Sebaceous glands

The **sebaceous glands** are associated with the hair follicles. They insert at the mid-portion of the hair shaft, around the top of the isthmus and just before the infundibulum. The gland is multilobular and composed of multiple acini where glands are found in abundance, and as a single acinus in sparsely distributed areas. Inside the lobules are germinative cells that line the periphery. The inner cells are called sebocytes and they produce a secretion (sebum) that is lipid-rich and composed predominantly of disintegrated cells (Figure 15.24). Distribution of these glands varies around the body—there are no sebaceous glands on the palms of the hands and soles of the feet; however, they are found in abundance on the face. Sebum is secreted through the excretory duct opening, and it is thought to act as an emollient on the skin surface, and may also have antibacterial properties.

15.6.5 Eccrine glands

The functional role of **eccrine glands** is one of thermoregulation. Their structure is coiled and tubular (Figure 15.25) and they are found in abundance on the palms of the hands and soles of the feet. This coiled structure twists through the epidermis and opens on the surface (Figures 15.26a and 15.26b). The morphology of the cells lining the duct is very similar to that in the apocrine gland. There are three cell types in the eccrine coil: clear cells, dark cells and

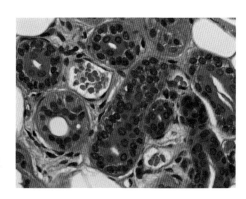

FIGURE 15.25
Eccrine gland; note its tubular nature (H&E stain, original magnification ×60).

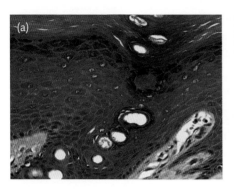 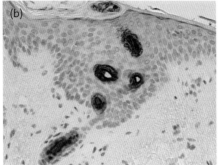

FIGURE 15.26

(a) Eccrine gland; note its twisting, coiled nature as it passes through the epidermis and eventually opens on the epidermal surface (H&E stain, original magnification ×60). (b) Immunocytochemical staining of the ductal opening of the eccrine gland (carcinoembryonic antigen [CEA] labelling, original magnification ×20).

myoepithelial cells. Clear cells are secretory in function and contain diastase-labile glycogen. They have a granular cytoplasm and round nucleus. The dark cells (or mucoid cells) contain neutral mucopolysaccharide and border the central lumen of the coiled structure. They also contain secretory granules that are PAS-positive. The spindle-shaped myoepithelial cells surround the coil structure, which in turn is surrounded by the basement membrane. The glandular excretory structure has three segments: a convoluted duct, a dermal component that is straight, and a spiral component called the **acrosyringium** that opens at the epidermal surface. The secretions are composed mainly of water which evaporates on the body's surface, effectively reducing body temperature. The eccrine gland is stimulated by non-myelinated sympathetic post-ganglionic nerve fibres, which can be seen surrounding the coil structure.

Acrosyringium
This is a component of the glandular excretory structure. It is lined by keratinocytes that are smaller than those of the surrounding epidermis, and are attached to the larger keratinocytes by desmosomal junctions.

15.6.6 Apocrine glands

Apocrine glands are located mainly in the axillae (armpits), mammary region, eyelids and genital areas. They are always associated with the pilosebaceous unit and arise from the point above the insertion of the sebaceous duct. The glands have a tubular, coiled secretory portion and a ductal excretory portion. The secretory portion (Figure 15.27) is considerably longer than the equivalent in the eccrine gland, and is composed of simple 'budding-type' epithelium. The excretory portion has a double cuboidal layer of cells. Under high-power magnification, it

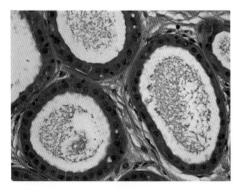

FIGURE 15.27
Secretory portion of the apocrine glands from armpit/axillary tissue (H&E stain, original magnification ×40).

is possible to identify microvilli on the surface of luminal cells. The cells appear highly eosino-philic due to a relatively high percentage of keratin filaments in the cell cytoplasm. The secre-tions from these glands are proteinaceous in nature but a functional role remains unclear.

Key points

The formation of body odour is believed to be associated with the effects of bacterial breakdown of apocrine gland secretions on the body's surface.

15.6.7 Hairs and nails

Hairs and nails represent some of the specialized adaptations of the adnexae. Most of the details relating to hairs have been discussed in the section on the hair follicles.

There are two types of hair on the body:

- *Terminal hair*, best described as the coarse, thick hairs of the body that arise from root bulbs deep in the dermis and subcutis.
- *Vellus hair*, which is fine, thin hair that only grows from the upper dermis.

Hair has three growth phases:

- Active growth (**anagen**)
- Regressive phase (**catagen**)
- Resting phase (**telogen**).

Hair can be sampled and examined in patients who have **alopecia** (Figure 15.28, Box 15.6).

A fact perhaps not so well known is the existence of mites that reside within hair follicles. There are two mites found in normal skin: ***Demodex folliculorum*** (Figures 15.29a and 15.29b) lives in hair follicles and ***Demodex brevis*** resides in the sebaceous glands. These mites are common observations in biopsies taken for pathological evaluation. They feed on sebum, the product of the sebaceous gland, and rarely cause adverse effects in the skin. Occasionally, however, they may cause localized pustular folliculitis and more widespread eruptions in **immunocompromised** patients.

Nails represent a protective covering over the distal portion of the fingers and toes. The nail is composed of a dense nail plate, which sits on stratified squamous epithelium. The plate is composed of hard keratin and rests on the nail bed. The nail plate is surrounded on three sides

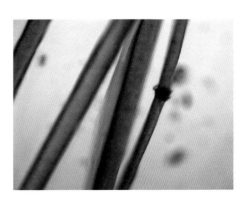

FIGURE 15.28
Hair removed from a patient under investigation for alopecia. Note the abnormal twist of the hair on the right (original magnification ×2).

BOX 15.6 *Alopecia*

Alopecia is abnormal hair loss. It is a very emotive subject and is increasing in incidence. Identifying the cause often involves extensive assessment of clinical details and also histological investigation of tissue biopsies and the hairs themselves. In many instances, defining the underlying causal mechanism is often the ideal approach, as this will give valuable information to the clinician, who is then better equipped to inform and manage the patient. Broadly speaking, alopecia is classified into two groups: scarring or non-scarring.

by skin folds, the left and right lateral and proximal nail folds. The proximal nail end (nail root) and the nail bed extend deep into the dermis and close to the bone joint. At the distal boundary of the proximal nail end is the lunula (the white half-moon). In this portion of the nail the keratinocytes are not fully keratinized and they retain their nuclei and therefore appear more opaque. Nails grow by proliferation and differentiation of the epithelium surrounding the nail root from the new cells that comprise the nail matrix. The nail plate then slides over the rest of the nail bed (Figures 15.30a and 15.30b).

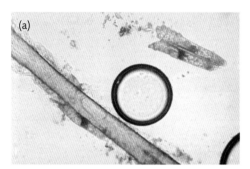
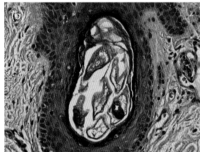

FIGURE 15.29
(a) Demodex mite attached to the side of a hair shaft, and additional mites separate from the hair shaft (original magnification ×20). (b) Normal skin hair follicle showing cross-sections of Demodex mites in the follicle centre (H&E stain, original magnification ×60).

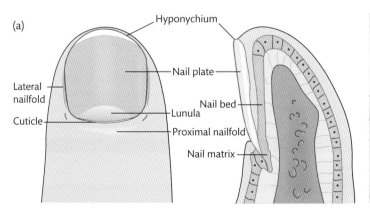
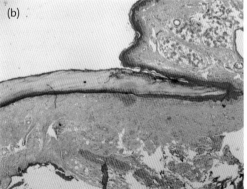

FIGURE 15.30
(a) Schematic presentation of the human nail and associated structures. (b) Nail and underlying nail bed, from a longitudinal section through normal toe tissue (H&E stain, original magnification ×2). Preparation by Ms Elizabeth Clayton.

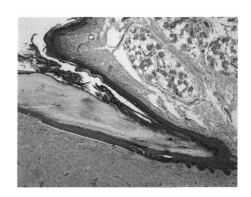

FIGURE 15.31
Toe nail and overlying cuticle skin (H&E stain, original magnification ×4).

The nail bed starts at the distal aspect of the lunula and ends at the skin under the free end of the nail, called the hyponychium. The nail-fold skin is highly keratinized and is called the eponychium (Figure 15.31).

Key points

The rate of nail growth is different for fingernails and toenails. Under normal circumstances, the rate of growth for fingernails is 0.1 mm per day, whereas for toenails it is approximately 0.05 mm per day.

SELF-CHECK 15.3

Describe the structures and functions of the components of the pilosebaceous unit.

15.6.8 Blood vessels, lymphatics and nerves

The circulatory system of the skin is slightly unusual and complex, due to varying functional requirements. It is composed of two plexuses that reside deep in the dermis and run horizontally to the epidermal surface. The deeper plexus is situated at the junction of the dermis and subcutis and is termed the cutaneous plexus. The second plexus is located just below the papillary dermis and is called the papillary plexus. The plexuses are interwoven arterioles and venules. The structure of the arterioles and post-capillary venules in two vascular plexuses is similar. In the papillary plexus, there is the formation of the capillary loops as arterioles and post-capillary venules meet the capillary network. The post-capillary venules possess a multi-layer basement membrane that has the ability to pull endothelial cells apart to allow the flow or exchange of cells and products between the blood and surrounding tissues, to increase vascular permeability and create inflammatory states. The primary role of the dermal vasculature is to provide nutrients to surrounding tissues and also increase blood flow to facilitate heat

CLINICAL CORRELATION

Pathologies of the nail

The nails are often affected in a host of pathological conditions. Most notably, autoimmune states such as lichen planus and psoriasis can disrupt normal nail growth. In addition, malignant melanoma can arise in the nail tissue and produce a particular subtype of melanoma termed 'acral lentiginous' malignant melanoma, which is the most common variant of melanoma found in the dark-skinned population.

loss from the skin surface in hot conditions, and, when required, reduce blood flow to the skin surface to conserve heat in cold conditions.

The lymphatic system is very similar to the vascular circulatory system. There are also two lymphatic plexuses. The upper plexus is at the base of the papillary junction and intertwines with the papillary vascular plexus. The deeper plexus intertwines with the deep dermal vascular plexus. The lymphatic vessels lack muscle coats (pericytes). They also have a discontinuous basement membrane which allows the exchange of lymph from the vessels into the tissues, and vice versa. The lymphatic vessels of the upper plexus also have lymphatic valves, which act to ensure that lymph flows in one direction.

Structures called glomus bodies, found mainly in acral skin (fingers and toes), are composed of highly convoluted arteriovenous structures in the reticular dermis. These structures are believed to be under sympathetic nervous system control and assist in thermoregulation.

The neural network of the skin is composed of myelinated and non-myelinated nerve fibres. There is a deep dermal nerve plexus at the junction of the dermis and subcutis. Large nerve bundles can be seen in the deep reticular dermis and subcutaneous fat. Smaller nerve bundles are found in the superficial nerve plexus, up to the papillary dermal compartment and epidermal ridges. Using immunohistochemical and ultrastructural methods, single nerve axons have been identified in the dermal papillae and epidermis.

There are also sensory nerve receptors, and broadly speaking these are either free or encapsulated. Free nerve receptors are found widely throughout the dermis and are also often associated with the epidermal appendages. These nerve receptors are unmyelinated nerve axons. There are free sensory nerve receptors within the epidermis and these are thought to be associated with Merkel cells (mechanoreceptors). The encapsulated nerve receptors are found within the dermis and include the Meissner corpuscles. These are found mainly in the papillary dermis and represent the mechanical receptors that respond to the sense of touch. Not surprisingly, they are found in greatest abundance within skin from the palms of the hands and soles of the feet, but also in the lips, eyelids and genitalia. Pacinian corpuscles (Figure 15.32) are mechanoreceptors that respond to vibration and pressure. They are located throughout the reticular dermis and also in the subcutis. As with Meissners corpuscles, they are also located in greatest numbers in a similar distribution pattern. Other encapsulated sensory receptors include the Ruffini corpuscles, Krause end-bulbs and the genital corpuscles.

SELF-CHECK 15.4

Describe the corpuscles found within skin. What are their functional roles?

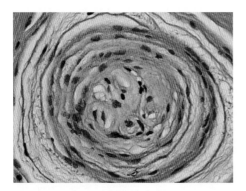

FIGURE 15.32

Pacinian corpuscle from dermal thumb skin (H&E stain, original magnification ×60).

15.6.9 Key cells of the dermis

Just as we have considered the key cells of the epidermis, we should also consider the equivalent cells of the dermis. For the purposes of this book, there are four main cell types that we need to consider: mast cells, fibroblasts, dermal dendrocytes and histiocytes.

15.6.9.1 *Mast cells*

Mast cells originate in the bone marrow from CD34-positive stem cells. They are located throughout the dermis but will be found in greatest numbers around blood vessels. Essentially, mast cells are specialized secretory cells that are not unique to skin tissue but are widely distributed throughout the body. These cells are easily identifiable histologically as they have a large oval nucleus and striking darkly stained cytoplasmic granules in an H&E-stained section. These granules can be seen as electron dense under the electron microscope (Figure 15.33). These cells synthesize a host of inflammatory mediators in granule form for quick release following the appropriate stimulus. The mediators primarily include histamine but also heparin, tryptase, chymase and neutrophil chemotactic factor. Mast cells are involved principally with increasing the permeability of vascular tissue following inflammatory responses. Mast cells are key to cutaneous allergic reactions. In addition, they have a role in eosinophilic activation and proliferation, and they also promote phagocytosis and are believed to be involved with connective tissue repair mechanisms. They are demonstrated histologically by virtue of the mediators they contain using tinctorial (e.g. toluidine blue) and also enzyme histochemical methods (e.g. chloroacetate esterase; Figure 15.34).

15.6.9.2 *Fibroblasts*

Fibroblasts are found in greatest number in the dermis. They are mesenchyme-derived cells responsible for the synthesis and breakdown of fibrous and non-fibrous connective tissue. The cells achieve this by modulating and synthesizing soluble mediators and are responsible

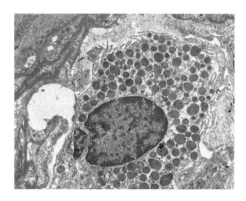

FIGURE 15.33

Mast cell demonstrating the darkly staining cytoplasmic granules. These granules release a host of inflammatory mediators under stimulation and burst free from the cells—a process termed mast cell degranulation (TEM, original magnification ×6000).

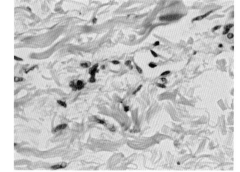

FIGURE 15.34

Enzyme histochemical staining (red) of mast cells (chloroacetate esterase [CAE] stain, original magnification ×60).

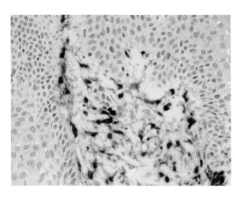

FIGURE 15.35
Immunocytochemical staining of dermal dendrocytes. These cells of the dermis are found in greater numbers in the papillary dermis and around hair follicles (anti-FXIIIa labelling, original magnification ×80).

for the production of dermal collagen, elastin and also the ground matrix substances of the dermis. Of great interest is the activity of fibroblasts in the process of regulating wound healing.

15.6.9.3 *Histiocytic cells*

Histiocytic cells (macrophages) are derived originally from bone marrow and leave the blood as monocytes and enter the skin tissue and differentiate into macrophages. These cells have a wide range of functions in skin as they do in all other tissue sites. Fundamentally, however, they are scavengers that phagocytose particles. They are also antigen-presenting cells with a role in antimicrobial defence and also antitumour responses and wound healing. They have a central role in the granulomatous responses of the skin and can multiply and merge to form giant cells characteristic of many granulomatous reaction patterns in the skin. They can be identified by a wide range of histiocytic antibodies (e.g. antibodies to CD68, Mac 387 and lysozyme).

15.6.9.4 *Dermal dendrocyte cells*

Dermal dendrocyte cells are best regarded as highly phagohistiocytic cells. Early studies suggested that they were a form of fibroblast; however, they have been shown to lack classical fibroblastic markers. They are found throughout the dermal compartment, but are present in greatest numbers in the papillary dermis and around hair follicles and associated sebaceous glands. They can be labelled with some markers of mononuclear macrophage lineage, but are characteristically positive for the marker factor XIIIa (Figure 15.35). In the skin, increased numbers of dermal dendrocytes are found commonly surrounding scar tissue. They are present in benign fibrohistiocytic tumours of the skin (e.g. dermatofibroma), and in increased numbers in inflammatory conditions such as eczema.

Key points

Anti-factor XIIIa is very useful in aiding the distinction between dermatofibroma, a benign fibrohistiocytic tumour that is FXIIIa-positive, and dermatofibrosarcoma protuberans (DFSP), which can show similar morphological features but is FXIIIa-negative and CD34-postive. It is important to realize that the use of antibodies raised against the histiocytic/fibrohistiocytic cell types does not always provide clear delineations between cells and their clear cell of origin, as there is considerable overlap between these cell types.

15.7 **Subcutis (subcutaneous fat)**

The subcutis is the final layer of skin tissue and is composed of fat-containing cells called adipocytes, which are held together in groups or lobules separated by connective tissue septa.

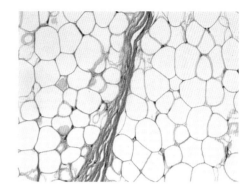

FIGURE 15.36
Lipid-laden adipocytes from the subcutaneous tissue deep within the dermis (H&E stain, original magnification ×60).

This layer of the skin will vary in thickness depending on a number of factors including location around the body, the age, gender and diet of each individual. Adipocytes contain large amounts of cytoplasmic lipid and when mature often show the nucleus pushed to one side of the cell. This layer of the skin has three functional roles: (i) it acts as an insulating layer that prevents heat loss; (ii) it acts as a buffer (body padding) to minimize damage following physical trauma; and (iii) the presence of lipid within adipocytes is an energy store (Figure 15.36).

15.8 Breast tissue

In basic terms, breast tissue is a specialized form of apocrine glands. The histology of the breast varies depending on a range of factors, most commonly including gender, tissue age, stage of the menstrual cycle, menopause, pregnancy and lactation. Many of these significant events are allied to significant hormonal changes that affect the tissue structure and architecture throughout life. The key hormones that affect the breast are oestrogen, which is required for ductal growth and branching, particularly during adolescence; progesterone, which is required for lobuloalveolar differentiation and growth; and testosterone, which stimulates the breast mesenchyme during fetal development. It is important to remember that there are a host of other steroid and peptide hormones that also have an impact on breast tissue regulation (e.g. glucocorticoids, insulin and prolactin).

The breasts are located on the anterior chest wall and overlie the pectoralis muscle. The breast tissue also extends and projects into the axilla (armpit) at the point termed the tail of Spence.

15.8.1 Female breast tissue

Female breast tissue is composed of ducts, ductules and lobular acinar units embedded in a fibrous and adipose tissue stroma (Figure 15.37).

The ductal/lobular units are organized into lobes that are poorly defined without obvious boundaries. Each lobe has a branching structure, with lobules draining into ducts that in turn drain into larger collecting ducts, often termed trunk ducts. These ducts then open onto the surface of the nipple; however, just before entering the nipple, the ducts expand to form a sinus—the lactiferous sinus.

The epithelium of the ductal/lobular system is bilayered and composed of an inner (luminal) cell layer and an outer (basal) layer referred to as the myoepithelial cell layer.

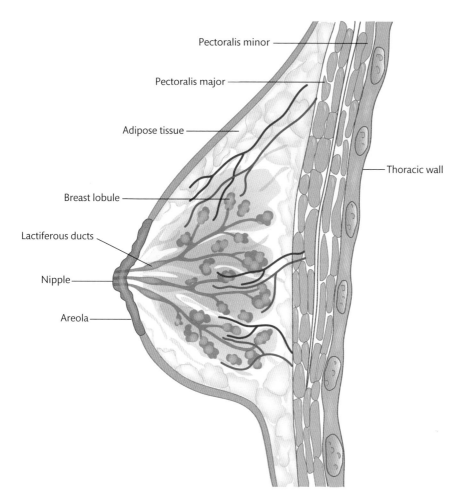

Pectoralis minor

Pectoralis major

Adipose tissue

Breast lobule

Lactiferous ducts

Nipple

Areola

Thoracic wall

FIGURE 15.37
Structural components of the normal female breast.
Patrick J. Lynch, Medical Illustrator/CC-BY.

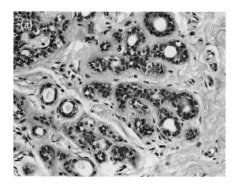
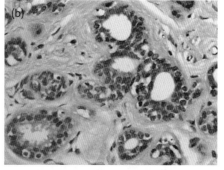

FIGURE 15.38
(a) Luminal cells from the ductal-lobular component of female breast tissue (H&E stain, original magnification ×40). (b) Luminal cells demonstrating cuboidal/columnar cells with oval nuclei and a pale eosinophilic cytoplasm (H&E stain, original magnification ×60).

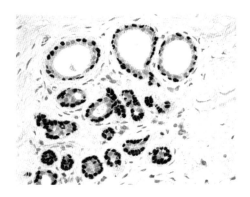

FIGURE 15.39
Immunocytochemical staining of myoepithelial cells surrounding the basal layer of the ductal-lobular structures (anti-P63 labelling, original magnification ×40).
(courtesy Janet freeman).

The luminal cells are normally cuboidal/columnar in appearance with oval nuclei and a pale eosinophilic cytoplasm (Figures 15.38a and 15.38b). Importantly, these luminal cells express a variety of low molecular weight cytokeratins (e.g. cytokeratins 7, 8, 18 and 19). Conversely, the outer basal layer of myoepithelial cells express high molecular weight cytokeratins, most notably cytokeratins 5/6, 14 and 17. In the majority of cases, the myoepithelial cells appear to be flat with compressed nuclei, but they can also resemble plumper epithelioid cells or even spindle-shaped cells, which appear similar to smooth muscle cells. As well as expressing high molecular weight cytokeratins, these cells are also positive for S100 protein, p63 (Figure 15.39), smooth muscle actin (Figures 15.40a and 15.40b), calponin and CD10 among a host of other, less well known markers.

These epithelial cells are secured to a basal lamina, which is composed largely of type IV collagen and laminin, and this sits outside the myoepithelial cell layer. Beyond this point, the breast tissue is composed of the stroma, which is made up of a mixture of fibroblasts, collagen, elastic fibres and adipocytes.

The lobule, in association with the terminal duct, is described as the terminal duct lobular unit and is regarded as the structural and functional unit of the breast. These structures are responsible for secretion and transport. The lobules have blind-ending sacs or ductules called acini, which have associated stromal tissue that contains cells such as lymphocytes, macrophages and mast cells.

There are four types of lobule, with types 1–3 commonly occurring simultaneously in normal breast tissue, and type 4 arising during pregnancy and lactation. The main structural difference between these types of lobule resides in the number of alveolar buds within them; type 1 has considerably less than types 2 and 3.

The nipple-areola structure shows increased pigmentation and contains many sensory nerve endings. The tip of the nipple has around 15–20 orifices. There are also elevations and undulations on the surface of the areola, termed tubercles of Montgomery. The nipple and areola are

FIGURE 15.40
(a) Immunocytochemical staining of myoepithelial cells (anti-smooth muscle actin [SMA] labelling, original magnification ×60).
(b) Immunocytochemical staining of myoepithelial cells demonstrating intense linear staining (anti-SMA labelling, original magnification ×40).
(courtesy Janet Freeman).

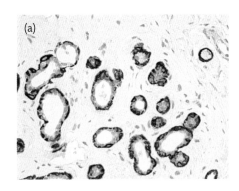

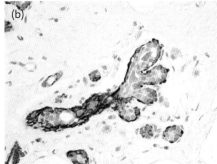

Breast cancer

Modern breast cancer treatment focuses on the assessment of receptors, most notably oestrogen and progesterone. More recently, considerable interest has focused on HER2 protein over-expression, p53 protein and associated mutations.

covered by keratinizing stratified epithelium. The nipple-areola area lacks pilosebaceous units; however, sebaceous and apocrine glands are present in the dermis of the nipple-areola tissue.

The breast undergoes considerable change as a result of hormonal or steroidal influences, thus pregnancy and lactation results in considerable epithelial proliferation, with a dramatic increase in the number of lobules and acinar units. During lactation there is distension of the lobular and acini units. In contrast, during the menopause there is involution and atrophy of the terminal duct lobular units, with associated reduction in acinar size and complexity, and reduction in collagen and adipose tissue in the surrounding stroma.

15.8.2 Male breast tissue

The male breast is composed of all the same components but generally has less stromal tissue. In addition, the epithelial elements normally consist of branching ducts without lobules.

Key points

The study of the epithelium of the breast ductal-lobular system is extremely important in the histological distinction between benign and malignant change in breast pathology.

15.8.3 Blood and lymphatics of the breast

The arterial supply to the breast is provided by the lateral thoracic arteries. The venous drainage largely follows the arterial system. There is a venous complex that runs from lateral to medial in the subcutaneous tissue.

Lymphatic drainage of breast tissue occurs via four routes: cutaneous, axillary, internal thoracic and posterior intercostal. The basin for lymphatic flow from the breast is the axilla.

SELF-CHECK 15.5

Describe the use of antibodies to evaluate normal and pathological breast tissue.

The lymphatic drainage system

An appreciation of the lymphatic drainage system of the breast is crucial in modern pathology. The principle of assessing metastatic spread of malignant cells from breast tissue by evaluating sentinel lymph node status (the first draining lymph node away from the primary tumour site) is central to how patient management regimes for breast disease are currently accessed.

15.9 Stains, antibodies and descriptive terms used in dermatopathology: a brief overview

Clinically, the study of skin disease falls to the dermatologists. Dermatologists in a clinical setting are almost unique in being able to make highly detailed visual assessments of any given skin lesion. This is not always the case with other disciplines in the assessment of deep tissue sites or other organ-specific pathologies. Therefore, what appears as a clinically defined disease process is confirmed dermatopathologically and supported by laboratory-based investigations. It is also important to remember that any given skin disease may change clinically and pathologically over time. Evidence for this comes from the evolving clinical changes one can see in many autoimmune states affecting the skin, as well as cutaneous lymphoma profiles. The subsequent laboratory tests will also reflect these clinical changes.

In order to introduce you to these laboratory tests, below is a table (Table 15.1) of the most common and widely used special stains employed in the histopathology laboratory for the study of

TABLE 15.1 Commonly used special stains employed in the investigation of skin disease.

Special stain	What demonstrated
Ziehl Neelsen	*Mycobacterium tuberculosis*
Alcian Blue	GAGS (mucin)
Chloroacetate esterase	Myeloid cells (mast cells)
Congo red	Amyloid
Gomori's reticulin	Reticulin fibres
Gram	Bacteria
Grocott	Fungus
Hale's dialysed Iron	GAGS (mucin)
Masson trichrome	Connective tissue
Masson Fontana	Melanin
Elastic van Gieson (EVG)	Elastic fibres
Periodic acid Schiff (PAS)	Basement membrane, fungus and glycogen
Diastase PAS	GAGS without glycogen
Perls' Prussian blue	Haemosiderin
Toluidine blue	Mast cells
Thioflavin T	Amyloid
Von Kossa	Calcium
Wade Fite	Leprosy bacillus
Warthin Starry	Spirochaetes (*Treponema pallidum*)

TABLE 15.2 Commonly employed antibodies used for the study of cell populations in skin disease.

Antibody	What demonstrated
CD1a	Langerhans cells
CD2	All T-lymphocyte cells
CD3	All T-lymphocyte cells
CD4	T-helper lymphocyte cells
CD8	T-cytotoxic lymphocyte cells
CD20	B-lymphocyte cells
CD45 (leucocyte-common antigen [LCA])	All lymphocytes (T and B)
CD34	Endothelial cells lining blood vessels
CD68	Monocytes/macrophages
Smooth muscle actin (SMA)	Smooth muscle cells
Carcinoembryonic antigen (CEA)	Ductal openings in glandular tissue
Desmin	Smooth and striated muscle
Epithelial membrane antigen (EMA)	Reacts with epithelial tissue
Factor XIIIa	Dermal dendrocytes
Ki67	Nuclear protein expressed in all cells during the cell cycle (except G_0 phase)
Melan A	Melanocytes
Neuron-specific enolase	Nerve and neuroendocrine cells (Merkel cells)
S100	Nerves, melanocytes, Langerhans cells, histiocytic cells
Laminin	Basement membrane zone
Lysozyme	Histiocytes
D2-40	Lymphatic endothelial cells
Vimentin	Intermediate filaments found in mesenchymal cells
MNF116	Broad-spectrum cytokeratin identifying epithelial cells
CK8/18	Low molecular weight cytokeratin expressed in glandular structures

skin tissue. In terms of specific cells and their immunophenotypic profiles, Table 15.2 includes some of the commonly used antibodies employed to delineate cell populations in skin disease.

The listed special stains and antibodies in Tables 15.1 and 15.2 appear in other volumes in this textbook series, most notably in *Cytopathology* and *Histopathology*. In these books, greater

TABLE 15.3 Definitions of key words and terms in dermatopathology.

Term	Definition/description
Acanthosis	Increase in the thickness of the spinous cells/layer of the epidermis
Anagen	Growth phase of the hair cycle
Apoptosis	Programmed cell death
Bulla	Fluid-filled blister
Colloid body	Remnants of keratinocyte that has died
Epidermotropism	Presence of lymphocyes within the epidermis
Exocytosis	Presence of inflammatory cells in the epidermis
Hyperkeratosis	Increased thickness of the stratum corneum
Microabscess	Mixture of neutrophils, fibrin and general cell debris in a cluster formation
Papillomatosis	A projection arising from the dermal papillae and epidermal tissue directly above it
Parakeratosis	Alteration in the usual appearance of cells in the stratum corneum, due to the retention of the cell nuclei
Pustule	Collection of neutophils and breakdown products mainly within the epidermis or hair follicles
Spongiosis	Intraepidermal oedema (swelling) widening the spaces between keratinocytes
Vesicle	Small fluid-filled cavity in the epidermis

attention is paid to the value of such stains and antibodies in the diagnosis of disease states throughout the human body.

As you read other textbooks in this series, which have a focus on disease processes, it is important to understand some of the more widely used words and terms in pathology. The final table (Table 15.3) in this chapter lists definitions of key terms and words used mainly to describe features seen in skin disease. This will provide you with a basic understanding of the descriptive nature of the science of pathology, particularly so in histopathology and cytopathology.

 Chapter summary

In this chapter we have:

● Described the structure and functions of the skin.

● Discussed the key features of the cells and structures that comprise the skin.

- Considered the importance and variation of skin as a structure and as the largest organ of the body.

- Discussed the regulatory systems and homeostatic mechanisms that regulate the skin.

- Considered the differential effects of normal and abnormal events that can occur in the skin.

- Defined key terminology associated with skin disease.

- Introduced a range of disorders of cutaneous disease processes.

Closing remarks

This is the last chapter in this book. You will have learned about the structure and function of tissue and cellular components of the human body, inside to out. Having read this you are now well equipped to read the discipline-specific volumes that comprise this series.

Anyone who has studied science will realize that it is a long and winding path of discovery and knowledge acquisition, often telling us more about our own personal abilities than anything else. What is important is that the knowledge gained can be applied constructively for the benefit of wider society.

 Further reading

- British Association of Dermatologists (www.bad.org.uk).

- British Skin Foundation (www.britishskinfoundation.org.uk).

- Burkitt HG, Young B, Heath JW, Wheater PR. *Wheater's functional histology. A text and colour atlas*. Edinburgh: Churchill Livingstone, 1993.

- Freedberg IM, Eisen AZ, Wolff K *et al.* eds. *Fitzpatrick's dermatology in general medicine* 5th edn, Vol 1. McGraw-Hill, 1999.

- Mills SE ed. *Histology for pathologists* 3rd edn. Philadelphia: Lippincott Williams and Wilkins, 2007.

- Moore G, Knight G, Blann AD eds. *Haematology*. Oxford: Oxford University Press, 2010.

- Orchard GE. Immunocytochemical techniques and advances in dermatopathology. *Curr Diagn Pathol* 2006; **12** (4): 292–302.

- Orchard GE, Nation BR eds. *Histopathology* Oxford: Oxford University Press, 2012.

- Shambayati B ed. *Cytopathology*. Oxford: Oxford University Press, 2011.

- Skin Cancer Foundation (www.skincancer.org).

- Suvarna K, Layton C, Bancroft JD eds. *Bancroft's theory and practice of histological techniques* 7th edn. Edinburgh: Churchill Livingstone, 2012.

- Weedon D ed. *Skin pathology* 2nd edn. Edinburgh: Churchill Livingstone, 2002.

Discussion questions

15.1 Discuss the structural features of the epidermis and describe how its structure relates to its function.

15.2 Describe the range of key cell types that reside within the dermis.

15.3 Discuss the structure of the BMZ, how important is this layer to the integrity of the skin as a functioning unit.

Answers to self-check questions and discussion questions, hints and tips are provided on the book's Online Resource Centre.

 Visit www.oxfordtextbooks.co.uk/orc/orchard_csf/.

Glossary

Accessory sex glands The collective name for the seminal vesicles, prostate, and bulbourethral glands.

Acinus Triangular area of liver with two portal tracts at the base and a single centrilobular venule at the apex.

Acrosyringium A component of the glandular excretory structure. It is lined by keratinocytes that are smaller than those of the surrounding epidermis, and are attached to the larger keratinocytes by desmosomal junctions.

Adenocarcinoma Cancer that originates from glandular epithelium.

Aetiology The cause of a disorder.

Allograft Transplant from one individual to another of the same species and genotype.

Alveolar macrophage Type of macrophage present in the pulmonary alveolus. Its function is the removal of foreign matter.

Ampulla The relatively long portion of the Fallopian tube leading from the infundibulum.

Ampulla of Vater The point of union of the common bile duct and the pancreatic duct.

Anaemia A haemoglobin concentration below the reference range for a particular gender and a particular age.

Androgen-binding protein A protein produced by Sertoli cells that binds specifically to testosterone.

Anterior pituitary The front lobe of the pituitary gland that secretes follicle stimulating hormone and other chemical messengers.

Antrum A cavity or chamber. May have specific reference to organs or sites in the body (e.g. ovarian antrum, gastric antrum).

Apoptosis An inflammation-independent mechanism through which controlled cell death is initiated.

Argentaffin cells Cells that take up silver stain.

Asphyxiation Death that can result due to lack of oxygen.

Asthma A common chronic inflammatory disease affecting the respiratory tract. It has many causes; symptoms include wheezing, coughing and shortness of breath.

Astral microtubules A type of microtubule; contributes to chromosome movement during cell division by pushing the spindle poles apart.

Atrophic A thinning or decrease in size of a tissue or organ.

Autonomic nervous system The involuntary part of the nervous system. It conducts impulses to cardiac muscle, smooth muscle and glands.

Autoradiography The process of exposing an X-ray film to the decay emissions of a radioactive substance—for example, the radioactively-labelled DNA probes used in Sanger sequencing —to produce an image for interpretation.

Axoneme The central core of a cilium, flagellum, or sperm tail consisting of a central pair of microtubules surrounded by nine other pairs ('9+2' formation).

Bartholin's glands A pair of mucin-secreting glands opening into the vagina.

Basal body An organelle formed from a centriole; gives rise to cilia and flagella.

Basal cells A type of reserve cell; relatively undifferentiated cells lying at the base of an epithelium, adjacent to a basement membrane.

Basement membrane Sheet of specialized tissue that forms an interface between other types of tissue; its primary function is to anchor epithelial tissue to connective tissue.

Beta-2 agonist A class of drug used to treat asthma. It acts on the β2-adrenergic receptor causing smooth muscle relaxation, resulting in dilation of bronchial passages.

Biliary tree (biliary tract) The path by which bile is secreted from the liver to the duodenum. It is formed of anastomosing branches and hence has a tree-like appearance.

Bipolar neurons Special neurons that have two extensions and are involved in the sense of smell.

Blood–testis barrier A physical barrier between the circulatory system and seminiferous tubules of the testis; controls the environment in which germ cells develop.

Bowman's gland Situated in the olfactory mucosa, beneath the olfactory epithelium.

Bronchial circulation The systemic vascular supply to the lung. It supplies blood to conducting airways down to the level of the terminal bronchioles.

Bronchopulmonary segments Each lung is separated by connective tissue into ten independently functioning compartments called bronchopulmonary segments. Each has its own supply of blood vessels and lymphatics.

Bulbourethral glands Two small glands located beneath the prostate gland; contribute fluids to semen during ejaculation.

Canal of Hering The intrahepatic bile ductules.

Canaliculus/canaliculi Thin tubes collecting bile secreted by the hepatocytes.

Carina Anatomical site where the trachea divides into the two primary bronchi.

Centriole A cylindrical-shaped structure within the cytoplasm consisting of nine triplets of microtubules ('9x3' formation); plays an important role in cell division.

Centromere Part of a chromosome that links paired chromatids.

Centrosomes Organelles in which microtubules are produced.

Cervix The lower part of the uterus.

Chromatid One of the two strands of a duplicated chromosome.

Cilia (singular cilium) Slender protuberances emanating from the body of a cell.

Cirrhosis Scarring as a consequence of damage to the liver.

Clara cells Non-ciliated epithelial cells present in terminal bronchi.

Clara cells secretary protein A substance produced by Clara cells.

Clitoris The female external sexual organ located where the labia minora meet.

Clusters of differentiation (CD) antigen The standardized term used to describe a range of molecules associated with the differentiation and maturation of cells.

Colloid oncotic pressure Osmotic pressure caused by proteins in plasma.

Columnar epithelium One or more layers of cells, which are longer than they are wide.

Conception Fertilization.

Connective tissue A diverse type of tissue that functions to support the body and bind other types of tissue together.

Contrast-enhanced chest CT scan During a CT scan, a radio-contrast agent or 'contrast medium' is administered intravenously in order to provide higher image quality.

Corpora cavernosa (singular corpora cavernosum) Two cylindrical-shaped columns of erectile tissue running along each side of the penis.

Corpus spongiosum A longitudinal column of spongy erectile tissue surrounding the urethra in the penis.

Cricoid cartilage Ring-shaped cartilage that surrounds the trachea.

Crossover Exchange of chromosomal segments between a matched pair of chromosomes during meiosis.

Crypt A narrow and deep invagination of epithelium.

Cryptic translocations The exchange of genetic material between chromosomes, where the morphology of the chromosomal structure and banding pattern are retained. Cryptic translocations may be missed when using banding techniques.

Decidual cells Enlarged and specialized endometrial stromal cells; facilitate nutrition of the developing embryo and assists gaseous exchange at the placental interface during pregnancy.

Decidualization Formation of decidual cells from endometrial stromal cells.

Diaphragm Sheet of muscle that extends across the bottom of the rib cage and separates heart, lungs and ribs from the abdominal cavity.

Diploid The condition in which a cell has two haploid sets of chromosomes.

Diverticulae (singular diverticulum) Sac-like outpouchings of a wall of a bodily structure. For example, the seminal vesicles are diverticulae of the vas deferens.

Ductal plate Primitive biliary epithelium that develops to form the intrahepatic bile ducts.

Dynein A protein molecule connecting the outer microtubule doublets in the axoneme of a cilium, flagellum, or sperm tail and responsible for their movement.

EBUS TBNA (Endobronchial ultrasound guided transbronchial needle aspiration) In this procedure, a bronchoscope is inserted via the mouth into the lungs. An ultrasound probe located at the tip of the bronchoscope provides images and guides the bronchoscopist to target suspicious areas for fine-needle aspiration.

Ectocervix Part of the cervix that extends into the vagina.

Effusion Pathological collection of serous fluid that collects in serous cavities.

Ejaculation Ejection of semen and sperm from the male reproductive tract.

Ejaculatory duct A duct formed from the union of the vas deferens and the duct of the seminal vesicle.

Embryo A developing human; the first eight weeks following fertilization.

Endocervical canal The narrow passageway of the cervix between the internal and external os.

Endocervix Endocervical canal.

Endocytosis A process by which cells absorb molecules such as proteins by engulfing them.

Endometrium The epithelial lining of the uterus.

Endomitosis The replication of chromosomes without the division of the cell nucleus. This leads to cells with, potentially, very high chromosome numbers.

Endoplasmic reticulum An organelle consisting of stacks of flattened compartments; synthesizes and stores various molecules within the cell.

Endothelial cells Found on the inner lining of blood vessels and lymphatic vessels, these cells have complex and unique functions, which you can read about in Chapter 9.

Epidermal appendages Structures including hair follicles, sweat glands, sebaceous glands and nails.

Epididymis A tightly coiled tube (duct) lined by pseudostratified epithelium; transports sperm from the testis to the vas deferens.

Epiglottis A thin flap of fibroelastic cartilage situated at the entrance of the larynx that keeps food from entering the lung.

Erectile tissue A framework of collagen and other fibres producing tissue spaces that may fill with blood.

Ethmoid bone One of the four bony structures that surround the nasal cavity.

Eukaryotic cells Cells containing a true membrane-bound nucleus and other membrane-bound organelles.

External os The opening of the cervix into the vagina.

Falciform ligament Attaches the liver to the anterior wall of the body. It is derived from the umbilical vein of the fetus.

Fallopian tubes A pair of tubes in the female reproductive system connecting the ovaries to the uterus.

False vocal cords A pair of thick folds of membrane that protect and sit slightly superior to the more delicate true vocal cords.

Ferritin An iron storage molecule containing the protein apoferritin combined with ferric iron.

Fertilization The fusion of male and female gametes to form a zygote.

Fetus A developing human; from eight weeks after fertilization to birth.

Fibrocollagenous tissue A type of connective tissue consisting of collagen and various other types of fibres.

Fibroelastic cartilage A flexible type of cartilage found in many areas in the body.

Fibrosis Formation of excess collagen or connective tissue as a response to injury.

Fimbriae (singular fimbria) Finger-like projections at the terminal portion of the fallopian tubes.

First polar body A non-viable cell that forms as a result of the first meiotic division of an oocyte.

Flagella (singular flagellum) Motile appendages which emanate from the body of a cell.

Follicle-stimulating hormone A hormone secreted by the anterior pituitary gland; stimulates ripening of ovarian follicles in females and spermatogenesis in males.

Frontal bone One of the four bony structures that surround the nasal cavity.

Gametes Reproductive cells (spermatozoa in males and ova in females) derived from germ cells and containing a haploid chromosome set.

Gametogenesis The process of cell division and differentiation that produces gametes.

Ganglia Structures in the peripheral nervous system that are equivalent to nuclei in the central nervous system.

Germ cells Specialized cells in the gonads, destined to become spermatozoa in males and ova in females.

Gestation The period of development from conception to birth.

Glans penis The rounded head of the penis.

Glisson's capsule Collagenous outer membrane of the liver.

Goblet cells Mucin-secreting columnar epithelial cells, so named because of their resemblance in shape to wine goblets.

Golgi apparatus An organelle formed by fusion of sections of endoplasmic reticulum; packages molecules ready for secretion or delivery to other organelles.

Gonads Gamete-producing organs; testes in the male and ovaries in the female.

Graafian follicle A mature ovarian follicle.

Granulosa cells Cells that surround the oocyte during oogenesis and secrete the sex steroids oestrogen and progesterone during various phases of the menstrual cycle.

Grey matter Commonly used term based on macroscopic appearance. Grey matter is neuron rich. (White matter lacks neurons, but has an abundance of myelin coated axons giving it the whiter colour.)

Haematocrit The proportion of whole blood occupied by red blood cells. This proportion is reported as a decimal fraction. For example, if red blood cells occupy 450 mL of 1000 mL blood, the haematocrit would be 45%, or 0.45 L/L.

Haemolytic anaemia The process whereby an individual develops a low red blood cell count and haemoglobin concentration as a consequence of reduced red blood cell survival.

Haemopoiesis The formation of the cellular components of blood

Haploid The condition in which a cell has only one set of chromosomes.

Helicine arteries Blood vessels in the penis that are involved in erection.

Helminth A parasitic worm commonly found within the intestine.

Hepatocyte Functional cell of the liver.

Hilum The part of an organ where blood vessels and nerves enter and leave; also organ-specific structures (e.g. in the lung: blood vessels, nerves, lymphatics and bronchus).

Histiocytes Macrophages that look like epithelial cells. Epithelioid histiocytes are seen in granulomatous inflammation.

Human immunodeficiency virus (HIV) A rotavirus and the causative agent in AIDS.

Human papillomavirus (HPV) A papillomavirus capable of infecting humans and causing warts and cancers.

Hyaline From the Greek, meaning glass-like.

Hybridoma A term used to describe the fusion of a B lymphocyte that has been stimulated to produce specific antibodies with a cancerous B-lineage cell called a myeloma cell. The fusion of these cells ensures continual production of identical immunoglobulins, and the fusion of the cancer cell ensures survival of the hybridoma under culture conditions.

Hydrodynamic focusing A technique used to deliver cells in single file to a set point within an enumeration analyser. The cells of interest are injected into the middle of a stream of fast-moving sheath flow. The sheath flow forms a wall, or barrier, to the injected cells and the different velocities of the fluids ensure they do not mix.

Hydrostatic pressure Pressure within the blood capillaries.

Hypothalamus A hormone-secreting region of the brain that controls a large number of bodily functions.

Immunocompromised The effect of compromised or absent immunity, which can be due to disease or the side effect of treatment.

Immunoglobulin Abbreviated to Ig, an immunoglobulin is a type of protein secreted into bodily fluids that binds to a particular antigen. Different classes of Ig include IgG, IgM, IgE, IgA and IgD. Each has a different structure and role within the immune response.

Infundibulum A funnel-shaped cavity; in the female genital tract it refers to the terminal portion of the Fallopian tube adjacent to the ovary.

Inhibin A hormone produced in the testis and ovary; inhibits the production of follicle stimulating hormone, thereby controlling spermatogenesis and oogenesis through a negative feedback mechanism.

Intercostal muscles Several groups of muscles that run between the ribs and help to move the chest.

Intermediate cells One or more layers of cells in stratified squamous epithelium lying between the superficial and parabasal cell layers.

Internal genitalia The reproductive organs within the body.

Internal nasal fossa The internal nasal cavity.

Internal os The opening of the cervix into the uterus.

Interstitium Supporting tissue surrounding the cells. In the respiratory tract it refers to the tissue surrounding the alveolar spaces.

Intramural segment A narrow section of the Fallopian tube continuous with the isthmus, leading to the uterus.

Ischaemia A reduction in blood flow to a tissue.

Isthmus Generally, a narrow connection between two larger structures; a short section of the Fallopian tube connecting the ampulla to the intramural segment.

Junctional complexes Complex attachment points between epithelial cells.

Kinetochore microtubules A type of microtubule; contributes to cell division by attaching to chromosomes before chromosomal segregation.

Kupffer cell Macrophage found in the liver, lining the sinusoids.

Labia The external visible portion of the vulva.

Lamina propria Thin layer of connective tissue under the epithelium.

Leukaemia The presence of cancerous haemopoietic cells in the bone marrow and peripheral blood.

Leydig cells Testosterone-producing cells found in the tissue spaces between the seminiferous tubules of the testis.

Lipofuscin A yellow/brown, granular, iron-negative lipid pigment found particularly in muscle, heart, liver and nerve cells; it is the product of cellular wear and tear, accumulating in lysosomes with age.

Liver transplantation The replacement of a diseased liver with a healthy liver from another human being (allograft).

Lobule Division of liver histologically. Roughly hexagonal, it has a single central venule and a portal tract at each exterior angle.

Lobules Small segments or divisions within an organ separated by septae; as in lobules of the testis.

Lymphatics Thin walled structures that carry lymph (also called lymphatic vessels).

Lymphoma Cancer that originates from lymphoid tissues.

Mallory–Denk bodies Inclusions found in the liver cells, typically, but not exclusively, in alcoholic hepatitis.

Mast cells Cells containing histamine and heparin which play a key role in the inflammatory process.

Maxillary bone One of the four bony structures that surround the nasal cavity.

May Grunwald Giemsa A Romanowsky stain.

Mediastinum The area between the lungs that contains the heart, aorta, oesophagus, trachea and lymph nodes.

Megakaryocyte The platelet progenitor cell derived as follows: mega = large, karyo = chromosomal complement, cyte = cell.

Meiosis A special type of cell division occurring only in germ cells; results in haploid daughter cells.

Menstrual cycle A cycle of physiological changes in the female reproductive system to facilitate fertilization and sexual reproduction.

Menstrual phase (menstruation) The stage of the menstrual cycle, lasting from two to five days, during which the endometrium is shed.

Mesothelial cells The lining cells of the serous cavities, which, in addition to the tunica vaginalis, include the pleural cavity (lining the lungs), the pericardial cavity (around the heart) and the peritoneal cavity (enveloping the abdominal organs).

Mesothelium Cells covering the serous membranes.

Metaplasia A change from one cell type to another in response to a stimulus. For example, squamous metaplasia of the cervix involves a change from columnar to squamous epithelium.

Microtubule organizing centre The site of microtubule formation.

Microtubules Hollow proteinaceous 'rods' that permit movement of various kinds within eukaryotic cells.

Microvilli (microvillus) Microscopic protrusions of the cell membrane that serve to increase the surface area of a cell.

Mitosis Nuclear and cellular division to produce two genetically identical diploid daughter cells.

Monoclonal antibodies Immunoglobulins that all have the same specific target protein (epitope).

Morphology The study of the shape, structure and appearance of an organism.

Mucociliary escalator A term that describes the movement of fluid and trapped microscopic particles over the surface of an epithelium, enabled by the synchronous beating of surface cilia.

Myeloid All cells of the bone marrow, with the exception of lymphocytes.

Myometrium The muscle layer of the uterus.

Naïve B cell A cell of B lineage that has yet to encounter an antigen.

Nasal conchae Bony structure in the nasal cavity.

Nasal vestibule The most anterior (front) part of the nose.

Necrosis The unprogrammed death of cells and tissues.

Negative feedback In hormone physiology, negative feedback refers to the inhibition of further secretion of a hormone when the levels of that hormone increase.

Neuroendocrine cells Neuroendocrine cells receive neural input and can release various messenger molecules into the blood. In the lung they may act as receptors for oxygen levels.

Nexin A linkage protein between microtubule doublets; prevents excessive movement of the doublets relative to each other.

Non-keratinizing stratified squamous epithelium Multilayered squamous epithelium without evidence of keratin formation.

Non-motile cilia Cilia that cannot beat and usually serves as a sensor.

Non-small cell lung cancer A broad division of lung cancer encompassing all epithelial cancers that are not covered by the small cell group of tumours.

Nuclei In the context of the central nervous system, the term nuclei is used to describe discrete functional groups of neurons, rather than the chromosome-containing structures of cells.

Oedematous An excessive amount of fluid in or around cells.

Oestrogen A hormone produced primarily by the granulosa cells of the ovary; involved in the regulation of the menstrual cycle and development of female sexual characteristics.

Olfaction Sense of smell.

Olfactory epithelium Specialized nasal epithelium involved in the sense of smell.

Oogenesis Female gametogenesis.

Oogonia (singular oogonium) Undifferentiated female germ cells.

Organelle A membrane-bound unit within a cell that has a specific function.

Oropharynx Back of the mouth.

Orthotopic transplantation Surgical procedure whereby the native organ is removed and replaced by a healthy donor organ, in the same anatomical location as the native organ.

Osteon The basic unit of bone, comprising bone lamellae, including embedded osteocytes, arranged in a concentric pattern with a central canal containing blood vessels. Perpendicular to the central canal, there are perforating canals that carry blood vessels into bone and the marrow space.

Ova (singular ovum) Female gametes.

Ovaries Female gonads.

Ovulation The release of a secondary oocyte from the ovary.

Ovulatory phase The stage of the menstrual cycle during which a secondary oocyte is released from the ovary.

Parabasal cells One or more layers of cells found in the deeper part of stratified squamous epithelium.

Paracetamol/acetaminophen An analgesic drug used to reduce pain and fever

Parasympathetic Parasympathetic stimulation causes bronchoconstriction.

Parietal layer The pleural membrane covering the wall of the cavity.

Penis The male external sexual organ.

Phagocytes From the Greek 'phagein' meaning to eat, and 'cyte' meaning cell. These cells ingest foreign matter.

Phagocytosis Describes the process whereby certain cells engulf and consume foreign material such as a bacterium or damaged tissue.

Pharynx From the Greek for 'throat'—the passage leading from the oral and nasal cavities in the head to the oesophagus and larynx.

Pleomorphism Variability in size and shape.

Pneumonia An infection of the lungs, usually due to infective agents such as bacteria, viruses or fungi.

Pneumothorax Presence of air in the pleural space. This leads to collapse of the lung, causing pain, discomfort and breathing difficulties.

Polar bodies Non-viable cells produced during meiosis.

Polar microtubules A type of microtubule; contributes to cell division by stabilizing the spindle apparatus and by pushing the spindle poles apart

Porta hepatis The entrance point of the hepatic portal vein and hepatic artery, and the exit point of the common hepatic duct.

Portal hypertension/hypotension High/low blood pressure in the portal vein system.

Portal tract Area of connective tissue at the edge of a lobule containing a branch of the hepatic artery and hepatic portal vein, which bring blood to the liver, and a bile duct to take bile from the liver.

Posterior lobe The most prominent part of the neurohypophysis, the posterior lobe (pars nervosa) of the pituitary contains neurosecretory axons and their endings.

Primary bronchioles Parts of the conducting portion of the respiratory system that are located at the end of the bronchi and terminate in the alveoli.

Primary follicle A structure within the ovary consisting of a primary oocyte surrounded by a layer of granulosa cells.

Primary oocyte A cell derived from an oogonium prior to completion of the first meiotic division.

Primary spermatocytes Cells produced from the mitotic division of spermatogonia.

Progenitor cells Cells that can become activated to replicate in response to cellular injury or cell death.

Progesterone A hormone produced by the corpus luteum following ovulation and by the placenta during pregnancy; numerous roles in the support and maintenance of pregnancy.

Prognosis Outcome of disease.

Proliferative phase (follicular phase) The stage of the menstrual cycle that follows menstruation and involves the maturation of ovarian follicles.

Prostate gland A gland beneath the bladder that secretes semen into the urethra during ejaculation; comprises approximately 30 per cent of the ejaculatory volume.

Pseudostratified [columnar] [epithelium] A single layer of cells that has a multilayered appearance because the nuclei of adjacent cells are at different levels

Pulmonary circulation Part of the cardiovascular system that carries oxygen-depleted blood away from the heart, to the lungs, and returns oxygenated blood to the heart.

Pyknosis describes a condensation or shrinking and thickening of a cell or nucleus.

Quenching Describes the process of reducing the fluorescence generated by a given substance.

Reactive mesothelial cells A term used by cytologists to describe normal mesothelial cells that are showing inflammatory changes.

Refractive index The ratio of the speed of light in one material compared to that in a second material of greater density. An illustration demonstrating refraction is provided in Figure 2.5. For each pair of materials, the figure calculated is a constant. In this ratio, the numerator represents the incident light and the denominator represents the refracted light.

Reproductive tract The part of the reproductive system that includes various cavities and ducts but excluding the gonads.

Reserve cells Non-specialized epithelial cells that divide and grow to form new tissue when the latter is damaged or ageing. Therefore, reserve cells are important for normal turnover, growth and repair of tissue.

Respiratory epithelium A type of epithelium found in the respiratory tract. Its function is to moisten the airway and form the first line of defence against pathogens and foreign matter.

Reti testis A network of channels carrying sperm from the seminiferous tubules to the epididymis.

Reticulin Supporting network for cells in tissue, composed of type III collagen.

Rigor mortis A post-mortem physical state of intense muscle contraction.

Romanowsky stain A biological stain comprising a mixture of dyes (methylene blue and eosin). Originally used to stain blood films, but later adopted for use in histopathology and cytology.

Salbutamol A short-acting β2-adrenergic receptor agonist used in the relief of bronchospasm.

Sarcoma Cancer that originates from connective tissues.

Sarcomere is the contractile unit of a myofibril, delimited by Z lines along the length of the myofibril.

Scrotum A skin-lined sac containing the testes.

Sebaceous Microscopic glands that produce an oily substance.

Sebaceous glands Oil-secreting glands opening on to the surface of the skin.

Second polar body A non-viable cell that forms as a result of the second meiotic division of an oocyte.

Secondary oocyte A cell derived from a primary oocyte between the first and second meiotic division.

Secondary spermatocytes Cells produced from the meiotic division of primary spermatocytes.

Secretory epithelial cells Cells that line bodily surfaces and structures and whose primary function is the secretion of various products of metabolism.

Secretory phase (luteal phase) The stage of the menstrual cycle following ovulation and preceding menstruation.

Semen A fluid produced by the accessory sex glands; facilitates transportation of sperm during ejaculation.

Seminal fluid Semen.

Seminal vesicles Glands that secrete semen into the ejaculatory duct; semen comprises approximately 70 per cent of the ejaculatory volume.

Septae (singular septum) Dividing walls or partitions between adjacent tissues.

Serous fluid A lubricating fluid produced by the mesothelium that permits the smooth movement of the lung during respiration.

Serous secretions Watery secretions produced by Bowman's gland.

Sertoli cells Cells within the seminiferous tubules supporting spermatogenesis.

Sexual intercourse Insertion of the penis into the vagina.

Sinusoid Type of blood vessel with fenestrated epithelium that facilitates exchange of nutrients, oxygen, carbon dioxide and waste products.

Small cell lung cancer A highly malignant tumour arising from neuroendocrine cells.

Soft palate Roof of the mouth.

Somatic cells Any cells of the body other than a germ cell or one of its derivatives.

Space of Disse Space between the sinusoid and hepatocytes. Contains plasma.

Spermatids Haploid male gametes produced from the meiotic division of secondary spermatocytes.

Spermatogenesis Male gametogenesis.

Spermatogonia (singular spermatogonium) Undifferentiated male germ cells.

Spermatozoa (singular spermatozoon) Male gametes.

Sphenoid bone One of the four bony structures that surround the nasal cavity.

Spindle apparatus System of microtubules that separates chromosomes during cell division.

Squamocolumnar junction An abrupt transition between the squamous epithelium of the ectocervix and the columnar epithelium of the endocervix.

Squamous carcinoma Type of cancer that originates from squamous epithelium.

Squamous metaplasia – respiratory tract Squamous metaplasia occurs as response to injury. It is commonly seen in smokers. The cells appear as sheets of polygonal-shaped cells, similar to that which occurs in the cervix.

Stellate cell Seen as part of the sinusoid lining, it produces collagen and extracellular matrix, and stores vitamin A.

Stem cells Cells that can differentiate into different cell types (also called progenitor cells).

Stroma Supportive tissue.

Stromal cells Cells of the connective tissue of any organ.

Superficial cells The uppermost layer of cells in stratified squamous epithelium.

Supporting cells Cells that provide structural support to the other epithelial tissues.

Supravital stains Used to stain live, unfixed cells. Usually, stains are applied to fixed cells, and as such are only used when the cells are no longer alive. An example of a supravital stain is new methylene blue.

Surfactant A lipoprotein secreted by alveolar cells. Its function is to reduce the surface tension of fluids in the lung.

Swell bodies The name given to venous plexuses in the nose. This area may become congested during allergic reaction or infection.

Sympathetic Sympathetic stimulation causes bronchodilation.

Terminal bronchioles The last part of the conducting portion of the respiratory tract.

Testes (singular testis) Male gonads.

Testosterone A hormone produced primarily by the Leydig cells of the testis; facilitates sperm production and proper development of male sexual characteristics.

The Royal Society The National Academy of Science in the UK. Founded in 1660, it has the mission statement 'to recognise, promote and support excellence in science and to encourage the development of science for the benefit of humanity'. The world's oldest scientific journal *Philosophical Transactions of the Royal Society* is published by this society.

Theca cells Cells that surround the granulosa cells in two layers (theca interna and theca externa) and provide the hormone precursors required for the production of sex steroids.

Thyroid cartilage The largest cartilage structure that surrounds the trachea.

Tight junctions Hold cells together and form a barrier that controls the passage of molecules and ions between cells.

Tinctorial staining A method that uses dyes to demonstrate components of cells and tissue structures.

Transbronchial fine-needle aspiration Aspiration through the wall of a bronchus.

Transformation zone Area of the cervix that undergoes squamous metaplasia.

Transitional epithelium See urothelium

Transthoracic fine-needle aspiration Aspiration through the chest wall or thoracic cavity.

Transudate Bodily fluid extruded from a tissue or passed through a membrane; low cell and protein content.

True vocal cords Two membranes stretched horizontally across the larynx that vibrate when air flows past them and which produce speech sounds.

Tubulin The protein component of microtubules.

Tunica albiginea A white membrane surrounding the corpora cavernosa of the penis; helps to retain blood in the penis during an erection.

Tunica vaginalis The serous membrane covering the testis.

Type 1 pneumocytes Type 1 pneumocytes are extremely thin squamous cells responsible for gaseous exchange. They occupy 95 per cent of total alveolar surface area.

Type 2 pneumocytes Type 2 pneumocytes have the potential to divide and differentiate into type 1 pneumocytes and also produce pulmonary surfactant

Urethra A duct carrying urine from the bladder to the external environment; also transports semen from the ejaculatory ducts.

Urethral meatus The external opening of the urethra.

Urothelium Specialized epithelium lining the majority of the urinary tract.

Uterus A hollow muscular organ located in the female pelvis; houses and nourishes the developing fetus.

Vagina A muscular tube connecting the uterus and cervix to the external environment.

Vas deferens A coiled duct that carries sperm from the epididymis to the ejaculatory duct.

Vascularized Includes many blood vessels.

Venous plexuses Congregation of multiple vessels.

Vibrissae Thick hairs inside the nostril.

Visceral layer The pleural layer covering the surface of the organ.

Vocalis muscle A band of muscle running parallel to the vocal cords that can affect the tension in the cords and thus influence the frequency of the sound produced.

Vulva Female external genital organs, comprising the labia and clitoris.

Wedge resection Surgical removal of a small wedge-shaped piece of tissue that includes the tumour and a small area of normal surrounding tissue

White matter Commonly used term based on macroscopic appearance. White matter lacks neurons, but has an abundance of myelin-coated axons giving it the whiter colour compared to that of grey matter. (Grey matter is neuron rich.)

Zygote The cell formed when two gametes fuse.

Index